Geographical Snapshots of North America

GEOGRAPHICAL SNAPSHOTS
OF NORTH AMERICA

Commemorating the 27th Congress of the
International Geographical Union and Assembly

Edited by
Donald G. Janelle

THE GUILFORD PRESS
New York   London

© 1992 The Guilford Press
A Division of Guilford Publications, Inc.
72 Spring Street, New York, NY 10012

Printed in the United States of America

This book is printed on acid-free paper.

Last digit is print number: 9 8 7 6 5 4 3 2 1

Library of Congress Cataloging-in-Publication Data
Geographical snapshots of North America: commemorating the 27th
    Congress of the International Geographical Union and Assembly,
    Washington, D.C., 9–14 August 1992 / editor, Donald G. Janelle.
        p.   cm.
    Includes bibliographical references and index.
    ISBN 0-89862-889-X.—ISBN 0-89862-030-9 (pbk.)
    1. North America—Geography. I. Janelle, Donald G., 1940–
II. International Geographical Union. General Assembly.
III. International Geographical Congress (27th : 1992 : Washington,
D.C.)
E40.5.G46   1992
917.3—dc20   92-6095
                                                            CIP

Contributors

Athol D. Abrahams
Department of Geography
State University of New York at Buffalo

Simon Baker
Department of Geography and Planning
East Carolina University

Klaus J. Bayr
Department of Geography
Keene State College

Stephen S. Birdsall
Department of Geography
University of North Carolina, Chapel Hill

E. M. Bjorklund
Department of Geography
University of Western Ontario

Anthony J. Brazel
Department of Geography
Arizona State University

Teresa L. Bulman
Department of Geography
Portland State University

David R. Butler
Department of Geography
University of North Carolina

James B. Campbell
Department of Geography
Virginia Polytechnic Institute and State University

George O. Carney
Department of Geography
Oklahoma State University

Patricia M. Chalk
Department of Geography
University of Western Ontario

Kathleen Christensen
Graduate School and University Center
City University of New York

Sandra L. Clark
Department of Geography
Arizona State University

Grady Clay
The Gazetteer Project
Louisville, Kentucky

William R. Code
Department of Geography
University of Western Ontario

Charles O. Collins
Department of Geography
University of Northern Colorado

Laura E. Conkey
Department of Geography
Dartmouth College

Susan L. Cutter
Department of Geography
Rutgers University

Donald W. Davis
Louisiana Geological Survey
Louisiana State University

Lary M. Dilsaver
Department of Geology and Geography
University of South Alabama

Ronald I. Dorn
Department of Geography
Arizona State University

Gary S. Elbow
Department of Geography
Texas Tech University

Deborah L. Elliott-Fisk
Department of Geography
University of California, Davis

Leonard J. Evenden
Department of Geography
Simon Fraser University

Bruce Fielding
Louisiana Department of Environmental Quality

K. Gajewski
Centre d'études nordiques
Université Laval

J. H. Galloway
Department of Geography
University of Toronto

Hugh J. Gayler
Department of Geography
Brock University

Philip J. Gersmehl
Department of Geography
University of Minnesota

Michael F. Goodchild
National Center for Geographic Information and Analysis
University of California, Santa Barbara

Warren E. Grabau
U.S. Army Corps of Engineers, Retired
Vicksburg, Mississippi

William L. Graf
Department of Geography
Arizona State University

Mark Gregory
Department of Geography
Oklahoma State University

William J. Gribb
Department of Geography and Recreation
University of Wyoming

Susan W. Hardwick
Department of Geography
California State University, Chico

Robert A. Harper
Department of Geography
University of Maryland at College Park

Richard Harris
Department of Geography
McMaster University

Douglas E. Heath
Geography and Earth Science Department
Northampton Community College

Briavel Holcomb
Department of Urban Studies and Community Health
Rutgers University

Donald G. Janelle
Department of Geography
University of Western Ontario

James H. Johnson Jr.
Department of Geography
University of California, Los Angeles

Laurence S. Kalkstein
Center for Climatic Research
Department of Geography
University of Delaware

Eugene C. Kirchherr
Department of Geography
Western Michigan University

Brian Klinkenberg
Department of Geography
University of British Columbia

Janet E. Kodras
Department of Geography
Florida State University

Victor A. Konrad
Canada-United States of America Fulbright Program

Arthur Krim
Society for Commercial Archeology

Mark P. Kumler
Department of Geography
University of California, Santa Barbara

Nina Siu-Ngan Lam
Department of Geography and Anthropology
Louisiana State University

Ary J. Lamme III
Department of Geography
University of Florida

Anthony J. Lewis
Department of Geography and Anthropology
Louisiana State University

Kam-biu Liu
Department of Geography and Anthropology
Louisiana State University

Ray Lougeay
Department of Geography
State University of New York, College at Geneseo

Glen M. MacDonald
Department of Geography
McMaster University

Alan G. Macpherson
Department of Geography
Memorial University of Newfoundland

George P. Malanson
Department of Geography
University of Iowa

Vincent H. Malmström
Department of Geography
Dartmouth College

Melvin G. Marcus
Department of Geography
Arizona State University

Harry L. Margulis
First College, Department of Urban Studies
Cleveland State University

Michele Masucci
Department of Geography
Auburn University

John R. Mather
Department of Geography
University of Delaware

Harold M. Mayer
Department of Geography
University of Wisconsin–Milwaukee

Patricia F. McDowell
Department of Geography
University of Oregon

Patrick McGreevy
Department of Geography
Clarion University of Pennsylvania

Thomas F. McIlwraith
Department of Geography
University of Toronto

John Mercer
Department of Geography
Syracuse University

E. Willard Miller
Department of Geography
Pennsylvania State University

Henry Moon
Department of Geography and Planning
University of Toledo

Keith W. Muckleston
Department of Geography
Oregon State University

Edward K. Muller
Department of History
University of Pittsburgh

Robert A. Muller
Department of Geography and Anthropology
Louisiana State University

Linda W. Mulligan
Department of Sociology
Ohio State University

Carolyn Murray-Wooley
Architectural Historian
Lexington, Kentucky

David J. Nemeth
Department of Geography and Planning
University of Toledo

Ann C. Noble
Department of Viticulture and Enology
University of California, Davis

T. R. Oke
Department of Geography
University of British Columbia

John E. Oliver
Department of Geography and Geology
Indiana State University

Clarence W. Olmstead
Department of Geography
University of Wisconsin–Madison

Amalie Jo Orme
Department of Geography
California State University, Northridge

Antony R. Orme
Department of Geography
University of California, Los Angeles

Risa Palm
College of Arts and Sciences
University of Oregon

Kathleen C. Parker
Department of Geography
University of Georgia

Anthony J. Parsons
Department of Geography
University of Keele

Allen K. Philbrick
Department of Geography
University of Western Ontario

Hong-Lie Qiu
Department of Geography and Anthropology
Louisiana State University

John P. Radford
Department of Geography
York University

Karl B. Raitz
Department of Geography
University of Kentucky

Rita Riccio
Department of Geography
San Diego State University

Bonham C. Richardson
Department of Geography
Virginia Polytechnic Institute and State University

Gisbert Rinschede
Institut für Geographie
Universität Regensburg

David A. Robinson
Department of Geography
Rutgers University

Thomas E. Ross
Department of Geology and Geography
Pembroke State University

James M. Rubenstein
Department of Geography
Miami University

Robert A. Rundstrom
Department of Geography
University of Oklahoma

Bruce Ryan
Department of Geography
University of Cincinnati

A. L. Rydant
Department of Geography
Keene State College

C. L. Salter
Department of Geography
University of Missouri–Columbia

Marcia Santiago
Statistics Canada
Ottawa, Ontario

H. P. Schmid
Swiss Federal Institute of Technology
Zurich, Switzerland

Darren M. Scott
Department of Geography
University of Western Ontario

David Seamon
College of Architecture and Design
Kansas State University

Marlyn L. Shelton
Department of Geography
University of California, Davis

Stephen J. Stadler
Department of Geography
Oklahoma State University

Michael Steinitz
Massachusetts Historical Commission

Frederick P. Stutz
Department of Geography
San Diego State University

Keith J. Tinkler
Department of Geography
Brock University

George Towers
Department of Geosciences
Southwest Missouri State

Daniel E. Turbeville III
Department of Geography
Eastern Oregon State College

Irina Vasiliev
Department of Geography
Syracuse University

H. Jesse Walker
Department of Geography and Anthropology
Louisiana State University

Stephen J. Walsh
Department of Geography
University of North Carolina, Chapel Hill

Barney Warf
Department of Geography
Kent State University

Barbara A. Weightman
Department of Geography
California State University, Fullerton

Charles F. J. Whebell
Department of Geography
University of Western Ontario

Emily E. Wheeler
Department of History and Geography
Georgia College

James O. Wheeler
Department of Geography
University of Georgia

Randy William Widdis
Department of Geography
University of Regina

Bobby M. Wilson
Department of Political Science and Public Affairs
University of Alabama, Birmingham

Marjorie Green Winkler
Center for Climatic Research
University of Wisconsin–Madison

Morton D. Winsberg
Department of Geography
Florida State University

Jennifer R. Wolch
Department of Geography
University of Southern California

Joseph S. Wood
Institute for Geographical Sciences
George Mason University

Donald J. Zeigler
Department of Political Science and Geography
Old Dominion University

Wilbur Zelinsky
Department of Geography
Pennsylvania State University

Foreword

The 27th International Geographical Congress celebrates 120 years of international cooperation among geographers. As in the 1870s, so in the 1990s, our quest is to understand the earth as human home. The task is never-ending, the challenge ever-changing, yet the spatial elements required to describe how people live and work and of how societies use their resources remain the same.

Geographical Snapshots of North America uses the concepts of point and line to interpret this remarkable continent in all of its geographical diversity. Points are concentrations or foci. Lines are paths of movement or boundaries. Together with areas that show the extent of things, they are the primitives upon which the logical structures of geography are built.

The dimensional primitives of geography provided the volume's authors with an umbrella of shade and comfort, no matter what their tools, techniques, methods, perspectives, or interests. All specialists, the contributors represent a kaleidoscope of subfields. Consequently, when you examine this collection of short essays you will encounter much of geography's lure, depth, texture, variety. Between its covers you will discover the breadth of analytical and philosophical approaches as well as topical interests of North American geographers. You will also find interpretations about North America that are of research and teaching value to geographers.

Like a family photograph, the book, organized around snapshot metaphors, instead of regions or topical specializations, will have a special meaning for Congress participants. But *Geographical Snapshots of North America* is no ordinary commemorative volume. It is a printing of peer-reviewed essays that will be of lasting value to all life-long explorers of the joy of geography.

GILBERT M. GROSVENOR
President, 27th International
    Geographical Congress
President and Chairman,
    National Geographical Society

ANTHONY R. DE SOUZA
Secretary General, 27th International
    Geographical Congress
Editor, *National Geographic Research
    & Exploration*

Preface

This book commemorates the fourth meeting of the international community of geographers to take place in North America in this century. It presents a sampling of short essays about the quadrant of Earth that stretches from Middle America and the Caribbean to the Arctic, and from the Atlantic to the Pacific. The volume is distinguished as much by the process of its development as by its contents. It is a contribution from a cross section of geographers. They range in experience from graduate students to applied geographers and to distinguished emeritus professors. Contributors represent the major branches of the discipline. They responded to an open call for submissions distributed through newsletters and convention fliers.

In spite of its inception as a random set of contributed manuscripts, the book represents the different perspectives used by geographers and its "snapshots" treat a broad range of North America's human and physical landscapes. Pulling together such diverse material under a single but nonetheless loose theme was exciting. And even though chapters are independent essays, with no deliberate linkage among them, the editors believe that the essential unity of the discipline suggests a principled structure for the book's contents. For those who sample broadly from these pages, the selections offer in their number an image of the greater whole. They offer a resource of ideas and imprints about North America, and insights on the practice and principles of geography.

Acknowledgments

The Publications Committee of the 27th International Geographical Congress thanks the contributors to this project. Anthony R. de Souza and the Organizing Committee were willing at some risk to break with the tradition of inviting essays from a few prominent scholars. Because of their support, we share the adventure and results of democratizing the procedure for crafting an important Congress publication.

We owe significant praise to the many authors who responded so willingly in preparing manuscripts in the uncertain but respected realm of peer review. They responded promptly to a demanding schedule and worked to make contributions accessible to a wide range of potential readers.

The Department of Geography, University of Western Ontario, provided a home and support for developing this book. The special efforts of Pam Brown, Trish Chalk, Marcia Santiago, and Darren Scott are noted. Pam handled the torrent of extra correspondence. Trish prepared the guidelines for graphics, and checked, completed, and revised maps and graphs. Marcia and Darren helped in the crucial late stages, preparing the final manuscript and checking the myriad details associated with a project of this scope.

Winfield Swanson, managing editor of *National Geographic Research & Exploration*, completed on short notice the task of copy editing the final manuscript. The cooperation and insights of Seymour Weingarten and the staff of The Guilford Press assured a quality production in time for the Congress.

I owe special personal thanks to the Publications Committee, whose members shouldered an immense amount of reading amidst already busy schedules. I also thank Barbara and Dan for their understanding support.

The Publications Committee, the Editor, and the authors share insights with colleagues from other regions who join us for the 27th International Geographical Congress. We hope, too, that other scholars, students, and armchair geographers, who seek appreciation and understanding of the world, will find our effort a credible reflection of the geographer's craft.

DONALD G. JANELLE
Chair, Publications Committee,
  27th International Geographical Congress
Professor and Chair, Department of Geography,
  University of Western Ontario

Contents

Introduction

A snapshot is a permanent record of a time and place. A geographical snapshot captures some essence of that record through the eyes and scholarly analysis of a geographer. In its elemental form, the geography of a region is an amalgam of points and lines that define its landscape features. Some points and lines are visible and easily spotted from the window of an aircraft or from an earth-bound vehicle. Some are but fleeting events that leave no permanent mark; while others, though abstract, underlie our geography. Geographers interpret these landscapes from a variety of theoretical and applied viewpoints and with a varied set of remarkable tools. They include maps, remotely sensed images, photographs, sketches, graphical analysis, mathematics, and literary skill.

The mandate for authors of *Geographical Snapshots of North America* was to focus on the symbolic or representative significance that selected points and lines have for interpreting North America. North America defines its geography by the lines and points that mark its jurisdictions, natural areas, paths of movement, junctions, and critical events. Some, of course, are more critical than others. Critical points in geographical space are products of events in time; for instance, Gettysburg, Yorktown, and Mount St. Helens. Some involve very short time spans, with each minute marked indelibly in the public consciousness: for example, the reactor core melt-down at Three Mile Island. Others are shaped only by the passage of decades, centuries, and millennia: the Mississippi River. Some extracted immense human labor: trans-continental railways, the Alaskan Pipeline, and the Trans-Canada Highway. Others are products of primarily natural forces: treelines and coastlines. Some, though visible, are often ignored: skid rows. Some are an experience: the journey to work. Some reflect social forces: breadlines. Though the events that define them may be short, some points assume a symbolic importance that transcends the succeeding years, decades, and

centuries. Thus, for example, Bunker Hill, scene of the first major battle in America's Revolution, is enshrined in the national culture. Some, though of questionable intrinsic durability, assume significance as symbolic places of philosophical or of social and political meaning: Walden Pond, Selma, and Wounded Knee. Some become national and international icons.

No single discipline can capture the full understanding of the richly varied landscape of a continent. The landscape itself changes, and no single snapshot can portray its existence and development. Nonetheless, the perspectives represented in this volume argue well for the role that geographers play in providing insight about North America's origins and directions.

The actual features discussed in this book are a mere fraction of the lines and points that define the continent's geography. Yet, they embrace a variety of regions, interregional linkages, landmarks, and barriers. They include structures of the artificial landscape and key elements of the natural system. The collectivity of these natural and human elements exposes a general sense of regional integration.

In part, the concentrations of professional geographers in certain regions, their propensities to submit manuscripts, and the current fashions that favor some research topics over others also contribute to the regional variations in coverage shown on the maps that introduce each of this book's ten parts. For instance, Arizona, Louisiana, and Ontario are more represented than Mexico or Alaska.

"Snapshot" implies brevity, informality, and a sense of the time. Many chapters share these traits, while others, also true to the snapshot metaphor, are studied compositions, with careful cropping of extraneous matter to evoke sharp images. Snapshot metaphors provide the organizational framework for this volume, but the Index also gives a framework by region and topical specialization.

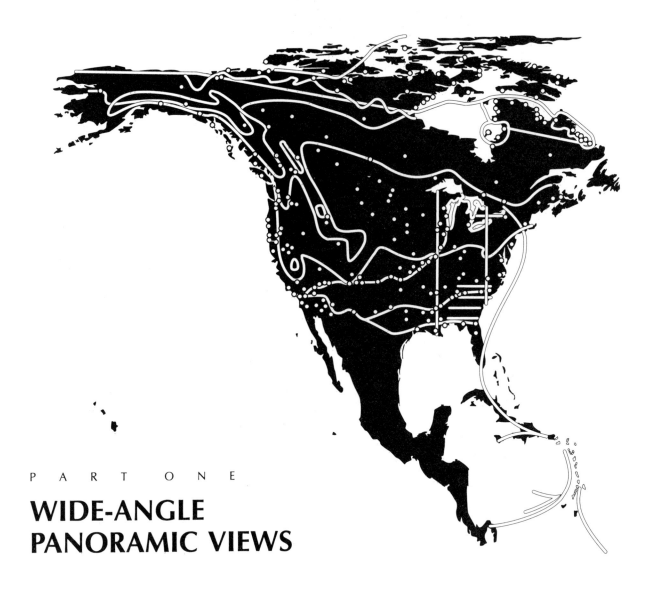

P A R T   O N E

WIDE-ANGLE PANORAMIC VIEWS

To capture the sweeping expanse of North America's geography, one needs a spacious exhibit. The transformation of its original Indian and Inuit populations and their territories requires vantage points through time and space. One needs a broad, open view to put in perspective the social and territorial evolution and organization of native North Americans. Similarly, historical depth of field adds substance and meaning to interpreting routeways of population and commerce through the islands of the Lesser Antilles. A contemporary traverse across the continent's natural north-south grain, from Ontario to Florida, meshes an understanding of the past with realities of the present. While many landscape features are simple lines on a map, these essays prove how geographical insight can tease out the hidden symbolism of major highways as well as the clues to environmental change represented by shifts in the continent's natural vegetation and precipitation boundaries. It is useful to start one's tour of North America with the focusing ability provided by a wide-angle lens.

—John R. Mather, University of Delaware

The Namerind Continent's Euro-American Transformation, 1600-1950

E. M. Bjorklund
University of Western Ontario

What happens to peoples with established cultural identity and spatial autonomy when "outsiders" intrude and persist over a long period? Commonly, geographical accounts of the emergence of "New World" culture in North America have Euro-centric perspectives (Meinig 1986). Equally important is what happened in the process from the viewpoint of indigenous cultures. Conventional interpretations of continent-wide colonization and occupancy have treated the events as if they occurred in a wilderness occupied only by "savages," "primitive people" who, whatever their number, constituted a threat. Furthermore, these groups were considered intrinsically inferior to the intruding population who gained territorial control over the continent. As a consequence, conquest and domination was justified as "inevitable." It is important to acknowledge documented evidence of the complete transformation of indigenous peoples and their homelands by the European migration into North America to provide some balance of geographical interpretation.

Nearly 200 tribes of native North American Indians (whose collective contracted-name is Namerinds) occupied the entire continent at the outset of European exploration and settlement. Starting nearly four centuries ago, their territories (nations) by mid-20th century either were eliminated entirely or reduced to small, discontinuous tracts and enclaves. Invading groups, referred to here collectively as Euro-Americans to reflect their varied European origins, penetrated the continent from four directions. Spanish explorer-colonizers focused on the southern and western faces of North America. French interests probed southern and northern parts. British and various North European attentions were directed at penetration from eastern and western coasts. Tentative toeholds expanded along coastlines and in-land along waterways and many existing Namerind land routes to form alignments of settlement that subsequently became frontal waves.

At first, only isolated localities were claimed. Later these expanded to colonial regions and national spaces. Most of the historic record describes this process from the developing Euro-American perspective (Meinig 1986; Morison 1971, 1974; Sauer 1971). Present discussion focuses on the less well-known Namerind perspective of this contact (Driver 1970; Marriot and Rachlin 1969; Oswalt 1966; Wax 1971).

Documentation about the actual transformation is fragmentary and scattered. Population of Namerind peoples is estimated at 18 million at the outset of European contact (Dobyns 1966; Dobyns and Swagerty 1983). Estimates used by the National Geographic Society (1972) suggest from 1 to 8 million. Whichever estimate is closer to the truth, the population pattern of indigenous people was altered irrevocably by an influx of some 45 million Europeans before 1950 (Wyotinsky and Wyotinsky 1953, 72). "Never before did so many people move so far within so short a period as in this great migration" (Philbrick 1963, 308).

Namerind Cultures, Circa A.D. 1600

Two sources are used to record the spatial pattern of Namerind peoples early in the contact process with Euro-Americans and for their present locations. Harold Driver's map (Figure 1) shows the approximate territories of Namerind nations circa A.D. 1600. The map depicts an entirely occupied continent—not vacant or only spottily inhabited as implied by the "wilderness" image so long perpetuated by Euro-Americans.

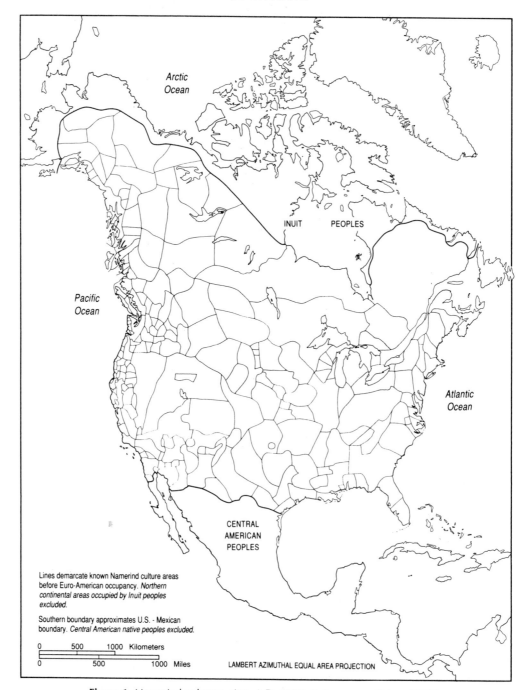

Figure 1. Namerind cultures, circa A.D. 1600. Data source: Driver (1970).

This provides a "before contact" view. Although Namerinds and contemporary scholars dispute particulars of Namerind territorial boundaries, for this account, those shown in Figure 1 may be taken as effective approximations.

As shown on the map, the original sizes of Namerind territories varied from a few hundred square kilometers, in the cases of several western and Pacific cultures, to tens of thousands of square kilometers, particularly in the continental interior. Whatever the size of their original culture area, each nation or tribe was characterized by a lifestyle based on resource-use regarded as sufficient for its needs and technology. Skilled cultivators, such as the people of the Cherokee Nation, were organized into town and village nuclei with attached cultivated tracts rimmed by wooded lands used in support of an agrarian way of life. By contrast, for example, people of the Ojibwa Nation based their economy on hunting-fishing-scavenging from woodland, lake, and river resources. They centered on multiple camp sites used intermittently.

The concept of boundary lines in a Western European sense did not exist in Indian territorial awareness. Namerind peoples were respectfully conscious of, and made use of, the varied resources throughout their respective regions. Territories were generally not rigidly bounded by formalized lines or limits of occupancy. There were, however, overlapping zones of contact among neighbors. Nonetheless, lines now drawn on maps representing these zones as edges are useful to suggest the regional extent of their attachment to territory and its resources for livelihood support. They attest to their beliefs in rights of land use rather than ownership, a concept that did not exist. They form a record of defensible space giving shape to places where Indian rules and codes regarding territory were persistently violated. In no sense was the pattern of A.D. 1600 construed as a static spatial arrangement of peoples—nor should it be. The pattern was dynamic, reflecting the ongoing inter-cultural contacts among the different tribal groups subject to adaptation over time.

Namerind Locations, Circa 1950

The map in Figure 2, based on information pieced together by Sol Tax (1961), records the locations of residual Namerind people as of 1950. It accounts for the tracts and vestigial patches allocated to Namerind survivors from prolonged, often harsh, hegemonic domination by generations of Euro-Americans. This gives an "after" view of post-European migration and settlement. In the United States and Canada, government-controlled reservations and reserves for Namerinds total roughly 2,424 million hectares (6 million acres) in Canada and 21,008 million hectares (52 million acres) in the United States (Driver 1970). This amounts to 1.3 percent of the original area of occupancy by Namerinds in 1600. The wide dispersal of allotted territories is strikingly apparent. Few bands, tribes, or nations occupy more than token parts of their original lands.

Statistical tests show that the spatial change in Namerind occupancy could not be accounted for by chance. Explanations of the contemporary pattern of Namerinds across the continent cannot exclude variations in the duration of Euro-American contact, differences in Euro-American perception, land-use patterns, or developmental policies regarding Namerind

peoples as significant factors to explain their contemporary circumstances. Though all sections of the continent have undergone marked transformation from the standpoint of Namerind occupancy patterns, the southeastern United States has been most radically altered by large agricultural enterprises. Generally, the elimination of Namerinds is most marked in areas taken over for Euro-American-style agriculture. The more intensive the contemporary land use, the more complete the destruction of Namerinds.

Dimensional Analysis

For each Namerind territory shown in Figure 1 the geometric centroid was selected and distances were measured from that point to current locations (shown in Figure 2) for which there are identifiable populations of that tribe today. The lines in Figure 3 connect origin centroids and destination points generalized as straight-line abstractions of routes used by Namerind peoples as they were moved from their homelands to the territories assigned by Euro-Americans. They represent paths of forced movements, long marches, flights from pursuing troops or settlers. In some cases these were retreats to highland "fortresses" and least preferred areas. The length of each line is an indicator of spatial displacement and alienation from their original domains. The number of lines to various destinations from any centroid is an indication of the degree of fragmentation of each Indian nation.

From this procedure two indices of transformation of the aboriginal way of life were derived from Figure 3, one of dislocation and the other of fragmentation. These indices reveal the degree of disaster and severity of disruption from the point of view of indigenous cultures. The net result of this point-to-point, line dimensional measurement of the Indian diaspora creates a single summary statement of the total stressful centuries-long "event." This event decimated virtually all the 200 Indian nations of North America. The transformation leaves only remnant groups dispersed to "least desired" resource-poor pockets.

Findings

Dislocation of Namerind cultures has taken the form of expulsions outside their original areas, shifts to much diminished areas within their homelands, or a combination of both. Tribal abilities to sustain their way of life were effectively weakened. Fragmentation, splitting tribal populations into separate groups and dispersing these widely, contributed to loss of language,

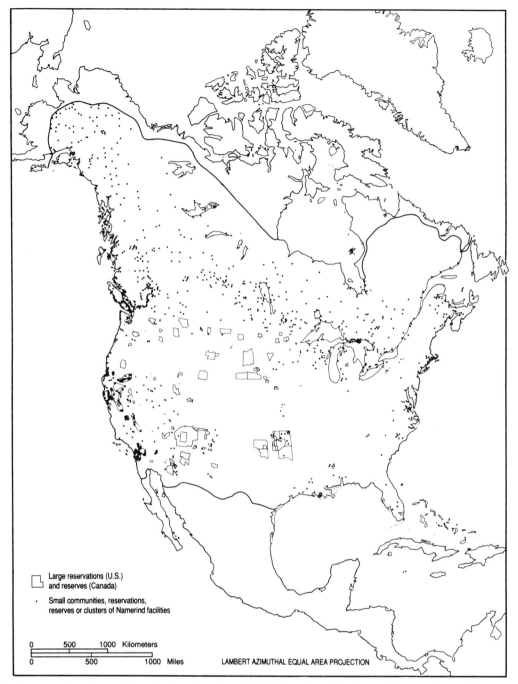

Figure 2. Contemporary locations of Namerind peoples, circa 1950. Pattern shows locations of Namerind populations on reservations and reserves, or clusters of families who retain some identity with their original culture. Data and map source: Tax (1960).

customs, and cultural cohesiveness. Where fragmentation has been coupled with close coexistence with other Namerind cultures, to say nothing of Euro-American culture, remnant groups have become transformed into hybrid populations. Dislocation and fragmentation, however, are not strongly correlated, which suggests that these were two powerful means

used separately and in combination to loosen the hold of Namerind peoples on their lands.

As territories and occupants were disturbed by Euro-Americans in search of new homelands for themselves, they created entirely new forms of spatial organization to ensure spatial integrity of the newly formed countries—United States and Canada. Each country, while it followed somewhat different policies and practices, had similar general objectives. These were elimination or removal of entire populations whenever their presence or resistance to intrusion was intense and could not be tolerated. Thus, as Namerind contact with the new cultures progressed, many indigenous ones were absorbed or obliterated (genocide), moved to locations outside their territories, or confined to limited localities within their former territories.

All these forms of territorial constraint had powerful impacts on Namerind population. Many were attacked by disease, starvation, harassment, and prolonged warfare, which decimated their numbers and demoralized any survivors. The Namerind nations' peoples,

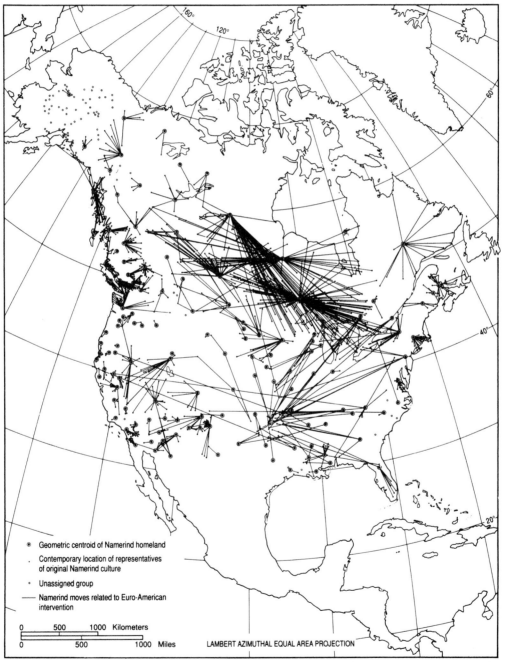

Figure 3. Dislocation and fragmentation of Namerind cultures by Euro-American occupancy. Geometric centroids of Namerind culture areas from Figure 1 are connected by lines to the locations assigned by Euro-Americans (from Figure 2).

fragmented into smaller groups and dispersed over wide areas (Figure 3), found contact virtually impossible to maintain. This weakened, if not destroyed, cultural integrity. Whatever the particular experiences were —far too complex to detail here—the cultural vitality of all groups was sapped, severely disrupted, and in many cases, obliterated.

The outcome of American and Canadian national policies has been a continental tragedy of enormous and enduring proportions for Namerind peoples. Yet, slowly, from the mid-part of this century onward, gaining some momentum at the present time, remnant populations of native peoples now rally to their own causes for cultural survival and territorial recognition and justice. Among the many Namerind groups seeking land claims settlements to rectify gross injustices in the past are the Abenaki in Maine, the Cree in eastern Canada, the Tsimshian on the Pacific coast, and the Hopi in the southwestern United States.

The dimensional measurements described in this paper assess cultural dislocation and fragmentation as the geographical basis of cultural genocide. They demonstrate the power of Euro-American strategies and tactics employed consistently against native peoples. Viable nations and tribal areas were reduced to isolated tracts (points).

Populations were dispersed to discontinuous localities (discontinuous, unconnected points), and were confined in enclaves (reservations, reserves) surrounded by a transformed continental area peopled by an unsympathetic, if not hostile or indifferent majority.

References

Dobyns, Henry F. 1966. Estimating aboriginal American populations: An appraisal of techniques with a new hemisphere estimate. *Current Anthropology* 7:395-416.

―――, and Swagerty, William R. 1983. *Their Number Thinned.* Knoxville, TN: University of Tennessee Press.

Driver, Harold E. 1970. *Indians of North America* (2nd edition revised). Chicago: University of Chicago Press. Maps 2, 38-45.

Marriot, Alice, and Rachlin, Carol K. 1969. *American Epic: The Story of the American Indian.* New York: G. P. Putnam and Sons.

Meinig, D.W. 1986. *The Shaping of America: A Geographical Perspective of 500 years of History.* Vol I. *Atlantic America, 1492-1800.* New Haven: Yale University Press.

Morison, Samuel Eliot. 1971. *The European Discovery of America. The Northern Voyages, AD 500-1600.* New York: Oxford University Press.

―――. 1974. *The European Discovery of America. The Southern Voyages, AD 1492-1616.* New York: Oxford University Press.

National Geographic Society. 1972. Indians of North America. Map Supplement # 6. *National Geographic Magazine* 42:739A.

Philbrick, A. K. 1963. *This Human World.* New York: John Wiley and Sons.

Oswalt, Wendell H. 1966. *This Land Was Theirs: A Study of the North American Indian.* New York: John Wiley and Sons.

Sauer, Carol Otrwin. 1971. *Sixteenth-Century North America: The Land and People as Seen by Europeans.* Berkeley: University of California Press.

Tax, Sol. 1961. *The North American Indians 1950 Distribution of Descendants of the Aboriginal Population of Alaska, Canada, and the United States.* Chicago: Department of Anthropology, University of Chicago.

Wax, Murray. 1971. *Indian Americans: Unity and Diversity.* Englewood Cliffs, NJ: Prentice-Hall.

Wyotinsky, W. S., and Wyotinsky, E. S. 1953. *World Population and Production,* Vol. I. New York: The Twentieth Century Fund.

Mapping the Inuit Ecumene of Arctic Canada

Robert A. Rundstrom
University of Oklahoma

Europeans and Euro-North Americans first came to terms with Arctic North America just as we do today—as an enigmatic area most commonly fitting one of three images: a sublime and ferociously peaceful habitat, a land of purity and whiteness, or a massive uninhabitable wasteland of rock, ice, and wind. Viewed from the south, the Arctic seems a vast featureless wilderness or frontier, without the reassuring point or linear referents to which we are accustomed, and offering little more substantive than isolated mineral resources and the promise of unending introspection for the periodic visitor.

I suggest another view of the Arctic, one vastly different, yet equally valid and compelling . . . and perhaps richer. I include the perspective of Arctic Canada's long-term residents—the Inuit and Inuvialuit (Eskimos)—while briefly sketching a human geography of their ecumene (inhabitable world). Unlike the southern view, the Inuit Arctic is filled with numerous points of habitation interconnected with lines of wildlife and human activity. Caribou paths, traplines, and an extensive network of travel routes for hunting, visiting, and recreation constitute the ordinary landscape of Inuit life. The Inuit world is one in which frequent movement along these lines is preferable to sedentary life in a government-planned settlement.

There are problems in bounding Arctic Canada. Whereas I dismiss "tree-line," the 10°C July isotherm, or permafrost limits with some misgivings, it is easy to dispense with the Arctic Circle or 60th parallel as a southern edge. Instead, I choose to use the limits set by the Inuit, and their patterns of movement, settlement, and land use (Freeman 1976). Implicit in this definition is the idea that Arctic Canada has limits in all directions, even northward; and wherever it is, it is humanly defined by the presence of a 1,000-year-old culture. Non-Inuit do not stay long in the Arctic.

The Inuit are the only ones who travel, hunt, and live season-to-season throughout this region. They know its resources and dangers intimately, and have a strong bond with their homeland.

Mapping the Inuit Ecumene

The map shows some features of the Inuit ecumene, and only a few that are important to Euro-Canadians (Figure 1). Regional and settlement names are printed in the native language, Inuktitut, using Roman orthography, although most are not officially accepted by the federal government in Ottawa (exceptions include Arviat, Iqaluit, Inuuvik, and all those in Nunavik). Wide, gray lines signify the crucial land claims and land-use boundaries. Scale varies on the map, because travel time and distance vary considerably with the season, local conditions, and purpose. Inuit travel among settlements by propeller-driven aircraft on occasion, but commonly use snowmobiles (winter) or four-wheel drive, all-terrain vehicles (summer) to traverse hundreds of kilometers on the ground.

Qallunaat (white, Euro-North American) cities are isolated points ambiguously located in "The South," and are shown close to the edge of the Inuit ecumene because of the speed with which jet aircraft can place Inuit in these cities. Most Qallunaat cities represent "outposts" for Inuit higher education, hospitalization, or bureaucratic services. Those in the United States are often portrayed on television. One of these, Detroit, is home to a television station whose broadcast is commonly viewed in Arctic living rooms.

Individual stations in the Distant Early Warning system appear on the map because Inuit laborers, paid by the U.S. government, built most of these isolated military outposts in the 1950s. Whereas Qallunaat

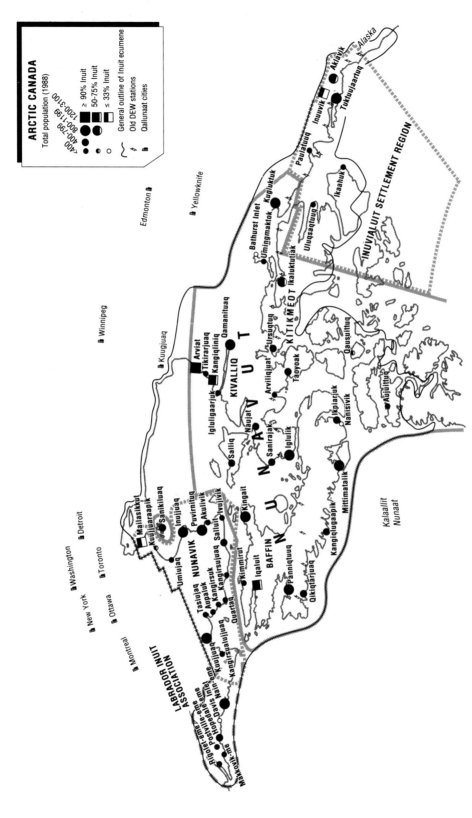

Figure 1. The Canadian Inuit ecumene, looking south from above the North Pole. Scale is larger in the north, and progressively decreases toward the south. Land claim and land-use boundaries are approximate, and numerous temporary camps and abandoned settlements are not shown. Data sources: Central Intelligence Agency (1978); Crowe (1990); Labrador Inuit Association; Müller-Wille (1987); Outcrop, Ltd. (1990); Statistics Canada (1988); and Tremblay (1989). © 1992 by Robert Rundstrom.

think of the Distant Early Warning Line as a linear window on the Soviet North, elder male Inuit recall the sites as points on the land associated with brief periods of cash employment. For younger Inuit, these sites confirm the increasing dominance and mystifying motivations of the Qallunaat now living among them.

Northern Labrador, northern Quebec (Nunavik), and two thirds of the Northwest Territories (Nunavut and the Inuvialuit Settlement Region) comprise the Canadian Inuit ecumene. It is twice the size of Alaska, almost a tenth of the continent, and its residents are more likely to claim a native heritage than anywhere else in North America. Seventy percent of Arctic Canadians claim a native culture, primarily Inuit, as their own. In the Kivalliq and Kitikmeot regions of Nunavut, 89 percent identify themselves as Inuit (Maslove and Hawkes 1990).

Points of Settlement

By any measure, Arctic Canada holds a small, dispersed population. Its 36,000 Inuit and non-Inuit residents are spread unevenly among 54 settlements. Two thirds live in Nunavut and the Inuvialuit Settlement Region, one fourth in Nunavik, and the rest in Labrador. An "average" settlement is 85 to 90 percent Inuit with only 650 people, but this statistic can be misleading (Labrador Inuit Association, personal communication, 25 February 1991; Outcrop, Ltd. 1990; Statistics Canada 1988; Tremblay 1989).

Significant Qallunaat minorities live in large towns such as Iqaluit, Kangiqliniq (Rankin Inlet), and Ikaluktutiak (Cambridge Bay), where government administration and services dominate. Although Inuit are the majority in these places, their ambience is akin to a small government town in Ontario. Mailasikkut (Chisasibi) and Aklavik have large Cree and Dene Indian populations respectively, and Inuuvik mixes Dene, Inuvialuit, and Qallunaat. A visitor notices that Inuit culture is best expressed in the smaller communities.

Inuit traditionally migrated among summer and winter camps out on the land and ice, developing an environmental schema that emphasized movement and linear routes among well-known points on the landscape. Their lifestyle changed dramatically beginning in the 1950s when the federal government relocated them to fixed hamlets and towns on the coast (Qamanituaq, or Baker Lake, is the single inland exception). Although some of these places were traditional seasonal campsites for the Inuit, many owe their existence to a local Distant Early Warning station (for example, Qikiqtarjuaq and Sanirajak), nearby mineral resource

(for example, Kangiqliniq and Qausuittuq), or a relocation decision made in Ottawa (for example, Umiujaq, Tikirarjuaq, and Aujuittuq). In one or two generations, the Inuit lifestyle has shifted from an emphasis on periodic linear movement away from camps to coping with a mainly sedentary urban existence. Since the 1950s, many elder Inuit and their children have adapted to urban life, becoming dependent on government welfare and the southern food their money can buy in the local Northern Stores (see Hudson's Bay Company) outlet. The youth are especially urbanized, knowing relatively little of the land, sea, and traditional routes of travel.

Like native people throughout North America, the Canadian Inuit population is growing fast. Doubling the 1960 estimate of 13,000 only took 20 years, and it should double again by 2000 (Irwin 1988). Approximately 37 percent are less than 15 years old (41 percent in Nunavik). In almost half the settlements, more than 40 percent are under 15 (Outcrop, Ltd. 1990; Statistics Canada 1988)!

Immigration within the ecumene easily exceeds emigration to the South. Inuit do not favor the population density, fast pace, and disorienting effects of Qallunaat cities. Moreover, there are real dangers for Inuit not attuned to metropolitan lifestyles, and who lack the local support groups typically available for other recent immigrants. Many long for their traditional foodways, and find the heat and humidity of a southern summer difficult to tolerate.

Northern jobs are scarce, however, and rarely last long. A few opportunities exist in emerging "cities" like Iqaluit and Kangiqliniq, where young families are increasingly attracted from smaller settlements. For example, during the 20 years from 1986 to 2006, Kangiqliniq is predicted to grow by 74 percent—from a population of 1,374 to 2,383. One third of this growth is expected to result from immigration from other Nunavut settlements as Inuit seek jobs associated with the local air force base, small college campus, hospital, and burgeoning administrative offices (Irwin 1988). But, inter-settlement migration undermines the social support normally drawn from the local extended family. Stresses have led to an increasing number of dependent, single-parent families.

Migration, high local birth rates, and an ineffectual government has created crises in all areas, notably housing. For example, the government of the Northwest Territories, comprising all of Nunavut and the Inuvialuit Settlement Region, strains to supply 300 new housing units per year when there is clear need for 10 times that many (Bell 1990). In 1990, the government built only 20 of the 212 housing units needed in Iqaluit, where homelessness has become a persistent and dangerous problem (Bell 1990). In short, fewer financial resources are available to meet the needs of an increasing population. Inuit society is becoming a tottering welfare state wholly dependent on external political decisions.

Although jobs are important, Inuit leaders recognize language retention as the fulcrum on which the future of Inuit society balances. Inuit use their language more than other native people use theirs (Maslove and Hawkes 1990). For example, 90 percent of Nunavik Inuit claim Inuktitut as a mother tongue in use at home; 70 percent usage is reported in Nunavut (Statistics Canada 1988). But rising numbers of wholly anglo- or francophonic Inuit families are now offsetting this tradition. Intermarriage is a major factor in the decline of Inuktitut-speaking households (Dorais 1989; Irwin 1988). In short, if English is spoken at home, the native language begins to disappear.

The New Need for Fixed Boundary Lines

In an attempt to gain control over their lives, Inuit began pressing land claims in the early 1970s. Canada never made treaties in the Arctic, and the Inuit have no cultural basis for bounding regions with lines. Rivers, seacoasts, traplines, caribou paths, and their own travel routes have been the crucial lines of their ecumene. All indicate movement within a complex social and physical landscape. Moreover, 200 years of contact with outsiders had not changed them appreciably in this respect until quite recently. Ironically, they must now adopt the Qallunaat practice of drawing firm boundary lines on maps—imposing stasis on the land—as a means to ensuring their cultural survival.

In 1975, the Inuit of Nunavik, a traditional-use area, finalized terms with the federal government and received simple title to 5,500 square kilometers, cash, and other concessions in exchange for extinguishment of aboriginal title. Next, the Inuvialuit to the west finally settled for 77,700 square kilometers in 1984 following protracted negotiations over important oil and gas resources in the Mackenzie Delta. They received surface and subsurface title to 45 percent of their lands, still a record in native land claims (Crowe 1990).

Nunavut, including the Sanikiluaq outlier, is home to approximately 17,000 Inuit, about half the population of Arctic Canada (Outcrop, Ltd. 1990; Statistics Canada 1988). Easily the largest at 314,000 square kilometers, the Nunavut claim would give Inuit title to just 20 percent of the traditional-use area. Finalization was scheduled for fall of 1991, with parliamentary legislation the following year. At this writing, a crucial boundary dispute with the Dene (Athapaskan Indian groups) to the west has stalled progress, and there is

concern that a final agreement may not be fashioned before an anticipated federal election and change of government in 1992 (Canadian Arctic Resources Committee 1990; Crowe 1990). When Nunavut is finally established, it will make Inuit the largest private landholders in the Americas.

The Labrador Inuit Association began pressing their claim in 1989. As the smallest in both area and population, the Labrador Inuit Association is perhaps the most disadvantaged. Now, they are by-standers to bickering between Ottawa and the provincial government of Newfoundland (which governs Labrador) over cost-sharing matters (Crowe 1990).

Of the four claims, Nunavut is unique because Inuit leaders have succeeded in getting a guarantee of self-government in addition to the claim settlement—that is, Ottawa has agreed to create a new official territory called "Nunavut." However, federal negotiators remain ambiguous on the timing of this innovation, and insist on settling the land claim first. Once authorized, Nunavut will allow the Inuit to establish Inuktitut as the official language of the territory. Their leaders see that event as absolutely crucial to cultural survival (Canadian Arctic Resources Committee 1990).

Conclusion

Europeans and Euro-North Americans have always seen Arctic Canada as an amorphous areal mass, ignoring the crucial points and lines familiar to its Inuit inhabitants. Traditionally, these consisted of innumerable temporary hunting and fishing camps crisscrossed by the wildlife paths, river systems, and human travel routes that linked them. Movement was the central theme of this network.

Today's Inuit struggle to give meaning to the points and lines of contemporary existence. As traditional knowledge recedes into the past, stresses of travel and hunting are replaced by stresses of sedentarism and dependency in fixed communities undergoing rapid, chaotic urbanization. The Inuit now recognize the need to use the land claims process to impose permanent lines of property ownership on their world. Points and lines were always a crucial part of Inuit life, but the meaning of such networks has shifted precipitously. What once indicated flexibility and movement, now represents containment and control. These are alien concepts emanating from a world not of their making, but a world in which the Inuit nevertheless must live. The Inuit have impressive adaptation skills; they will continue to struggle with this latest and most crucial test of their survival.

Acknowledgments

Many thanks to David Craven for cartographic assistance, and to the National Film Board of Canada for permission to compile from their base map.

References

Bell, J. 1990. The crowded Arctic. *Arctic Circle* July-August:22-30.

Canadian Arctic Resources Committee. 1990. Nunavut revisited. *Northern Perspectives* November-December:2-28.

Central Intelligence Agency. 1978. *Polar Regions Atlas.* Washington, DC: Central Intelligence Agency.

Crowe, K. 1990. Claims on the land. *Arctic Circle* November-December:14-23.

Dorais, L.-J. 1989. Bilingualism and diglossia in the Canadian Eastern Arctic. *Arctic* 42:199-207.

Freeman, M., ed. 1976. *Report, Inuit Land Use and Occupancy Project.* Ottawa: Department of Indian Affairs and Northern Development.

Irwin, C. 1988. Lords of the Arctic: Wards of the state. *Northern Perspectives* January-March:2-12.

Maslove, A., and Hawkes, D. 1990. *Canada's North: A Profile.* Ottawa: Minister of Supply and Services Canada.

Müller-Wille, L. 1987. *Gazetteer of Inuit Place Names in Nunavik.* Inujjuaq: Avataq Cultural Institute.

Outcrop, Ltd. 1990. *Northwest Territories Data Book: 1990-1991.* Yellowknife: Outcrop.

Statistics Canada. 1988. *Census of Canada 1986.* Ottawa: Queen's Printer.

Tremblay, C. 1989. Housing in the North: Ten years of progress. *Rencontre* June:14.

The Lesser Antilles

J. H. Galloway
University of Toronto

The Lesser Antilles form a cultural region that has played a far greater role in the geography of North America than its scant area and small population would suggest. There are hundreds of islands in the region, tidied into a neat line that traces a broad arc through 1,000 kilometers from the Virgins in the north to Grenada in the south (Figure 1). The total area is a mere 7,000 square kilometers with a population today of approximately 1.7 million. Guadeloupe has the most space (1,702 square kilometers) and Martinique the most people (400,000). Many of the islands are of volcanic origin and rise steeply from the sea to sharp peaks; the others are low coral formations. They lie in the path of the trade winds, which roll impressive Atlantic breakers onto their windward beaches, pour orographic rain onto their hills, and moderate the tropical heat to a comfortable norm. The combination of topography and trade winds means that where rainfall is abundant, there is scant level ground for cultivation; where the land is level, as on the coral islands, water is often in short supply. The beauty of the islands, their meagre resources, and their shared record of human occupation in pre- and post-Columbian times form the roots of a strong regional identity.

The islands first appeared in human history as a route by which aboriginal groups migrated from the mainland of South America into the heart of the Caribbean. In colonial times, they again played a prominent role as the hearth of the sugar industry in North America. During the last 150 years, they have been a source of migrants to many places around the Caribbean and even farther afield. They now have in common economic and political problems.

The Pre-Columbian Routeway

Migrating populations might have made their way most easily to the Caribbean islands from the surrounding mainland via three access points: from the southern tip of Florida to Cuba, a distance of 145 kilometers; across from the Yucatán Peninsula to Cuba, a distance of 200 kilometers; and up from South America along the line of the Lesser Antilles. This last route is particularly attractive to sailors of simple craft because the trade winds and ocean currents favor a northerly direction of movement from Trinidad and the Venezuelan coast; distances between the islands are short; and each island is in sight of its neighbors, excepting the anomaly of Barbados over the horizon to the east. The archaeological record so far unearthed gives no evidence of any mass movement of people either from the Yucatán or from Florida to Cuba; rather, contacts between Cuba and the mainland were of a chance nature such as the arrival of craft blown off course in storms. The evidence, however, leaves no doubt as to the importance of the third route: all three aboriginal groups present in the Caribbean when Columbus arrived—the Ciboney, the Island Arawak, and the Carib—had entered the Caribbean from South America via the Lesser Antilles. The Ciboney were the first, beginning their migration along the island chain by about 3770 B.C. according to carbon dates. They were followed by the Arawak who by 1492 had pushed the Ciboney into the farther reaches of eastern Cuba and southern Hispaniola.

It was the Ciboney and the Arawaks who introduced agriculture and the tropical South American complex of crops to the Caribbean. Maize was a later arrival and probably was brought into the Caribbean via South America rather than across from Yucatán. The Caribs were the last to begin the journey along this routeway. Reputedly cannibal—although the extent of this practice is unknown—they were warlike, killed the Arawak men, and married their women. By 1492 they had completed the conquest of the Lesser Antilles and had begun to invade the larger islands. A consequence of the use of this easy migration route meant that the

Caribbean belonged culturally to northern South America rather than to Mesoamerica or to Florida (Rouse 1966).

Although routeways are for passage rather than for settlement, the Lesser Antilles did support a permanent population and indeed at the time of the Spanish arrival in 1492 there may have been as many as 175,000 inhabitants. This figure results from multiplying the area of the islands by what could well have been the average density of population in the Caribbean at that time, 25 per square kilometer (Watts 1987). This population quickly fell victim to the Spanish need for labor, particularly in the gold fields of Hispaniola. Slavers worked through the Lesser Antilles between 1512 and 1520, leaving them bereft of population save for a few Caribs who eluded capture in the mountainous interiors of some of the islands (Watts 1987). During their long occupation, the aborigines necessarily changed the landscape of the islands, clearing land for their crops, and felling trees for houses and fuel, but they placed relatively little pressure on the natural resources and passed the islands on to their conquerors in sound ecological condition.

Culture Hearth

The emergence of the Lesser Antilles and of Barbados in particular as the hearth of the sugar industry in North America was perhaps an unexpected development and requires some explanation. With the incorporation of the Lesser Antilles into the Atlantic world of commerce, the locational context of the islands changed. The link to South America ceased to matter while the links to the rest of the Caribbean and to Europe became all important. The Spaniards' appreciation of the geography of the Caribbean evolved as their New World conquests proceeded so that within a few decades of the arrival of Columbus the islands, both great and small, were of little concern to them other than as a protective screen for their prime enterprise, the mainland empire. Their strategic focus shifted northward, to the Straits of Florida, which the prevailing winds and ocean currents made the natural exit for their fleets returning to Spain from Mexico and Peru.

Abandoned by Spain, the Lesser Antilles were nevertheless attractive to the north Europeans when they began to look for islands to colonize in the 1620s. Uninhabited and undefended, they were easy to occupy. Their location upwind of all other territory in and around the Caribbean made them relatively safe from Spanish attack but at the same time a good base from which to trade with or to threaten the Spanish pos-

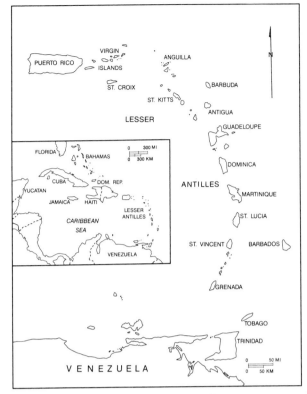

Figure 1. The Lesser Antilles.

sessions. A further important consideration was the easy passage back to Europe. Ships from the Lesser Antilles could run north before the trades to pick up the westerlies in mid Atlantic, in contrast to ships out of a port within the Caribbean, which had an arduous tack to a passage between the Antilles or had a long reach round the tip of Cuba to the gulf stream and so home to Europe. Into this changing geopolitical scene intruded the problem of a suitable cash crop for the colonies.

By the early 1600s, the sugar industry, which Spain had once attempted to foster on Hispaniola, finally failed, in good measure a victim of Spain's declining interest in the economic possibilities of its Caribbean islands. However, for the English and French sponsors of the new colonies in the Lesser Antilles, disillusioned by the 1630s with the poor returns they received from tobacco and cotton, sugar cultivation seemed a promising alternative, but they lacked the necessary capital and technology to make the transition. The Dutch, who at this time were being gradually driven from their plantations in Brazil and were looking for new sources of sugar for their refineries, willingly gave this assistance. On Barbados, in the 1640s, small holdings began to give way to plantations and European laborers to be replaced by African slaves. This "sugar revolution" spread soon afterwards along the line of the Lesser Antilles (Galloway 1989).

The culture hearth took on very distinctive characteristics. The small size of the islands ensured planters made careful, efficient use of the scarce resources. They took care of the soil by manuring, preserving the moisture content and controlling erosion. Fuel-efficient furnaces reduced the demand for fuel; bagasse (crushed cane) supplemented and then replaced timber as deforestation advanced. These innovations, introduced first in Barbados, diffused quickly through the Lesser Antilles and later onto Jamaica, Cuba, and St. Dominique (Haiti) (Galloway 1989). Profits lay in producing sugar, and planters hence emphasized the cultivation of sugar cane even at the expense of food crops with the consequence that the Lesser Antilles soon came to depend even for staple foods on imports from Europe and North America.

They also ran a deficit in human beings. Disease and the arduous nature of the agricultural work meant that the slave population was not self-sustaining but had to be continually replenished with fresh arrivals from Africa. The gradual increase in the population of the islands over the decades is a testimony to the scale of the slave trade. By the mid-years of the 18th century, the Lesser Antilles' role as a culture hearth had come to an end. They had ceased to be a center of innovation in the sugar industry and the Greater Antilles had become more important sources of sugar. The legacy for the Lesser Antilles of this period of prominence was a deforested landscape, intensive commercial land use, a slave society with a small white planter class, and a greater density of population than anywhere else in colonial America. In a few favored localities in Massachusetts, New Jersey, and Pennsylvania there were by the end of the 18th century as many as 20 people per square kilometer; Barbados, by contrast, had a population of 78,000 or 181 per square kilometer (Galloway 1989).

Exodus

As the 18th century passed into the 19th, the circumstances of the Lesser Antilles changed yet again, for the better in some respects, for the worse in others. On the positive side of the ledger, slavery was abolished and conditions of health improved. Immigration, whether forced or free, was no longer necessary to maintain the population. Indeed, the islands began to register net increases, and built towards a surfeit of people (Curtin 1990). On the negative side was the decline in the economic importance of the islands. The new world of the nineteenth century offered greater scope for investment in other places and other industries than in the sugar plantations of these tiny territories. The colonial powers' neglect of these islands was made all the more easy by the emergence of an alternative source of sugar in the sugarbeet industry of Europe. Without new economic activity to give employment, without land which the growing population could farm, many of the people of these islands had to leave to earn a livelihood. Migration characterizes the long third phase in the historical geography of these islands. This phase lasted well into the 20th century.

Migration began on an appreciable scale in 1838 when the British government ended apprenticeship, the post-slavery form of labor that had continued to bind former slaves to plantations. Immediately the Lesser Antilles became a routeway again, but this time with the flow of movement in the opposite direction, from north to south, as former slaves made their way by schooner from the densely populated islands of St. Kitts, Antigua, and Barbados to the new colonies of Trinidad and Guiana where land was unoccupied and labor was in demand. Between 1839 and 1849, Trinidad received 10,278 immigrants from these islands and British Guiana another 7,582 (Galloway 1989).

In later years the movement southward extended to Surinam, Cayenne, and Brazil. Islanders took jobs on ships and roved the world well beyond the Caribbean looking for work. In 1892, some Barbadians even contracted to work in the Belgian Congo and went there accompanied by a few mates from Martinique and St. Lucia. Between the years 1861 and 1903, the net emigration from Barbados may have been as high as 103,500 (Galloway 1989; Richardson 1985). In the early 20th century, Panama became a major destination where there was work in digging the Canal (Richardson 1985); and the developing oil industry attracted workers to Lake Maracaiba, Curaçao, and Aruba.

Patterns of seasonal migration evolved as islanders moved to work as cane-cutters in the sugar harvests of Cuba and the Dominican Republic. Over the decades many thousands made their way to the United States, Canada, Great Britain, and France. This diaspora has been of economic benefit to the islands in that the remittances of the migrants have helped sustain their families at home.

Questions

The islands today make little impact on the rest of North America except as tourist destinations. They are no longer a routeway; their strategic role has gone. However, they are the scene of two ongoing struggles: the search for a viable economy and the search for a viable political status. The once dominant sugar industry is now in full retreat with only two of the islands, St. Kitts and Barbados, retaining a significant level of production. The low prices the crop commands means that the future of sugar, even on these two islands where it has been grown for more than 300 years, is not secure.

Profitable, alternative forms of land use are hard to find. Cattle ranching does not provide nearly the same level of employment as sugarcane; bananas are already an important crop and the market for such specialty crops as nutmeg (grown on Grenada) and arrowroot (grown on St. Vincent) is limited. The mass tourist industry is a consequence of cheap, rapid air travel that came with the introduction of jet passenger planes in the 1960s. It is a mixed blessing, providing some, mostly unskilled, work as well as foreign exchange, but it has failed to provide a major market for locally grown produce because the tourists prefer North American food. It has also raised the cost of land to the local population and has strained the freshwater resources, particularly of the drier islands. A question mark therefore remains over the path the island economies might take.

A second question hangs over the political future of the islands. The common heritage and regional consciousness has so far not proved sufficient to overcome old colonial attachments; even among the English-speaking islands, political federation is an idea for the future. Guadeloupe and Martinique are overseas departments of France, and the social services of the French state are an inducement to remain so. The Americans, Dutch, and British still administer clusters of small islands as dependent territories. The largest of the British islands have become independent since the 1960s and are now engaged in an experiment of making mini-statedom work that may afford a model for mini-states elsewhere.

References

Curtin, Philip D. 1990. *The Rise and Fall of the Plantation Complex: Essays in Atlantic History.* Cambridge: Cambridge University Press.
Galloway, J. H. 1989. *The Sugar Cane Industry. An Historical Geography from Its Origins to 1914.* Cambridge: Cambridge University Press.
Richardson, Bonham C. 1985. *Panama Money in Barbados, 1900-1920.* Knoxville: University of Tennessee Press.
Rouse, Irving. 1966. Mesoamerica and the eastern Caribbean area. In *Handbook of Middle American Indians,* Vol. 4, ed. Robert Wauchope, pp. 234-242. Austin: University of Texas Press.
Watts, David. 1987. *The West Indies: Patterns of Development, Culture and Environmental Change Since 1492.* Cambridge: Cambridge University Press.

Crossing the American Grain, or the Happiness of Pursuit

Grady Clay
The Gazetteer Project
Louisville, Kentucky

Growing up in Atlanta, I was taught that it was indeed THE CENTER OF THE UNIVERSE, and its Peachtree Street the model for all MAIN DRAGS. Living in Cambridge taught me to be neighborly toward Boston, THE HUB OF THE UNIVERSE. Living in Alaska reinforced the knowledge of being OUTSIDE, a long way from STATESIDE. Curiosity about places seized me early.

Here I have selected a slice of placeful reality, carved out of landscapes caught in passing. I record a series of cross-section jaunts north and south across the geographic grain of North America. Although offered in the format of a single trip, it is rather a pastiche, derived from many trips and treks.

This runs counter to the historical east to west traverses of North America along the "Route of the Pioneers" and the movement of ideas that followed. I find that cutting north to south across that grain offers a more liberating—and often disconcerting— way to experience the continent and its vagaries.

In such an oblique light, I patch-work my way down-country from Canada to Florida within a swathe roughly 485 kilometers (300 miles) wide. It lies within the 82nd and 86th meridians, via Ontario, Ohio, Indiana, Kentucky, Tennessee, Georgia, Florida (Figure 1).

Across our route lie great and timeless flows: climatic, runoff, piped, and electronic. Weather, manufactured across Siberia and the Pacific Ocean, sweeps into our path from the west, meeting billows of moisture swelling up from the Gulf of Mexico. Once we are past the Great Lakes, runoff generally flows south and west, slowly carrying surface soils seaward. The great flow of electronic traffic runs east-west. And underneath

it all pulse vast networks of pumped and piped water, sewage, gas, and petroleum.

This cross section also takes us across nine historic overlays of interventions by the federal government, each leaving behind unique and often visible traces. They are: (1) the national or Jeffersonian grid established by the Land Ordinance of 1785; (2) the 19th-century network of mostly disused or abandoned navigation canals, and the successor lake-lock-and-dam systems for river navigation; (3) the original National Road (now U.S. 40), and other U.S. highways and the post-1944 interstate network; (4) hundreds of airways of aircraft and flyways of birds, most of them federally designated and monitored; (5) two national borders: between the United States and Canada, and between the North and the old Confederacy, with discernible traces of Civil War fortifications; (6) sites of 15 Indian territorial cessions made under duress, and the former lands of six Indian nations; (7) military bases of World Wars I and II, especially air bases dependent on southern flying weather; (8) the electric power grid of the Tennessee Valley Authority and other utilities; and (9) six state lines, with diverse oddities at their borders, such as local zoning difference, "Last Chance" filling stations for liquor or gasoline, sudden flurries of marriage parlors for under-21 youths, and border shifts in highway surface and maintenance.

Beyond these divisions other empires have set up their boundaries: federal administrative regions, with headquarters in Atlanta, Chicago, Cincinnati, Cleveland, Memphis, Richmond, St. Louis. No two bureaucracies divvy up their subject territories in quite

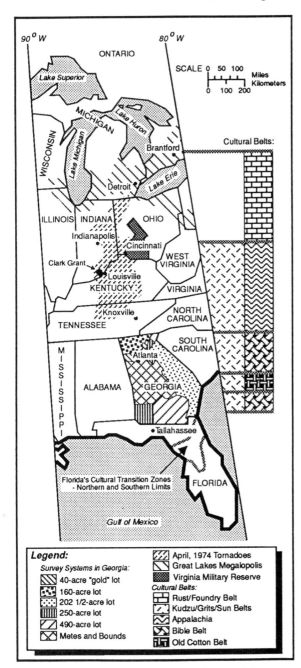

Figure 1. Grain-crossing map of the Canada-United States cross-section described in this article. Drafted by Amy Ahner, from sketches by Charles William Brubaker in Clay (1980), and other sources.

140 to 240 days, so that southward the shade turns into a generic place essential to summer comfort. By mid-Georgia, crape myrtle becomes a tree tall enough to shade small houses, as does the banana farther south. At Tallahassee, we drive along our first legally designated network of "canopy roads," their trees protected to shadow historic local routes.

On a flight from Detroit to Toronto, looking down at the snow-covered field-and-woods pattern of south-central Ontario, I see dark strips of woodland neatly aligned along what in the United States are called the back forties, distant from the farms' frontages on local roads. But soon all falls into disarray; the roads turn higgledy-piggledy, the back forties an unruly tangle. Ontario's maps suggested that original surveys, which took off at right angles from the shores of Lakes Erie and Huron, converged inland in the Brantford region, with ensuing adjustments or compromises that went every-which-way (Figure 2).

In Ontario, one inescapable overlay is the presence of the Great Lakes Megalopolis, studied with intensity by the Greek ekistician, Constantinos Doxiadis (1966-1969). This belt of urbanization and heavy traffic extends some 3000 airline kilometers from Chicago-Detroit to Toronto-Ottawa. Canadian geographers call it "The Anglo-American Heartland." Jerome F. Pickard (1972) identified an adjacent configuration as the Lower Great Lakes Metropolitan Belt. Herman Kahn coined another term, "Chipitts," for the Chicago-Pittsburgh urbanization.

Within that Heartland lies the Sunparlour, Canada's southernmost climatic zone. Its sunny flank extends in an arc along the north shore of Lake Erie. It offers scenic stretches of well-tended vegetable farms, vineyards, and orchards intensively producing food and drink. It echoes eastward along the shores of Lake Erie in the Niagara Fruit Belt. Both belts get some government protection from urbanization, not readily apparent at first glance.

In crossing Lake Erie, one also crosses one of the world's more peaceable national borders, its current placidity resting on history and the latest free-trade

the same way. Competing radio-TV stations and other media carve up our route into scores of listener or market areas.

North America's climatic zones shape ends and means along the way. At our Ontario beginning, rainfall comes to only 81 centimeters (32 inches). It doubles in south Florida. The growing season expands from

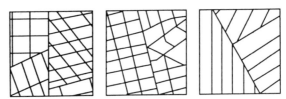

Figure 2. Three sketches of odd geometric jumbles where three or more surveys come together in central Ontario, derived from a 1912 Ontario map. Drafted by Amy Ahner.

treaty between Canada and the United States. But this border-crossing still throws up mental blocks. Most map makers of the United States and Canada still tend to go blank beyond their own national border as well as at state line.

For some 485 kilometers (300 miles) from Central Ontario into southern Ohio, our route crosses what Joel Garreau (1982) has called "the Foundry," an international region earlier known as the Industrial Belt. U.S. portions were re-dubbed the Rust Belt following the migration of industry southward after World War II. In search of a less pejorative term, pundits came up with the now-familiar generic Frost Belt.

Northern Ohio has become an investment center for Japanese capital, a concentration zone for new steel and auto production and parts plants. Of 66 Japanese-owned or joint-venture steel plants in the United States, the greatest concentration lies in Ohio with suppliers festooned along the "just-in-time" transport routes of Kentucky and Tennessee (Kenney and Florida 1991; Perugi 1991)

Our traverse keeps us mostly on Eastern Standard Time except for Indiana; five counties in the northeast corner stay on Central time all year around; five in the southwest corner are on Eastern Standard Time but refuse to change to Daylight Savings Time in summer. The Eastern Standard Time and Central Standard Time zones split Kentucky and Tennessee in midsection. Our route also moves along the edges of several settlement zones, successive frontiers of the 19th century, an area where barn types migrated westward from New England and Pennsylvania.

By the time our route reaches the vicinity of Franklin, Indiana, it will cross the westward-moving path of the U.S. center of population. For 50 years, from 1890 to 1940, that center sauntered across Indiana before reaching Illinois. There it now reposes, awaiting headcounts of new Pacific Rim and Hispanic migrants into the South and West that will someday shift the center of population to—Missouri?

From the northern coasts of Indiana-Ohio southward, our route stays within the Jeffersonian grid of 1-mile square sections into which the territory was surveyed and sold off under the Northwest Ordinance of 1785. In mid-Ohio, we confront that aberration known as the Virginia Military District, where land was surveyed by metes-and-bounds, to be dispersed after 1790 among Revolutionary veterans (Thrower 1966)—providing fees for title lawyers ever since.

In both Indiana and Ohio, we traverse that generic zone called Tornado Country. 148 tornadoes occurred within 24 hours, on 3-4 April 1974, the largest number on record (Fujita 1975). We cross traces of many destructive tornado paths, mostly running southwest-to-

northeast. Less spectacularly, in tribute to prevailing winds of winter, shelterbelts of evergreen trees and shrubs extend around their north and west sides to protect farmsteads along our route.

Now through the Corn Belt—also known as the Corn-Hog-Soybeans Belt, extending three states westward from Ohio. In northern Indiana the national grid left its imprint in The Great Wet—what once was the vast swamp of the Kankakee River Basin, the greatest wildlife wetland in the Midwest. Ditchers-and-drainers of the 19th century dried out the basin and carved it up squarely for soybeans and corn, between recurrent floods (Figure 3). In downtown Indianapolis our path crosses the alignment of the great National Road, now U.S. 40, that carried streams of settlers westward. The road is now overshadowed by interstates.

The pace speeds up between Indianapolis and Louisville as we encounter many belts, borders, and transitions, including deicing line for planes between flights. Deicing is routine in Indianapolis on winter days when it is not needed in Louisville, 195 kilometers (120 miles) farther south. Not far south of Indianapolis, corporate and association names in local telephone books begin to fatten up with the prefixes "Southern" and "South." "Dixie" makes an early appearance, with "Dixie Highways" aligned north-south across southern Illinois, Indiana, Ohio, and Kentucky. Confederate flags, flown or bumper-stickered, soon make a showing.

Along in here is the "grits line," where hominy grits appear, asked-for or not, at breakfast. In the vicinity of Seymour, Indiana, we pass through what Air Corps pilots, in training during World War II, called Thunderstorm Alley (Stuart Alexander, interview, 1987). The "kudzu line" appears hereabouts, the northern limits of that rambunctious legume (Figure 4).

Figure 3. Near Naked City, Indiana, the slow, historic flow of swampwaters is interrupted. Pipe-toters who have been at work here since the mid-19th century converting swampland to cropfields. Photograph by G. Clay.

Figure 4. Kudzu vine festoon the Kentucky-Tennessee State Line, one of the migrant's way-stations en route northward. The condition of the sign indicates that travelers are also gun-toters. Photograph by G. Clay.

By the time our southward course reaches the Ohio River, highly visible clues tell us that we are passing out of the Corn-Hog-Soybean belt into the Tobacco Belt: large black or unpainted barns with vertical air-dry vents, and tobacco patches, bright green-to-yellow in summer. Along the river lies an intermittent but measurable "Fog Belt."

To cross the Ohio River involves a cultural leap over what Wilbur Zelinsky (1973) has called a "first-order cultural boundary." The traveler is now in the South, "by far the largest of the three primordial American cultural areas" (Zelinsky 1973). We also make a linguistic leap. Somewhere in central Indiana-Ohio we have crossed the "Guess-Reckon Line," with "reckon" more often used south of the line (Beveridge 1990). "Angleworms" abound in northern speech, but give way to "earthworms" as we move south. The crudity "anymore" (as in "He's hard of hearing anymore") comes on strong in Kentucky; while "babushka," the immigrant's shawl along with "Bossy" (for cow), barely make it south of the Mason-Dixon Line. "Adam's house cat" (as "I don't know him from Adam's house cat") flourishes south of Kentucky. By middle Georgia, the traveler will have taken leave of "Adam's off-ox" (Cassidy 1985); streams will have become branches, lima beans turn into butterbeans; and we encounter a "Bubba [for brother] Belt" that runs through Georgia-South Carolina speech.

At the Ohio River we also cross what, during the Civil War, was a contested national border that separated victor from vanquished for generations afterward. Geographer William Warntz once identified the Ohio River as the sharpest "income front" in the United States. It shows the two greatest drops in "wealth density per square mile," a measure of capital formation, between contiguous states. These breaks occurred be-

tween Kentucky and Illinois-Ohio (Warntz 1865). In 1941, when wartime defense contracts were transforming the nation, the dollar drop at the Ohio River was precipitous—from $1,173.9 million in Ohio contracts to $123.3 million worth in Kentucky (*The Courier-Journal* 1941).

At Louisville where we cross the Ohio River, no telephone book exists to fully embrace the two-state Louisville metropolitan area. The short kilometer and a half between north and south river banks until 1991 remained, technically speaking, "long distance." Also at the Ohio River, land settlement patterns change from the Jeffersonian grid to metes-and-bounds common to the old Colonial states, Kentucky having been originally part of the Virginia colony.

Now our traverse leaps some 485 kilometers (300 miles) across Kentucky-Tennessee to encounter the Ducktown or Copper Basin at the Tennessee-Georgia-North Carolina border. The Copper Basin is the largest man-made desert in the Eastern United States, the possible West Coast contender being the Owens River Valley desertified by Los Angeles water swindlers beginning 1904 to 1905 (Clay 1980). This gaudy, red-yellow-ochre landscape was created by a century of logging, and open roasting of copper ore, the sulfide fumes killing all vegetation. This left some 12,000 hectares (30,000 acres) of deeply eroded gully-fields where oft-replanted pines struggle to clothe the brilliant, bare, and harsh subsoils (Figure 5).

Underneath all this lies the systematic land survey—not Mr. Jefferson's upright grid but a Tennessee variation skewed a few degrees off true north-south. Tennessee had the wisdom to pick up the systematic structure of townships, fractional townships, and 65-hectare (160-acre) school quarter section, so that the Ducktown school district still gets royalties for minerals extracted from the schools' 65 hectares (160 acres).

Figure 5. Gullyfields cover all traces of the original survey grid covering the Ducktown Basin of southeast Tennessee. Looming on the skyline is a water tank that bears the name, "Sahara," a local hotel. Photograph by G. Clay.

Immediately south, at the Georgia state line, we fall off the Ducktown grid into metes-and-bounds, with local variations of the grid southward. Eastern Georgia was settled by metes-and-bounds during its Colonial period. But middle to western Georgia, settled after the Revolution, and increasingly after the war of 1812, still shows the imprint of five different rectangular surveys, aligned roughly north-south.

Somewhere in middle Georgia we cross the "Northern Limit of the Southern Magnolia," and "the Southern Limit of the White Birch." Around Macon the coastal palmetto enters as a civic street tree, and two hours south of Macon we encounter the Spanish moss. The skyline lifts appreciably as we enter the tall-treed "Pecan Plantation Belt." Here too we enter the remnants of the famous Cotton Belt, its farms now highly mechanized. Soon enough, we reach the home territory of the copyrighted, litigated, and much-imitated Vidalia Onion named for the town Vidalia and its unique regional mix of soil and climate.

South Georgia is also the historic locale of the Pulpwood Belt, still expanding. Georgia Pacific Company moved its headquarters from the Pacific Coast to Peachtree Street, Atlanta, and owns large hunks of former cotton land in middle Georgia. And not so coincidentally, our route also weaves in and out of several smell-and-smoke plumes from Southern pulp-paper plants, noxious signals of the national appetite for paper.

In south Georgia, depending on the season, one crosses what I have termed a "Drop Zone"—a predictable zone where specific items, goods, or appendages are discarded. Here, the castoffs are large chunks of ice, clinging to the undercarriage of truck-trailers from up North in winter. By south Georgia most of them drop off to become "road kills."

South Georgia is also an opportune place to observe the westernization of the Eastern United States. In parts of Georgia and Illinois-Indiana-Ohio, large-scale mechanized farming, with quarter-mile irrigation rigs, and farmsteads a kilometer off the highway, has made the landscape resemble the West. Depopulation has continued as tenant farmers, black and white, have been dispossessed or escaped their old bonds and tumbledown houses which dot the backwoods (Figure 6). More than 6.5 million blacks moved north between 1910 and 1970, crossing the grain toward the North.

De-settlement is an unsettling act. Abandonment disturbs deep psychic roots. "We have been abandoning places since 1800 and the impact of withdrawal has always been painful" observed the late Kevin Lynch (1986).

Our route has touched wide segments of abandoned lands. None are identified on standard maps, and most

Figure 6. Far back from well-traveled frontages, abandonment is writ large across many farmsteads. This one, in the Appalachian outliers of southern Indiana, hovers between decrepitude and dereliction. Photograph by G. Clay.

are in transition among absentee owners, speculators, and tax assessors. Even as I write, the future abandonment of many a Southern military base, recently declared surplus, hangs in the Congressional balance.

In southwest Georgia, our traverse passes through "Quail Shootin' Country," a unique concentration of old cotton-turned-pineywoods plantations. These were converted, safely after the Civil War, to absentee Yankee ownership for seasonal quail shooting. The ownership and the hunts continue across State Line to the outskirts of Tallahassee (Paisley 1978).

Tallahassee itself is booming in the 1990s, its new suburbs hotly contested in hearings required by Florida's growth-management law of 1985. But in another sense, Tallahassee is The End—of the southern red-clay belt, the last observable undulations of the Piedmont terrain. At Cody Scarp, a few kilometers south of the state capital, the land level drops off, the soil changes, and a traveler descends to flat, scrawny, sawgrass-palmetto-and-sand country. For good reasons—poor soil only one of them—the 11 counties around Tallahassee are the least developed in Florida. Between Tallahassee and central Florida (beyond our traverse), a continuing traveler will cross six cultural border zones (Lamme and Oldakowski 1982) that separate the Tallahassee-Panhandle region from the rest of super-growth Florida to the south, where the state bird is jokingly identified as the construction crane.

A good observation post for ending our traverse is at the top of the State Office Building in downtown Tallahassee. Overhead, higher in the sky, the airways run with our route—north to south. Out on the highways the flow goes up and down-country, carrying "Snowbirds," tourists and retirees flocking south in autumn, and north in the spring.

Along this 1,855-kilometer (1,150-mile) route, we have been testing generic places in the marketplace of language against everyday experience. These genera are parts of the grammar by which we negotiate changing worlds. I see them as a great kindred to which we all belong—with family rules we disregard at considerable risk. It is easy to say "It's chaos and to hell with it," or to fall back on old cliches. But cutting across the grain of old preconceptions makes it clear that the generic place-naming and descriptions offer a way to find ourselves again—back for awhile on familiar territory. And that is not a bad generic place to be.

References

Beveridge, Charles. 1990. Olmsted Centennial address in Louisville, Kentucky.

Cassidy, Frederic G., ed. 1985. *Dictionary of American Regional English*, Vol. 1, A-C. Cambridge: The Belknap Press of Harvard University Press.

Clay, Grady. 1980. *Closeup: How to Read the American City*. Chicago: University of Chicago Press.

Courier-Journal. 1941. Map of defense contracts by states. 22 November.

Doxiadis Associates (Athens, Greece). 1966-1969. *The Great Lakes Megalopolis*, Vols. 1-3, Studies for the Detroit Edison Company. Detroit, Michigan.

Fujita, T. Theodore. 1978. Mapping and cartography. *Landscape Architecture Magazine* September:cover.

Garreau, Joel. 1982. *The Nine Nations of North America.* New York: Avon.

Kahn, Herman, and Weiner, Anthony J. 1967. *The Year 2000*. New York: Macmillan.

Kenney, Martin, and Florida, Richard. 1991. How Japanese industry is rebuilding the rust belt. *Technology Review* February/March:29.

Lamme, Ary J. III, and Oldakowski, Raymond J. 1982. Vernacular areas in Florida. *Southeastern Geographer* 22(2):99-109.

Lynch, Kevin. 1986. About wasting. Unpublished manuscript, chapter 8, p. 2. Cambridge: circulated for review by MIT Press.

Paisley, Clifford. 1978. *From Cotton to Quail: An Agricultural Chronicle of Leon County, Florida, 1860-1967.* Gainesville: University of Florida Press.

Pickard, Jerome. 1972. Report, federal commission on population growth and the American future. *The New York Times* 6 February.

Perugi, Deborah. 1991. Map from "How Japanese industry is rebuilding the rust belt." *Technology Review* February/March:29.

Thrower, Norman J. S. 1966. *Original Survey and Land Subdivision.* Chicago: Rand McNally.

Warntz, William. 1965. *Macrogeography and Income Fronts*, Monograph Series No. 3, p. 28. Philadelphia: Regional Science Research Institute.

Zelinsky, Wilbur. 1973. *The Cultural Geography of the United States*, pp. 118-119. Englewood Cliffs, NJ: Prentice-Hall.

Main Streets of the Sunbelt: I-10 and I-20

Donald J. Zeigler
Old Dominion University

Jackson, Mississippi, bills itself as "a Sunbelt community that is vintage Deep South." In Columbia, South Carolina, you can buy a radiator hose at Sunbelt Auto Parts. And in the Mesilla Valley of New Mexico, Sunbelt Realty's billboards try to entice prospective buyers to "Sunset Hills—A New Neighborhood." Stretching across the continent from Florence, South Carolina, and Jacksonville, Florida, to Santa Monica, California, these and other communities along Interstate Highways 10 and 20 exploit the Sunbelt theme to sell their wares, to characterize their lifestyles, and to sharpen their regional identity (Figure 1). As dual Main Streets of the U.S. Sunbelt, which meet and marry in western Texas taking the I-10 name, these great east-west highways and the corridors they traverse call attention to many of the forces shaping the post-industrial geography of the United States and its southern flank. They have changed the places where we live, work, and shop, and even more importantly, they have rearranged the map of growth and decline. Five of the nation's 20 largest metropolitan areas, for instance, are now along I-10 and I-20, and these are among the fastest growing urban regions in the country.

The interstate highway system, of which I-10 and I-20 are a part, comprises a 68,600-kilometer network of woof and warp. These masterpieces of civil engineering strengthen the fabric of national unity by breaking down the barriers of isolation and speeding the diffusion of national trends. The interstates constitute only slightly more than 1 percent of public roads and streets in the United States, yet they carry a fifth of the traffic. The majority of the network in urban areas carries well over 40 vehicles per minute, and a fifth carries more than one vehicle per second. Most Americans use the interstate system daily, and all Americans benefit from the economic efficiencies it provides in moving goods and people around the country. Consider the bumper sticker seen along I-20, which read: "If you bought it, a truck brought it." Lettuce from California gardens, for instance, makes its way along the interstate system to the tables of the old garden state, New Jersey, in 72 hours, and at a price that everyone can afford.

Just as there was nothing indecisive about America in the post-World War II period, with its growth-oriented, mass-production mind-set, there was nothing indecisive about the highways it built. The interstate system had materialized in cartographic form by 1947, and in the federal budget by 1956 (Seely 1987). Despite references to "defense highways," the interstates were conceptualized with one purpose in mind, to link the nation together in the most efficient way possible. According to A. Q. Mowbray they symbolized the American Dream: "to drive from coast to coast without encountering a traffic light" (Clapp 1984, 176).

Highway Names and the Sense of Place

Even what we call these roads betrays their efficiency, for their names are numbers, and the numbers, because of the conventions they follow, are quickly placed on our mental maps. The name I-10, for instance, is quickly recognized as an east-west highway because it carries an even number, as a major transcontinental route because it ends in naught, and as farther south than I-20 because its number is lower. Few people would recognize it by its honorary sobriquet, the Christopher Columbus Transcontinental Highway, for the only place where that is made known to the traveler

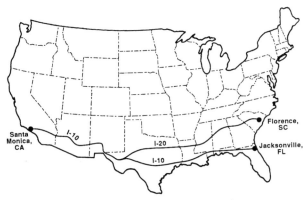

Figure 1. Interstate Highways 10 and 20.

is at its origin in Santa Monica, California. In this respect the interstates differ from their predecessors, and even from the first-generation of federally funded highways, the U.S. Highway System which began in 1921. These two-lane highways carried numbers too —U.S. 80 across the present-day Sunbelt, for instance—but they also carried names. U.S. 80 was known as the Dixie Overland Highway in the southeast and the Old Spanish Trail and Bankhead Highway farther west, each with a trail marker of unique design.

Just as I-10 and I-20 came to replace old U.S. 80, they also came to solidify the tradition of using numerical monikers rather than names, further damaging Americans' sense of place. As Gertrude Stein might put it, no matter where you are on the interstate system, "there is no there." Yet, it is not only the sense of place that has been sacrificed, it is the places themselves. Small towns were by-passed and large cities pierced by highways that limited access and eliminated intersections. The extent to which they captured the national spirit was epitomized by Lewis Mumford when he declared "our national flower is the concrete cloverleaf" (Clapp 1984, 177). Indeed, there are no real places along these corridors of unimpeded connectivity, only single-function pull-offs with no permanent inhabitants or no inhabitants at all. Witness the welcome centers that try so hard to be something other than homogenized places, the commercial monopolies most common along toll-road sections, and the self-service pit-stops that are announced by the generic "Rest Area" signs. Imagine such service centers a half century ago without at least a hamlet. The new places that did materialize arose at the ends of off-ramps, where they took numerical monikers of their own—"That's at Exit 6"—and where they siphoned growth away from places along the old U.S. highways. What we have witnessed along the interstate system is a collapse of "in-between space," often to the detriment of "in-between places."

Spatial and Environmental Interactions

Traveling on the interstates often mutes the environmental stimuli that set our minds to work. While the view from the interstate is often spectacular, it is more general, more void of details, and more fleeting, particularly when compared with the secondary roads. As the interstate highways become more like coast-to-coast by-passes, travel becomes more like reading a *Reader's Digest* condensed book, or the blurb on the book jacket. The honeysuckle and morning glories of the Dixie Overland Highway, the smell of turpentine in Bienville National Forest, the cluck of chickens riding to their doom in ghostly white school buses in Morton, Mississippi, are some of the sights, smells, and sounds that are so much a part of the "blue highways" (Least Heat Moon 1982) and so lacking along the network of "green highways" we know as interstates. Stopping to eat, to spend money, to relax, to talk to people along the way represent more active forms of interaction with the environment. Slower travel invites such interactions; faster travel discourages them. One particular endemic disease is easy to contract on the interstate system. It is what Peter Jenkins (1979, 37) calls "mileage craziness." He defines it as becoming "overly concerned with arriving" and "obsessively placing more importance on how many miles are traveled than on the real reason for traveling."

As technological advances in transportation have rapidly overcome the friction of distance, spatial interactions between places have multiplied exponentially, often at the expense of human-environment interactions with places along the way. The interstate highways have promoted spatial interaction at the expense of ecological interaction (Figure 2). The two discontinuities in Figure 2 (and there may be more) represent conditions that dramatically reduce chances for interaction with the environment. For example, once people envelope themselves in their vehicular shells, as at point A, they become less aware of their environment: the chances of face-to-face contact are practically nil; topography loses the meaning it has to the cyclist; and climate becomes immaterial when the heater or air conditioner is turned on. Likewise, when the mode of transit leaves the ground, as at point B, relationships with people and places along the way are reduced to zero; in fact, to travel by plane means that you do not even have to know what "the way" is. The same may be said for the interstates, for

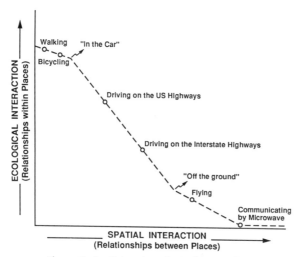

Figure 2. Spatial and ecological interaction.

how can you say that you have been to Mississippi if you have only skimmed the surface on I-20 with a brief refueling stop at Exit 17?

As the interstate system was aborning in the early 1960s, a report was commissioned by the Automobile Manufacturer's Association, which predicted that the interstate highways would have three major effects (Wilbur Smith and Associates 1961). First, they would improve accessibility and thus the attractiveness of central business districts. Second, they would stimulate the planned development of metropolitan areas. Third, they would become instrumental in the redevelopment of older central cities. From the perspectives of the 1990s, as only a few more (controversial) sections of the interstate system have yet to be built, we can see that their effects have been the antithesis of projections. They seem to have stimulated suburban and exurban development, often unplanned, at the expense of the central city. Rather than leading "auto-mobile" Americans to the downtown, they have led them away— and with people has gone business to more accessible locations around freeway interchanges (Moon 1987). In fact, if the megalopolitan northeast is seen as the nation's downtown and the Sunbelt as its suburbs, the interstates seem to have diminished the former's role in favor of the latter.

In large measure, the interstate highways have rearranged the cardinal place geometry of the nation (Elliott 1984), and with it the space-searching behavior of the American people. At the metropolitan scale this rearrangement may be seen in the Atlanta-Birmingham corridor, a region along I-20 in Georgia and Alabama, where the national competition between Sunbelt and Frostbelt is being played out in microcosm. Whereas Atlanta has emerged as the business hub of the "new South," Birmingham represents an outlier

of declining Frostbelt industry, proving that the economic climate is not equally sunny everywhere south of the 37th parallel. Birmingham, in fact, was tagged "the most unsouthern of southern cities" by the Great Depression Writers' Program (1941, 4), and the same could be said for the other cities of northern Alabama, notably Anniston.

Anniston appears on the map as the only metropolis between Atlanta and Birmingham in the I-20 corridor. Judging from travel along the interstate, however, one might never know that Anniston, anchor of a metropolitan area of 116,000, even exists. Instead, Oxford, a village-turned-city in Anniston's southern suburban fringe, seems to have eclipsed the mother city in a competition to become the capital of the Atlanta-Birmingham corridor. Oxford has sapped the growth of Anniston as surely as the Sunbelt has sapped the growth of the Frostbelt.

Oxford considers itself to be one of the pearls along Alabama's I-20 and bills itself as the "Gateway to the Future" and the "Crossroads of Alabama." Indeed, with the completion of I-20, one of Oxford's advantages is its accessibility. Whereas U.S. Highway 78 passed right through the heart of Anniston on its way across the South, its interstate replacement by-passed the city in favor of Oxford, miles to the south. Anniston of the industrial past—producing iron, steel, textiles, and chemicals—contrasts sharply with nearby Oxford of the post-industrial future with its expanding amenity base backed up by light manufacturing. Oxford is attempting to restructure the cardinal place geometry of northeastern Alabama, first in people's perceptions and then in reality. When the region's residents need higher-order goods and services, Oxford wishes to be perceived as their nearest larger neighbor.

Probably the most telling evidence of the competition between the two is to be found in Oxford's promotional literature. Proximity to Anniston is never mentioned though it does boast of its location in "one of the state's Standard Metropolitan Areas." Furthermore, the locator map (Figure 3), shows Anniston to be equidistant between Oxford and Gadsden. In fact, Oxford is only 6 kilometers away from Anniston, while Gadsden is 53 kilometers away (Office of the Mayor ca. 1985). This competition between new and old, accessible and inaccessible, is repeated all along the I-10–I-20 corridor. The result is the emergence of a new set of cardinal places.

Conclusions: Looking Back and Looking Ahead

We watched as wagon trails replaced Indian paths, and as corduroy roads replaced wagon trails. We watched as a system of U.S. highways improved upon the dirt

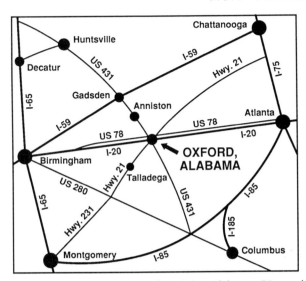

Figure 3. Oxford, Anniston, and Gadsden, Alabama: Distorted distance relationships.

roads of the 19th century, and again as the modern interstate network nudged aside its predecessor U.S. highways. In the region south of the 37th parallel, Interstate Highways 10 and 20 have emerged as the major east-west thoroughfares, speeding the movement of goods, people, and ideas across what has become known as the Sunbelt, and helping to establish growth (albeit far from ubiquitous) as one of the region's most widely recognized characteristics. These ribbons of concrete and asphalt have transformed transportation habits and the way in which Americans think about distances. They have abetted the decline of some older metropolitan centers, of many "old downtowns," and of countless small places that did not merit an exit of their own. Yet, they have also spawned a new and larger archipelago of growth poles and "new downtowns" around interstate highway interchanges and nearby commercial strips.

Speed and efficiency have been the trademarks of interstate highways, as they have multiplied exponentially the potential for people and places to interact with each other. Perhaps unfortunately, the interstate system has also fostered a tendency for us to travel more and see less, for the view from the interstate is a fleeting view, a distant view, and a biased view of the American landscape. In short, the interstate system has encouraged spatial interaction at the expense of human-environmental interaction. It has contributed to a demise of the many unique characteristics of places which have given the American landscape its local color. Landscapes made of local crafts have given way to landscapes made of interchangeable parts.

What will replace the interstate highways in our never-ending quest to conquer the distances that sep-

arate us? Is it already in operation and will geographers recognize its presence on the landscape? Perhaps, it is the technology that enables us to replace space-bound transport corridors with an invisible interstate network of electromagnetic waves. In our emerging post-industrial society, information is becoming the raw material that powers growth, and information can be transported independently of the transportation system. Computer networks, often proprietary and usually invisible, will have as much influence on restructuring the map of growth and decline in the next 30 years as the interstate highway system had on the last 30.

While geographers may not be able to see the emerging generation of space-conquering technology as lines on the land, they can certainly see its results. The revolution in telecommunication is already reshaping I-10 and I-20 landscapes. Its effects are visible in the vertical development of downtowns, indicators of growth in the quaternary economy, in the transactional nodes that are embedded in the suburbs, often around interstate interchanges, and in the satellite dishes that punctuate the countryside. They are also hinted at in the growth of franchise empires, and in the billboards that advertise "over 400 locations nationwide," for such networks of lodging and eating facilities would be impossible were it not for an ability to store, transport, and coordinate information with a minimum of effort. Satellite dishes and coaxial cables may not seem as if their impact on the landscape is as profound as the tens of thousands of kilometers of interstate highways, but let not the invisible nature of the next generation of space adjusting technologies fool us into thinking it is unimportant.

References

Clapp, J. A. 1984. *The City.* New Brunswick, NJ: Center for Urban Policy Research, Rutgers University.

Elliott, H. M. 1984. Cardinal place geometry in the American South. *Southeastern Geographer* 24:65-77.

Jenkins, P. 1979. *A Walk Across America.* New York: Fawcett Crest.

Least Heat Moon, W. 1982. *Blue Highways.* New York: Fawcett Crest.

Moon, H. E. Jr. 1987. *Rural Development Perspectives* 4(October):35-38.

Office of the Mayor. ca. 1985. "Oxford, Alabama 'The Crossroads . . .'"

Seely, B. E. 1987. *Building the American Highway System.* Philadelphia: Temple University Press.

Wilbur Smith and Associates. 1961. *Future Highways and Urban Growth.*

Writers' Program. 1941. *Alabama: A Guide to the Deep South.* New York: Richard R. Smith.

Route 66: Auto River of the American West

Arthur Krim
Society for Commercial Archeology

Route 66 extended from Chicago to Los Angeles, from the Midwest to California, from Lake Michigan to the Pacific Ocean, from State Street to Wilshire Boulevard, from deep winter to endless summer. It served as an auto river of the American highway frontier across two thirds of the United States during the mid-20th century. While today its official line is distinguished by five interstate highways, the original U.S. Highway Number 66 was a key link from the settled core of the urban Midwest to the opportune frontier of southern California and to the emergent settlement of the Southwest (Figure 1).

The logic of Route 66 lay in its location as the major connecting path around the southern flank of the Rocky Mountains from the Great Plains to California, a relatively easy gradient of high plateaus through New Mexico and Arizona where winter passes were generally opened to auto travel (Snyder 1990). The diagonal traverse from Chicago to Los Angeles was made across a recent American frontier, in former French and Spanish territory. The area was hardly American before 1850, and even in 1900, was still an active frontier of settlement where new railroads cut across recent Indian territory (Meinig 1971). After 1920, the automobile laid claim to this Southwest landscape and Route 66 proved to be a primary corridor incorporating the region into the American cultural realm.

Oklahoma Origins

Route 66 was conceived in the newly admitted state of Oklahoma as an effort by local boosters to link the former Indian Territory with its two urban poles in Chicago and Los Angeles (Krim 1990a). The intent was to create a diagonal line of traffic through Oklahoma that would divert travel from the established routes through Kansas City and Denver from the Midwest to California. This project was advanced by Cyrus S. Avery, Highway Commissioner in Tulsa, Oklahoma, as part of the Good Roads movement before World War I (Scott and Kelly 1988). With an oil boom in Oklahoma and local prosperity of Tulsa, such a diagonal seemed logical and necessary to tie the Sooner State with the new automobile trails then being marked in the Midwest.

Avery was appointed as Oklahoma Highway Commissioner after the War and was a member of the Joint Board on Interstate Highways that created the U.S. road-numbering system in 1925. Avery, with colleagues from Missouri and Illinois, designated the Chicago-to-Los Angeles highway as U.S. Number 60. However, opposition from Kentucky forced the choice of a secondary number in the national grid, and Number 66 was selected as a compromise in 1926 (Krim 1990b). Thus, Route 66 was designated as the national highway from Chicago to Los Angeles, through Tulsa. Avery wrote to Washington officials, "We assure you that it will be a road through Oklahoma that the U.S. Government will be proud of" (Scott and Kelly 1988, 17).

After designation as U.S. Highway Number 66, the logic of the diagonal line, direct from Chicago to Los Angeles, became more obvious to automobile travelers. Furthermore, the alliterative sound of the number "66" sparked a commercial campaign with the formation of the U.S. 66 Highway Association in 1927 and the designation of Phillips 66 gasoline, both again in Oklahoma (Scott and Kelly 1988; Wallis 1988). These commercial campaigns promoted Highway 66 as the Main Street of America in the last years of the 1920s' boom prosperity.

30

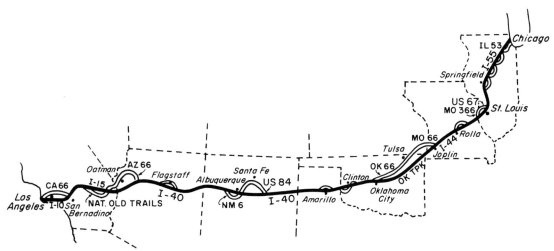

Figure 1. Route 66, as an auto river of the American west, showing the current interstate highway in black and the relic meander loops of early route sections in white. Current state highway numbers are given for relic sections. Source: Krim (1990a), with the permission of Iowa State University Press.

Route Locations

The advantage of U.S. 66 lay in its routing from St. Louis directly southwest across the Ozark Mountains to Oklahoma and then west into the Panhandle of Texas, at Amarillo. From here the original highway looped north from Santa Rosa, New Mexico, into Santa Fe and then down the Rio Grande to Albuquerque, later achieved with a short-line across the mountains at Clines Corner in 1937. From Albuquerque the highway followed the main line of the Santa Fe Railroad all the way into Los Angeles, through Grants and Gallup, New Mexico, to Flagstaff, Arizona, skirting just south of the Navajo Indian Reservation and the Grand Canyon National Park. West of Kingman, Arizona, U.S. 66 turned up into the Black Mountains at Oatman, then down to the Colorado River at Needles, California. This tortuous circuit was later eliminated following the railroad around the mountain flank in 1952, but not before the Oatman grade became infamous as the most dangerous section of Highway 66. *The Guide Book to Highway 66* describes the route as follows:

> Now you start up the Gold Road Hill, possibly the steepest grade you will encounter on US 66. The east side of this grade rises 1400 feet in 9 miles, starting with a long gentle slope and becoming more difficult as you near the summit. In the last half mile there are two or three quick hairpin turns. The west side of this grade is the steepest, but will present no problem to the westbound driver, who should, however, keep his car in second gear going down (Rittenhouse 1989 [1946], 110).

From Needles to Barstow, U.S. 66 paralleled the Santa Fe Railroad past tank stops at Bagdad and Amoy, mostly traveled at night to avoid the searing heat of the deadly Mojave Desert. Then the highway exited the confines of Cajon Pass into the bountiful bloom of the San Bernardino Basin and along Foothill Boulevard into Los Angeles, through the orange groves of Azuza, Arcadia, and Pasadena. Eventually U.S. 66 was extended to the Pacific Ocean at the Santa Monica Pier, linking the waters of the East with the waters of the West (Scott and Kelly 1988).

Highway Mythology

A significant period in the development of Route 66 was marked by the Dust Bowl migration of the 1930s. The highway served as the escape route from the Ozarks and from the Panhandle counties of Texas and Oklahoma to the expected opportunities in California. Route 66 served as the direct line of travel from the Dust Bowl and proved to be an effective funnel for families dispossessed by the dry years and devastation of the plains (Gregory 1989). California novelist John Steinbeck captured the drama of the migration in his classic *The Grapes of Wrath* after he followed the Dust Bowl trek in 1937 along U.S. 66 from Chicago to Barstow (DeMott 1989):

> Highway 66 is the main migrant road. 66—the long concrete path across the country, waving gently up and down on the map, from the Mississippi to Bakersfield—over the red lands, twisting up into the

mountains, crossing the Divide and down into the bright and terrible desert, and across the desert to the mountains again, and into the rich California valleys (Steinbeck 1939, 160).

World War II was a crucial period in the history of Route 66, marking its use as an economic artery that linked the industrial centers of the Midwest to the aircraft and naval bases of southern California (Scott and Kelly 1988). Along its route were strategic facilities, training fields, and explosives depots in the Mojave Desert and atomic research facilities in New Mexico. Thus, when the war ended, the expectations of prosperity were captured in the song-map of the Chicago-to-Los Angeles journey, "Get Your Kicks on Route 66," written in early 1946 by Bobby Troup and his wife Cynthia on their transcontinental trip from Pennsylvania to California (Krim 1990a):

If you ever plan to motor west,
Travel my way, take the highway that's the best.
Get your kicks on Route Sixty-six!
It winds from Chicago to L.A.
More than two thousand miles all the way.
Get your kicks on Route Sixty-six!
Now you go through St. Louis and Joplin, Missouri;
Oklahoma City looks mighty pretty;
You'll see Amarillo; Gallup, New Mexico;
Flagstaff, Arizona; Don't forget Winona;
Kingman, Barstow, San Bernardino
(© by Londontown Music, Inc.).

The Troup song-map proved to be a popular jazz hit of the postwar period, recorded first by Nat King Cole in 1946 with later covers by in the rock-and-roll era by Chuck Berry and The Rolling Stones. More importantly, its rhyming town names served as a practical geography of the long-distance stop-over points. The Troup song-map thus became a cartographic guide to motorists moving to the California frontier (Krim 1990a).

Historic Preservation

The postwar expansion of the Southwest defined Route 66 as the primary channel of traffic through the region. Along the segment from Oklahoma to Arizona, growth centers such as Amarillo, Albuquerque, and Flagstaff developed as major regional cities (Meinig 1971). Route 66 also gave access to national tourist attractions (for instance the Grand Canyon) during the national prosperity of the 1950s. However, with the designation of the interstate highway system in 1957, federal upgrading and relocation of U.S. 66 was initiated. From

this date on, the singular numbering of Route 66 was replaced by a multiple of interstate freeways, including I-40 along the primary segment from Oklahoma City to Barstow. Gradual replacement of two-lane U.S. 66 sections with four-lane interstates occurred in the 1960s. The final segment of original U.S. 66 was by-passed around Williams, Arizona, in 1984 and the federal number retired the next year (Scott and Kelly 1988).

While official designation of U.S. Highway 66 no longer exists, its role as a primary auto road to California immortalized the highway in the popular imagination. Like the Mississippi of Mark Twain, U.S. 66 was a river of migration and opportunity to the western frontier. And like the Mississippi, Route 66 has left relic meanders and former ox-bows that serve to mark its continual flow (Figure 1). These abandoned sloughs are now historic segments of the famous highway. State preservation organizations and recently enacted federal legislation have marked a national historic corridor along the path of old Route 66 (Wallis 1990).

Symbolic Survival

Amplified by the popular imagery in *The Grapes of Wrath* and "Get Your Kicks on Route 66," the highway has achieved international fame well beyond its immediate path. It has become a signet of the American auto road and its route, a geography of myth. Indeed, Route 66 was unique among early transcontinental auto highways. It symbolized the role of the automobile in shaping the mass culture of the United States to a degree not found in other modern cultures by the mid-20th century. The result is an expanding awareness that Route 66 was more than a simple highway number; it became the auto river across the American frontier. It was a transmission line of popular culture across the expanse of desert and high mountains separating the settled East from the emergent Southwest. It helped to open the frontier to development. By this process, Route 66 became embedded in the landscape. It continues to trace a primary line of travel across the western United States and enhance a historic memory that is worthy of national preservation.

References

DeMott, Robert, ed. 1989. *Working Days: The Journals of "The Grapes of Wrath."* New York: Viking Penguin.
Gregory, James N. 1989. *American Exodus: The Dust Bowl and Okie Culture in California.* New York: Oxford University Press.

Krim, Arthur. 1990a. Mapping Route 66: A cultural cartography. In *Roadside America*, ed. Jan Jennings, pp. 198-208. Ames, IA: Iowa State University Press.

_____ . 1990b. Highway reports: First numbering of Route 66 discovered in Missouri. *Society for Commercial Archeology Journal* 11:10-11.

Meinig, D. W. 1971. *Southwest, Three Peoples in Geographical Change.* New York: Oxford University Press.

Rittenhouse, Jack D. 1989. *A Guide Book to Highway 66: A Facsimile of the 1946 Edition.* Albuquerque: University of New Mexico Press.

Scott, Quinta, and Kelly, Susan Croce. 1988. *Route 66: The Highway and Its People.* Norman: University of Oklahoma Press.

Snyder, Tom. 1990. *The Route 66 Traveler's Guide.* New York: St. Martin's Press.

Steinbeck, John. 1939. *The Grapes of Wrath.* New York: Viking Press.

Wallis, Michael. 1988. *Oil Man: The Story of Frank Phillips.* New York: Doubleday.

_____ . 1990. *Route 66: The Mother Road.* New York: St. Martin's Press.

The Northern Treeline of Canada

Glen M. MacDonald
McMaster University

K. Gajewski
Université Laval

The northern treeline of Canada stretches over 4,800 kilometers from the Alaskan border to the Atlantic Ocean and represents one of the most important geographic boundaries in North America (Figure 1). The presence of trees on the landscape is both influenced by and influences environmental factors such as climate, permafrost distribution, and hydrological regime. In addition, the treeline's location has significantly influenced the cultural geography of Canada and now coincides with political boundaries in two regions.

At the time of European contact, areas north of treeline were occupied by Inuit of the Eskimoan language family (MacMillan 1988; Figure 2). The arctic cultural tradition of the Inuit was built around resources available from the tundra and adjacent coastal regions. The treeline served as the northern boundary of lands traditionally occupied by tribes of the Athapaskan language family in western and central Canada and of tribes of the Algonkian family in eastern Canada. The subarctic cultural tradition of the Namerind living along the southern edge of treeline was based on resources available from the boreal forest.

In recent years, the traditional boundary between the lands occupied by Inuit and Indian peoples has become a matter of political importance (Figure 2). Agreement was reached in 1987 to divide the Northwest Territories into two separate political entities (Government of the Northwest Territories 1987). The northeastern area is largely occupied by Inuit and the southwestern area is occupied by Indians, Métis, and non-natives.

In recognition of traditional patterns of native land occupation, a large segment of the proposed political boundary follows the treeline (Figure 2). Along the Hudson Bay coast of Québec, the James Bay and Northern Québec Agreement divides the territory into Cree and Inuit administration areas along the 55th parallel, which roughly follows the northern limit of the boreal forest (Gouvernement du Québec 1984).

Despite the enshrinement of its current location as a political boundary, it is important to consider that the treeline is also a dynamic boundary that has shifted geographically in response to variations in climate. This relationship to climate has made the treeline an important feature for studying climatic change. Here, we focus on the physical-geographical consequence of the northern treeline of Canada.

Biogeography

The northern treeline is not a sharp border between forest and tundra, but rather a zone of transition, from forest to boreal woodland to forest-tundra and finally to tundra (Figure 1). In strict terms, the treeline is the boundary beyond which tree species do not grow taller than 5 meters. However, the treeline is generally mapped as the northern limit of the forest-tundra (Payette 1983). Small individuals and krummholz (shrub-form) stands of coniferous tree species may be found north of the mapped limit of the forest-tundra, but these are diffusely scattered (Elliot-Fisk 1988; Payette 1983).

The boreal woodland zone consists of open stands of coniferous trees with some denser stands of deciduous and coniferous trees in more protected locations (Rowe 1972). Lichens such as *Cladonia stellaris* and *Stereocaulon paschale* are a diagnostic ground cover in well-drained areas, while poorly drained sites often support *Sphagnum* peatlands. Coniferous trees on well drained sites include white spruce (*Picea glauca*) and jack pine (*Pinus banksiana*). Balsam fir (*Abies balsamea*) is found on mesic sites east of Hudson Bay, but it is absent from the boreal woodland in the west. Deciduous trees of the

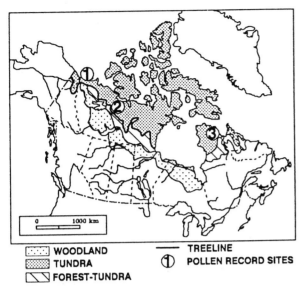

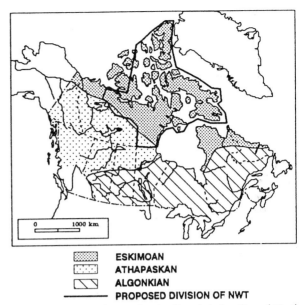

WOODLAND
TUNDRA
FOREST-TUNDRA

TREELINE
① POLLEN RECORD SITES

Figure 1. The distribution of treeline zone vegetation across northern Canada (after Rowe 1972). The mapped treeline corresponds with the northern limit of forest-tundra vegetation.

ESKIMOAN
ATHAPASKAN
ALGONKIAN
PROPOSED DIVISION OF NWT

Figure 2. The boundaries of major linguistic families of North American Indians and Inuit in northern Canada (after MacMillan 1988) and the proposed boundary for subdivision of the Northwest Territories (after Government of the Northwest Territories 1987).

woodland zone include paper birch (*Betula papyrifera*), balsam poplar (*Populus balsamifera*), and aspen poplar (*Populus tremuloides*). Poorly drained sites are occupied by black spruce (*Picea mariana*) and larch (*Larix laricina*).

The forest-tundra is characterized by stands and individual conifers interspersed with areas of tundra, with the extent of tree cover decreasing northward. The treeline is formed most often by white spruce in the west, and white and black spruce in the east, although other species such as larch can locally form treeline (Elliot-Fisk 1988; Payette 1983; Rowe 1972). Paper birch and poplar are encountered less frequently in the forest-tundra zone, although balsam poplar is the northernmost tree found in Labrador. Dominant shrubs in the tundra areas include shrub birch (*Betula glandulosa*), alder (*Alnus* spp.), willows (*Salix* spp.) and members of the heath family (Ericaceae). Wet sites support *Sphagnum* peatlands and sedge (Cyperaceae) meadows. A moderately diverse flora of herbs with arctic and boreal affinities is found on open sites.

Climatology

The location of treeline is controlled largely by climate. The position of treeline is commonly correlated with the distribution of temperature, for example the 10°C mean July isotherm corresponds generally with the northern treeline in Canada (Halliday and Brown

1943). Tree growth in Canada is also limited to regions that receive more than 90 kilocalories per square centimeter per year of absorbed radiation (Hare and Ritchie 1972). The geographic position of the Canadian treeline corresponds roughly with the summer position of the arctic front, which is the zone of transition between dry, cold arctic air masses and warm, moist, southern air masses (Bryson 1966; Figure 3). This dominance

ARCTIC-PACIFIC FRONTAL ZONE
BAND OF HIGH FRONTAL ACTIVITY

Figure 3. The modal position of the arctic front in July and the band of high frontal activity in summer (after Bryson 1966).

of dry, cold arctic air is responsible for the low summer temperatures and dry conditions that typify the climate north of treeline. The presence of this airmass boundary is responsible for the steep climatic gradients that are typical of the treeline zone (Hare and Ritchie 1972).

The physiological relationships between tree growth and climate are complex, and it is probably a combination of factors that is responsible for the location of treeline. Low availability of radiation energy decreases growth rates and the physiological energy available for reproduction. The shorter growth season and meteorological variability in the north affect the phenological cycle. Establishment of trees by sexual reproduction is rare or episodic at the current northern limits of spruce and periods of tree establishment are correlated with warm summers (Black and Bliss 1980; Elliot-Fisk 1983; Scott et al. 1987). The spatial structure of the forest-tundra—alternating patches of tundra and forested land—appears to have resulted from the failure of reproduction following fires during the past several thousand years (Nichols 1976; Payette and Gagnon 1985). Harsh conditions in winter likely contribute to the small stature of northern trees by destroying needles and apical buds that extend above the snow cover. Permafrost at treeline causes mechanical and thermal damage to root systems that impedes tree establishment and growth (Brown and Péwé 1973).

At the same time, differences in albedo and retention of snow caused by the presence of coniferous trees may be important in influencing local climate (Hare and Ritchie 1972) and possibly global climate as well (Wilson et al. 1987).

Permafrost

In general, the southern limits of continuous permafrost corresponds with the northern limits of forest-tundra (Brown and Péwé 1973; Figure 4). The presence of continuous permafrost profoundly affects the hydrological regime, as groundwater flow is limited to the active layer (Brown and Péwé 1973). The southern limit of the boreal woodland coincides in a number of areas with the widespread occurrence of discontinuous permafrost. Although primarily a function of air temperature, permafrost distribution is also affected by characteristics of the land surface. Factors related to treeline—including the presence and type of forest vegetation, depth of the litter layer, and soil type—can influence permafrost development (Bonan and Shugart 1989).

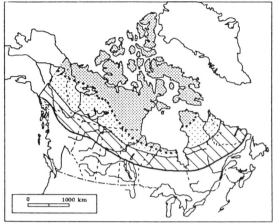

CONTINUOUS PERMAFROST ZONE
•••• SOUTHERN LIMIT OF CONTINUOUS PERMAFROST ZONE
DISCONTINUOUS PERMAFROST ZONE
SOUTHERN FRINGE OF DISCONTINUOUS PERMAFROST ZONE
—— SOUTHERN LIMIT OF DISCONTINUOUS PERMAFROST ZONE

Figure 4. The distribution of permafrost (after Brown and Péwé 1973).

Climate Change and the Postglacial History of Treeline

The close relationship between the arctic front and the geographical position of treeline has stimulated studies of past climate changes using fossil pollen, dendroclimatology, and historical records. Understanding the dynamic nature of treeline is important, as shifts in treeline imply changes in hydrologic and geomorphic processes that can significantly influence conservation and resource use in the north.

In our research, we use fossil pollen deposited in lake sediments located at or beyond modern treeline to reconstruct postglacial treeline history. These pollen data record large-scale regional changes in the vegetation distribution. Chronological control is provided by radiocarbon dating. Deglaciation occurred by 12,500 BP (radiocarbon years before present) in northwestern Canada, between 9,000 BP to 7,000 BP in central Canada, but not until 8,000 BP to 6,000 BP in Québec-Labrador (Dyke and Prest 1987). Significant changes in treeline vegetation have occurred since deglaciation (Figure 5). Large amounts of spruce pollen in the sediments of Sleet Lake (Figure 5, site 1) indicate that forest cover extended north of the modern treeline in northwestern Canada between 9,500 and 6,000 BP (Spear 1983). The fossil pollen record for spruce at Queen's Lake in central Canada (Figure 5, site 2) suggests a northward expansion of treeline occurred between 5,500 BP and 3,500 BP (Moser and MacDonald 1990). Fossil pollen from RAF II (Richard

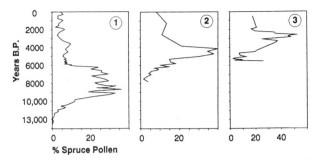

Figure 5. Fossil pollen records of spruce from the sediments of (1) Sleet Lake, Northwest Territories (Spear 1983), (2) Queen's Lake, Northwest Territories (Moser and MacDonald 1990), and (3) RAF II Lake, Québec (Richard 1981). Locations of pollen sites are shown on Figure 1.

1981) in Québec (Figure 5, site 3) suggests that forest cover was more dense in the forest-tundra between 4,000 to 2,000 BP.

This fossil pollen evidence indicates that treeline has been a dynamic boundary during the postglacial. Shifts in the treeline probably resulted from shifts in the position of the arctic front. As treeline moved, so did other hydrologic, geomorphic, and cultural boundaries, although the response time of these other boundaries may vary. In addition, changes in treeline appear to have been asynchronous in western, central, and eastern Canada. Our observations of past treeline changes suggest that the predicted global warming could also result in asynchronous adjustments of the vegetational and physical environmental conditions to complex climate changes.

References

Black, R. A., and Bliss, L. C. 1980. Reproductive ecology of *Picea mariana* (Mill.) BSP, at tree line near Inuvik, Northwest Territories, Canada. *Ecological Monographs* 50:331-354.

Bonan, G. B., and Shugart, H. H. 1989. Environmental factors and ecological processes in boreal forests. *Annual Review of Ecology and Systematics* 20:1-28.

Brown, R. J. E., and Péwé, T. L. 1973. Distribution of permafrost in North America and its relationship to the environment: A review, 1963-1973. In *Permafrost: North American Contribution, Second International Conference*, pp. 71-100. Washington, DC: National Academy of Sciences.

Bryson, R. A. 1966. Air masses, streamlines and the boreal forest. *Geographical Bulletin* 8:228-269.

Dyke, A. S., and Prest, V. K. 1987. Late Wisconsinan and Holocene history of the Laurentide ice sheet. *Géographie Physique et Quaternaire* 41:237-264.

Elliot-Fisk, D. L. 1983. The stability of the northern Canadian tree limit. *Annals of the Association of American Geographers* 73:560-576.

————. 1988. The boreal forest. In *North American Terrestrial Vegetation*, eds. M. G. Barbour and W. D. Billings, pp. 33-62. Cambridge: Cambridge University Press.

Government of the Northwest Territories. 1987. *Boundary and Constitutional Agreement*. Yellowknife: Government of the Northwest Territories.

Gouvernement du Québec. 1984. *Le Nord du Québec, Profil Régional*. Québec: Direction Générale des Publications Gouvernementale du Ministre des Communications.

Halliday, W. E. D, and Brown, A. W. A. 1943. The distribution of some important forest trees of Canada. *Ecology* 24:353-373.

Hare, F. K., and Ritchie, J. C. 1972. The boreal bioclimates. *Geographical Review* 62:333-365.

MacMillan, A. D. 1988. *Native Peoples and Cultures of Canada*. Vancouver: Douglas and McIntyre.

Moser, K. A., and MacDonald, G. M. 1990. Holocene vegetation change at treeline north of Yellowknife, Northwest Territories. *Quaternary Research* 34:227-239.

Nichols, H. 1976. Historical aspects of the northern Canadian treeline. *Arctic* 29:38-47.

Payette, S. 1983. The forest tundra and present tree-lines of the northern Québec-Labrador peninsula. *Nordicana* 47:3-23.

Payette, S., and Gagnon, R. 1985. Late Holocene deforestation and tree regeneration in the forest-tundra of Québec. *Nature* 313:570-572.

Richard, P. J. H. 1981. Paléophytogéographie postglaciaire en Ungava par l'analyse pollinique. *Paléo-Québec* 13. Québec: Université du Québec.

Rowe, J. S. 1972. *Forest Regions of Canada*. Ottawa: Canadian Forestry Service Publication 1300.

Scott, P. A., Hansell, R. I. C., and Fayle, D. C. F. 1987. Establishment of white spruce populations and responses to climatic change at the treeline, Churchill, Manitoba. *Arctic and Alpine Research* 19:45-51.

Spear, R. 1983. Paleoecological approaches to a study of treeline fluctuations in the Mackenzie delta region, Northwest Territories: Preliminary results. *Nordicana* 47:61-72.

Wilson, M. F., Henderson-Sellers, A., Dickinson, R. E., and Kennedy, P. J. 1987. Investigation of the sensitivity of the land-surface parameterization of the NCAR community climate model in regions of tundra vegetation. *Journal of Climatology* 7:319-343.

The North American Snowline

David A. Robinson
Rutgers University

Careening down a mountainous slope, loathing a long, if not impossible, commute home, worrying about the availability of moisture for the germination of crops, remembering a time when "it was up to my knees from December to March." These are just a few of the diverse thoughts, opinions, and memories people have of snow. Falling snow or snow lying on the ground exert an impact on human activities as diverse as engineering, agriculture, travel, recreation, commerce, and safety (Rooney 1967; Snyder *et al.* 1980). Snow influences hydrologic, biologic, chemical, and geologic processes at and near the surface of the earth (Jones and Orville-Thomas 1987). Empirical and modeling studies also show snow cover to have an influential role within the climate system (Barnett *et al.* 1989; Walsh *et al.* 1985). Global models of anthropogenically induced climate change suggest enhanced warming in regions where snow cover is currently ephemeral (Dickinson *et al.* 1987). For this reason, it has been suggested that snow cover might be a useful index for monitoring global climate change (Barry 1985).

Just where and when does snow lie on the ground over the North American continent? I explore this by examining the boundary separating snow-covered from snow-free land. I document the annual march of this snowline equatorward and back to the Arctic using satellite-derived data from the past two decades and station observations to examine the variability of the midwinter snowline over the central United States since the turn of this century.

Delineating the Snowline

The location of the snowline may be determined by observing surface conditions from the ground or from an elevated platform. Most commonly, the former includes integrating observations from a network of official climatologic stations, and the latter involves interpretation of one or a suite of satellite images.

Ground observations of snow cover are commonly made daily, although in some regions (particularly mountainous areas) measurements are made only once or twice a month. While practices vary, if the standard open, level observing site is greater than 50 percent snow covered, snow depth is measured and recorded. With daily observations and a network of stations to draw from, the location of the snowline may be quite accurately determined. However, a series of caveats must be ascribed to the preceding statement. Problems with the delineation of the snowline from surface observations arise: when observers fail to make measurements or do not maintain standard observation practices (Robinson 1989); when drifting makes an accurate assessment of cover difficult; where a sufficiently dense network of stations is absent (for example, mountainous terrain and arctic lands); where significant topographic diversity makes interpolation of snow conditions between stations virtually impossible; and where snow cover is exceedingly patchy. Station observations of snow cover are available over the United States and Canada for the past century, although the number and coverage of these sites increased significantly towards the middle of this century and has remained relatively stable for the past 50 or 60 years.

The delineation of the snowline from aircraft and satellites is achieved by several means. Local snowlines are examined from aircraft photographically as well as by instruments that measure surface-emitted gamma radiation. Regional and continental snowlines are gleaned from visible satellite observations of solar radiation reflected off of the earth's surface and from microwave radiation emitted by the surface. The visible approach has the benefits of being directly interpretable by the human eye and by having imagery with a resolution of 1 kilometer available daily over the entire North American continent. Disadvantages of a visible approach include the inability to monitor surface conditions where clouds are present and where dense vegetation precludes reliable observations of the underlying surface. Low solar illumination is not a significant liability, since most high-latitude regions are snow covered before the diminution of light and remain covered until spring. Charts of snow cover from visible imagery have been available continuously since the late 1960s.

Microwave radiation penetrates winter clouds, permitting an unobstructed signal from the earth's surface to reach a satellite. The discrimination of snow cover is possible mainly because of differences in emissivity between snow-covered and snow-free surfaces. Spatial resolution is on the order of several tens of kilometers, making a detailed delineation of the snowline difficult, particularly where snow is patchy. It is also difficult to identify shallow or wet snow using microwaves, and recognition of snow is a problem where vegetation masks the surface. Microwave-derived snow products have been available since the late 1970s.

The accuracy of delineating the snowline over North America also depends upon the condition of the snow pack. The situation is simplest when snow cover is fresh, as the transition between full cover and snow-free ground commonly occurs over only several kilometers. Charting the snowline where the snow pack has not been replenished for days to weeks is more difficult. With time, the coverage of snow becomes patchy, as a result of drifting and preferential melting on open fields and south-facing slopes. Thus, the width of the transition zone between full cover and snow-free ground may be several tens of kilometers.

In summary, the snowline produced from satellite or station observations will at best be accurate to within a few kilometers (for instance, fresh snow and visible satellite charting), and may be difficult to discern within a range of 50 or perhaps even 100 kilometers (for instance, old snow and a network of stations located tens of kilometers apart). However, such limitations are not of great concern when analyzing the snowline over large regions, particularly when averaging on monthly or longer time scales.

North American Snowline: 1971-1990

The snowline over the North American continent is in a constant state of flux. Its movement may be tracked using data from U.S. National Oceanographic and Atmospheric Administration hemispheric snow charts.

The National Oceanographic and Atmospheric Administration weekly charts depict the snowline based on a visual interpretation of photographic copies of visible satellite imagery by trained meteorologists. Charting began in 1966, but only since the early 1970s has the accuracy been such that charts are considered suitable for continental-scale studies (Wiesnet *et al.* 1987).

The snow season begins in North America in September, when snow cover becomes established over the high Arctic and on lofty mountain peaks. By October, the mean snowline (defined here as the isoline denoting a 50 percent frequency of snow cover) at lower elevations has advanced beyond the Arctic Circle and encircles the Canadian Rockies (Figure 1). The snowline continues its equatorward trek throughout the fall and early winter, reaching its southernmost position in January or occasionally in February (Figure 1). At this time, the mean snowline lies close to the 40th parallel across the United States, taking a dip towards 35° in the southern Rockies. All of Canada and Alaska, except for maritime regions and the western Canadian prairie, are virtually assured a continuous mid-winter snow cover, while snow may cover the ground as much as 10 percent of the time as far south as 35°.

In April, the mean snowline in the center of the continent retreats across the United States-Canada border and is located near the prairie-boreal forest

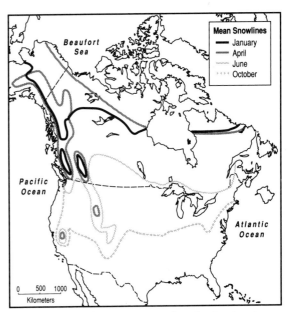

Figure 1. Mean North American snowlines for October, January, April, and June, derived from National Oceanographic and Atmospheric Administration weekly snow charts for January 1971 to June 1990. The mean snowline is defined as the isoline denoting a 50 percent frequency of snow cover.

boundary in central Canada (Figure 1). Toward the east, the snowline lies between 45° and 48°. The Rockies, Cascades, and Sierras normally remain snow covered in mid spring. The mean snowline in June lies near the taiga-tundra boundary in Alaska, the Keewatin and northern Mackenzie Districts of the Northwest Territories, and in northern Québec and Newfoundland (Figure 1). The Canadian Rockies also remain snow covered. By the end of June, the North American snow season has almost run its course. Melt has progressed well into the islands of the Canadian archipelago, and by mid July all lands are snow free.

The seasonal position of the snowline has varied considerably from year to year over the past two decades. This is well expressed by the area of North America covered with snow in each season (Figure 2). Snow areas have varied from year to year by 2 to 3 million square kilometers in each season, the extreme maximum varying from 1.2 times the minimum in winter to 1.8 times the minimum in summer. No trends in fall, winter, or spring snow cover are discernible over the period, although the springs of 1987 through 1990 exceeded or were close to previous minima. Summer cover was appreciably lower in the early and late 1980s than at other times in the past two decades.

Central United States Snowline: 1900-1989

A network of 143 stations with daily snow data since the turn of the century is available over eleven central U.S. states, enabling an investigation of snowline dynamics over this period (Robinson 1988). The inhomogeneous spacing of the stations and occasional

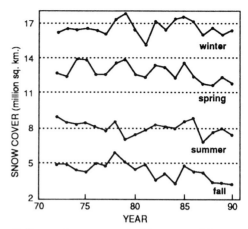

Figure 2. Seasonal snow cover (million square kilometers) over North America (including Greenland) from winter 1971 to 1972 to fall 1990. Data from National Oceanographic and Atmospheric Administration weekly snow charts.

data gaps in the network requires grouping of station reports into 1° latitude by 4° longitude divisions. Division values are an average of all available reports, which range from one to eight depending on day and region. Of the past nine decades, the January snowline was at its southernmost in the 1970s and was at its northernmost, roughly 3° or 4° poleward of the 1970s line, in the 1900s (Figure 3). A division is considered to be snow covered in a given decade when more than half of the days have a cover more than or equal to 2.5 centimeters for at least five Januaries. A decade-by-decade count of the number of divisions meeting this criterion shows the first five decades of this century to have less January snow cover than the most recent

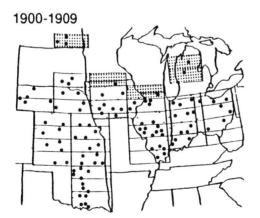

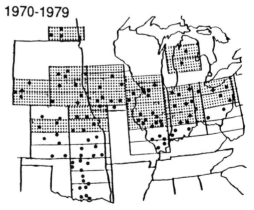

Figure 3. January snow cover over the central United States from 1900 to 1909 and 1970 to 1979. Study divisions with five or more years with more than half of the days in the month with a snow cover greater than or equal to 2.5 centimeters. Dots = study stations.

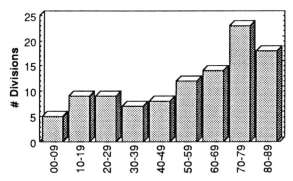

Figure 4. Decadal summary of study divisions in the central United States (cf. Figure 3) with more than half of the days in January having a snow cover of greater than or equal to 2.5 centimeters in at least five years. Pre-1980 data from the station network; 1980 to 1989 data from National Oceanographic and Atmospheric Administration weekly snow charts.

four (Figure 4). This is particularly notable in the central Great Plains.

Conclusions

Should model projections of future climate warming prove accurate, the mean North American snowline will retreat poleward and toward mountain peaks in all seasons. This will have wide-ranging consequences on water supplies, agriculture, tourism, and many other processes and activities. The significant year-to-year variability in the snowline over past decades has provided glimpses of snow-drought effects that would become more common with warming.

Despite a recent decrease in spring snow cover, there is no indication of any wide-scale retreat of the North American snowline in the past several decades. In fact, if the January snowline in the central United States is any indication (caution: it may not be, particularly in other seasons), a precipitous retreat of the line would have to occur before conditions exceeding the first half of the century were surpassed. Clearly, we must continue to observe the snowline from ground and satellite platforms to gain a better understanding of its role in future climate change and the identification of such change.

Acknowledgments

I thank M. Hughes and F. Keimig for technical assistance. This work is supported by the National Oceanographic and Atmospheric Administration under grant NA90AA-D-AC518 and by the National Science Foundation's Climate Dynamics Program under grants ATM 89-96113 and ATM 90-16563.

References

Barnett, T. P., Dumenil, L., Schlese, U., Roeckner, E., and Latif, M. 1989. The effect of Eurasian snow cover on regional and global climate variations. *Journal of the Atmospheric Sciences* 46:661-685.

Barry, R. G. 1985. The cryosphere and climate change. In *Detecting the Climatic Effects of Increasing Carbon Dioxide*, DOE/ER 0235, eds. M. MacCracken and F. Luther, pp. 109-148. Washington, DC: U.S. Government Printing Office.

Dickinson, R. E., Meehl, G. A., and Washington, W. M. 1987. Ice-albedo feedback in a CO2-doubling simulation. *Climatic Change* 10:241-248.

Jones, H. G., and Orville-Thomas, W. J., eds. 1987. *Seasonal Snowcovers: Physics, Chemistry, Hydrology.* Dordrecht, Holland: D. Reidel.

Robinson, D. A. 1988. Construction of a United States historical snow data base. In *Proceedings of the 45th Eastern Snow Conference*, ed. J. Lewis, pp. 50-59. Lake Placid, NY: Eastern Snow Conference.

———. 1989. Evaluation of the collection, archiving, and publication of daily snow data in the United States. *Physical Geography* 10:120-130.

Rooney, J. F. Jr. 1967. The urban snow hazard in the United States. *Geographical Review* 57:538-559.

Snyder, J. R., Skold, M. D., and Willis, W. O. 1980. Economics of snow management for agriculture in the Great Plains. *Journal of Soil and Water Conservation* 35:21-24.

Walsh, J. E., Jasperson, W. H., and Ross, B. 1985. Influences of snow cover and soil moisture on monthly air temperature. *Monthly Weather Review* 113:756-768.

Wiesnet, D. R., Ropelewski, C. F., Kukla, G. J., and Robinson, D. A. 1987. A discussion of the accuracy of NOAA satellite-derived global seasonal snow cover measurements. In *Large Scale Effects of Seasonal Snow Cover, Proceedings of the Vancouver Symposium*, IAHS Publication 166, eds. B. Goodison, R. Barry, and J. Dozier, pp. 291-304. Vancouver, BC: International Association of Hydrologic Science.

P A R T   T W O

FILTERS AND LENSES

To geographers, boundaries are filters as well as barriers. They screen the commerce of ideas, goods, and people, and demarcate differences in origins and lifestyles. Borders between Canada and the United States and between Mexico and the United States have served these functions with varying degrees of closure. Their permeability has depended on the social, political, and economic circumstances of the time. Recent movements of Central American refugees to Canada via the United States and negotiations over free trade have evoked strong arguments over the meanings and roles of these national frontiers. But even at local levels within countries, boundaries are important filters. Thus, the mismatch of soil, vegetation, property, and other boundaries can play havoc with implementation of agricultural and other policies by public agencies. The crossing of water barriers by new bridges can open access to formerly isolated areas and elicit conflicts between insiders and outsiders. Mental constructs are also filters. Thus, the philosophical perspectives of geographers themselves serve as sieves for sorting out ideas about the interpretation of landscapes. Through the lens of one's personal frame of reference, the messages evoked from environmental settings, architecture, paintings, and writings mesh into patterns of meaning. It's a good idea to keep a selection of lenses on hand.

—*Victor A. Konrad, University of Maine and Canada-U.S. Fulbright Program*

A Canadian Geographer's Perspective on the Canada-United States Border

Randy William Widdis
University of Regina

Canadians are the most smug people in the world. Americans are the most arrogant people in the world. That is why the world's longest undefended border separates them. Canadians secretly like the Americans being so arrogant because it enables them to be so smug.

—FOTHERINGHAM (1989, 84)

Geographers study boundaries both objectively and subjectively for what they reveal about landscape elements of neighbouring states and as part of larger investigations into relationships between those adjacent political units. The scientist classifies and defines while the humanist considers the symbolic meanings of an artificial creation. To the empiricist, the Canadian-American border is understood by pertinent facts: At 8,890 kilometers, it is the longest continuous boundary line in the world; and with 130 crossing stations and more than 100 million people and almost $200 billion in goods and services crossing each year, it is the busiest border on the planet (Colombo 1986; Macdonald 1989). To the humanist, this boundary can only be understood as an interpreted emotional experience, a symbolic marker defining a Canadian community, at least an Anglo-Canadian community.

The border itself is not a living entity, imbued with meaning; it is the people who live along it who give the border symbolic significance. The fact that 75 percent of all Canadians and only 12 percent of Americans live within 160 kilometers of the border explains to some extent its greater influence on the lives of those who live north of the "line" (Colombo 1986, 8). Yet more important than geographical propinquity to a geopolitical boundary is what the border represents

to Canadians for the development of a "historically and geographically specific social system" (Pred 1984, 281).

Canada as a "historically contingent society," developing within the context of its own internal evolution, has always framed its "becoming" in its changing political, economic and cultural relationships with the United States. That the relationship with the United States functions as a barometer by which Anglophone Canadians measure their evolving identity is not surprising given the complex and varied nature of the ties linking various transborder regions. (French Canadians have developed their own cultural boundaries within the country, shields serving to strengthen their own sense of national identity.) Historical geography, distinguished by its concern with understanding data in context and a subsequent focus on places and the processes that create them, a mode of inquiry that combines functional description with structural explanation, and a synthesis of subject and object whereby a humanist perspective provides the skeleton on which the researcher adds the flesh of empirical data, serves as a useful basis upon which to consider Canadian-American relations and the different meanings applied to the border at various spatial and temporal levels. It is from this particular disciplinary viewpoint that I offer the following comments on the symbolic significance of the border in Canadian-American relations.

That different meanings have been offered regarding Canadian-American relations is evident upon examination of metaphors used to describe the border. Canadians, particularly Anglophone Canadians, in their fiction and popular culture have tended to view the border as a dividing line or shield, protecting a fledgling culture from a dominating presence. The metaphor of the "border as a shield" symbolizes that for Canadians, our relationship with the United States has played a major part in developing what symbols we do have, an important consideration given the reality of living in an environment dominated by American symbols, icons, and myths.

A discontinuous and disjointed settlement experience, combined with the overwhelming American presence, have restricted efforts to create national symbols. Quebec exists largely as a nation because of its unique culture, reinforced by language, and its association with a distinctive historical geography, but Anglophone Canada has always struggled to find its niche within the continent. Anglo-Canadians can draw only on a meagre reservoir of symbols and myths for guidance. Symbols help us to interpret who or what we are and what we can be and myths are particularly important because they transform secular history into sacred legends. Anglophone Canada has had relatively few myths around which emotions, beliefs, memories, and nostalgia have been ritualized, the exceptions being the trek of the Loyalists northward in 1784 and the idea of "Canadian heroes standing up to Yankee bullies" during the War of 1812 (Kilbourn 1988, 18). But even the aggrandizement of the War of 1812 and the myth of the Canadian "David" standing up against the American "Goliath" must be weighted against the reality of Canada's being defended largely by British troops and Indian loyalists and the irony that trade between regions on both sides of the border continued largely unabated even as armies crossed that same line.

The most striking aspect of the border as shield metaphor is its oppositional character. As Anthony Cohen (1985, 58) wisely states, "boundaries are relational rather than absolute; that is, they mark the community in relation to other communities." The border serves as the basic reference point for historical, literal, symbolic, and psychological interpretations of an Anglo-Canadian identity. Yet identity is a problematic concept; it is heavily contextual, difficult to measure, and changes over time and according to milieu. What is constant in the Canadian experience is that all groups in different regions and at different times have interpreted their identity vis-à-vis their relationship with the United States. And in this context, the border is the emotional and ideological focal point for the never-ending debate over the nature of these relationships.

This is the basic premise of Patrick McGreevy's recent examination of the contrasting meanings assigned to the Niagara border region by citizens from both countries. McGreevy (1988, 307) metaphorically describes the border as a "wall of mirrors, reflecting back different meanings to Canadians and Americans, meanings that in turn reflect different ideologies of nationalism." Canadians take much better care of their side of this border region in comparison to Americans because Niagara symbolizes the "front entrance to the house," the place where English Canadian nationalism begins; whereas for Americans, this place represents "the back alley where the trash is kept," the place where American's manifest destiny ended with the War of 1812 (McGreevy 1988, 313). Yet as McGreevy readily admits, the meanings assigned to the border by citizens living on both sides are not so clearly distinguishable at other places because of different historical geographical experiences.

There was no similar militaristic conflict along the St. Croix River (despite the antics of the drunken Fenians in 1866) or the 49th parallel. The strong commercial links between the Maritimes and New England and the east-west flow of trade and migration into the west paralleling the transcontinental railway resulted in a different transborder experience for these Canadian regions and generated different interpretations of the significance of the border. Regional separation within Canada resulting from cultural plurality and geographical isolation combined with different kinds of relationships with American border regions to create a variable settlement experience, produce different levels of attachment to the idea of Canada, and elicit different interpretations of the symbolic meaning of the border. This regional dimension of Canadian life has been and continues to be a major factor in the development of Canadian identity, a fact noted by many observers but nowhere more eloquently expressed than by the recently deceased giant of Canadian literary criticism, Northrop Frye.

Through a lifetime of study, Northrop Frye became one of the foremost students and interpreters of Canadian culture and in doing so, it might be argued, he unconsciously adopted a historical geographical approach to the study of Canadian identity. He views the elusive question of identity as nothing more than an expression of culture including the human imagination. Since the imagination—that is, the ideas by which we live—is so shaped by personal experience and perception, Frye maintains that in a country as large and diverse as Canada, identity is not a "Canadian" question but a "regional" question. Frye insists that unity and identity in Canada are quite different concepts but are often confused in the minds of Canadians:

Identity is local and regional, rooted in the imagination and in words of culture; unity is national in reference, international in perspective, and rooted in political feeling (Frye 1971, ii).

The tension between national unity and regional identity, Frye (1971, 220) believes, means that the important question perplexing Canadians is not "Who am I?" but rather "Where is here?" He emphasizes the fact, later elaborated upon by R. C. Harris (1982), that there was no temporally and spatially continuous settlement experience as in the United States. Small communities and regions, geographically isolated from one another, ensured a development of what Frye calls a "garrison mentality" and Harris terms an "island archipelago."

Both Harris and Frye express their views of the historical geographical essence of Canada in the form of metaphors, the former seeing Canada as a collection of islands in a stormy sea called Confederation and the latter comprehending the country in the form of a cartographical metaphor, that is using the legend and conventions of a map (Canadian Broadcasting Corporation 1975). To Frye, each voyageur (Canadian) in search of the national image (Here) is involved in a journey that has no arrival; the map is not yet complete. The individual identifies and interprets ideas, events, and experiences largely within a regional frame, which enables him to orient himself in time and space. He is asking himself "Where is here?" but this leads him still into uncharted territory. In this quest the voyageur must recognize that there is a body of cultural assumptions, framed by a regional consciousness, involved in the appreciation of the nature of Canada which acts as a filter through which the imagery passes. Frye reasons that an individual's Here is neither static nor complete but continually evolves as new ideas, events, and experiences permeate his or her consciousness.

The regional frame the voyageur uses to orient himself in territory is bounded east and west by the rest of Canada and north and south by his transborder relationship with neighboring American regions. These contradictory east-west and north-south forces have created in essence a border parallax (Gwyn 1985). A parallax is defined as: "an apparent change in the position of an object resulting from the change in the direction or position from which it is viewed" (*Webster's New World Dictionary* 1970, 1030). The ideological position from which the border is viewed plays a major role in its interpretation. This is certainly evident in the recent debate between two noted Canadian historical geographers, Cole Harris (1990a), who maintains that the emergence of Canadian regions and regional identities had more to do with the east-west transcontinental expansion of trade and settlement than proximity to American regions; and Victor Konrad (1990, 127), the major spokesperson for the "borderlands thesis," who despite claims to ideological neu-

trality (McKinsey and Konrad 1989, iv), takes an ideologically full position that "North America runs more naturally north and south than east and west" (Konrad 1990, 127).

Regional borders in Canada, Harris insists, are more the result of distinctive European encounters with different Canadian settings than simply being peripheries of American core regions. To Harris, the border is a shield against continentalism and a historical geographical interpretation of the settlement of Canada, he insists, "gives us a particularly clear fix on all of this, situating modern Canada in its own remarkable, if symbolically diffident past" (1990b, 1). The present situation (that is, free trade, constitutional crisis) leads him to believe that we are in great danger of balkanizing, of becoming "a mosaic of antagonistic regional and ethnic enclaves, sitting ducks for the continental pressures that have always borne on this country, and do so now more than ever" (Harris 1990b, 2).

The "borderlands thesis" views borderlands as regions of cultural synthesis where cross-border flows of people, goods, and ideas acted as integrative elements contributing to a continental dynamic (McKinsey and Konrad 1989, 7-8). Proponents maintain that several distinct borderlands based on regional affinities have developed over time, reflecting different types of interactions taking place across the border, interactions between complementary areas that dovetail (McKinsey and Konrad 1989, 29). In their view, the debate over free trade exaggerates the significance of its economic and political components as the flow of goods and services is mediated by subnational structures rotted in regional cultures which transcend the border, traditional structures ensuring that political and cultural sovereignty continue to exist (McKinsey and Konrad 1989, 15).

It seems to me that this cultural synthesis/east-west perspective debate rests in part upon two dialectically opposed visions of the core-periphery paradigm: the neoclassical perspective that views core-periphery exchanges as mutually beneficial because of the trickle-down mechanisms of the marketplace; and the neo-Marxian view that sees core-periphery relations as unequal and exploitative because of the unequal exchange mechanisms inherent in capitalist markets. Both sides would deny any adherence to such polar ideological positions but the rhetoric of the debate demonstrates that, in spite of their attempt for objectivity, it is impossible for Canadian scholars to consider the symbolism of the border and the question of Canadian-American relations without being political.

My interpretation of history is that Canadians have built a distinct society with which they identify at a number of levels despite regional tensions and continentalist forces. Canada is a fragile but continuously evolving east-west society based on a historical development of staples along the St. Lawrence-Great Lakes axis and the creation of instruments bridging what Harris (1982) calls the island archipelago: an extensive nationalized transportation linkage, a complex of government sustained cultural agencies, and redistributive policies designed to reduce regional and social disparity and "humanize" the marketplace. The fact that the present government is implementing policies that threaten those east-west instruments bridging regional/cultural islands and, by implication, national sovereignty, is a cause of great concern for many Canadians. In short, elements of our historical-geographical development and a different political culture have served to balance the very powerful north-south pull that connects American and Canadian border regions in many ways. Perhaps the best way to examine Canadian-American relations and the significance of the border is to attempt to control one's political ideology as much as possible and take an intermediary position between Harris's east-west and Konrad's north-south arguments; the truth lies somewhere in the middle of this dialectic.

McGreevy is right; the Canadian-American border is a wall of mirrors reflecting different meanings, meanings that in turn reflect different ideologies of nationalism (1988, 307). For Anglophone Canadians in particular, by examining the nature of our relations with America, we in turn see ourselves. What we see in the mirror is largely influenced by what we want to see, because ultimately nationalism, regionalism, and continentalism are territorial ideologies.

The vision I see reflected in the mirror disturbs me profoundly. The islands are shrouded in the fog of indifference and intolerance and the bridges are in danger of collapsing under the weight of constitutional and economic troubles. Or, as was the case of the Mercier Bridge, the bridges are closing because of indifference rather than from collapsing because of too little structural support. And the voyageurs grow more weary with each passing day.

References

Canadian Broadcasting Corporation. 1975. *Journey Without Arrival* (film).

Cohen, Anthony. 1985. *The Symbolic Construction of Community*, Key Ideas Series, No. 1. Chichester: Ellis Horwood.

Colombo, John Robert. 1986. *1001 Questions About Canada.* Toronto: Doubleday Canada Ltd.

Fotheringham, Allan ("Dr. Foth"). 1989. Different—In a matter of speaking. *Maclean's* 3 July:84.

Frye, Northrop. 1971. *The Bush Garden: Essays on the Canadian Imagination.* Toronto: Anansi.

Gwyn, Richard. 1985. *The 49th Paradox: Canada in North America.* Toronto: McClelland and Stewart Ltd.

Harris, R. C. 1982. Regionalism and the Canadian archipelago. In *Heartland and Hinterland*, 1st edition, ed. L.D. McCann, pp. 459-484. Scarborough, ON: Prentice-Hall.

———. 1990a. The Canadian archipelago and the borderlands thesis. In *Association of American Geographers Annual Meeting Program and Abstracts*, p. 96. Washington, DC: American Association of Geographers.

———. 1990b. Editorial. *Newsletter for Canadian Historical Geographers* 5:1-2.

Kilbourn, William. 1988. The peaceable kingdom still. *Daedalus* Fall: 1-30.

Konrad, Victor. 1900. Borderlands: A concept for reinterpreting North America. In *Association of American Geographers Annual Meeting Program and Abstracts*, p. 127. Washington, DC: Association of American Geographers.

Macdonald, Marci. 1989. Fields of force. *Maclean's* 3 July:26-30.

McGreevy, Patrick. 1988. The end of America: The beginning of Canada. *The Canadian Geographer* 32(4):307-318.

McKinsey, Lauren, and Konrad, Victor. 1989. *Borderlands Reflections: The United States and Canada*, Borderlands Monograph Series No. 1. Orono: Canadian-American Center, University of Maine.

Pred, Alan. 1984. Place as historically contingent process: Structuration and the time geography of becoming places. *Annals of the Association of American Geographers* 74(2):279-297.

Webster's New World Dictionary, 2nd college edition. 1970. Toronto: Nelson, Forst, and Scott.

Urban Expressions of Cultural Duality

John Mercer

Syracuse University

Despite global tendencies toward homogeneity in the "advanced" market economies, Canada and the United States are not on the same path to a similar social condition. Rather, as S. M. Lipset (1990) aptly observes, the two societies are more like a pair of trains traveling along parallel railway tracks. Many kilometers (and some centuries) later, they are far from their origins but still separate. This is most evident in a comparison of American and Canadian cities. Three quarters of the population in both countries live in cities that constitute a most crucial set of points within North America. I contend that there are two distinct urban systems in North America, two highly differentiated sets of cities bounded by the border. These cities are the products of two differing cultural contexts; the cultural motifs of the societies are embedded in their urban forms and in the lives of their urban populations. Equally, the cities, some certainly more important than others, have been significant in socioeconomic and political terms in the organization of the two distinct cultural contexts (Goldberg and Mercer 1986). One of the most basic yet concrete representations of the border's division of North America lies in the nature of urban places in Canada and in the United States. The urban landscapes, urban forms, social geographies, and urban local governments all bear witness to the power and significance of this division, symbolized in the international border.

While the border demarcates and defines the territorial edge of two societies, Canada and the United States do share certain basic characteristics. They are both capitalistic, with private ownership of the means of production dominant and the institution of private property reflected in high levels of home ownership, private markets in rental housing, and low levels of public-sector housing. As a result, social relations are,

broadly speaking, capitalist in nature, and struggles over working conditions and the distribution of surplus value created in the labor process have been widespread in both countries. Both societies also have federal systems of governance and adhere strongly to the principles and practices of representative democracy.

That said, one can immediately recognize that important differences exist even with respect to these commonalities. Canada has a level of state ownership in many industrial, utility, and transportation sectors that differentiates it from the United States. Many commentators have argued, J. B. Cullingworth (1987) perhaps most convincingly, that private property and land use are more strictly regulated by local governments in Canada than in the United States. The federal systems of government are differentiated sharply by contrasting divisions of powers between federal and state or provincial authorities, by a greater role accorded the judiciary in the United States, and by an enduring commitment to different forms of government—one congressional, the other parliamentary in nature. Each country's urban areas are governed within federal frameworks but there the similarity ends. Central-local relations for Canadian cities mean relations with the provinces, the federal government being of lesser importance. In contrast, the federal relationship with cities in the United States has been paramount; city-state relations have been highly variable and generally of lesser importance, despite attempts to increase the states' roles and reduce that of federal authority, especially during the Reagan presidency.

The cultural context for urban development possesses two other important features in addition to capitalist and inter-governmental relations. The countries differ sharply in their social composition, captured in two fundamental dualities—one, American, is principally expressed in racial terms; the other, Canadian, is chiefly encountered in socio-linguistic terms. A full detailing is beyond the scope of this essay. There is little doubt however that a hallmark of American cities is the remarkable persistence still of a high level of racially based residential segregation. Because of the particular geographical location of the francophone and anglophone groups, residential segregation in metropolitan areas is less the concern (except in Montreal) than are regional tensions and inter-provincial relations, and relations between Quebec and the government of Canada. Nevertheless, given the location of highly urbanized southern Quebec in the urban-centered core of the Canadian space economy, this is a critical element in understanding Canadian society and urban structure in Maurice Yeates' heartland region, *Main Street* (Yeates 1975). These long established dualities are being challenged by the outcome of an important geographic and social process. Immigration from regions beyond the traditional European sources is altering the composition of labor forces and urban populations although a notable cross-national difference again emerges, there being a far higher proportion of foreign-born persons resident in Canadian cities than in American ones. Indeed, the noted travel writer Jan Morris has dubbed Toronto as "the emblematic immigrant destination of the late twentieth century" (Morris 1984, 44).

A second crucial feature of the cultural context for urban development is that constellation of values and beliefs in a society that defines appropriate ways of doing things (including the making of cities). There are several key differences here. Although complex, the notion of an American individualism and Canadian collectivism captures the essential difference. This distinction can only be sketched here. American political philosophy is grounded in universalistic norms, sustaining notions like liberty and equality. Liberal philosophy reaching its apogee in America is strongly connected to the privileging of the individual, competition, and free enterprise. Other intellectual and philosophical traditions molded Canadian ideologies, as did a more modest liberalism. A conservative tradition sustains the maintenance of community order and the public peace over individual rights (only recently constitutionally guaranteed). An important socialist tradition emphasizes social justice, public sector intervention for the common weal, and cooperative enterprises. Not surprisingly then, there is an acceptance and willingness by many Canadians of having governments and their agents act directly in various market contexts in the provision of goods and services. Though not viewed universally as beneficial, the support in Canada for public enterprise is much higher than in the United States where greater skepticism and

doubt over the desirability and efficacy of public authorities exists.

Private Cities versus Public Cities

I assert that Canadian cities are more public in their nature and American ones more private. These are not sharply drawn polarities but should be thought of as overlapping zones on a public-private continuum.

The conception of the American city as "private" is not entirely original, following as it does S. B. Warner (1968). It expresses a strong commitment to individualism and individual freedoms; to the protection of private property rights under the Constitution; to the use of private mechanisms and individual user fees in the provision of infrastructures and goods and services; to homeownership, especially of the single detached residence; and to the concept of autonomy in local government. In a privatized society, problems are solved in a highly personalized fashion, withdrawal being prominent. In America, conditions of life in many cities have led to withdrawal to the safer ground of suburban jurisdictions with the power to exclude. It is equally necessary to recognize an active federal government in urban affairs, a full panoply of land-use powers in the hands of local governments, and a wide range of public authorities delivering services such as education, housing, and transportation. A central question is this: Whose interests are best served by these public agents? A popular and much encouraged view is that government serves all the people in given jurisdictions; but a more caustic view suggests that special interests have captured federal administrations (Greenstone 1982). The land-use powers of local governments, especially in the suburban domain, have often been exercised causally in exclusionary practices to benefit the interests of existing residents who may constitute a community or simply be an aggregation of private interests seeing their dwellings as commodities. In short, the private city is an expression of power relationships within urban areas.

The public city is more attuned to Canadian values, ideologies, and current practices. It expresses a stronger commitment to variously defined collectivities; to the maintenance of social order rather than individual freedoms; to the preservation of ethnic identities; to a greater trust and belief in the competence of government and its bureaucracies; and to the idea of active intervention in the form of urban planning by city, suburban, and innovative forms of metropolitan governments. The public city exhibits a higher quality of urban development consistent with high servicing standards set by local authorities: a well-developed and high-quality public transportation system; an extensive system of community and recreation centers with quality parks and open spaces; all publicly provided; and a public school system that is not seen as having virtually collapsed in its central jurisdiction. There is in Canadian cities a multifaceted presence of a more extensive welfare state, of health care and hospital services, of policing, and of a wide array of Crown corporations. There is, however, a private counterweight to the public city and its equation with a Canadian identity. Privatism exists in Canada, expressed in housing and urban development as plainly as in other walks of life. The commitment to homeownership and the use and exchange values associated with private property are on a par with those in the United States. Canadian municipalities reveal exclusionary practices. Residential covenants and individual discrimination in urban housing markets were and are not unknown. While government has been an important actor in Canadian economic development, the private sector remains the driving force. More narrowly, the driving force for Canadian *urban* development is essentially private, as in the United States. This does not discount the development activities of public agents but it recognizes that theirs is not generally the primary role (Logan and Molotch 1987).

Conclusion

The Canada-United States border sets apart two distinct cultural contexts, creating divided ground. The expression of these different contexts can be clearly seen in the distinct landscapes of the public and private cities and, equally importantly, in the "landscapes of the mind" of those who occupy a powerful continental duality—Americans and Canadians.

References

Cullingworth, J. B. 1987. *Urban and Regional Planning in Canada*. New Brunswick, NJ: Transaction Books.
Goldberg, M. A., and Mercer, J. 1986. *The Myth of the North American City: Continentalism Challenged*. Vancouver: University of British Columbia Press.
Greenstone, J. D., ed. 1982. *Public Values and Private Power in American Politics*. Chicago: University of Chicago Press.
Lipset, S. M. 1990. *Continental Divide*. New York: Routledge.
Logan, J. R., and Molotch, H. L. 1987. *Urban Fortunes: The Political Economy of Place*. Berkeley: University of California Press.
Morris, J. 1984. Flat city. *Saturday Night* 99(6):38-46.
Warner, S. B. 1968. *The Private City*. Philadelphia: University of Pennsylvania Press.
Yeates, M. 1975. *Main Street: Windsor to Quebec City*. Toronto: Macmillan.

The Pacific Coast Borderland and Frontier

Leonard J. Evenden
Simon Fraser University

Daniel E. Turbeville III
Eastern Oregon State College

It took at least a century for the boundary between the United States and British North America, later Canada, to be established. In 1818, the 49th parallel of latitude was established as the boundary in the west, but this agreement did not apply beyond the Rockies. Finally, in 1846, it was agreed by treaty that this line should be extended to the coast at the Strait of Georgia (Figure 1). Between these dates, however, settlement in the area known to Americans as the Oregon Country was open to people of both powers—a "joint occupation."

The establishment and subsequent character of this segment of the boundary illustrates the distinction between boundary and frontier: the one conventionally understood as a line of containment, the other a dynamic zone of predominantly outward movement (Kristof 1959). To extend the boundary to the Pacific coast was imperative, a task made difficult by remoteness and the loose character of settlement in the west, ". . . that shadowy belt of American occupation . . . where it met the corresponding nebulous outskirts of the far-away Canadian state on the St. Lawrence River" (Semple 1911, 211). The traumatic memory from the War of 1812, of fixing the boundary at the "Niagara frontier," was cautionary. Canadians considered the War of 1812 a victory in that the boundary at Niagara was maintained in the face of the advancing American frontier and this experience may have represented a significant American frustration (McGreevy 1988). But the circumstances of the boundary establishment in the far west differed from those at Niagara, and "The United States . . . won the race for the effective settlement of the Oregon country" (Thomson 1966, 261).

Frontier and Focus

The steady growth of American settlement in the Willamette Valley, the "garden of Oregon," stretching south from the lower Columbia, forged the focal point of the territory, a counterpoise to the (British) Hudson's Bay Company post on the lower Columbia. Americans were ambitious, not only to engage in the fur trade, but to own and reside on the land, develop farms, introduce American forms of government, and gain access to the Pacific. In 1843, in an increasingly tense atmosphere, the Hudson's Bay Company relocated its headquarters to a newly established fort at the southern tip of Vancouver Island, to be created a British crown colony in 1849. This fort was the nucleus around which would form the city of Victoria, capital of the colony. Thus the two principal foci had emerged, of land-based American settlement and the British coastal fort and colonial capital (Figure 2).

Not long after, in 1858, thousands of Americans were attracted to Victoria and the Fraser River by the discovery of gold in the interior of British Columbia. The British responded by sending a regiment, the Royal Engineers, to the area and by establishing the mainland colony of British Columbia. The Engineers built routes to the interior and established townsites, the best known being New Westminster, positioned on the high right bank on the north side of the unbridged Fraser, a site protected by the expanse of the river from the possibility of invasion from the south.

The Americans felt no similar fear of land-based invasion. What they feared was the power of the Royal Navy, a substantial consideration, given that the boundary in Georgia Strait was still being contested. Hence they established a number of coastal forts, including one at Port Townsend at the entrance to Puget Sound. Just as New Westminster had become the dominant focus in the new mainland colony, so Port Townsend was ambitious to become the main economic center on the American side. As tensions eased, both places lost out to larger centers, namely Vancouver and Seattle, which lay at a distance from the border. Thus a nascent settlement system, focused at the border for political and military reasons, yielded to a network

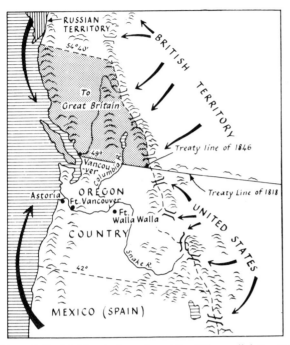

Figure 1. Establishing the boundary at the 49th parallel. Source: Watson (1963), reprinted with the permission of Longmans UK Ltd.

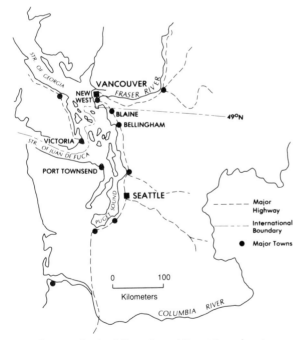

Figure 2. Strait of Georgia and Puget Sound region.

of places that grew up in response to economic development and that drew the focus away from the border. This second network spatially anchors today's settlement.

Unity and Separation at the Border: Contemporary Perceptions

To a Canadian who lives within a few minutes' drive of the border, the crossing represents not only an impediment but also an opportunity of access to a market with a wide range of favorably priced consumer goods. Recreational land is also comparatively cheap, considering its proximity to Vancouver, and development regulations have been permissive. To an American, the mounting pressures attending this spillover of people as shoppers, holiday residents, and curious neighbors is both welcome and irritating. Some development opportunities are stimulated in this region of historically marginal prosperity, but traffic line-ups in the small border towns at times render the streets almost impassable to residents, and queues at the checkout counters in the stores cause delays. Comparatively relaxed regulations concerning drinking and entertainment establishments can also result in behavior that local authorities are ill-equipped to handle (Simmons and Turbeville 1985). As for purchasing recreational land "Canadians now . . . are . . . condemned as alien interlopers and land-grabbers" (Rutan 1977, 7). The boundary is clearly permeable, and the border region, most importantly in Whatcom County, is a place of mixing.

First-year university students in Vancouver, with only rudimentary geographical education, were asked in 1978 to draw an outline map of Canada from memory. Results are conventional but, in a few cases, the far western boundary segment is modified or transgressed (Figure 3). Such perceptual permeability of the boundary perhaps suggests regional unity in the popular mind. In contrast, a study of popular regions based on the choice of regional names for metropolitan enterprises may suggest the opposite (Figure 4). In Zelinsky's study, the term "Northwest" coincides in large measure with the older notion of the Oregon territory. The boundary of "West" extends into Canada and intriguingly points out the region of historic Willamette Valley settlement. "Pacific" identifies not only distant California but also adjacent Canada. Thus, literally nominal evidence points to popular regional boundaries that are spatially constricted in the border zone, emphasizing separation. Openness and closure appear to coexist in boundary perceptions in this region.

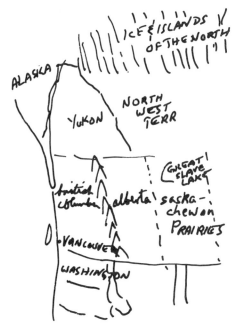

Figure 3. Perception of boundary permeability by a Canadian student.

Perceptual ambiguity at the personal and popular levels, however, is complemented at the state level by spatially wide-ranging perceptions of regional unity. In the mid-1970s the elected legislators of Washington state and British Columbia, without reference to national contexts, met in Bellingham, the compromise

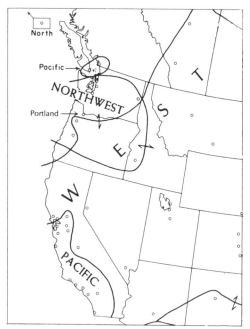

Figure 4. Popular regions: names of metropolitan enterprises. Source: Zelinsky (1980), reprinted with the permission of the Association of American Geographers.

border location, to discuss regional issues of common interest (Forward and Gerhold 1974). Following the passage of the free-trade agreement (Figure 5), this discussion has been reopened around the idea of forming a new political and economic geographic unit. The Pacific Northwest Economic Region comprises five states and two provinces and its aim is to promote cooperation in the region to become more competitive globally (Figure 6). Its core area comprises the Strait of Georgia and Puget Sound lowlands; the wider region combines the favored coast and an extensive resource hinterland. About one third of its 15 million people live in the politically bisected core. How the aims of economic cooperation can be achieved must be negotiated, as has been the case for two centuries in matters pertaining to the border.

Unity and Separation at the Border: Contemporary Behavior

The United States, as the dominant power, is a looming presence in Canadian life. Paradoxically, the lesser power has a comparatively strong presence in the border

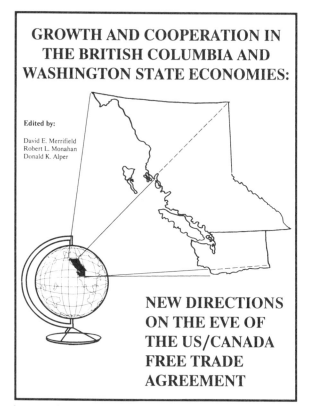

Figure 5. The region in continental and international perspective. Design by Greg Foy, reprinted with the permission of the Western Washington University Center for Canadian and Canadian-American Studies, and Center for Economic and Business Research.

Figure 6. Logo of the Pacific Northwest Economic Region. Reprinted with the permission of the Pacific Northwest Economic Region.

regions. Nowhere does this imbalance have greater local effect than at the Pacific end of the 49th parallel, that is in the lower Fraser Valley of British Columbia, including metropolitan Vancouver, and in adjacent Whatcom County, with its county seat of Bellingham.

Prices for consumer goods are generally lower on the American side, attracting Canadian shoppers, but Vancouver provides the closest major urban "downtown" for American border region residents. Cross-border tourism has grown and Whatcom County has become a popular place for Canadians to purchase recreational property. Canadian travelers use Bellingham airport for flights to destinations in the United States, and in this way the airport is developing an overflow function for Vancouver. Both Seattle and Vancouver have hosted world fairs, and it is now conventional to point to Expo '86 in Vancouver as the moment when a turning point was reached in the development of the border economy on the American side. Economic planning in the hitherto rural Whatcom County is now largely predicated upon the existence of the adjacent, numerous, and reasonably well-paid Canadian population. Some Canadian firms, for example in the wood-products industries, having found it increasingly difficult to compete in traditional American markets under current terms of trade, have relocated to the south side of the border to take advantage of domestic status and lower production costs, especially labor.

"Day-trippers" going to Bellingham and the border communities to shop have increased in number from 6.6 million in 1986 to more than 13 million in 1990. Spending in this region, by British Columbian residents, increased from nearly 67 million to almost 200 million Canadian dollars during the same period. Popular purchases include gasoline, groceries, clothing, and electronic items. The unprecedented demand for these goods is transforming the border towns from service centers for the traditional agricultural and fishing economies to suburban shopping centers in a metropolitan economy, albeit centers dislocated from Vancouver by the spatial intervention of the international boundary. This "boom" in the development of shopping facilities on the south side of the border is matched by a decline in demand for such outlets and goods on the north side. Indeed, in 1990, certain British Columbian merchants, with failing businesses located near the border, brought a lawsuit against the Canadian government for failing to collect all payable duty on goods purchased by travellers in the United States.

With such pressures, and with a recent agreement that binds the two countries to the principle of mutual "free trade," it is difficult to foresee any abatement in these trends. Overall, nearly 9 million automobiles crossed this segment of the boundary during 1989, representing a threefold increase since 1985. This has prompted efforts to speed up cross-border traffic. But to expedite the flow of traffic is not universally regarded as desirable. The issue is difficult in both countries, for it means that local conditions must be understood in distant capitals. By the time national priorities are weighed and decisions made, the effect at the local level can be quite uncoordinated.

Together yet Apart

In recognition of the peaceful settling of the boundary disputes in this area, Sam Hill of Seattle, a Quaker and entrepreneur, initiated the construction of the Peace Arch as a gesture of good will. Built astride the boundary, it was dedicated in 1921 (Figure 7). A

Figure 7. Ceremony at the Peace Arch, 1930s. Reprinted with the permission of the Western Washington University Center for Pacific Northwest Studies.

plaque on the east side of the structure displays the image of the *Mayflower*, recalling the long cultural memory of America to this point on the boundary. This is balanced on the west side by a plaque bearing the image of the *Beaver*, the first steamship on the northwest coast, in the service of the Hudson's Bay Company. These plaques are a study in the contrasts of the wider influences brought into focus at this point. The arch itself, its iron gate fixed in an open position, and its classical imagery perhaps more American than Canadian, has come to symbolize the meeting of the two countries and local border communities in a spirit of friendship, even of kinship. Indeed, its south-facing inscription reads "Children of a Common Mother."

Since the Peace Arch was built, this region has come to face the Pacific world to a greater degree than ever before. It does so with increasing self-consciousness and perhaps self-assurance. The full ramifications of any movement to regional economic cooperation will not be known for some time, but it will force issues of border regional development and cooperation. No matter how separate or similar the national experiences, however, the essential unity of the region can never be completely denied. If it were to be composed today, the inscription on the Peace Arch might well be modified to read "Partners in a Common Destiny."

References

Forward, Charles N., and Gerhold, George A., eds. 1974. *Environment and Man in British Columbia and Washington*, Symposium on Canadian-American Relations. Bellingham: Western Washington State College.

Kristof, Ladis D. 1959. The nature of frontiers and boundaries. *Annals of the Association of American Geographers* 49:269-282.

McGreevy, Patrick. 1988. The end of America: The beginning of Canada. *Canadian Geographer* 32:307-318.

Rutan, Gerard F. 1977. The ugly Canadian: Canadian purchase and ownership of land in Whatcom County, Washington. *Social Science Journal* 14:5-15.

Semple, Ellen C. 1911. *Influences of Geographic Environment*. New York: Henry Holt.

Simmons, Terry, and Turbeville, Daniel E. III. 1985. Blaine, Washington: Tijuana of the north? In *Current Research by Western Canadian Geographers: The Cariboo College Papers, 1984*, Occasional Papers in Geography, B.C. Geographical Series No. 41., ed. Edgar L. Jackson, pp. 47-57. Vancouver: Tantalus Research.

Thomson, Don W. 1966. *Men and Meridians*, Vol. 1. Ottawa: Queen's Printer.

Watson, J. Wreford. 1963. *North America: Its Countries and Regions*. London: Longmans.

Zelinsky, Wilbur. 1980. North America's vernacular regions. *Annals of the Association of American Geographers* 70:1-16.

The Overground Railroad: Central American Refugee Flows to Canada

Gary S. Elbow
Texas Tech University

Boundaries and Pathways

Boundaries are lines along which countries make physical contact (Glassner and de Blij 1989). They mark the limits of national jurisdiction and control the passage of goods and people between neighboring states. The boundary between the United States and Mexico stretches across approximately 4,000 kilometers of desert and semi-arid country between northern Mexico and the southwestern United States. Roughly half of the length of the border is marked by the Rio Grande, a river that, despite its name, usually carries so little water that it can be waded. The remainder of the boundary follows straight lines traced across the landscape in total disregard of terrain features. This boundary marks the official separation between Anglo-American and Latin American cultures and, more important, it is the only place in the world where a wealthy, highly industrialized region comes into direct physical contact with a relatively poor region with far less productive economies.

Mexicans and Central Americans can travel overland cheaply and easily to the frontier areas. From there, they can wade across the Rio Grande or cross the boundary somewhere along its unguarded course through the desert. With such easy access and the promise of relatively high paying jobs, it is hardly surprising that Mexicans have crossed into the United States for decades. Since about 1980, they have been joined by many Central Americans, many of whom are political refugees.

If the boundary presents few physical limits to passage between Mexico and the United States, its legal significance is enormous. For Central Americans seeking entry to the United States, the boundary is a symbolic wall that blocks their pathway to the north. From the United States government's perspective, the boundary is an ineffective barrier across which pour thousands of Central American and other aliens each year. This latter view of the boundary is reflected in United States government responses to Central American refugees who seek asylum in the country.

The Canadian boundary with the United States (the only country with which Canada shares a land boundary), on the other hand, has not been a serious obstacle for refugees. Until the law was changed in 1988, any alien who could reach the border and present himself or herself as a refugee would receive permission to reside in Canada until the case could be adjudicated (Nash 1989). For Guatemalans and Salvadorans, because of their countries' political problems there is priority for awarding of temporary refugee status (Nash 1989). Far from being a barrier, the Canadian boundary was virtually open, at least until 1988.

This paper discusses how one movement in the United States, the overground railroad, capitalized on these differences in the barrier effect of international boundaries to facilitate the permanent resettlement of Central American refugees in Canada.

Refugees and the Central American Crisis

For the past decade and a half Central America has suffered a devastating series of crises. These problems are especially acute in El Salvador, Guatemala, and Nicaragua. El Salvador is gripped in a civil war that began in 1978. Thousands have been killed, between

20 and 25 percent of the population has been displaced from homes, and the economy has been severely damaged (Ferris 1987). In Guatemala, guerrilla warfare and military repression have cost thousands of lives and devastated large areas. Once a relatively strong economy, Guatemala has suffered inflation and growing international debt (Ferris 1987). Nicaragua had a bloody revolution in 1979, resulting in the death of more than 50,000 persons and destruction of its economic infrastructure. United States efforts to destabilize the Sandinista government in the 1980s led to thousands more deaths and weakened the economy to near collapse by 1990.

One reaction to these crises is an unprecedented flow of refugees from Central America during most of the decade of the 1980s and continuing to the present. Some of these refugees sought sanctuary in neighboring countries and a few fled to Canada, but the most popular destination, by far, was the United States. For 1985, the United States Bureau of the Census estimated the number of Salvadorans in the country at 280,000, of whom only 80,000 were legal immigrants. The number of Guatemalans was estimated to be 205,000, of whom 70,000 were legal residents, and there were thought to be 185,000 Nicaraguans, 50,000 of whom had legal status (Peterson 1986, 44-46). These are conservative estimates; independent sources place the numbers of Salvadoran nationals living in the United States at two to three times the official estimates, while estimates for Guatemalans and Nicaraguans fall fairly close to the Peterson figures (Zolberg et al. 1989, 212). In addition to those who established themselves in the United States, either legally or illegally, thousands were apprehended at the Mexican border or within the country, detained, and subsequently deported.

The Overground Railroad

In the United States certain groups of citizens became convinced that government policies for granting political asylum were unfair. They feared that many refugees, who faced persecution or even death if they returned to their home countries, were being illegally deported. These concerns, as it turns out, were well founded, and recent court decisions in the United States mandate readjudication of the cases of thousands of Guatemalans and Salvadorans who were denied asylum (*The New York Times* 1990).

The overground railroad is one response to these concerns. It is a name that refers to both a movement and an organization. The overground railroad movement is dedicated to aiding qualified Central American

refugees in the United States to obtain asylum, either in the United States or, more frequently, Canada, where granting of refugee status is far more likely. To accomplish this goal, the overground railroad bonds refugees out of detention centers, provides support while applications are being processed, and facilitates transfer to destinations in Canada when they are accepted by that country.

The name "overground railroad" evokes parallels with the underground railroad that smuggled escaped slaves from Confederate states to the Union and to Canada before and during the Civil War (1861 to 1865) in the United States. The adjective "overground" stresses that this movement works within the legal systems of the United States and Canada, making it acceptable to religious congregations sympathetic with the plight of Central American refugees but opposed to violating immigration laws. The term railroad is symbolic; refugees are transported to Canada most frequently by air, and never on a railroad. Still, the name suggests the idea of real, if indefinite, transportation lines linking sources of refugee origin with their ultimate destinations. The name "overground railroad" was also adopted by one of the organizations that helps to place refugees with churches or other support groups while they are awaiting action on their asylum applications.

Differential Treatment of Refugees

Inconsistency in the handling of refugees from different Central American countries is a primary point of conflict between the government of the United States and citizens who oppose its Central American refugee policy. The United States government claims that Guatemalan and Salvadoran migrants respond primarily to economic factors, while concerned citizens argue that many of the illegal entrants are legitimate political refugees who fled their homes out of concern for their personal safety (see especially Jones 1989 and Stanley 1987). This issue is of far more than academic interest because it affects the response of the United States government to the awarding of refugee status or political asylum.

Refugees from El Salvador and Guatemala, nominal democracies supported by the United States, are generally considered to have economic motivations for migrating and, therefore, are considered to be ineligible for political asylum (Dreifus 1987; Zolberg et al. 1989). On the other hand, the United States government opposed the former Sandinista government of Nicaragua, and migrants from that country during most of the 1980s found it relatively easy to qualify for political asylum because immigration officials assumed that they had legitimate fear of persecution at home (*U.S. News and World Report* 1989).

Thus, during the 1980s only a tiny fraction of Guatemalans and Salvadorans who applied for asylum—two to three percent—were approved and allowed to stay in the United States (Dreifus 1987; Quammen 1986). The vast majority, those who were denied asylum, faced deportation or went underground to avoid being returned to their home country. Meanwhile, about 35 percent of asylum applications from Nicaraguans were approved. This occurred despite reports that government repression in Nicaragua was much less than in Guatemala and El Salvador. This permissive attitude toward Nicaraguans changed during 1989, when the economy collapsed and thousands presented themselves at the Mexico-United States border to apply for asylum (Fins 1989).

The argument for economic causes of migration is supported because most migrants from Guatemala and El Salvador arrive overland via Mexico. If these people were legitimate political refugees, reasons the U.S. Department of Justice, they would stop in the nearest country to their homeland instead of travelling hundreds of miles to the United States. On the other hand, well-to-do Nicaraguans, who could afford to fly, were considered to have primarily political motivations for emigrating, since they came directly to the United States.

Central American Refugees in Canada

Canada has never been a popular destination for Central American immigrants. The most obvious reason for this is that most refugees travel overland through Mexico to the United States. Going to Canada would add well over 1,500 kilometers to their trip and they would have to cross through the United States, an unknown country whose language most do not speak. Added to the barrier of distance is the intervening opportunity of employment in the United States. Also, refugees from the tropics may be discouraged by images of Canada as a northern country with a harsh and forbidding climate. Finally, the large Spanish-speaking population in the United States, coupled with possible contacts through family or friends who preceded the immigrant, are further inducements for the potential refugee to think first of the United States rather than Canada as a destination.

For these reasons, refugee flows to Canada from Central America are only a fraction of the flows to the United States. This may be why the government of Canada has taken a more liberal approach in granting asylum to migrants from Central America. However,

Canada also takes pride in its historic role as a refugee sanctuary (Nash 1989). The first people who might be considered refugees were the United Empire Loyalists. These supporters of the English crown were forced to leave the United States during and after the Revolutionary War. Subsequently, Canada provided sanctuary for escaped American slaves in the mid-1800s and for Russian Mennonites in the 1870s. This pattern continued into the modern era when, between 1947 and 1987, Canada admitted more than 400,000 refugees. New legislation passed in 1988 greatly restricted access via direct border crossings, but left largely intact the broader body of Canadian refugee law (Nash 1989). Also worthy of note is the existence of both public and private programs of language training, resettlement, and employment for refugees in Canada (Universidad para la Paz 1987; Weiner 1987).

Facilitating Refugee Flows

Early in the 1980s, groups in the United States became aware of the relative ease with which Central American refugees could gain admission to Canada. They formed the overground railroad to take advantage of the situation and developed programs to facilitate the legal movement of refugees from the United States to Canada.

The principal actors in the overground railroad movement are Jubilee Partners, an independent Christian group in rural Georgia; the overground railroad, headquartered in a Mennonite church in Evanston, Illinois; and Mennonite International Refugee Assistance. Jubilee Partners was founded in 1979 to aid Asian refugees who could gain entry to the United States if American citizens volunteered to sponsor them (Smith 1986). By 1982, leaders at Jubilee Partners were aware of the dilemma of Salvadoran and Guatemalan refugees in the United States, and they developed a unique strategy to help them.

Working with Mennonite organizations that had connections in Latin America, the United States, and Canada, Jubilee Partners screened refugees detained in centers at the United States-Mexico border and identified those who were likely to qualify for asylum in Canada. Those refugees were bonded out of the detention centers, helped with the Canadian refugee application process, and supported in the United States until they gained admission to Canada. Refugees who applied for asylum in the United States were given temporary legal status while their cases were pending. This generally provided them with more than enough time to qualify for refugee status in Canada.

The overground railroad located church congregations or other groups that could support refugees while they awaited adjudication of their asylum claims. Prospective refugees who are released from detention usually must be provided with housing, food, and health care because they rarely speak adequate English; they have no money; and they may not qualify for work permits in the United States. Even for those who do have permission to work, employment is difficult to find and likely pays only minimum wages, so help with living arrangements, transportation, learning English, and other matters is essential.

Origins and Destinations of Refugees

The overground railroad movement has focused on refugees from El Salvador and Guatemala. Estimates are that 75 to 80 percent of the refugees placed by Jubilee Partners and the overground railroad are Salvadoran. Another 20 percent are Guatemalan. Hondurans and Nicaraguans comprise fewer than 5 percent of the total. Between its inception in 1983 and mid-1989 the overground railroad movement placed around 1,300 refugees in Canada (Rivas 1989). This estimate represents about 5.4 percent of the nearly 24,000 Latin American refugees who were admitted to Canada between 1982 and 1987 and comprises less than half of the total of refugees admitted from El Salvador, Guatemala, and Nicaragua (Nash 1989). They are an even smaller percentage of the hundreds of thousands of Central American refugees who have arrived at the United States border during this time. Nevertheless, the movement is significant, both for the services it provides to the refugees who are helped and, perhaps even more, for the awareness of conditions in Central America that it promotes.

Refugees accepted for residence in Canada are normally placed in one of the major urban areas. Of the Central American refugees resettled by the overground railroad in 1989, about 60 percent were placed in Ontario, mainly in Toronto; 10 percent went to Montreal, 10 percent to Vancouver, and the remainder were scattered out among other provinces (D. Janzen, personal communication, 9 March 1990). Jubilee Partners reports that placement of their refugees has been approximately equally divided with one third in Toronto, one third split between Vancouver and Winnipeg, and the remainder going elsewhere in Canada (D. Lawrence, personal communication, 16 February 1990). No refugees from Jubilee Partners have gone to Yukon or Northwest Territories nor to the provinces of Newfoundland, Prince Edward Island, or Québec.

Conclusions

The overground railroad movement developed in response to the refugee policies of the United States, which transformed the linear United States-Mexico boundary into a selective barrier to legal entry of certain Central Americans. Under these policies refugees from a country with a government opposed by the United States, such as Nicaragua, had a far greater chance of being granted political asylum than someone who came from a country whose government was supported by the United States. Responding to the arbitrary nature of United States policy in blocking certain refugee flows while facilitating others, organizations such as the overground railroad took advantage of an open and consistent policy in neighboring Canada. There, the boundary is not a barrier. In fact, for favored groups such as Salvadorans and Guatemalans, entry is facilitated. In this way, the overground railroad serves a role similar to its Civil War namesake, the underground railroad. For oppressed people, it provides pathways to sanctuary across the forbidding barriers formed by boundary. On the domestic scene, these organizations also provided a focus for efforts to reform United States immigration policy and eliminate the inconsistent treatment of refugees from different countries. This action is aimed at reducing the barrier effect of the border with respect to Central American refugee flows.

References

Dreifus, C. 1987. No refugees need apply. *Atlantic Monthly* February:32-35.

Ferris, E. G. 1987. *The Central American Refugees.* New York: Praeger.

Fins, A. N. 1989. The battering ram at the golden door. *Business Week* 6 February:52-53.

Glassner, Martin Ira, and de Blij, Harm. 1989. *Systematic Political Geography,* 4th edition. New York: John Wiley.

Jones, R. C. 1989. Causes of Salvadoran migration to the United States. *The Geographical Review* 79:183-194.

Nash, A. 1989. *International Refugee Pressures and the Canadian Public Policy Response,* Discussion Paper 89.B.1. Ottawa, ON: Institute for Research on Public Policy.

The New York Times. 1990. For refugees the door swings open. 24 December:22.

Peterson, L. 1986. *Central American Migration: Past and Present,* Center for International Research, Staff Paper No. 25. Washington, DC: U.S. Bureau of the Census.

Quammen, D. 1986. Seeking refuge in a desert. *Harpers* December:58-64.

Rivas, M. 1989. Overground railroad. *The Dallas Morning News* 23 July:41A,43A

Smith, K. S. 1986. Where welcome waits: Jubilee's night ride to Canada. *Commonweal* 14 February:67-68.

Stanley, W. D. 1987. Economic migrants or refugees from violence? A time-series analysis of salvadoran migration to the United States. *Latin American Research Review* 22:132-154.

Universidad para la Paz, Universidad Nacional. 1987. *Los Refugiados Centroamericanos.* Heredia, Costa Rica: Universidad para la Paz, Universidad Nacional.

U.S. News and World Report. 1989. As the huddled masses roll across El Norte. 30 January:10-11.

Weiner, G. 1988. Canada's refugee policy. In *Human Rights and the Protection of Refugees under International Law*, ed. A. Nash, pp. 233-240. Halifax, Canada: Institute for Research on Public Policy.

Zolberg, A. R., Suhrke, A., and Aguayo, S. 1989. *Escape from Violence: Conflict and the Refugee Crisis in the Developing World.* New York: Oxford University Press.

Maquiladora: Labor-Transport Cost Substitution along the Border

Frederick P. Stutz
San Diego State University

A setting for joint cooperative economic development is the 3,225-kilometer (2,000-mile) international border separating the most powerful and technologically advanced First World nation on earth—the United States—from a poor, Third World nation—Mexico. The *maquiladora* program was initiated by the Mexican government in 1965 as a replacement for the Bracero guest-worker program, which ended the year before. The program allows foreign companies to establish complete ownership of assembly plants in Mexico to produce goods for export markets. Ninety percent of approximately 1,700 maquiladora manufacturing plants in Mexico are within a few kilometers of this international border. "*La linea*" as it is called in Mexico, separates not only two large and vastly different countries, but also separates the American companies from a great labor-saving advantage.

Under U.S. tariff laws, U.S.-made components can be sent to Mexico for fabricating in the branch plants and returned to the United States for final assembly and distribution at much lower import duty levels. While the maquiladora program is early in its development, it is growing 20 percent per year and has surpassed the most optimistic expectations in both countries. The number of employees in "twin plants" in Mexico was approximately 500,000 in 1990, and is expected to grow to almost 900,000 by 1995 (Table 1). Without the maquiladora option, many labor-intensive U.S. companies could not maintain domestic plants and would not be competitive in the world market.

Fifty randomly selected maquiladora plants were queried since the fall of 1989 to decipher location patterns within the Tijuana metropolitan area of approximately 2 million people. The most important locational consideration for these plants was direct access to unskilled female labor, but city services were also important.

Labor-Supply Theory

Trade models in international economics argue that manufacturing will expand in areas where there is a pool of available and inexpensive labor. The labor, which reduces the total cost of manufacturing and leads to higher profits, acts as a magnet for growth (Conroy 1975, 44). Further, manufacturing expands where a small proportion of existing labor is engaged in manufacturing. Per capita income typically rises in such areas after many years, as the industrialization process continues to cycle upward and agglomeration occurs, bidding up the price of labor. Immigration to the region then begins, as has occurred in the Mexican border area. Tijuana, the most rapidly growing of Mexican cities, with average disposable incomes twice the national average, is an example. Migrants, from the interior of Mexico and from Central America have augmented its total labor supply.

Historically, labor shortages and economic fluctuations created large flows of Mexican labor north across the border (Alba 1982). Elimination of the Bracero program in 1964 resulted in a loss of approximately 200,000 jobs for Mexican workers, greatly harming the Mexican economy and leading to general unemployment along the northern border (Baerresen 1971). It is difficult for Mexicans to obtain green cards, that permit them to work legally in the United States. Mexico, therefore, created the maquiladora program

Table 1. Growth of the Maquiladora Industry in Mexico

Year	Number of plants	Number of employees	Value added ($million U.S.)
1970	120	20,327	N/D
1976	448	74,496	366
1980	620	119,546	772
1986	890	249,833	1,295
1990*	1,780	469,863	2,239
1996*	3,557	932,509	4,563

* Projections of Grupo Bermudez.

Sources: National Institute of Statistics, Secretariat of Programming and Budget, Mexico, June 1988; Grupo Bermudez Industrial Developers, Research Department, 1988; "The Mexican Economy 1988-1989." Ciudad Juárez, Chihuahua, Mexico: Grupo Bermudez.

to attract capital from the United States and to use labor in Mexico.

While low wage rates are important and average $0.90 per hour compared with $15.00 per hour in the United States ($2.50 in Taiwan and $4.00 in Korea), the location of plants in Mexico offers additional advantages for U.S. labor intensive industries, for example, the ability of managerial and technical personnel to live in the United States while managing twin plants in nearby locations across the border (Baerresen 1971; Grunwald 1985). Although geographical proximity of the plants along the international border is important, it does raise serious questions about redistribution of economic and social benefits in both countries. While R. South argues that maquiladoras are clearly not "relocated runaway manufacturers" (South 1990, 564), W. Wilson (1987) regards the location of maquiladoras in Mexican border towns as largely responsible for the high rates of joblessness among black males and for the emergence of an urban underclass in large American cities.

The Maquiladora Program Today

The benefits of the maquiladora program for both the United States and Mexico are substantial (Kolbe 1988; South 1990). U.S. firms become more competitive in world markets by combining advanced technology with low-cost Mexican labor, even given higher transportation costs to distribute their products. Employment opportunities on the north side of the border increase to meet needs for warehousing, administration, product finishing, and distribution. Additionally, about 30 percent of the wages paid by U.S. firms in Mexico are spent on the purchase of United States products

in the border cities of Texas, New Mexico, Arizona, and California.

The benefits for Mexico include higher levels of employment opportunity and more income for the border area's population. As Mexican workers train in industrial processes and new technologies, their improved skills add to the general economic development of Mexico. Export to the United States of maquiladora products brings needed foreign exchange into Mexico, especially important since the substantial devaluation of the Mexican peso. Mexican wages fell from among the highest in the Third World to among the lowest, in dollar terms, when the peso dropped from about $0.05 in the early 1980s to less than $0.003 in the late 1980s.

Devaluation stimulated the growth of the maquiladora industry; it now represents the second largest source of foreign exchange earnings for Mexico, after the oil industry. With the glut of oil on world markets and the price plunge of a barrel of crude from $38 in 1980 to $18 in 1990, maquiladoras are a bright spot in a beleaguered economy. Additionally, Mexican plants can import foreign components from the United States and further manufacture them without payment of custom duties. Table 1 summarizes some of the important benefits to Mexico from the rapid growth of maquiladora investment, gains that are most apparent in border communities such as Tijuana.

Intraurban Location Factors

Do factors of industrial location in Tijuana, Mexico, differ from locational processes at the intraurban level in the United States? I posed this question to 50 managers or management-level employees of maquiladora plants. They confirmed that site selection in Mexico is extremely important and can make the difference between success or failure in the business venture. The most important locational decision is based on direct availability of labor. Wage rates and worker turnover can also be reduced by a wise locational choice. Ninety percent of the products manufactured require large quantities of labor in assembly. Six-day work-weeks are the norm.

The most important characteristic of a good site in Tijuana is proximity to female labor. Mexican labor is less geographically mobile than American labor. Most Mexican workers do not own automobiles, and public transit and urban freeways are not nearly as generally available as in the United States. An ideal location allows workers, usually unorganized and non-militant females, to walk to work. This could reduce the stress on the labor force and allow for lower wages, lower fringe benefits, and decreased turnover of workers. However, worker turnover remains high by American standards.

The housing shortage in Tijuana is a major contributor to the high labor turnover rates. This problem, coupled with the employer preference for women workers, negatively affects family formation and family structure in Mexico's border towns. All too frequently, female workers have no alternative to maquiladora jobs and they become easily exploited.

The availability of infrastructure and city services is also an important locational dimension in Tijuana. Electricity, sewage, power, housing, and telephone service are not generally available or accessible in all manufacturing locations. Care must be taken to obtain a guarantee from public utility agencies that service can be provided inexpensively for an extended period.

The final important locational dimension is zoning. In the past, it was easy to get zoning variances and this led to many industrial buildings within residential neighborhoods, with incipient excess demand on local services and resulting pollution. Some manufacturing plants are purposely located in residential areas and are not well marked, thereby attempting to escape identification and taxation by local authorities. Currently, a new plant permit is not issued unless the site is properly zoned for industrial use. However, within residential areas, firms can recover the higher costs of constructing buildings and amenities because of greater employment satisfaction, labor savings, and lower worker turnover.

Conclusion

Mexico's maquiladora program has been in existence for 25 years. The program was intended to create employment in Mexico, generate foreign exchange, and produce a transfer of technology. Currently the nearly 1,700 maquiladoras in Mexico employ more than 500,000 workers.

The program allows foreign countries to establish 100 percent foreign-owned plants in Mexico. Further, there is favorable duty treatment for maquiladoras. U.S. firms can send machinery, equipment, raw materials, and supplies to Mexico without paying any duties because those items represent only temporary imports. The main benefit to foreign firms is access to low cost Mexican female labor. Most other costs, such as land, construction, leasing, and utilities are also lower in Mexico than in the United States.

Problems of high labor turnover rates, lack of housing, and inadequate infrastructure in Mexico, and moral questions concerning the exploitation of laborers persist. Additionally, the possible negative effects of the maquiladora assembly plant system on U.S. workers, especially minority workers who are concentrated in large central cities, has yet to be fully understood. While the border no longer separates American capital from Mexican labor, la linea still represents a formidable political and social divide between Third World and First World North Americans. The central question now is how to negotiate a free trade agreement so that the relationship between Mexico and the United States can be made more mutually beneficial. A key social issue is how the costs to U.S. and Mexican labor of accelerating and formalizing economic integration through free trade agreement can be reduced to tolerable limits by sound policies.

References

Alba, F. 1982. Mexico's northern border: A framework of reference. *National Resources Journal* 22:749,763.

Baerresen, D. W. 1971. *The Border Industrialization Program of Mexico.* Lexington, MA: Heath Lexington.

Conroy, M. E. 1975. *The Challenge of Urban Economic Development.* Lexington, MA: Lexington Books.

Grunwald, J. 1985. The assembly industry in Mexico. In *The Global Factory, Foreign Assembly in International Trade,* eds. J. Grunwald and K. Flamm, pp. 137-139. Washington, DC: The Brookings Institution.

Kolbe, J. T. 1988. The Wharton Study. *The Congressional Record,* pp. 1162-1182.

South, R. 1990. Transnational "maquiladora" location. *Annals of the Association of American Geographers* 80:549-570.

Wilson, W. 1987. *The Truly Disadvantaged.* Chicago: University of Chicago Press.

The Urban Colonias of El Paso County, Texas

George Towers
Southwest Missouri State University

The United States-Mexico border divides the two great geopolities, the First World and the Third. Such global terminology suggests a great gulf between the two nations. However, at the urban conglomerations that bridge the border, social, economic, and political currents wash from one world to the other, blending to create new social and spatial forms. This essay examines how two of these processes—Mexican immigration and American land-use politics—have joined in the formation of the "colonias" of El Paso County, Texas.

Colonias

Colonias are irregular, unserviced, and, until very recently, largely unregulated residential subdivisions in unincorporated areas near border cities in Texas and New Mexico. A typical colonia covers about 10 hectares and is divided into roughly 100 equally sized private lots. In El Paso County, approximately 250 colonias house some 70,000 people (U.S. General Accounting Office 1990) (Figure 1).

Colonias developers convert agricultural land for residential use by simply marking off lots and bulldozing dirt roads. Colonias lots are not connected to public sewer or water systems and all housing must be provided by the buyer. The lots' emptiness, however, ensures their affordability; lots are typically sold for about $10,000, paid in monthly installments of $100 or $150.

Colonias residents are drawn from El Paso's poor. In 1988, the median annual household income in El Paso's colonias was $11,497, less than half the U.S. median of $24,500 (Rand McNally 1990; Texas Department of Human Services 1988). Residents are almost exclusively of Mexican descent and 53 percent of colonias households are headed by a Mexican native

(Texas Department of Human Services 1988). Among colonias household heads, three fifths do not have a high school diploma and half do not read or speak English (Texas Department of Human Services 1988).

Because of the unserviced nature of colonias development, life in these areas is difficult and dangerous. Colonias residents must not only build their own housing, often living in shelters nearly indistinguishable from those in the shantytowns surrounding Latin American cities, but must also provide their own sewage facilities and drinking water. While a few colonias households use outhouses, most dispose of sewage via self-built, often substandard septic tanks or cesspools. Drinking water is laboriously hauled in or is obtained from private wells. Because the groundwater from these wells is often contaminated by private sewerage and the chemicals deposited in the soil by years of agricultural use, colonias residents suffer extremely high rates of water-borne diseases—tuberculosis, hepatitis, and shigellosis. For example, in San Elizario, an unincorporated community in southeastern El Paso County, two thirds of the children have been infected with hepatitis (*El Paso Times* 1988).

Socially, colonias reflect Mexico's demographic and economic pressures; but, spatially, colonias result from American urban land-use politics. The object of examining these two sources of colonias growth is to reveal the merging of the Third World with the First in the urban landscape of the United States-Mexico border.

Mexican Emigration

The demand for colonias housing is due, in large part, to immigration from Mexico, an outcome of the relationships structuring the world economy. As pe-

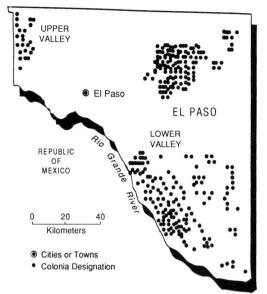

Figure 1. Colonia locations in El Paso County, Texas. Source: U.S. General Accounting Office (1990, 23).

ripheral, predominantly agrarian societies attempt to urbanize and industrialize within the capitalist world economy, rural populations are loosed from the land, labor surpluses are generated, and masses emigrate to the core countries.

Postwar Mexican economics and demographics exemplify this process. In seeking development, Mexican policies were designed to increase agricultural productivity and generate an urban industrial economy. As farms became larger and farming more mechanized, owners of small farms were displaced and agricultural labor made redundant. Mexican industry, however, has not been able to absorb this dislocated work force. Capital-intensive industrial development results in slow job creation and a demand for industrial skills that few in the expanding labor force possess.

Rapid population growth puts additional pressure on the Mexican labor market. B. S. Bradshaw and W. P. Frisbie (1983) estimate that for every 100 persons who left the labor force between 1970 and 1980, 357 entered it. This "labor force replacement ratio" was expected to rise to 407 in the 1980s and fall to only 330 in the 1990s. In absolute terms, the Mexican labor force was expected to grow by 17.5 million from 1970 to 2000 (Bradshaw and Frisbie 1983).

Mexican migration has resulted in El Paso's rapid growth. Between 1960 and 1990, the city's population nearly doubled in size, growing by 277,540 residents. Persons of Spanish origin, the census category that best approximates El Paso's population of Mexican descent, accounted for 99 percent of this growth, 274,626 people. Similarly, from 1960 to 1980, the

number of El Pasoans born in Mexico increased by 48,444. In relative terms, in 1960, 44 percent of El Paso's population was of Spanish origin and 11 percent Mexican born; by 1990, 70 percent was of Spanish origin and, in 1980, 17 percent was Mexican born (U.S. Department of Commerce 1960, 1980, 1990). These figures demonstrate that immigration from Mexico drives population growth in El Paso and, with population growth, the demand for low cost colonias housing.

The Politics of Urban Land Use

Colonias also reflect processes within American society, particularly the nature of American land-use politics. The formation of some 250 unserviced, irregular subdivisions adjacent to a large American city is, at first glance, an incongruity. But, this sort of development is entirely predictable. Students of the "urban growth machine" claim that municipal governments are strongly influenced by coalitions of pro-growth land interests. Their fortunes dependent upon the city's economic growth, landowners and speculators involve themselves intimately with the workings of local government (Logan and Molotch 1987). City government is equally dependent on economic growth to maintain fiscal stability and elected officials rely on growth coalitions for campaign funds and decision-making cooperation (Elkin 1985). These relationships cause government to protect the commodity status of urban land, and to avoid interfering with growth (Logan and Molotch 1987).

Growth, however, does have costs and countereffects. Development often brings congestion, environmental hazards, and service shortfalls. These undesirable by-products of growth—which are typically suffered disproportionately by the least powerful segments of the population—can return to foul the growth machine as citizen groups oppose the pro-growth coalition (Gottdiener 1985). The political repercussions of citizen resistance may redirect growth and restructure the growth coalition itself (Elkin 1985).

Colonias Development and El Paso Government

The history of colonias development in El Paso illustrates these general observations for two distinct periods. First, from the early growth of the colonias in the 1960s until the late 1980s, the three arms of local government involved with colonias development— the City of El Paso, El Paso County, and the El Paso

Public Service Board—avoided the problems created by colonias development, unwilling to regulate colonias growth or to extend services. Second, with growing public concern over colonias conditions and opposition to colonias development, in the last few years local government has increased its control over colonias development and its efforts to service the colonias.

The City of El Paso has authority over development in the city's "extra-territorial jurisdiction," an 8-kilometer-wide band encompassing some 775 square kilometers adjacent to the city limits. In the 1970s and 1980s, although the city had the power to prohibit subdivisions without water or sewage systems, colonias were tolerated in the extra-territorial jurisdiction. The City Plan Commission and City Council claimed that the extra-territorial jurisdiction would eventually be annexed and that the city must ensure orderly growth. The city took the position that given its limited resources for enforcement and its need to maintain a record of development beyond the city limits, to insist on serviced subdivisions would only lead to the growth of more unregistered, unmanageable colonias.

El Paso County government was effectively powerless to control colonias development beyond the extra-territorial jurisdiction until the late 1980s. Previously, the county could only require that developers file plans at the county courthouse. Without the power to regulate service provision, county commissioners approved dozens of colonias subdivisions annually.

The El Paso Public Service Board secures water supplies for the city. Throughout its history, the Public Service Board has been reluctant to send water beyond the city limits without acquiring compensatory water supplies. Federal funding encouraged the Public Service Board to extend water to the extra-territorial jurisdiction during the 1970s, but in 1979, the Public Service Board cut off water to the extra-territorial jurisdiction. Sharing the city's concern with orderly growth and easy annexation, the Public Service Board claimed that water provision provoked growth beyond the city limits, growth that was becoming increasingly disorderly and unmanageable. Although the Board failed to curtail disorderly development, it kept the colonias waterless.

Development interests are well represented in city and county government and on the Public Service Board. The City and County Plan Commissions, boards of appointees to advise City Council and County Commissioners Court on land-use policy, have been dominated by land interests and have delayed stricter regulation of colonias development (E. Morales, Assistant County Attorney, El Paso, interview, 19 November, 1990). Commission members endorse the developers' position that additional government interference is dysfunctional and serves only to price both developers and lot buyers out of the market (*El Paso Times* 1990). In 1987, three of the Public Service Board's five members were developers and in 1988, a colonias developer was named chairman (*El Paso Herald Post* 1987b). Critics of the Public Service Board's refusal to provide the colonias with city water have long charged that this policy limits standard development to those places in which Board members have speculative interests (*El Paso Herald Post* 1987a).

The inability and unwillingness of government to address the lack of water and sewer service for the colonias has been assailed by colonias residents and their supporters. El Paso Interreligious Sponsoring Organization, a community group formed by Saul Alinsky's Industrial Areas Foundation, has effectively brought the colonias' hazards to public attention. By coordinating voter registration drives and staging candidate forums, El Paso Interreligious Sponsoring Organization has made a place for the colonias and colonias residents in local politics (Towers 1991). Now, many leading local politicians, including the current county judge and her predecessor, have committed themselves to improving colonias conditions.

Local government's responsiveness to the colonias is evinced in real efforts to prohibit further colonias development. In January 1988, the city enacted a new Subdivision Ordinance that lots smaller than 0.4 hectares must be supplied with public water service and lots smaller than 0.2 hectares with sewer service. These regulations, enforced since mid-1989, enable the city to prohibit colonias development in the extra-territorial jurisdiction.

In 1990, the county gained the authority to regulate developers. Under the auspices of Proposition 2, a 1989 referendum securing state funds for water service to the colonias, Texas border counties were required to adopt rules limiting the growth of unserviced subdivisions. Thus, El Paso County prohibits the sale of lots smaller than 4.05 hectares without public water service and lots smaller than 0.4 hectares without public sewage. These rules criminalize colonias development outside the extra-territorial jurisdiction.

The Public Service Board, too, has changed its course. In 1988, the Board reached an agreement with the El Paso County Lower Valley Water District Authority, formed in 1985 to acquire water service for the colonias, to pump drinking water to Lower Valley Water District Authority customers beyond the city limits in exchange for the water rights to agricultural land in outlying El Paso County. In 1990, the Public Service Board generalized this policy and agreed to provide water to the remainder of the county.

The history of colonias development exemplifies the findings of the urban-growth literature regarding urban land-use politics. El Paso government, under the direct and indirect influence of the development community, turned a blind eye to colonias development, opting not to interfere with growth. Colonias development, however, has brought about hazards and hardships. The sufferings of colonias residents has engendered public resistance to this form of urban growth and has redirected development politics in El Paso.

Summary

Colonias are a new landscape unique to the United States' border region, a landscape developed from American land-use politics and peopled by Mexican immigrants. They are found adjacent to such Texas border cities as El Paso, McAllen, Laredo, and Brownsville. This new spatial form uneasily joins the economic and demographic pressures jarring Mexican society with the problematic politics of American urban growth and land use. The colonias are spatially and socially representative of the meeting of the First World with the Third along the United States-Mexico borderline.

References

Bradshaw, B. S., and Frisbie, W. P. 1983. Potential labor force supply and replacement in Mexico and the states of the Mexican cession and Texas, 1980-2000. *International Migration Review* 17:394-409.

Elkin, S. L. 1985. Twentieth century urban regimes. *Journal of Urban Affairs* 7(2):11-28.

El Paso Herald Post. 1987a. Drinking neighbors' sewage. 3 January:B1,B6.

———. 1987b. El Paso's 10,000: "Third World" lurks on outskirts of Sun City. 7 April:Special Section.

El Paso Times. 1988. Health care poor in San Elizario. 7 June:B1,B2.

———. 1990. Its health vs. cheap housing in colonias flap. 1 July:A1,A4.

Gottdiener, M. 1985. *The Social Production of Urban Space.* Austin: University of Texas.

Logan, J., and Molotch, H. 1987. *Urban Fortunes: The Political Economy of Place.* Berkeley: University of California.

Rand McNally. 1990. *1990 Commercial Atlas and Marketing Guide*, 121st edition. Chicago: Rand McNally.

Texas Department of Human Services. 1988. *The Colonias Factbook.* Austin: Texas Department of Human Services.

Towers, G. 1991. Alinsky organizing in El Paso and San Antonio: EPISO, COPS, and Mexican-American politics. *Border Issues Series*, Working Paper No. 1. Tucson: Udall Center for Studies in Public Policy, University of Arizona.

U.S. Department of Commerce. 1960, 1980, 1990. United States census of population. Washington, DC.

U.S. General Accounting Office. 1990. *Rural Development: Problems and Progress of Colonia Subdivisions near Mexico Border.* Washington, DC.

The Hosmer Silt Loam in Montgomery County, Illinois, and American Farm Policy

Philip J. Gersmehl
University of Minnesota

The Food Security Acts of 1985 and 1990 are landmark legislation. For the first time in the history of U.S. farm policy, soil conservation is a prerequisite for governmental aid. Farmers who plow fragile land, drain wetlands, or continue to farm erodible land without a government-approved conservation plan will lose their eligibility for federal assistance.

Administering this far-reaching law is difficult in many parts of the country. In this paper, I explore ways in which farm policy is affected by the intersections of five kinds of lines—boundaries between geomorphic or vegetation regions, lines on soil survey maps, political borders between counties, and fence lines on individual farms. To summarize one small part of a project that involved interviews with farmers and soil conservation officials in 320 counties across the nation (Gersmehl *et al.* 1989), I have selected a small area in south-central Illinois as a case study. This area illustrates how a variety of different kinds of map lines can interact to affect the implementation of the Conservation Reserve, a provision of the Food Security Act that pays farmers to reestablish protective cover on land that is officially classified as "highly erodible."

A Case Study

Bond, Montgomery, and Shelby Counties, Illinois, are near the southern edge of the Prairie Peninsula, an eastward extension of the midcontinental grassland prior to agricultural conversion (Figure 1A). More than half of Bond County was forested; Montgomery and Shelby Counties had more grassland, with forest only on sloping land near river valleys. The soil pattern reflects this vegetation history, with dark-colored mollisols on broad uplands and light-colored alfisols on the slopes.

Several middle Pleistocene glaciers covered all three counties, but the most recent Wisconsinan glacier stopped in Shelby County. Slightly more than half of that county has a layer of young glacial till, which is more fertile than the older till south of the terminal moraine. A layer of silty loess covers both tills, to a depth of several meters to the west and progressively thinner toward the southeast. Soil taxonomists have used arbitrary lines to divide loessial soils into several thickness groups; two of those lines cross the case study area (Figure 1B).

Climate is almost uniform throughout this study area, and crop productivity depends primarily on the kind of glacial till, the thickness of loess, and the pre-agricultural vegetation cover. Average yields of corn, for example, are 150 bushels per acre on the young till north of the moraine, 130 bushels per acre on thick-loess grassland soils on older till, and 80 bushels per acre on thin-loess forest soils on older till. Roughly in the middle of the yield range is the Hosmer silt loam, the taxonomic category of soils that form on sloping forest land within the zone of medium-thick loess over older till. Yields from Hosmer soil are about 110 bushels per acre, and erosion is a severe hazard. Long-term productivity depends on careful management, and the land manager usually faces a rather difficult choice among several unpleasant alternatives: excessive erosion, expensive terraces, or profit-reducing crop rotations.

Highly Erodible Land and Its Users

For the purposes of the Food Security Act, the Hosmer silt loam is classified as highly erodible on all slopes steeper than 4 percent and on long slopes between 2 and 4 percent. Because of its marginal yields and erosion

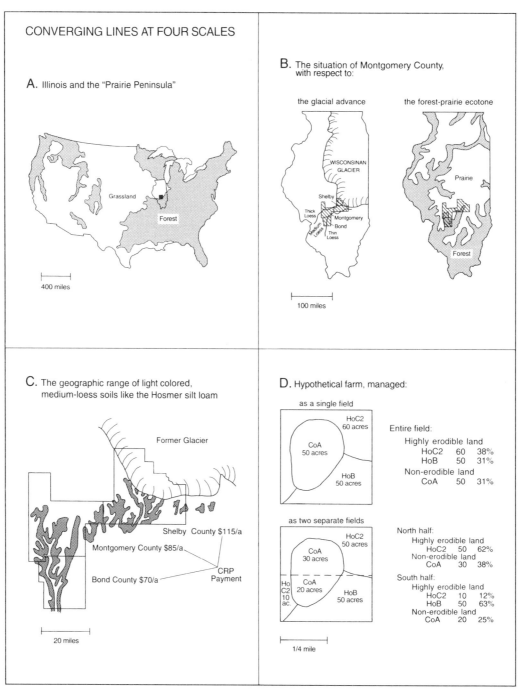

Figure 1. Map lines at four scales. The Hosmer silt loam is the category of soils that formed on forested slopes in the zone of medium-thickness loess (100 to 150 centimeters [40 to 60 inches]). A is adapted from Paullin and Wright (1932); B and C are adapted from Fehrenbacher *et al.* (1984); D is a hypothetical landholding, based on patterns described in Downey and Odell (1969). I have simplified the soil map for clarity—an actual field of this size in this environment would probably have between 10 and 20 separately delineated soil areas.

hazard, few farmers plant annual crops on steep Hosmer silt loam. Those few tend to be either economically desperate, ignorant of the erosion hazard, unscrupulous enough to trade future productivity for present income, or willing to take the expensive steps (such as terracing or no-till farming) necessary to grow annual crops without excessive erosion. The farmers who have fields of Hosmer silt loam but choose *not* to raise annual crops on it also fall into several different categories: risk-averse, well-to-do, part-time, or livestock-oriented (who prefer to use steep land for pasture or hay rather than annual crops).

Heterogeneity of the farmer population is one reason it is difficult to devise equitable and efficient policies for controlling erosion. A policy intended to help the financially desperate will almost inevitably reward the unscrupulous. Likewise, a policy to promote new tillage practices among risk-averse farmers will usually penalize those who innovate at their own expense. These value-laden issues tend to cloud any discussion of American farm problems; in this paper, I will temporarily ignore them in order to focus on technical problems in classifying highly erodible land. These problems are especially acute with medium-depth, medium-productivity soils such as the Hosmer silt loam.

Resource Heterogeneity within a Farm

The first technical problem is the fundamental arbitrariness of the lines on a county soil survey map. These lines represent the best judgment of a technician who typically has time to sample the soil in only a few sites per square mile. In the words of one soil surveyor I spoke with near Montgomery: "We don't really map soils; we map the landscapes that we have learned are usually associated with various kinds of soil, and we drill holes to check those inferences every once in a while."

That kind of resource mapping is more than acceptable for general planning. But the rules of the Food Security Act require the district soil conservationist (a county-level government official) to determine how much of a field has erodible soil mapping units (such as Hosmer silt loam, 4 to 7 percent slope). If these soils occupy more than 33 percent of a field, the field must have a conservation plan; if they pass the 67 percent threshold, the field is declared eligible for enrolment in the Conservation Reserve.

The arbitrariness of this percentage (67 percent) does not pose a problem in regions like central Montana or west Texas, where nearly all soils are highly erodible. It is also of little concern in level areas of Ohio,

northern Illinois, or central Iowa, where few locally important soils fall into the highly erodible category. But in the region of the Hosmer silt loam, the exact position of the lines drawn by soil surveyors is very important, because of the way these lines interact with property lines and with the fence lines established by individual landowners.

For example, consider a hypothetical landholding of 160 acres (about 64 hectares, the acreage conveyed under the Homestead Act of 1867). If this farm were located in a hilly part of Montgomery County, Illinois, it would probably have a fertile prairie soil on the upland and some lighter-colored and erodible soil like Hosmer on the slopes. It also would have straight borders, because land in this region was surveyed according to the rules of the U.S. Public Land Survey. With straight property lines, farmers usually find rectangular fields easier to manage, and therefore the interior fences of many Illinois farms are also straight north-south or east-west lines.

The soil surveyor might examine the south part of this hypothetical farm and choose to map much of it as Hosmer B (2 to 4 percent slope), because the slopes there are mostly between 1 and 3 percent and the individual patches of darker soil or more level land are generally smaller than 2 acres (the minimum size for a separate delineation on a soil survey map; see Figure 1D). The north part, by contrast, has a fairly large level area flanked by much steeper land. The transition between the level area and the slope is narrow enough, and close enough to the border between dark and light soils, that the surveyor chose to map the field as two separate areas of level (A-slope) grassland soil and fairly steep forest soil (C-slope). (The dividing lines between slope classes are set arbitrarily within each county, and in fact the B- and C-slopes for the Hosmer series in Bond County are 2 to 5 percent and 5 to 10 percent, rather than 2 to 4 percent and 4 to 7 percent as in Montgomery County. This is one way in which a county line can affect our perception and classification of the soil resource. The Montgomery-Shelby County line poses an even bigger version of the same problem, because the Shelby County soil survey is not yet finished, and the surveyors who are working on it have apparently decided that there is not enough Hosmer soil in the county to warrant a separate category on the maps. Their compromise solution is to combine several kinds of forested loess soil and to map them all as Ava silt loam, a thin-loess forest soil with a slightly lower yield potential and a slightly higher erosion hazard than Hosmer.)

If our hypothetical 160 acres were managed as a single field, all of it would be eligible for the Conservation Reserve, because the survey map shows about 69 percent of the area in mapping units that are classified as highly erodible, and the Hosmer C soil occupies the most area. Now, suppose an east-west fence divided this land into two fields. The north field, with the steepest land and the most serious erosion hazard, would *not* be eligible for the Reserve, because the map shows only 62 percent of the field as highly erodible Hosmer C soil. Meanwhile, most of the southern field has a B-slope (2 to 4 percent) and the district conservationist must go out to the field and measure the length of the slopes in order to classify the field. A few long slopes would tip the entire field into the highly erodible land (HEL) category, even though it is actually much less vulnerable to erosion than the north field. In short, strict application of the rules has resulted in a misclassification of both fields. This kind of problem is far from rare—more than 200 of the 320 district conservationists in my interview sample reported difficulty in classifying medium-sized fields in areas with intermediate slopes or complex soil patterns. (For clarity, my example has only three soil mapping units on 160 acres; similar-sized fields with 20 or 30 separately mapped kinds of soil are common in many regions, and soil surveyors, aware of the arbitrariness of their survey lines, generally view as ludicrous the idea of measuring all of these soil areas to a cumulative precision of a few percentage points.)

The arbitrary nature of some soil map lines is compounded by the arbitrary way in which different farmers manage a field. For example, an erosion-conscious farmer might choose to build contour fences around the best 140 acres of our hypothetical 160-acre field and use 20 acres of the worst land for pasture. An ignorant or unscrupulous farmer, meanwhile, might be willing to plow the entire 160 acres, including the steep land.

The cropland field of the careful farmer, with shorter slopes and less than 60 percent erodible land, would not be eligible for the Conservation Reserve. The farmer who plowed the entire field, however, could enrol all of it (including the flat upland) in the Reserve and collect an annual payment from the government. In effect, a farmer who chooses to use land carefully has less chance to participate in the Reserve.

This would not be a problem, if the only lines of concern were field borders and soil map lines, and if Conservation Reserve payments were figured solely on the basis of the productivity of individual fields. In this part of Illinois, however, the vegetation border, the glacial moraine, and the county lines also play important roles, because the payment levels were established on a county-unit basis, as described below.

Average Crop Yields and Payment Ceilings

The government used the average yields in each county to determine the maximum payment for land that farmers put in the Conservation Reserve Program. In Shelby County, with its area of fertile soils on young glacial till, the 1987 agricultural census reported an average corn yield of 140 bushels per acre, and the maximum Conservation Reserve Program payment was set at $115 per acre per year. In Bond County, with a large area of forested slope and generally thin loess, the average corn yield was only 106 bushels per acre, and the payment ceiling is $70. And in Montgomery County, with less forest than in Bond County but no young till as in Shelby County, average corn yield was an intermediate 122 bushels per acre, and the maximum Conservation Reserve Program payment is $85. These payment ceilings are reasonable attempts to accommodate differences in average productivity in three different pedogenic environments. However, the borders between soil regions do not correspond to county lines. Soils like the Hosmer silt loam occur in all three counties, and, in fact, they are precisely the kinds of soil most likely to be enroled into the reserve in each of the counties (Figure 1C).

At this juncture, all of these different kinds of lines interact to cause a major problem. Soils like the Ava or Hosmer silt loam have essentially the same yield potential in each of the three counties in my case study; the 1987 cash lease value of this kind of cropland was about $60 per acre. A careful farmer probably would not grow continuous corn or other annual crops on Ava or Hosmer soil. But for those desperate or unscrupulous farmers who were using these soils for corn or soybeans in 1981 to 1985, the payment to put the land in the Conservation Reserve would be $115 in Shelby County and only $70 in Bond County. In other words, one farmer would be paid 64 percent more than another with the same kind of land, simply because of the location of the county line. To compound the effect, a poor steward of soil in Shelby County can collect nearly $20,000 per year by enroling 160 acres in the Conservation Reserve. Meanwhile, a more careful owner of a similar tract of land may have chosen to leave a small amount of steep land in pasture. This person would be unable to enrol any land in the Reserve and would have to settle for about half the income (and much more uncertainty) from cash leasing. The incentives are clearly perverse, but, given the verbal reticence of typical farmers, the motivational effects of this kind of program tend to be obvious in rural coffee shops and almost invisible in tabulations of questionnaire results in academic journals (Gersmehl 1989).

Conclusion

A convergence of map lines in southern Illinois clearly illustrates one of the pervasive inequities associated with the Conservation Reserve Program (and, indeed, the entire structure of paid diversions and price supports that is at the core of American agricultural policy). The purpose of the program is to assist hard-pressed farmers, but in tens of thousands of places throughout the country, the landholding borders, fence lines, and administrative lines do not follow the boundaries between resource regions. This has allowed people (often but not always justifiably described as unscrupulous) to clear fields and choose crops on the basis of vagaries in local administrative rules rather than the innate capability of the soil. The numbers are not trivial— on the basis of my interviews and field data, I estimate that governmental programs have altered the size of fields on about half a million farms and biased the selection of crops away from the optimum on perhaps 130 million to 160 million acres of land, more than one third of the total cultivated land in the country. And that distortion, in turn, has made American agriculture simultaneously less efficient and less equitable than it should be.

The solution to the problem outlined in this paper, unfortunately, requires a radical change in American farm policy. That change must address two problems, one mechanical and one motivational. The mechanical problem of interactions between soil boundaries and field lines can be addressed by using a flexible system to measure crop yield and soil erodibility for individual fields. Gathering this kind of data is a time-consuming task, but not much more so than the present system of measuring the areas of soil mapping units that fall into the arbitrarily defined category of "highly erodible."

The motivational problem arises because the program systematically rewards the poorest land steward. The benefits of the Conservation Reserve Program are generally not available to farmers who have taken care of their land. This aspect of the problem will probably prove very difficult to solve. It was described as serious (in response to an open-ended question about problems) by more than 96 percent of the farmers and field soil scientists in our sample. At the same time, its importance was denied by more than two thirds of the policy-makers I interviewed in Washington, DC. Their response is that they were authorized to devise a program to protect land, not to reward good farmers. In the words of one who later requested anonymity, "this is a soil program, not a farmer program." Unfortunately, defining the scope of their responsibility in those terms reveals a deplorable ignorance of the ongoing process of land transfer—if one systematically penalizes land stewardship, one should hardly be surprised to find a downward trend in the proportion of land operated by good stewards. In fact, one truly heartening fact about American agriculture is that so many farmers continue to practice good stewardship is spite of the wrong headed federal policies. Future historians may well conclude that the celebrated individualism and independence of the American landowner was the factor that bought us enough time to fix the misguided incentives that plagued American farm policies in the 1980s.

Acknowledgments

This research was supported, in part, by National Science Foundation grant number SES-88-21432. I thank Tanya Mayer and Greg Chu for drafting the illustrations. I also acknowledge my great debt to Dwight Brown, Fraser Hart, Glenn Johnson, and the hundreds of district soil conservationists, other local governmental officials, and farmers who were willing to share their time and expertise so freely.

References

Downey, C. E., and Odell, R. T. 1969. *Soil Survey of Montgomery County, Illinois.* Washington, DC: U.S. Department of Agriculture.

Fehrenbacher, J. B., Alexander, J. D., Jansen, I. J., Darmody, R. G., Pope, R. A., Flock, M. A., Voss, E. E., Stout, J. W., Andrews, W. F., and Bushue, L. J. 1984. *Soils of Illinois,* Bulletin 778. Champaign: University of Illinois, College of Agriculture.

Gersmehl, P. J. 1989. Bonanza for the land miners: A contemporary Western saga. In *The American West,* ed. R. Kroes, pp. 81-98. Amsterdam: Free University Press.

———. Baker, B., and Brown, D. A. 1989. Effects of land management on "innate" soil erodibility: A potential complication for compliance planning. *Journal of Soil and Water Conservation* 44:417-420.

Paullin, Charles O., and Wright, John K. 1932. *Atlas of the Historical Geography of the United States.* Washington, DC: Carnegie Institution.

Phillips, D. B., and Goddard, T. M. 1983. *Soil Survey of Bond County, Illinois.* Washington, DC: U.S. Department of Agriculture.

The Chesapeake Bay Bridge: Development Symbol for Maryland's Eastern Shore

Michele Masucci
Auburn University

Maryland's eastern and western shores were connected by the Chesapeake Bay Bridge in 1952 (Figure 1). Prior to construction of the bridge, access to the Eastern Shore was only possible by ferry or by driving the perimeter of the Chesapeake Bay. Once considered remote to outsiders, the region is now accessible in 5 to 7 minutes by the Bay Bridge.

It is difficult to overstate the impact of the Bay Bridge on the Eastern Shore. Along regional highways large billboards proclaim the Eastern Shore the "Land of Pleasant Living." Several area chambers of commerce have adopted and promoted this slogan to tourists. Yet, the occasional bumper sticker protests: "There is no life west of the Eastern Shore," or, more directly, "Bomb the Bridge." The first bumper-sticker slogan could be misinterpreted as an affirmation of positive regional consciousness, because its local meaning harbors sentiments of resentment and mistrust. The second offers a solution to the problem of intrusion.

For Eastern Shore natives, the Bay Bridge represents change that has not reflected their needs and preferences. The metaphor of bridge construction and the specific symbolism of the Chesapeake Bay Bridge sharpen the contrast between insider- and outsider-perspectives on regional development. In this essay, the emerging landscape is considered through the themes of discovery and development, place meanings, and bridges in context.

Developing the "Remote"

One consequence of improved accessibility to the Eastern Shore is the contrast between prospering and declining communities. Shore settlements are traditionally located at river headwaters or on bay inlets, reflecting nearly three centuries of reliance on water transportation in the region (Walsh 1988). Many of these communities are now by-passed because they are inefficiently arranged for road transportation. Cities farther inland, such as Easton, Cambridge, and Salisbury, have experienced rapid residential, commercial, and industrial growth because they are located where interstate highways cross navigable rivers.

The Bay Bridge anchors an additive development process, affirmed by the construction of a second bay bridge in 1973. Subsequently, traffic congestion increased along commercial segments of Route 50 and on two-lane bridges. Eventually, the older structures were upgraded in Cambridge (the Malkus Bridge, completed in 1987), Vienna (the Vienna Bridge, completed in 1989), and Kent Island (the Kent Narrows Bridge, completed in 1990).

Uneven use of the Bay Bridge underscores a more intriguing impact on the region's developmental status. On summer Friday afternoons, four out of five lanes over the Chesapeake Bay direct traffic toward the Eastern Shore. On Sunday afternoons, this pattern is reversed. Seasonal differences are also notable. For example, in January 1990, 301,106 non-commuter automobiles crossed the Bay Bridge; during July of the same year, automobile traffic rose to 651,130 (Maryland Department of Transportation 1991). The emergence of the Eastern Shore as a tourist destination has led to conflicting place sentiments between insiders and outsiders.

The bridges on Route 50 confirm tourist discovery of the region because their purpose is to link Ocean

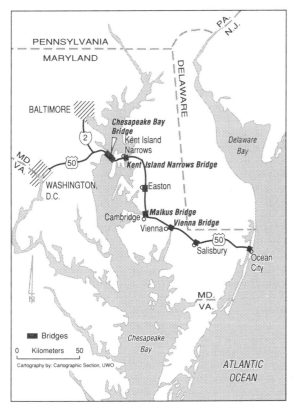

Figure 1. Route 50 links Maryland's Eastern and Western Shores, making ocean resorts accessible to metropolitan populations.

City to metropolitan Baltimore and Washington. Eastern Shore residents can question who is best served by this connection. Inland economic prosperity has also meant encroachment on prime agricultural land, while waterfront land values have soared. Ocean City's construction boom during the 1970s and 1980s produced a generic skyline of commercial strips, high-rise, flat-roofed condominiums, and stilt-framed beachfront homes. Most residences, businesses, and streets are vacant during the off-season, as a few thousand year-round residents anticipate hundreds of thousands of summer visitors.

The transformation of Eastern Shore landscapes is an ongoing struggle over place meanings (Buttimer 1980; Wilson 1980). Favored locales are subject to the risks of development because of the collective impact of recreating tourists (Dovey 1989; Pearce 1988). Myths and symbols have become substitutes for lost place meaning (for example, Allen 1976; Tuan 1990). Since the construction of the Bay Bridge, symbol and myth formation regarding place labels, local artisans, and regional life-styles has been one way of preserving the image of the Eastern Shore as it was (for example, Buttimer 1980; Fowler 1987).

The Struggle over Place Meanings

However, myths and symbols are not a substitute for true regional knowledge. Regional subtleties have been obscured by the quest for authenticity. No summer vacation is complete without a pilgrimage to Phillips Crab House, where 10 dollars buys a dozen steamed crabs. Shore natives also enjoy steamed crabs, purchased by the bushel or "all-you-can-eat" at the Red Roost, a broiler house converted into a restaurant. Whitehaven and Cherry Hill are among the Wicomico County fishing areas that outsiders never seem to find. While tourists sunbathe on Assateague Island, Eastern Shore residents favor boating along the Nanticoke, Wicomico, and Pokomoke rivers.

"Bay Country," a popular label for the region, can be interpreted differently by insiders and outsiders (for example, Mudimbe 1988; Pearce 1988). Moreover, Eastern Shore natives are not ignorant of its meaning from the outsider's perspective. This caption attracts trade from tourists seeking an implied environmental setting (Duncan 1978). The Bay Country Shop, in Cambridge, exemplifies how commercialized symbols appear to be authentic (Duncan 1978; Pearce 1988). It houses local and imported decoys, collectors' editions of books describing the region, labrador retriever memorabilia, and clothing for afternoon outings in the country. However, for insiders, the regional label "Bay Country" is less useful lacking reference to the state of Maryland, Delaware, or Virginia, which comprise the peninsula. Natives either identify themselves as from the Eastern Shore, from Delmarva, or from one of its subregions or towns.

Outsiders are familiar with well-known images of Eastern Shore locales, such as: watermen and seafood-processing plants, ponies on Assateague Island, Frances Kitching's kitchen on Smith Island, Blackwater National Wildlife Refuge, the Wye Oak, broiler houses two stories tall, and Ocean City's boardwalk. These images of the region have been popularized through film and literature (Harvey 1989; Horton 1987). Typical, but as yet unrepresentative, scenes also include: "for sale" signs along the bays and rivers, both settled and unsettled; neo-colonial riverfront estates; yachts; and the discernable architecture of docks and carefully bulkheaded shores. New subdivisions named Nithsdale and Foxchase boast faux castle entrances, and lakes are created so that estates can be located on waterfront land. Replaced drawbridges and suspended ferry schedules result from the forgotten meaning of water as transportation. Former farms boast corrugated metal churches in place of barns and broiler houses.

The elevation of decoy carving to the status of high art is a symbol no less remarkable than the "Country

House" phenomenon. Decoys, formerly used by hunters to attract waterfowl to marshes, are now collected and created as naturalistic art. Antique and prize-winning works were exhibited at the Smithsonian Institute in 1986. The Ward World Championship Carving Competition (held in Ocean City each April) and the Ward Exhibition (held in September at Salisbury's Civic Center) are sponsored by the Ward Foundation, named in honor of Crisfield carvers Lem and Steve Ward. Each November, Easton hosts its Waterfowl Festival. The "Country House" phenomenon refers to a Salisbury business that has capitalized on Eastern Shore cultural symbols and artifacts with impressive commercial success. Given that backwater communities on the Eastern Shore are home to more than 500 decoy artists, perhaps the oddest of the items to be purchased at the "Country House" are the mass-produced rustic decoys, meant to augment the perfectly balanced country home decor.

Bridges in Context: A Perspective on Local Meanings

Outsiders may interpret "bridge" as a noun: a form of land over water. The Bay Bridge is the physical link between peninsular and continental space. According to this view, bridge building implies a discrete division between land and water such that water impedes transportation and development. But in a bay estuary ecosystem, land and water are not discernable in absolute terms. There are varying degrees of the interface between land and water: marshes and swamps, tidal streams and shores (streambeds are exposed during low tides), and subtle distinctions in the importance of elevation related to water-table heights.

This delicate ecosystem is overpowered by features such as the Kent Narrows Bridge. It replaces a drawbridge at that site with a six-lane structure. Its height is necessary to accommodate the river traffic below; but the land is flat, and terrain and vista are not impeded. The bridge looms in the horizon and from a distance surfaces as the largest feature, either cultural or physiographic, of Kent Island. Bridge pilings fill wetlands, and nearby, luxury condominiums parallel dredged river channels.

T. Horton (1987) notes local skepticism about the purposes of bridge building on the Eastern Shore. The new bridge connecting Hooper's Island to Dorchester County (completed in 1980) is visible from across the Chesapeake Bay. It has been described as a "monument to man's stupidity, man's waste, and governmental interference and inefficiency" by local residents, who insisted that the old bridge be carefully saved in sections

to be reconstructed as a private pier (Horton 1987, 204-205). A more dramatic example of local sentiment about bridges can be found in Cambridge, where the old Choptank Bridge remains partially intact and serves as a park; the middle portion has been removed so that boats can navigate the river. Unlike the new bridge, the former was a transportation link *and* a meeting place, favored fishing and crabbing haunt, and pedestrian route.

The discontinuity in place meanings between insider and outsider images of the Eastern Shore is notable for its extraordinary pervasiveness. When the Chesapeake Bay Bridges of regional landscapes acquire symbolic significance, it is because they are commonly recognized as tangible artifacts of change. In the midst of change, culture and heritage are lost, and from the perspective of those who have lost a personal struggle to maintain the continuity of place experiences, that loss can be difficult to reconcile.

Acknowledgments

I thank Robert Myers and Kenneth Bindas for their helpful comments and Michelle Stewart for help with graphics.

References

Allen, J. L. 1976. Land of myth, waters of wonder: The place of the imagination in the history of geographical exploration. In *Geographies of the Mind: Essays in Historical Geography*, eds. D. Lowenthal and M. J. Bowden, pp. 41-61. New York: Oxford University Press.

Buttimer, A. 1980. Social space and the planning of residential areas. In *The Human Experience of Space and Place*, eds. A. Buttimer and D. Seamon, pp. 166-187. New York: St. Martin's Press.

Dovey, K. 1989. The quest for authenticity and the replication of environmental meaning. In *Dwelling, Place, and Environment: Towards a Phenomenology of Person and World*, eds. D. Seamon and R. Mugerauer, pp. 33-49. New York: Columbia University Press.

Duncan, J. S. 1978. The social construction of unreality: An interactionist approach to the tourist's cognition of environment. In *Humanistic Geography: Prospects and Problems*, eds. D. Ley and M. S. Samuels, pp. 269-282. London: Croom Helm.

Fowler, P. J. 1987. The contemporary past. In *Landscape and Culture: Geographical and Archaeological Perspectives*, ed. J. M. Wagstaff, pp. 173-191. Oxford: Basil Blackwell.

Harvey, D. 1989. *The Condition of Postmodernity: An Enquiry into the Origins of Cultural Change*. Cambridge, MA: Basil Blackwell.

Horton, T. 1987. *Bay Country*. Baltimore: Johns Hopkins University Press.

Maryland Department of Transportation. 1991. *SHA Traffic Inventory.* Baltimore: Maryland Department of Transportation, State Highway Administration, Office of Traffic.

Mudimbe, V. Y. 1988. *The Invention of Africa: Gnosis, Philosophy, and the Order of Knowledge.* Bloomington, IN: Indiana University Press.

Pearce, P. L. 1988. *The Ulysses Factor: Evaluating Visitors in Tourist Settings.* New York: Springer-Verlag.

Tuan, Yi-Fu. 1990. Realism and fantasy in art, history, and geography. *Annals of the Association of American Geographers* 80:435-446.

Walsh, L. S. 1988. Community networks in early Chesapeake. In *Colonial Chesapeake Society,* eds. L. G. Carr, P. D. Morgan, and J. B. Russo, pp. 200-241. Chapel Hill: University of North Carolina Press.

Wilson, B. M. 1980. Social space and symbolic interaction. In *The Human Experience of Space and Place,* eds. A. Buttimer and D. Seamon, pp. 135-147. New York: St. Martin's Press.

A Diary Interpretation of Place: Artist Frederic Church's *Olana*

David Seamon
Kansas State University

This article explores a unique North American place—*Olana*, the 101-hectare (250-acre) home of landscape artist Frederic Church (1826-1900). In the mid-19th century, Church was America's pre-eminent artist and painted such panoramic, wall-size landscapes as *Niagara* (1857) and *The Heart of the Andes* (1859). Largely designed by Church himself, Olana is 48 kilometers (30 miles) south of Albany, New York, directly across the Hudson River from the Catskill home of artist Thomas Cole, Church's teacher and founder of the "Hudson River School" of landscape painting. Today, Olana is a New York State Historic Site and attracts more than 250,000 visitors annually. The Persian-style mansion (Figure 1) and spectacular views of the Hudson River and the Catskill Mountains (Figure 2) have been praised by lay public and critics alike. Architectural and landscape historians argue that Church's design efforts at Olana are as powerful aesthetically as his landscape paintings (for example, Carr 1989; Huntington 1966; Ryan 1989; Scully 1965).

To identify and to describe Olana's unique sense of place, I used three separate interpretive approaches: first, studying several hundred of Church's own paintings, sketches, and letters dealing with the Hudson, Catskills, and Olana; second, examining scholarly and laypersons' descriptions of Olana and Church's work and life; and, third, writing a diary account of Olana as I experienced it firsthand (Seamon 1989).

In this article, I discuss the third approach, which I call firsthand explication. Through being on the site and walking, looking, writing, sketching, photographing, and reflecting on past experiences, I sought to empathize with and to articulate the architectural, environmental, and human qualities that make Olana a special place, at least for me as a representative 20th-century visitor. At the same time, I compared these personal recognitions with discoveries that arose from my study of laypeoples', scholars', and Church's own descriptions of Olana. The eventual result was my identifying the six broad themes summarized in Table 1 (Seamon 1989).

In this article, I present representative portions of the diary that I kept to clarify my Olana experiences. To suggest how these firsthand descriptions contributed to identifying the six themes of Table 1, I include bracketed numbers at the end of passages that pointed toward a particular theme.

•   •   •

September 25, 1988

I remember my first visit to Olana in the summer of 1968. After a trip to Bard College, five friends and I were driving back to Albany State University, where we were all undergraduate students. We passed Olana's entrance, and one friend, an American Studies major who liked Church's paintings, asked if we could stop. We drove up the long, winding entry road, parked the car, and walked to the house, which immediately struck me emotionally. I sensed that this place was special—made by a person of genius, though I knew little about Church at the time [theme 6]. I recollect most vividly the breathtaking quality of the panoramic view [3]. Someone pointed out that Church intentionally built the lake below the house to echo visually the bend in the Hudson that one sees at this point [5]. This fact touched me deeply, perhaps because it was the first time that I really saw how design could enhance an already extraordinary place.

Figure 1. Frederic Edwin Church, Olana, south façade, Hudson, New York, 1982. Main house, 1870-1874, designed with structural assistance from architect Calvert Vaux; studio addition, 1888-1891, designed by Church. Reprinted with the permission of Friends of Olana, Inc., and New York State Office of Parks, Recreation and Historic Preservation, Taconic Region, Olana State Historic Site.

September 30, 1988

As I spend time at Olana and try to experience it intentionally, I realize more and more deeply that, as a place, it is much more than a house [1]. Just as important is the relationship of the house to the larger surrounds of the Hudson River and Catskill Mountains. These surrounds are uneven in that the most important parts incorporate the view from the house to the west, south, and, especially, the southwest.

This view, however, is *not only visual*; it is not simply scenery or panorama or landscape as, say, one sees when looking through a pay-for-view telescope at some dam or city skyline [3,4]. The view is visceral, affecting, touching. It seems to speak of something besides itself. Perhaps, in this view, we discover something about ourselves [2].

October 5, 1988

I spent time at the top of the house tower today. It is normally not accessible to the public, but the Site Manager understood my wish to study the views from that spot. I'm not sure that anyone knows whether Church went up on the tower often or at all. I would expect not, however, since the platform is only about

eight by ten feet in area and is enclosed by a delicate balustrade about a foot high. The significant thing about the view is that one can see in all directions. The north and east views become visible, but they are *undistinguished*. The eastern view toward the Berkshire Mountains of Massachusetts is monotonous— without any unusual or harmonious geographical features to catch the eye. The northern view is dominated by Mt. Merino, a large hill that hides the river and the town of Hudson. But what is it about the western and southern views that is so exhilarating? Why do some views and panoramas at Olana seem to touch us much more than others? [3,4].

October 17, 1988

I'm walking on the Pond Road that runs around Church's artificial lake. I look up at the house far

Figure 2. Olana's recessed veranda that Church called the "ombra," from which there is a spectacular view of the Hudson River and the Catskill Mountains. Reprinted with the permission of Friends of Olana, Inc., and New York State Office of Parks, Recreation and Historic Preservation, Taconic Region, Olana State Historic Site.

Table 1. Olana's Sense of Place: Six Interpretive Themes

1. **A Home on High.** Olana was first of all a home where Frederic Church and his family could live a comfortable and private life. By its very siting on a high hill, Church could literally obey the Christian injunction: "To breathe a clear, refined atmosphere."

2. **Nature as the Word of God.** If the views are crucial in understanding Olana, it is not because Church saw them as simply "beautiful scenery." Like many Americans of his time, Church deeply believed that nature reflected spiritual meaning, which he interpreted through his strong Protestant faith. For Church, Olana's landscapes and views were a continual source of religious inspiration and prophecy.

3. **Seeing the Essence of Landscape.** Church's way of portraying landscape involved a union of science and art. He used careful observations of things and scenes in nature to create large-scale landscape paintings. In one sense, the panoramas seen from Olana are a real-world example of Church's aim in painting: A controlled view that portrayed the essence of a particular landscape—in the case of Olana, the Hudson River and Catskill Mountains.

4. **Touching Mountains, Water, and Sky.** The mountains, water, and sky play an essential part in understanding the views at Olana. Perhaps the sky is most important because its light and atmosphere most closely symbolized God for the Hudson River painters. In the last 30 years of his life, Church frequently sketched and painted the Olana sky and regularly watched the sunsets.

5. **A Landscape in the Romantic Style.** To complement and enhance the beauty of Olana's natural site, Church manipulated the landscape, constructing the lake, planting trees, and building paths and roads. He was influenced by the romantic style of landscape gardening, which emphasized an informal arrangement of the landscape, including winding roads, irregular clusters of trees, and lake with meandering shores. These qualities can be seen in Olana's landscape.

6. **A Home of Eclectic Style.** The house at Olana was a recombination of contrasting elements and styles into a new artistic unity. For both architecture and interior design, Church drew together motifs and objects from many different times and places. The result is an architectural style difficult to categorize. It has variously been called "Persian-Moorish-Eclectic," "Italianate-Eastern-Picturesque," and "Aesthetic Movement."

Source: Seamon (1989).

above me. From this location, the house is literally a "home on high" [1]. The building rests beautifully on the hill: anchored, upright, sturdy, and secure. It seems to meet its surroundings graciously—comfortably holding its walls to touch the views that it looks out upon. Yes, it *proudly* faces the distances before it. I feel how the house looks out and greets the clouds and river and mountains. The house is the center of this place, the circumference of which is the surrounding land, water, and sky. How extraordinary, when one thinks of it, that a building and its wider landscape can be so much of a piece! An environmental whole of center and horizon.

I walk up the farm road toward the house. A sunny, windy day and the height of autumn color. There is a ridge of cumulus clouds above the hill on which the house sits, and another ridge along the west bank of the river. I sit on the stone wall to the southeast of the house and ponder a premise from Huntington (1980), who argues that, in Church's paintings, one finds rather than loses oneself spiritually; nature becomes a symbol of God (in contrast to luminist painters of the Hudson River School, where one *does* lose oneself) [2].

Do I find rather than lose myself here in the view from Olana? For me, the sense is on-looking. The world is before me, rather than I in the midst of the world. The world below and around is too far away, and I can only look out upon it. I can't be in it. I move a short distance away to the terraced lawn on the south side of the house. I realize that, from this location, the view of the river and mountains is the most perfect—exactly arranged as if a picturesque painting in a picture frame [3]. In this location, especially, one also feels surrounded by the sky [4]. There are some dramatic lighting effects as I sit. White cumulus clouds that at times become black. A more fragile, almost cirrus-like cloud higher up that covers the sun at one point and radiates light like some glowing white cloth. I think one *can* apply Huntington's premise to Olana: one does not become lost in this view but, rather, feels oneself to be a kind of privileged spectator who gives conscious attention to the natural world before him or her [1,2].

But if my position in relation to the surroundings automatically makes me feel an on-looker, I must also emphasize that the elements of the landscape are important—the water, mountains, and sky. They hold my attention because they seem so perfect. I'm wondering what people notice in what order. As I observe my own experience, I note that my eye seems first of all to move to the river—to its sinuous curves, which, in turn, carry my attention to the smooth rhythmic curves of the mountains behind. From there, attention is drawn to the cumulus-filled sky and back to the river, then left to the lake and back to the river again [3,4].

Why does the view hold my attention so? Always when I'm here, I feel that I could look for hours. First, of course, because it is such a panoramic view—almost one-hundred degrees. One feels he or she can see the whole world from this place. In the center is the river, which widens and curves and makes what Church called "the bend in the river." This afternoon the sun strikes the river and the surface sparkles like illuminated metal foil. The contrast between the luminous water and dark west bank dramatizes the curve.

There is also Church's human-made lake. Sooner or later the eye comes to that. I note how the sharp curve of the lake completes the unfinished left bend in the river further below [5]. The water in the landscape is a soft element; it somehow mediates the "hardness" of the land and mountains. I try to imagine this valley without the river and lake. I realize that the view would be much less enticing—the water softens the view and provides a resting point so that the eye wants to linger [4,5].

And above both water and mountains is the sky, which is still filled with a few cumulus clouds that begin to catch the orange rays of a setting sun. In my height as viewer, I also look out on the sky. I look out at the four elements—earth, water, air, and fire—and they are arranged in such a gracious, harmonious way [2,3]. Balance and breath come to mind. In some way, each is in a right proportion, and one feels a sense of wholeness and reality. Taken separately, the mountains, river, and valley are not that special. The Great Smokies are more beautiful and the Mississippi much more impressive. But the specific combination of geographical elements before me somehow join together in a whole much greater than each part alone. There is environmental and experiential synergy [3,4].

• • •

Commentary

The importance of placement and surroundings in contributing to Olana's sense of place is frequently mentioned in scholarly writings (for example, Carr 1989; Huntington 1966; Ryan 1989; Scully 1965). One value of my firsthand explication of Olana was that my understanding of its spatial and geographical context became deeper and more meaningful than if I had mastered this knowledge intellectually through other commentators' accounts that I, the reader, could only know cerebrally and vicariously. The direct experiential awareness that I gained from these efforts to encounter and to know Olana directly, particularly as many realizations were later corroborated by the other two interpretive approaches, provided an understanding more resonant and thick than what I might have gained from outside facts and other peoples' accounts alone.

A major aim of humanistic and phenomenological geography is to develop ways of seeing, thinking, and describing that lead to a richer, more appreciative, understanding of the natural and built worlds (Relph 1989). One practical vehicle is extended, first-person presence in place and efforts to understand that place through direct experience that is thoughtfully recorded and examined for wider meanings. In one sense, first-hand explication is nothing new: earlier geographers like Paul Vidal de la Blache and Carl Sauer regularly enlisted careful firsthand observation to study landscapes, environments, and places. In this past fieldwork, however, the observational process involved careful attention to the material, empirical, and, therefore, objective environment. A phenomenological approach, including the method of firsthand explication, widens the range of these earlier studies by legitimatizing the use of so-called "subjective" dimensions of geographical field work—the researcher's feelings, intuitions, awarenesses, experiences, inner states, and personal understandings in regard to the environment, landscape, or place.

Clearly, the use of firsthand explication is risky in that one can see too little or too much; one can miss crucial dimensions of place or imagine dimensions that are not really there. This weakness illustrates the need for complementary means by which to see and to understand the place—for example, thoughtful observation, careful drawing and photographing, reflections on other individuals' portrayals of the place, interviews with residents and visitors, and so forth (Alexander *et al.* 1977; van Manen 1990).

In my Olana work, I sought to minimize misreading and embellishment by studying the "texts" of Church's letters, drawings, and paintings; and by pondering other individuals' interpretations of Olana's uniqueness as a place. I believe that the resulting distillation, summarized in Table 1, provides one accurate portrait of the underlying qualities that contribute to Olana's powerful sense of place.

In seeking to describe any place or landscape, geographers need methods and languages whereby places and landscapes are rendered faithfully, in words and pictures that they themselves would use if they had the voice to speak (Relph 1989). Toward this end, humanistic and phenomenological perspectives offer incisive conceptual and methodological tools (Seamon 1987). One promising approach is firsthand explication of place complemented by other descriptive bases. Carefully chosen, created, and arranged, these materials can offer a comprehensive ground for thorough interpretation and accurate generalization.

Acknowledgments

This research was supported by grants from the National Endowment for the Arts and from Kansas State University. For their assistance, I thank the New York State Office of Parks, Recreation and Historic Preservation, Taconic Region, and Olana's staff, especially James A. Ryan and Robin Eckerle.

References

Alexander, C., Ishikawa, S., and Silverstein, M. 1977. *A Pattern Language*. New York: Oxford University Press.

Carr, G. 1989. *Olana Landscapes: The World of Frederic Edwin Church*. New York: Rizzoli.

Huntington, D. C. 1966. *The Landscapes of Frederic Edwin Church*. New York: George Braziller.

————. 1980. Church and luminism: Light for the elect. In *American Light: The Luminist Movement, 1850-1875*, ed. J. Wilmerding, pp. 155-190. Washington, DC: National Gallery of Art.

Relph, E. 1989. Responsive methods, geographical imagination and the study of landscapes. In *Remaking Human Geography*, eds. A. Kobayashi and S. MacKenzie, pp. 149-163. Boston: Unwin Hyman.

Ryan, J. 1989. Frederic Church's Olana: Architecture and landscape as art. In *Frederic Edwin Church*, ed. F. Kelly, pp. 126-156. Washington, DC: Smithsonian Institution Press.

Scully, V. 1965. Palace of the past. *Progressive Architecture* May:185-189.

Seamon, D. 1987. Phenomenology and environment-behavior research. In *Advances in Environment, Behavior and Design*, Vol. 1, eds. G. T. Moore and E. Zube, pp. 3-27. New York: Plenum.

————. 1989. *Recommendations for a Visitors' Center at Frederic Church's Olana Based on the Approach of Christopher Alexander's Pattern Language*, Final Report, Design Arts Program, National Endowment for the Arts, Grant No. 87-4216-0152.

van Manen, M. 1990. *Researching Human Experience*. Albany, NY: SUNY Press.

POSITIVES
AND NEGATIVES

North Americans harbor positive feelings about their continent and the constructive use of its bounty—the grandeur of its physical beauty and the richness of its resources. They hold pride in the creation of dynamic cities and productive agricultural landscapes. They boast of forethought in the designation of the world's first great national parks. But there are downsides to North America's rapid development and affluence. Negatives abound in the polarization between rich and poor, the plight of an increasing homeless population, and deleterious social breakdown in large cities. Threats to health and to pristine natural sites stem from air and water pollution and from toxic waste sites and radiation sources. Serious loss of prime agricultural lands to the undisciplined growth of urban centers, and destructive over-use of national parks add to national concerns over balanced development and environmental quality. Geographers in the forefront of exploring these social and environmental problems share their perspectives and research findings in the chapters of this section. It is useful to keep a number of negatives in mind when printing positive pictures.

— James H. Johnson Jr., University of California, Los Angeles

Bankers' Nirvana and New Calcutta: New York City in the 1980s

Barney Warf
Kent State University

F ew places in North America are as culturally diverse or evoke as rich an ensemble of images and symbols as New York City. Wall Street, Greenwich Village, the Statue of Liberty, Empire State Building, United Nations, Carnegie Hall, Rockefeller Center, and Central Park are household names to millions of Americans and magnets that draw countless tourists. New York is the ultimate "post-modern" conurbation, simultaneously nirvana and cesspool, which powerfully illustrates the interpenetration of global capital and the poignant worlds of everyday life. The city offers dramatic contrasts in wealth and poverty in close proximity to each other, and exemplifies the positive and negative extremes of advanced corporate capitalism. In this respect, New York as prescient vision of the future substantiates both rosy scenarios of the post-industrial city as well as desolate, yet compelling, visions of a world bifurcated between the idle rich and a swollen permanent underclass.

In contrast to the mid-1970s, when it was decimated by the evacuation of 300,000 manufacturing jobs and the loss of almost 1 million people, New York in the 1980s enjoyed a prolonged period of economic growth (Sassen 1989; Warf 1988b). Its resurgence rested heavily upon its status as global banking and financial center, a position seriously rivalled only by Tokyo and London (Scanlon 1989). Finance is to New York what steel was to Pittsburgh: the engine that drives the regional economy. New York's financial community, centered on Wall Street, employed more than 530,000 people in 1989. In the late 1970s banking grew rapidly with the shift to floating exchange rates and the influx of petrodollars; in the 1980s, it continued to grow as corporate and federal government borrowing soared and a wave of highly leveraged corporate mergers and

buy-outs brought lucrative fees. Even the threat of Third World debt, which depressed earnings in money center banks, has been mitigated by write-offs, debt-equity swaps, and sales on the secondary debt market. The growth of banking has been greatest among foreign firms: in 1989, 360 foreign banks, with assets over $330 billion (one half of all foreign bank assets in the United States), used New York as their base of operations to penetrate the North American market. Today, 20 percent of New York's banking employment is in foreign-owned firms, especially Japanese banks such as Dai Ichi Kangyo, the world's largest corporation in terms of assets (Warf 1988a).

The securities industries also expanded rapidly during the 1980s with the New York stock exchange's famous "bull market," in which the Dow Jones reached unprecedented heights. Deregulation of capital markets unleashed new sources of investment funds, particularly pension and mutual funds, while trade exploded on the foreign markets, including not only stocks and bonds but currencies, options, and futures, all of which became linked 24 hours per day by telecommunications systems to similar centers in Europe and Asia. Driven by computerized trading, the New York stock exchange currently trades more than 150 million shares per day, up from 12 million per day in the 1970s. One consequence of this trend is much higher volatility, as exemplified by the stock market "crashes" that occurred in October 1987 and October 1989. As with banking, New York's securities industry witnessed the growth of foreign firms, particularly heavily capitalized Japanese companies such as Nomura and Daiwa. After 1988, however, the securities industry underwent a pronounced contraction, exemplified by the collapse of Drexel Burnham Lambert, and its employment declined by almost 25 percent. The on-going collapse of the financial sectors in the early 1990s has left the New York region in severe economic recession, as manifested in its depressed real estate market and acute shortages of tax revenues.

New York also boasts of being the "communications center of the world," with one of the world's largest telecommunications networks. Most jobs in New York involve the collection, production, processing, transmission, or consumption of information in one capacity or another. Far more than the national average, New Yorkers work in offices and at computers; more word processors are found on Manhattan than in all of Europe combined (Warf 1988b). Roughly 400,000 New Yorkers are employed in business services, including one third of the nation's largest law firms, 70 percent of the its advertising and public relations companies (particularly on Madison Avenue), and seven of the "Big Eight" accounting firms. Manhattan is also the navel of a vast entertainment industry, including Broadway, Times Square, and 42nd Street (Figure 1), that attracts 18 million tourists annually, one quarter of which

come from foreign nations, particularly Japan. By spending $2.5 billion annually in the New York area, tourists form a significant source of revenue that sustains many of the city's 100,000 hotel rooms and 25,000 restaurants.

With 32 million square meters (340 million square feet) of office space, the New York metropolitan region is the largest conglomeration of offices in the world. One fifth of all office space in the United States is found within a 50-mile radius of its downtown. Manhattan—the geographically smallest, most densely populated county in the United States—is differentiated into two distinct office markets—midtown, predominantly occupied by business services, and downtown, home to a large complex of financial services. In downtown, the World Trade Center—famous for its twin 110-story-high towers that lumber over the city's skyline—forms the world's largest office complex. The sheer density of Manhattan offers powerful agglomeration economies to services firms (and 60 headquarters of Fortune 500 companies), which rely upon face-to-face interaction to minimize linkage costs with suppliers and clients.

During the construction boom in New York in the 1980s, foreign investment grew significantly. Currently, 20 percent of Manhattan's office space, or 6 million square meters (66 million square feet), is foreign-owned. Canadians, who own the World Finance Center, headquarters to American Express, form the largest single group of foreign investors by square footage. However, Japanese firms, with purchase of a controlling share of Rockefeller Center in 1990, currently constitute the largest group in terms of dollar values, particularly since New York's real estate still remains considerably less expensive than Tokyo's. Other foreign landlords include investors from the Netherlands, Germany, Britain, and the Philippines.

New York was once the largest manufacturing center in the world, home to an enormous complex of textile, shipbuilding, and metal-working firms and countless other small industrial plants. Despite four decades of job loss in manufacturing, the de-industrialization of New York is not yet complete. More than 360,000 New Yorkers are employed in manufacturing, far more than most metropolitan areas in the United States. Large parts of Brooklyn, Queens, and the Bronx continue to be employed in garment production, tool and die making, food processing, wood working, and electronics. In the suburbs, aerospace, chemicals, and "high technology" firms are often tied to military contracts. Plagued by high taxes and labor and land costs, New York City continues to lose blue-collar jobs twice as fast as the rest of the United States, with significant effects on its occupational and residential structure.

Figure 1. 42nd Street, heart of Manhattan. Photograph by Barney Warf.

As financial and business services have grown, the city's income distribution has become progressively polarized, inducing a growing schism between rich and poor (Sassen 1989). Today, New York exhibits one of the greatest class bifurcations in North America, a harbinger, perhaps, of the future service economy elsewhere.

The recent changes in New York's economy have had dire effects on its housing markets. Spurred by the well-paying jobs in financial and business services, a tide of gentrification has engulfed New York, turning much of Manhattan into a "yuppie dormitory." The Upper West Side, for years a low-income community of ill repute, has been transformed into one of the most chic and costly parts of the metropolis (Wilson 1987b), as have communities such as Chelsea (Wilson 1987a), western Brooklyn (Warf 1990), parts of northern New Jersey, and sections of Harlem, heart of the region's black community (Schaffer and Smith 1986). Residential real estate prices have risen rapidly; average housing prices in 1990 in the metropolitan region reached $190,000, while many condominiums and brownstones in New York City sold for three times that amount, far beyond the reach of ordinary residents. Gentrification, commercial construction, rent control, and insufficient construction have conspired to lower the residential vacancy rate below 2 percent. The tight housing market is particularly difficult for the poor, many of whom are confined to tenement buildings and welfare hotels. Within a few miles of one another live some of the wealthiest and most poverty-stricken residents of America: New York is First World and Third World simultaneously.

If the resurgence from the 1970s brought New York renewed prosperity, it also exhibits a dark side. The city suffers from a seemingly insuperable array of problems. Despite its high tax rates, many service-dependent poor people and an inefficient municipal government have led to chronic public revenue shortages and cutbacks in public services. The educational system faces serious shortages of teachers, an often violent and alienated student body, alarmingly high high school dropout rates, and deteriorating buildings. The medical system is on the verge of total collapse, besieged by inadequate funding, an influx of poor desperate for health care, and the AIDS epidemic (1 in 20 New Yorkers is infected with the HIV virus, including large proportions of the gay and intravenous drug-using populations). The city's violent crime rate, which has given it a notorious reputation nationally, climbed by 20 percent annually in the 1980s, particularly in connection with the growing trade in illegal drugs. Its streets, bridges, tunnels, and public transportation systems, which moves one third of all mass transit riders

in the United States, are sadly dilapidated: New York is a First World city with a Third World infrastructure.

The booming service economy has done little for New York's large and impoverished underclass, which is denied access to well-paying jobs by a lack of marketable skills (Bailey and Waldinger 1984). In the city's notoriously poor education system, half of the students fail to complete high school. In tens of thousands of decrepit apartment buildings in Harlem, the Bronx, and Brooklyn live more than 2 million people mired in poverty—one quarter of the city's population—including entire families with malnourished children in buildings often without heat or water. More people live in public housing in New York than in the entire city of New Orleans. The most vulnerable members of this rabid competition for shelter are the homeless: almost 100,000 New Yorkers sleep on the streets or in emergency shelters, earning the city the popular epithet "New Calcutta" (Marcuse 1986).

New York's ills are disproportionately concentrated among its minority population, especially blacks, who comprise one quarter of the city and suffer the highest rates of unemployment, lack of affordable housing, crime, low educational attainment, and dependency on public services. A drug-related epidemic of violent crime has made homicide the leading cause of death for young black men. In Harlem—the center of American black intellectual and artistic life since the 1920s—male life expectancy is lower than in Bangladesh. Brooklyn's Bedford-Stuyvesant, the largest single black community in the United States, suffers similarly (Quimby 1979). Despite widespread racism and severe economic disadvantages, blacks contribute mightily to the city's cultural and political life (including the newest mayor, David Dinkins).

Language fails to do justice to the astonishing ethnic diversity of New Yorkers, who form a kaleidoscope of cultures. Ellis Island continues to serve as a reminder of New York's long history as a major entrepôt for immigrants from around the world. Today, older neighbourhoods of Ukrainians, Poles, Germans, Greeks, and Italians jostle side-by-side with newer ones formed by recent arrivals from the Third World (Marshall 1987). The Jewish population, which comprises one third of New Yorkers, dates back to German and East European Ashkenazi arrivals in the late 19th century, who congealed around the garment industry in Manhattan's infamous "lower east side". Recent Jewish immigrants are often Sephardim from Arabic countries. In southern Brooklyn, 50,000 Russian Jews inhabit "Little Odessa," the largest Russian population outside the Soviet Union.

In the 1950s, New York received its first airborne wave of immigrants, from Puerto Rico. Today almost

400,000 New Yorkers are of Puerto Rican ancestry, part of a complex migration system that intimately connects the city with the Caribbean island. In large parts of the Bronx and Brooklyn, once home to the Jewish and Irish middle class, little English is spoken. Frequently poverty stricken, Puerto Ricans have made significant employment gains in medical care, clerical services, and retail trade.

While New York continues to lose native-born residents through net emigration, it gains residents by net foreign immigration. Today, one half of the city's 7.5 million inhabitants come from overseas. Between 1980 and 1986, immigration to New York was dominated by Dominicans, Caribbeans, Chinese, Colombians, and Indians (Figure 2). In New York's public school system—the nation's largest—more than 100 languages are spoken. More than 20,000 Albanian-Americans cluster together in the Bronx. Three separate Chinatowns (in Manhattan, Queens, and Brooklyn) form part of a larger and rapidly growing Asian community, fueled in part by the exodus from Hong Kong; Chinese tongs and Vietnamese youth gangs have engaged in frequent clashes over turf and access to the lucrative drug trade. Brooklyn's Arab district includes Egyptians, Yemenis, Morrocans, Syrians, and Palestinians. In Brooklyn, 700,000 immigrants from Jamaica, Haiti, Trinidad, Barbados, and other Caribbean islands make New York the largest West Indian city in the world (Kasinitz 1987). Inevitably, immigrants form new neighborhoods and transform old ones, creating textbook examples of residential invasion and succession. Immigrants lend more to New York than its distinctively international flavor, they play an important role in the local economy (Bogen 1987). Unskilled workers, predominantly from Third World nations, offer an inexhaustible source of cheap labor. Others, such as Korean-owned fruit and vegetable vendors, have erected new businesses (Kim 1981). Chinese and Dominican immigrants have penetrated the city's textile industry (Waldinger 1986).

For better or worse—or, perhaps, for better *and* worse—New York is the essence of the world's places collapsed to the head of a pin. The city illustrates the most accomplished and dynamic facets of American society in its monuments to finance, its universities, museums, and artistic life, as well as some of the most deplorable in its many suffering poor. It is abundantly clear that its linkages to global capital have not benefited all of its residents equally: the gains from financial and business services, which fueled the transformation of its commercial and residential landscapes, have done little for New York's impoverished minorities. Numerous social pathologies, from homelessness to crime to the overburdened public service system, have gotten progressively worse in the last decade, not better. Few more explicit examples exist of the obsessive tendency of capitalism to continuously create new landscapes and new social formations, to generate new predicaments while resolving old ones (Berman 1982). In its melange of peoples, occupations, incomes, and lifestyles, New York illustrates with great clarity the diversity of the changing North American panorama.

References

Bailey, T., and Waldinger, R. 1984. A skills mismatch in New York's labor market? *New York Affairs* 8(3):3-18.

Berman, M. 1982. *All That Is Solid Melts into Air.* New York: Penguin.

Bogen, E. 1987. *Immigration in New York.* New York: Praeger.

Kasinitz, P. 1987. The minority within: The new black immigrants. *New York Affairs* 10:44-58.

Kim, I. 1981. *New Urban Immigrants: The Korean Community in New York.* Princeton: Princeton University Press.

Levine, R. Young. 1990. Immigrant wave lifts New York economy. *New York Times* 30 July:8.

Marcuse, P. 1986. Abandonment, gentrification, and displacement: The linkages in New York City. In *Gentrification of the City*, eds. N. Smith and P. Williams, pp. 153-177. Boston: Allen and Unwin.

Marshall, A. 1987. New immigrants in New York's economy. In *New Immigrants in New York*, ed. N. Foner, pp. 79-101. New York: Columbia University Press.

Quimby, E. 1979. Bedford-Stuyvesant: The making of a ghetto. In *Brooklyn USA*, ed. R. Miller, pp. 229-238. New York: Columbia University Press.

Sassen, S. 1989. New trends in the sociospatial organization of the New York City economy. In *Economic Restructuring and Political Response*, ed. R. A. Beauregard, pp. 69-113. Beverly Hills: Sage.

Scanlon, R. 1989. New York City as global capital in the 1980s. In *Cities in a Global Society*, eds. R. Knight and G. Gappert, pp. 83-95. Beverly Hills: Sage.

Schaffer, R., and Smith, N. 1986. The gentrification of Harlem? *Annals of the Association of American Geographers* 76:347-365.

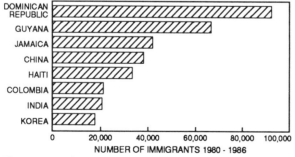

Figure 2. Total immigration to New York City, 1980 to 1986. Source: Levine (1990).

Waldinger, R. 1986. *Through the Eye of the Needle: Immigrants and Enterprise in New York's Garment Trades.* New York: New York University Press.

Warf, B. 1988a. Japanese investments in the New York Metropolitan Region. *Geographical Review* 78:257-271.

———. 1988b. The New York Region's renaissance. *Economic Development Commentary* 12(2):13-17.

———. 1990. The reconstruction of social ecology and neighborhood change in Brooklyn. *Environment and Planning D: Society and Space* 8:73-96.

Wilson, D. 1987a. Institutions and urban revitalization: The case of Chelsea in New York City. *Urban Geography* 8:129-145.

———. 1987b. Urban revitalization on the Upper West Side of Manhattan: An urban managerialist assessment. *Economic Geography* 63:35-47.

Metropolitan Points of Power and Problems in the United States

Robert A. Harper
University of Maryland

Traditionally, the northeastern region of the United States has been the center of power in the country. However, current evidence presents a different picture. Economic and cultural dominance, as well as the country's social problems are concentrated in five separate, but closely integrated urban regions (consolidated metropolitan statistical areas, or CMSAs). These five—New York, Los Angeles, Chicago, San Francisco, and Philadelphia-Wilmington—are not only the largest population centers, but they function primarily at national and international levels, not at regional scales.

Moreover, the ties of the five largest urban centers are, first of all, with one another, as suggested by domestic airline flights (Table 1). All but one of the busiest connections (the second-order link between Dallas and Houston) involve the Big Five and two thirds of these routes are between one Big-Five center and another. The Big Five are the origin or destination for more than three quarters of the busiest domestic routes in the country. These data suggest that the five centers, in many ways, function as if they are one dominant central point.

Spread as they are from coast to coast—two along the Atlantic coast, one in the Midwest, and two on the Pacific Coast—they provide evidence that traditional regional dominance has been replaced by a network connecting the largest centers of the three most important regions economically, culturally, and politically. In addition to intensive transport connections, they are joined by sophisticated communications technologies for managing the flow of money, data, and information.

The Big Five account for more than 21.2 percent of the country's population (Table 2). Individually, their populations dwarf those of most states. The New York CMSA, with 18.1 million people in 1990, is larger than all states except California (29.8 million). Only three states (California, New York, and Texas) have more people than the Los Angeles CMSA; only six are larger than the Chicago CMSA; and only 12 have more people than the San Francisco or Philadelphia CMSAs. The gap between the smallest of the Big Five (Philadelphia) and the next largest CMSA (Detroit) is more than 1.3 million. The New York CMSA includes within its boundaries 4 metropolitan statistical areas (MSAs), besides New York City, with more than 1 million people each; Los Angeles has two satellite MSAs of more than 1 million; and San Francisco has one.

These CMSAs contain some of the world's most influential businesses and entrepreneurs and a large portion of the country's poorest people (Table 2). They are home to 207 of the 400 richest people in the country, whose estimated total wealth in 1988 was $140 billion (Forbes 1990). At the same time, they contain more than 5.5 million people who live below the poverty level and 1.3 million welfare recipients (U.S. Bureau of the Census 1986).

Economic Dominance

Economically the five largest metropolitan areas account for a major share of the country's manufacturing output, wholesale trade, and financial dealings. According to the latest figures (U.S. Bureau of the Census 1986), they produce almost a quarter (24.1 percent) of the value added by manufacture; almost a third (31.8 percent) of wholesale sales. They also control nearly a third (32.9 percent) of the country's total bank deposits.

Table 1. Major Domestic Airline Connections, 1989

Cities	Orders*: 1	2	3	4	1	2	3	4
		Between the Big Five			With other U.S. cities			
New York	1	2			2	1	7	10
Los Angeles	2		1				2	6
Chicago		1	1	2			3	12
San Francisco	1	1		1			1	4
Philadelphia				1				1
Other U.S. cities						1	3	13

* First order: greater than 250,000 passengers in sample survey; second order: from 200,000 to 250,000; third order: from 100,000 to 200,000; fourth order: from 50,000 to 100,000.

Source: Air Transport Association of America (1990).

Most important of all economically, they are the dominant business management centers for the country and for U.S. interests overseas. They were host headquarters to 186 of the 500 largest corporations in total assets in 1989 (Table 3)—including 57 of the largest 100 (Forbes 1989). The large publicly held companies headquartered in these five CMSAs control almost half (49.6 percent) of the $6.6 trillion in total assets of the entire 500. Almost two thirds (62.1 percent) of those assets are controlled by 83 of the country's largest financial institutions headquartered in these CMSAs. These banks, insurance companies, and brokerage houses funds for much of the country's economic expansion. More than 100 of the other large companies based in the Big Five invest in mines, timber, farm products, and industrial capacity, and market an important share of the country's output.

Cultural Dominance

The cultural impact of the five largest CMSAs is equally great. They host the headquarters of the major com-munications companies—all of the television networks and an important share of production companies that provide television material. They house the largest concentration of the country's book (both trade and textbook), magazine, music, and record publishers. Here are the leading symphony orchestras, theatrical producers, ballet companies, museums, and galleries, and some of the major professional sports teams. They host the leading private foundations that support cultural achievement, social causes, and medical and scientific research. They are sites of some of the country's leading universities and research organizations and the headquarters for major religious institutions. Here, too, are the leading change agents—the radicals and critics concerned with such crucial issues as environmental conservation and social justice.

Concentrations of Social Problems

These five largest centers of power and wealth are also the chief destinations in the country for the poor, the destitute, and the foreign. As such, they have the

Table 2. Rich and Poor Residents in the Five Largest CMSAs

	Percentage of U.S. population (1990)	Number of 400 wealthiest Americans (1989)	Percentage of U.S. poor (1979)
New York	7.3	100	8.0
Los Angeles	5.8	38	4.9
Chicago	3.2	20	3.2
San Francisco	2.5	21	1.8
Philadelphia	2.4	28	2.2
Total	21.2	207	20.1

Source: U.S. Bureau of the Census (1991).

Table 3. Number of Headquarters of the 500 Largest U.S. Public Corporations in the Five Largest CMSAs, 1988

	Energy	Multicompanies (conglomerates)	Manufacturing	Financial
New York	4	6	26	40
Los Angeles	3	3	2	17
Chicago	2	3	7	10
San Francisco	1	1	2	8
Philadelphia	2		5	8
Total	12	13	42	83

	Trade	Transport and communication	Other	Total (all hdqts)
New York	3	8	5	92
Los Angeles	1	3	3	32
Chicago	2	5	3	32
San Francisco			2	14
Philadelphia			1	16
Total	6	16	14	186

Source: Forbes (1989).

largest concentration of racial and ethnic minorities and its results—forces of segregation and racial tension—in the country. The last census (1980) showed that these five centers contained more than one fourth (16.2 percent) of the country's Afro-Americans and close to half (42.6 percent) of the country's Hispanic population (U.S. Bureau of the Census 1986). Since that time the proportions are said to have increased. Until 50 years ago the center of America's melting pot, these urban centers are the scene of a new ethnic mix that includes Koreans, Chinese, various Southeast Asian groups, Indians, and Pakistanis; Nigerians, Ethiopians, and other Africans; as well as Hispanics from a number of Latin American countries. These centers still house some of the largest Polish, Italian, Russian, and Greek populations in the world and now have significant Native American populations.

The Big Five: New York and the Others

The five largest metropolitan areas function as the principal linkages of the U.S. urban network. They dominate the country economically and culturally. Only the political function, headquartered in Washington, DC, a metropolitan area scarcely half the size of the smallest of these centers, is centered in the Big Five. Yet the businesses, institutions, and wealthy

families in these centers wield influence in Washington as well.

The basic functions of the Big Five show a high degree of complementarity, particularly for financial and other business management and cultural decisions. Businesses in one center depend on those in others. As a result, the highest density flows of information, money, and people in the country move among them, as do domestic air flights.

In their influence on the country, the Big Five should really be called New York and the Other Four. New York, with only slightly more than one third of the population total of the Big Five centers, accounts for at least half the totals of the five in most economic activities—wholesale trade (50.6 percent), bank deposits (60.5 percent), and, most importantly, in total assets controlled by large public corporations (59.0 percent). Among the large corporations, New York dominates in the four leading categories in assets: banking (60.5 percent of total assets), manufacturing (67.9 percent), multi-companies or conglomerates (81.1 percent), and energy companies (52.0 percent). New York's importance as a business center is unquestioned. New York also has more than twice as many major air connections—domestic and international—as any other Big Five center (Table 1). It has the most major air links to other places in the country including half the first- and second-order routes and it is by far the leading international connecting point in the United States.

Conclusions

When the physical size of the United States and the global dominance of its economy are considered, the importance of just five large urban places is remarkable. Together, the Big Five serve the three leading population and economic regions of the country: New York and nearby Philadelphia, the Northeast that has been the country's dominant region throughout its history and its European gateway; Chicago, the industrial-agricultural Middle West heartland; and Los Angeles and San Francisco, the most rapidly growing area of the 20th century and the gateway to the rapidly expanding Pacific Rim. However, the regionalities of these centers are no longer the source of their importance. Today, as with Rome, Constantinople, Venice, Amsterdam, Paris, and London in the past, it is the long-range, inter-regional, international connections that are of first importance. Today, these long range connections are global. As business and cultural centers, the Big Five centers operate on a national and international scale quite apart from the regions in which they are located. They are prime examples of the multimillion urban centers built on global connections with other large world urban centers such as London, Tokyo, and Paris. Such centers attract the wealth and power of the world along with an increasing share of the world's poor, who see them as centers of hope for a new life.

References

Air Transport Association of America. 1990. *Origin-Destination Survey of Airline Passenger Traffic 1989: Domestic City Pair Summary.* Washington, DC: Air Transport Association of America.

Forbes. 1989. *The Forbes 500.* New York: Forbes.

Forbes. 1990. *The 400 Richest People in America.* New York: Forbes.

U.S. Bureau of the Census. 1986. *State and Metropolitan Data Book.* Washington, DC: U.S. Government Printing Office.

U.S. Bureau of the Census. 1991. *Preliminary Reports 1990 Census.* Washington, DC: U.S. Government Printing Office.

The Unplanned Blue-Collar Suburb in Its Heyday, 1900-1940

Richard Harris
McMaster University

Many scholars have treated the sprawling middle-class suburb of owner-occupied homes, for better or worse, as a peculiarly American landscape (Fishman 1987; Jackson 1985). They have assumed that blue-collar suburbs emerged on a large scale only with the mass affluence of the 1950s. But if the United States (along with Australia) can plausibly claim to be the world's first suburban nation it is because American workers have, for many decades, been able to realize the dream of owning their own suburban homes (Harris and Hamnett 1987). Blue-collar suburbs date back to the 1800s and had their heyday in the early decades of this century. Arguably they, more than the affluent enclaves of suburban mythology, have been a distinctive feature of the American urban scene (Harris 1988).

In the early 1900s, districts at the urban fringe were often unregulated and unplanned. In such a setting, low-income households, and especially blue-collar workers, accommodated themselves by building their own homes. The suburbs that they made expressed personal achievement in an urban setting in a manner that had few parallels in the Old World, reflecting the particular opportunities of a place and time.

The Making of Homes

In North America before World War II many families built their own homes. This was especially true in rural areas, and above all on the frontier, where settlers had little or no choice in the matter. Owner-building also played a significant role in the development of urban areas. We do not have precise national data but, from their national survey of building inspectors, R. Whitten and T. Adams (1931) imply that about

one quarter of new single-detached homes in the 1920s were owner-built. Owner-builders could be drawn from almost all classes of people. Indeed, there seems to have been a minor middle-class fad for owner-building in the early 20th century, one associated with a movement back to the land. But most people who built their own home did so from necessity and, disproportionately, were employed in blue-collar occupations. A case study of Toronto, Ontario, found that between 1901 and 1913 about 35 percent of all new homes were owner-built. This average disguised a variation from less than 10 percent among owners and managers to more than 40 percent among skilled blue-collar workers and more than 70 percent among the unskilled (Harris 1990). In practise, then, owner-builders were most likely to be working class, and suburban owner-building helped to transform the social geography of many North American cities after the turn of the century.

Workers built homes in a variety of ways. Some planned and built their homes with assistance from no one except immediate family. Starting with a small suburban plot, often mortgaged, they bought lumber in small quantities and worked evenings and weekends to put up a simple shelter. This might take the form of a one- or two-room shack, or a windowless basement. Later, as time and income allowed, improvements were made. The original home might be extended with an addition, an extra story or a basement. Alternatively, a new dwelling might be built and the old shack moved to the back of the lot to serve as a shed or garage. In some cases—we have no real way of knowing how many—owner-builders helped, and received help from, their neighbours. They bartered services, participated in building "bees," or contracted out the more difficult tasks. They might build and rebuild their home over a period of many years. Acting independently of one another, they developed lots in a haphazard fashion, so that developed sites might abut vacant lots for years or even decades.

The Earlscourt area of Toronto, which was subdivided in the 1900s, is a case in point (Harris 1992) (Figure 1). A photograph of part of this area, taken by amateur photographer John Boyd in October 1916, shows a varied landscape of modest homes, some no more than shacks (Figure 2). Vacant land was used by children for games, by their parents for growing vegetables, or, judging from the worn path in the right foreground of Boyd's photograph, by anyone who wanted to take a shortcut to the next street.

As the unsurfaced road and sidewalk in this photograph also suggest, settlements of this type were at first ignored by public authorities. Zoning and building regulations did not exist or were laxly enforced. Families built whatever they wished, in whatever manner and style, and virtually without regard for neighboring property. Streets, sewers, water mains, street lighting, fire protection, telephones, postal service, and public

Figure 2. A general view of Earlscourt in 1916. Owner-building created a landscape used for gardens, games, and shortcuts. Photograph by John Boyd, National Archives of Canada/PA 69932.

transit were often absent. In its early years Earlscourt had none of these services. People walked everywhere: a mile or so to the end of the streetcar line or to the store where they could shop and pick up mail, to the well at the end of the block for water, to the end of

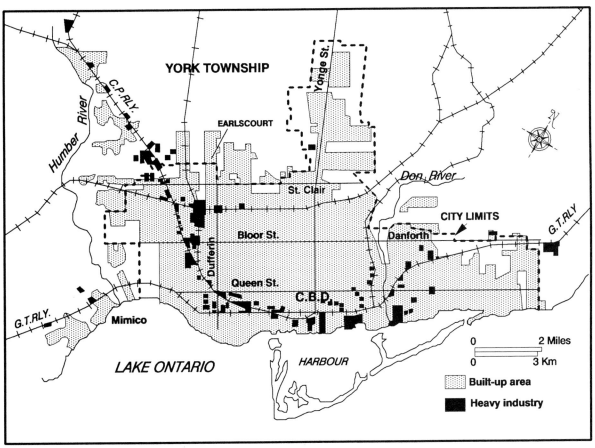

Figure 1. Earlscourt in relation to the built-up area of Toronto, 1915.

the garden for the "outhouse"; many walked miles to work. Women found it difficult to maintain homes in such isolated places but they, or at least their husbands, often resisted later attempts to install public services: more services meant higher taxes (Zunz 1982). Owner-built suburbs, then, grew up through the independent actions of many thousands of households.

The Shaping of Unplanned Suburbs

The suburban owner-builders did what made sense to themselves, largely without reference to the actions of others. Certainly, once the land was subdivided, no person or group attempted to design and coordinate the overall development process. This distinguishes North America from most European countries. There, the rarer examples of self-building were coordinated by municipalities which installed services and regulated and assisted the construction process. In the 1930s, for example, the city of Stockholm initiated a "magic house" program through which they regulated and gave technical assistance to owner-builders (Oxholm 1935).

Although the American suburbs were unplanned, they were shaped by a particular combination of circumstances. From the early days of European settlement, North American cities were characterized by a free market in land, one which was unfettered by feudal restrictions and patterns of ownership. Wood was—and indeed still is—relatively inexpensive and the development of balloon frame construction in the 19th century placed a cheap and easy method of construction at the disposal of many who had no training and few building skills (Doucet and Weaver 1985). Even in the earlier phases of industrial urban development, these factors often helped to give workers the option of suburban self-building as an alternative to living in cramped quarters downtown. Around the turn of the century, however, several developments combined to enhance the opportunities for owner-building.

First, a restructuring of American industry went hand in hand with a decentralization of factories to the outskirts of the major cities and to industrial satellites. At one time the costs of commuting might have kept workers close to centrally located jobs; now the same consideration drew them increasingly towards the fringe. Second, at about the same time, the struggles of the labor movement won for workers a shorter day which, incidentally, made more time for men to spend, if they chose, working upon their homes. Wages also rose, albeit slowly, and helped bring streetcars, and then cars, within the reach of the blue-collar worker.

This widened his—and his family's—residential options. Third, and by no means least, central cities stopped annexing their suburbs (Teaford 1979). Most urban growth after the turn of the century took place outside the boundaries of the central cities in suburbs that became, and have remained, independent political entities. This was important because although most city governments had building regulations by World War I, most suburbs did not (Fairlie 1901). Thus at the time that jobs were drawing workers out of the city, cheap land and lumber, coupled with permissive public authorities, created inviting opportunities for them to build their own homes at the suburban fringe.

These forces counted for more in some cities than in others. They were probably most significant in the midwest, where cities were growing more rapidly than back east and where manufacturing played a large part in urban employment. Certainly, the clearest demonstrations of self-building activity in the urban literature on this period are for the midwestern cities of Detroit, Milwaukee, and their closest Canadian counterpart, Toronto (Harris 1990; Simon 1976; Zunz 1982). The evidence for eastern cities such as New York and Washington is more fragmentary, but in part this is because the historians of these cities have been predisposed to focus upon the inner city tenements and courts rather than upon the suburbs (cf. Barrows 1981). Contemporaries often noted, and almost as frequently deplored, the owner-built suburb. In his book *Neglected Neighbors*, for example, Charles Weller (1909) noted the existence of "shanties" in all parts of Washington but especially at the suburban fringe. I believe that, although regional variations were important, the unplanned working-class suburb was a common and distinctive feature of North American cities in this period.

The Legacy

The unplanned suburb rapidly declined after World War II. In the interwar years a growing number of suburban municipalities had adopted building, zoning, and subdivision regulations that made self-building more difficult. During the 1930s, in both Canada and the United States, the federal government initiated changes in the system of mortgage finance. These changes made it easier for workers to buy homes from speculative builders and developers, rather than having to rely on their own labor. In most cities, by the 1950s, the unplanned suburb was history.

Unplanned suburbs built in the first half of this century have left traces on the urban landscape (for example, Barnett 1978). In most places, however, these are fast disappearing. Owner-built homes, even after years of improvement, were often quite modest. Most were not built to last. They rarely have the architectural features that would encourage a local historical board to designate them as worthy of preservation. As a result, in many cases they have been demolished. A few have persisted. In the area of Toronto once, but no longer, known as Earlscourt, there are still a good number of owner-built homes that date from the early 20th century. Although in a state of disrepair, they speak across the decades of a distinctive era of city-building (Figure 3). In many more cases, the original homes still exist but under an accretion of improvements. One of the reasons why Earlscourt fell into disuse is that in the 1950s the neighborhood was taken over by Italians. With so many Italians in the construction trades, the homes of the area were given a face-lift. Sometimes the changes are modest and obvious (Figure 4). Curved arches protrude from a frontage that still boasts an original gable and characteristically local type of false front. Here, with a little imagination, the past can be reconstructed. But in many more cases the present exterior offers few clues to the history of the home.

The current usage, and indeed the likely persistence, of owner-built homes depends to some extent on the economic fortunes of the central city. The fringe suburbs of the early 20th century have become the accessible inner suburbs of today. Where, as in Toronto, em-

Figure 4. Gothic traces of the first British settlers in Earlscourt are being obscured by the romanesque improvements of later Italian immigrants. Photograph by Richard Harris (1988).

ployment in the central area has continued to grow, the owner-built homes of the 1900s or the 1920s constitute a valuable stock of affordable homes. This can pose a threat: as land prices rise it becomes economic to buy up these homes only to tear them down and build new "monster" homes more in keeping with the gentrifying city. In other cities, such as Detroit, where employment trends are downward these homes may be left to rot. Either way, the prognosis is not good. In another generation very little direct evidence may be left of one of the more distinctive elements in the historical geography of the North American city.

References

Barnett, R. 1978. The libertarian suburb: Deliberate disorder. *Landscape* 22(3):44-48.

Barrows, R. G. 1981. Beyond the tenement: Patterns of American urban housing, 1870-1930. *Journal of Urban History* 9:395-420.

Doucet, M. J., and Weaver, J. 1985. Material culture and the North American house: The era of the common man, 1870-1920. *Journal of American History* 72(3):560-587.

Fairlie, J. A. 1901. *Municipal Administration.* New York: Macmillan.

Fishman, R. 1987. *Bourgeois Utopias: The Rise and Fall of Suburbia.* New York: Basic Books.

Harris, R. 1988. American suburbs: A sketch of a new interpretation. *Journal of Urban History* 15(1):98-103.

———. 1990. Self-building and the social geography of Toronto, 1901-1913: A challenge for urban theory. *Transactions, Institute of British Geographers* 15(4):387-402.

Figure 3. In Earlscourt, many homes speak across the decades of a distinctive era of city-building. Photograph by Richard Harris (1988).

———. 1992. "Canada's all right:" The lives and loyalties of immigrant families in a Toronto suburb, 1900-1945. *The Canadian Geographer* 36(1):13-30.

———, **and Hamnett, C.** 1987. The myth of the promised land: The social diffusion of home ownership in Britain and North America. *Annals of the Association of American Geographers* 77(2):173-190.

Jackson, K. T. 1985. *Crabgrass Frontier: The Suburbanization of the United States.* New York: Oxford University Press.

Oxholm, A. 1935. *The Small Housing Scheme of the City of Stockholm.* Washington, DC: Forest Products Division, Foreign and Domestic Commerce, U.S. Department of Commerce.

Simon, R. D. 1976. Housing and services in an immigrant neighborhood: Milwaukee's Ward 14. *Journal of Urban History* 2(4):435-58.

Teaford, J. C. 1979. *City and Suburb: The Political Fragmentation of Metropolitan America, 1850-1970.* Baltimore: Johns Hopkins University Press.

Whitten, R., and Adams, T. 1931. *Neighborhoods of Small Homes.* Cambridge, MA: Harvard University Press.

Weller, C. 1909. *Neglected Neighbours.* Philadelphia: J. C. Winston.

Zunz, O. 1982. *The Changing Face of Inequality: Urbanization, Industrial Development and Immigrants in Detroit, 1880-1920.* Chicago: University of Chicago Press.

White-Collar Home-Based Work

Kathleen Christensen
Graduate School,
City University of New York

Home-based work constitutes one of the most controversial labor issues in the 1990s. Advocates argue that working at home increases autonomy on the job, enhances flexibility in balancing the demands of work and family, and protects the basic right of U.S. citizens to work where they want. Critics maintain that working at home creates an invisible work force that is easily exploited, forces women back into the home, and precludes the development of a national policy of child care and elder care supports.

The public debates on homework, both industrial as well as white-collar, have been intense and vociferous. Yet, according to the most recent data collected by the U.S. Bureau of Labor Statistics, only 1.9 million Americans work exclusively in their homes—representing only a small fraction of the entire American work force (Horvath 1986). Why has such a minority of workers been able to claim so much U.S. political, economic, and social attention?

One of the most likely reasons is that the move toward white-collar homework illuminates larger trends that are propelling American firms and families to rethink the traditional division between the home and work place. The purpose of this paper is to address how the home has become a focal point of work and family life.

Although home-based work is not new to U.S. society, the emergence of white-collar office homework is. Until the Industrial Revolution, agricultural and cottage industries dominated the economy, and, throughout history, certain groups such as scholars, writers, crafts-people, and artists have worked at home, seeking a measure of solitude. However, since the advent of the industrial era, other needs such as supervision, communication, and the cooperative use of resources and equipment have predominated, leading to the centralization of the work place, first in factories, then in offices. The notion of work and family as separate and relatively autonomous behavioral spheres grew out of this physical separation of the centralized work place and the home.

Current changes in the economy and the family are precipitating changes in attitudes toward the separation of the home from the work place. The move toward "lean and mean" corporate labor forces and the shifting boundaries between work and family are resulting in new work arrangements for men and women, including work at home.

A number of terms are currently in vogue to cover gainful employment in the home; the most inclusive being "home-based work" and "homework." These terms can be used interchangeably to characterize any paid work done in the home regardless of the employment status of the worker. In addition, they can cover work that is either done exclusively in the home or based out of the home. For example, authors usually work in their homes, whereas sales representatives spend much of their time on the road, working out of their homes.

More restrictive terms than home-based work and homework also exist to cover computer-mediated homework. Futurist Jack Nilles of the University of Southern California coined "telecommuting" to describe computer work done by company employees at home that allows them to substitute their computers for their commutes. In effect, both technology and the employment status of the workers are embedded in the label.

The notion of the electronic cottage, originally forged by futurist Alvin Toffler in *The Third Wave* (1980), focuses on the technological, rather than employment, aspect of homework. Wired for electronic work, the home often sits in an electronic network made possible by advanced telecommunication technology. Although the electronic cottage houses both the employed and self-employed, the term most often refers to the self-employed.

Much of the media's attention to homework has been limited to telecommuting and to the electronic cottage, implying that the technology causes work to be done at home. Yet, contrary to conventional wisdom, which holds that computers will enable millions of people to work in their homes, I argue that the causes for any large-scale movement to white-collar homework, whether or not it is computer-mediated, will have more to do with prevailing conditions in the economy and the family than in the availability of computer technology.

Structural Changes in the Economy: Core and Ring Workers

Firms in the United States have undergone profound changes in the last decade as a result of down-sizing, mergers and acquisitions, and the need to stay competitive in an increasingly global economy. In fact, between the beginning of 1980 and the end of 1987, the Fortune 500 companies reduced their work forces by 3.1 million, going from an aggregate of 16.2 million employees to 13.1 million.

This type of internal labor-market turbulence has prompted many companies to rethink their overall staffing attitudes and practices. The traditional attitude that firms took toward their white-collar work force entailed permanent, or at least relatively secure, employment, with some notion of career advancement. Yet current staffing practices challenge that stance. Many companies now think of their personnel in much the same manner as they do their inventories, striving for a just-in-time staffing strategy to parallel their just-in-time inventory systems that keep supplies and materials just sufficient to meet current demand.

This desire for elasticity in staffing has resulted in an ad hoc two-tiered work force in many U.S. firms (Belous 1989; Christensen and Murphree 1988; Polivka and Nardone 1989). The first tier comprises a core of salaried employees on the company payroll toward whom the traditional attitude still holds. These core employees are accorded a relatively high degree of job security, perquisites, health and pension benefits, and opportunities for training and skill upgrading. The second tier includes a cadre of workers, many of whom are not on the company payroll, hired as self-employed independent contractors, temporaries, or casual part-timers. These workers have weak ties to the company,

are generally hired for finite periods, often in a non-systematic fashion, and receive no health coverage, pension plans, or other benefits. Many in the second tier previously worked for the firm as core employees.

This second tier work force goes by many names, including contingent work force, peripheral or secondary workers, and even reserve work force. Although this tier has always existed in the United States and elsewhere, there is evidence that its numbers are growing (Belous 1989). Examples abound in the publishing, television, and advertising industries as professional and technical employees are laid off from the core tier and hired back as independent contractors, euphemistically referred to as free-lancers, management consultants, or entrepreneurs.

Home as Site for Contingent Workers

The home becomes an important work site for many contingent workers hired on contract bases (Christensen 1986) due to tax advantages, access to families, and limited alternative work-sites. The practice of contracting out work to be done at home, however, has become controversial because, according to the Internal Revenue Service, many of these contractors should be classified as employees and should receive the benefits and protection that come with employee status (General Accounting Office 1989). As a result of their self-employed status, they are often deprived of the health and safety net they rightfully deserve thus, creating a potentially large group of vulnerable, second class workers.

Telecommuting for Core Employees

Not all people who work at home for companies are part of the contingent work force, however. According to a recent survey I conducted for The Conference Board, a business research and information institute located in New York City, 29 of 521 of the nation's largest firms offer telecommuting to core employees, often as a way to attract and retain the best people for their remaining core positions (Christensen 1989). Changing demographics have influenced U.S. firms' ability to recruit and retain these high-quality workers. According to the 1987 Work Force 2000 report commissioned by the U.S. Department of Labor, U.S. population growth has leveled off, particularly among the educated middle class. This means that, by the end of this century, the United States will have fewer workers trained for jobs that require education and technical skills. The consequences of the changing balance between supply and demand are twofold—a tightening of local labor markets and an increasing mismatch between available skills and new job requirements.

U.S. labor markets are feeling the pinch. A recent survey of more than 700 human resources executives by the American Society for Personnel Administration reveals that 43 percent report problems finding qualified executives; 66 percent cite difficulties finding technical help (*Wall Street Journal* 1989). To attract workers, the study finds, higher wages are being offered by 58 percent of these companies, tuition aid by 52 percent, and better health benefits by 31 percent. All of these recruitment incentives are costly. Some firms are turning, therefore, to less expensive incentives such as flexible schedules, which include professional part-time, job sharing, and telecommuting as effective tools for recruiting and retaining the core employees they want.

When properly designed, telecommuting can meet the needs of both employers and employees. Specific examples of such successful telecommuting programs include Mountain Bell, Pacific Bell, and J. C. Penney, all of which view work-at-home as a scheduling device to attract and retain valuable employees, to reduce absenteeism, and to cut the costs of office space. What is important about these corporate programs is that their home-based employees maintain their employee status, are paid exactly what they would be paid if they worked in the office, and receive all of the health and pension benefits they would get as on-site employees. In addition, they are considered for promotion and training. The companies pay for all equipment and telephone costs. Further, to ensure that the home-based employee maintains a high profile in the company, Pacific Bell requires employees to come into the office at least one day a week.

Firms such as these three may augur well for the type of enlightened and strategic planning that U.S. business increasingly needs as the United States moves toward the turn of the century. For these firms, a move toward work-at-home alternatives may be just one of many ways to cultivate good workers who need or want flexibility.

Structural Changes in the Family

The traditional family in which the father goes out to work and the mother stays home has undergone rapid change, and this change has contributed to increased demands for flexible scheduling, including work-at-home. Fewer than one tenth of U.S. families (7 percent) fit the traditional model. In fact, the norm is much more likely to be the dual-earner family in which both spouses have paying jobs or the female-headed family where the woman is the sole breadwinner. By March 1985, nearly 17 percent of all families, approximately 10.5 million, were headed by women who were divorced, separated, widowed, or never married.

One result of these changes in the family is that the traditional boundaries between women's work and men's work have changed. Significantly, no longer is women's work solely unpaid labor in the home and men's work solely paid labor outside the home. Women's entrance into the work force in large numbers over the last several years has profoundly reconfigured the boundaries between work and family, in effect between production and reproduction.

Yet, the burden of these changes has fallen primarily on women who now work the double day: paid labor one shift, unpaid domestic work another. The vaunted ideal remains flexibility—an ability to set hours or stretch days in such a fashion that both shifts can be accomplished.

For many women, the home presents itself as an ideal work place, particularly if they can work as employees for a firm. Realistically, however, few companies currently allow telecommuting, although increasing numbers are exploring the option (Christensen 1989). Therefore, if women are serious about working at home, they will more than likely do so on a self-employed basis.

According to the U.S. Small Business Administration, women-owned businesses are the fastest growing segment of the small business population. Between 1977 and 1982, the number of female nonfarm sole proprietorships grew at an annual rate of 6.9 percent, nearly double the overall annual rate of 3.7 percent.

Home as Site for Women Balancing Work and Family Demands

Evidence indicates that the home serves as the work place for many women who are in business for themselves. According to the Bureau of Labor Statistics, of those home-based workers who worked 35 hours or more at home, nearly 70 percent were self-employed in unincorporated businesses (Horvath 1986).

Much of the public debate implies that mothers with children under 18 constitute the prime candidates for home-based businesses. Yet, Bureau of Labor Statistics figures reveal that of the women who work 35 or more hours a week at home, approximately 259,000 have children under 18, while an almost equal number are women without children (Horvath 1986). Circumstances propelling women to work at home are varied. Those with children do it to be near their families, while those without often are trying to reenter the labor force after years of raising their families. Older women approaching retirement need supplemental income, while others find that they have increasing responsibility for elderly family members. Some are part of the "sandwich" generation, women caught between caring for both elderly family members and children.

By the year 2000, the Census Bureau estimates that the "oldest old," those Americans aged 85 or older, will number 5 million and are likely to be women who have little or no retirement benefits. According to Dana Friedman of the Families and Work Institute, a nonprofit research and advocacy organization in New York City, nearly 80 percent of the oldest old live outside nursing homes and thereby require some type of assistance. For example, in 1986, The Travelers Corporation of Hartford, Connecticut, surveyed their employees over the age of 30 who provide some type of care to an elderly parent. Most of those who needed care were widowed mothers or mothers-in-law, while most of those who provided care were women.

For many women, the home appears to be an ideal site in which to meet the demands of both work and family. Yet my research indicates that, at least for women with pre-school-aged children, homework does not eliminate the need for other forms of child care (Christensen 1988). In a national survey of 7,000 home-based workers, I found that two thirds of the women doing professional and managerial work used child care; only one third of women doing clerical work had help. These women typically worked late at night, while maintaining full responsibility for the house and family. Under these conditions, few women saw work-at-home as an ideal solution to balancing the needs of work and family (Christensen 1988).

Conclusion

Since the Industrial Revolution, the home has represented a private point on the American landscape.

Technological, economic and social changes, however, are facilitating the development of paid labor in the home, thereby creating homes that are both public and private points. This change will have long-standing consequences for what we think of as the division between work and family, making what has always been a rigid boundary a more permeable one.

References

Belous, R. 1989. *The Contingent Economy: The Growth of the Temporary, Part-Time and Subcontracted Workforce*, NPA Report 239. Washington, DC: The National Planning Association.

Christensen, K. 1986. Testimony before the Committee on Government Operations, U.S. House of Representatives. *Pros and Cons of Home-Based Clerical Work*, pp. 27-36. Washington, DC: U.S. Government Printing Office.

————. 1988. *Women and Home-Based Work: The Unspoken Contract*. New York: Henry Holt.

————. 1989. *Flexible Staffing and Scheduling in U.S. Corporations*, Research Bulletin 240. New York: The Conference Board.

————, **and Murphree, M.** 1988. *Introduction in Flexible Workstyles: A Look at Contingent Labor*, pp. 1-4. Washington, DC: U.S. Department of Labor, Women's Bureau.

General Accounting Office. 1989. *Tax Administration: Information Returns Can Be Used to Identify Employers Who Misclassify Workers*. Gaithersburg, MD: General Government Business.

Horvath, F. 1986. Work at home: New findings from the current population survey. *Monthly Labor Review* 109(11):31-35.

Polivka, Anne, and Nardone, Thomas. 1989. On the definition of contingent work. *Monthly Labor Review* 112(12):9-16.

Toffler, A. 1980. *The Third Wave*. New York: William Morrow.

Wall Street Journal. 1989. Labor shortages are getting tighter and tighter, companies say in labor letter. 7 February:1.

Breadlines

Janet E. Kodras
Florida State University

People are in those lines because the food is free.
—U.S. Attorney General Edwin Meese (1985)

The breadlines that form outside the nation's soup kitchens, extending down the steps of the Haven of Rest Mission and across the lawn of the Good Shepherd Catholic Church, illustrate a fundamental paradox in the United States—poverty in the midst of plenty, hunger in the world's most powerful agricultural nation. The lines were longest, and the paradox strongest, during the 1930s and 1980s. The variable length of the breadline may be taken as symbolic measure of American social failure.

I explore here the reasons for the paradox, examining these periods of failure within the framework of "regimes of accumulation," first advanced by the French regulation school (Aglietta 1979). This perspective holds that each regime in the development of capitalism is dominated by a particular type of production (craftswork, mass-production manufacturing), which is facilitated by a set of regulatory mechanisms (crafts guild insurance funds, state social programs). These eras are interspersed by periods of crisis, during which problems in the previous regime undermine its continuation and give rise to new organizations of production. Importantly, the process of change from one mode of production to another during crisis generates new social problems, such as an increase in hunger, but these must ripen into specific societal conflicts before regulatory mechanisms can be designed to address them (Kodras and Jones 1990).

Regulation theory treats the 1930s and the present period, beginning in the mid 1970s, as eras of crisis. During both periods, the economic base was shifting from one dominant type of production to another, but regulatory mechanisms lagged behind. Designed to facilitate previous modes of production, they were insufficient, and inappropriate, to address new forms of deprivation and hunger resulting from the emergent system. At such times breadlines, the last resort of those in need of a meal, lengthen outside the nation's charitable organizations.

The 1930s Crisis

In his novel *The Grapes of Wrath*, John Steinbeck paints a most vivid portrait of the contradiction of hunger in the midst of agricultural abundance during the 1930s Great Depression:

And in the growing year the warmth grows and the leaves turn dark green . . . the year is heavy with produce. And first the cherries ripen. Cent and a half a pound. Hell, we can't pick 'em for that. . . . And the pears grow yellow and soft. Five dollars a ton. We can't do it. And the yellow fruit falls heavily to the ground and splashes on the ground. . . . Men who can graft the trees and make the seed fertile and big can find no way to let the hungry people eat their produce. . . . The works of the roots of the vines, the trees, must be destroyed to keep up the price, and this is the saddest, bitterest thing of all. Carloads of oranges dumped on the ground. The people came from miles to take the fruit, but this could not be. How would they buy oranges at twenty cents a dozen if they could drive out and pick them up? And children dying of pellagra must die because a profit cannot be taken from an orange. . . . In the eyes of the hungry there is a growing wrath. In the souls of the people the grapes of wrath are filling and growing heavy, growing heavy for the vintage (Steinbeck 1939, 382-385).

It was the awareness of this absurdity that most threatened the legitimacy of the economic system during the 1930s. Oranges set on fire, milk dumped in ditches, pigs slaughtered for lack of a market, were simply more visible than the "unproduced potential of the industrial sector" during the Depression (Poppendieck 1986, xiii).

What was the nature of the economic system that generated such a paradox?

Throughout the early 20th century, a new form of industrial production was gaining momentum, based on large-scale mass production. This system of production would emerge to dominate and to define the postwar economy. But as yet, large segments of the labor force were still employed in agriculture or crafts industry, and the types of social assistance then in place were designed to address indigence and hunger resulting from failures in those types of production. Agriculturists called upon resources from the extended family during difficult periods, while craftsworkers developed trade associations that provided mutual insurance funds to assist members during the slack season (Piore 1987). The punitive system of public relief operated within the English Poor Law tradition of local responsibility for poverty, and by degrading and isolating those few who did receive support, it legitimated the belief that poverty was a sign of individual failure to compete in the labor market (Piven and Cloward 1979). Similarly, the gains made during the Progressive Era to address new forms of suffering, derivative of ascendant industrialization, were as yet partial and unevenly realized. Unemployment funds established by labor unions covered less than 1 percent of the labor force at that time (Bernstein 1970).

These forms of social assistance were clearly inadequate protection against the enormous deprivation arising after the Crash of October 1929. The number of unemployed shot up from less than half a million in October 1929 to 15 million—one third of the labor force—in 1933 (Bernstein 1970). A survey of school children found that fully one fourth suffered from malnutrition, and 100 cases of death by starvation were reported in New York City hospitals alone during 1931 (Bernstein 1970; Piven and Cloward 1979). Individuals and private organizations devised innumerable schemes to meet the growing problems of hunger:

> Breadlines proliferated; some eighty-two separate lines were operating in New York City alone. . . . Gangster Al Capone opened a breadline in Chicago. St. Louis society women distributed unsold food from restaurants. Someone placed baskets in New York City railroad stations to enable commuters to donate vegetables from their gardens. Harlem radio personality Willie Jackson opened a penny restaurant where meals were sold for one cent per dish (Poppendieck 1986, 24-25).

Not all of the food queues formed outside churches and private charitable organizations: "Lines form each day at the garbage dumps from eight in the morning to five in the afternoon. Men and women come there to see if they can't find food to carry back home with them. They get some, if they come early enough" (Hallgren 1933, 100). The situation was worst in rural areas, where the American Red Cross was often the only formal mechanism of relief. It was in the largest cities, however, that the growing masses of the hungry first began to challenge the notion that poverty was a reflection of individual failure. The concentrated magnitude of suffering demonstrated that larger, macroeconomic forces were at work. As growing awareness of the source of the problem spread during the early years of the Depression, mob looting of food stores and demonstrations under the Communist banners "Fight—Don't Starve" increased in frequency and violence (Leab 1967; Piven and Cloward 1979).

Throughout the early years of the Depression, President Herbert Hoover had remained steadfast in his conviction that recovery was imminent and that local public and private forms of assistance were the proper conduit of such aid as was needed. But hunger in the midst of agricultural overproduction was everywhere evident, and Hoover could not remain persuasive for long. Economic disaster translated into political realignment. Basing his campaign on recovery for "the forgotten man at the bottom of the economic pyramid," Franklin D. Roosevelt trounced Hoover in the 1932 Presidential election (Piven and Cloward 1979). Beginning with emergency recovery measures, the national government began to take a more active role in social assistance with a series of programs known as the New Deal. Only in the face of growing disorder in the cities and countryside, and the unmitigated failure of municipal and private efforts, did the federal government assume responsibility, and even then, its role was seen as a temporary solution during crisis. Nevertheless, the national programs established during the Depression set precedents for a federal role in social assistance during the postwar period.

The Postwar Regime

Economic stability was eventually restored, as the emergent industrial system geared up to support the war effort and to meet domestic and international demand for U.S. manufactured products following the war. New mechanisms of social assistance were shaped by macroeconomic Keynesian policy, a social contact between capital and labor that traded worker compliance for a share in industrial profits, and a welfare state that moderated the effects of periodic economic downturns (Kodras and Jones 1990). The welfare state in the United States was underdeveloped relative to those in most Western European industrial nations.

Furthermore, the mechanisms did not extend to all segments of the American population, but they were evidently sufficient to facilitate the continued growth of the economic system as it gained global hegemony in manufacturing. State intervention in the labor process (a combination of labor laws, redistribution through taxation, and outright provision) had come to replace previous types of social regulation, such as mutual aid within crafts associations and extended agricultural families (Cox 1989).

Significantly, state efforts to address hunger were not an initial component of the developing welfare state. Domestic food policy was oriented toward production controls, maintaining agricultural prices by cutting back on output (Poppendieck 1986). Strong economic performance at the height of this regime held down the incidence of hunger, although it did not eliminate it altogether. Insufficient food consumption was greatest in the deep South: approximately one quarter of all households in Mississippi, Arkansas, and South Carolina consumed inadequate diets in the early 1960s (U.S. Bureau of the Census 1970). Agricultural production controls and the increasing export of surpluses concealed, but did not eliminate, the underlying contradiction of hunger in the midst of agricultural abundance.

Not until the early 1960s did the national government institute efforts to increase food consumption among the poverty population. The food stamp program was part of President Lyndon Johnson's War on Poverty. Ironically, the U.S. Department of Agriculture, whose role it is to serve the interests of the farm economy, was put in control of the food stamp program. The Department of Agriculture exercised its authority to ensure that the program operated more to the benefit of the agricultural sector than the poverty interests. Thus, the postwar governmental regulation of the food system was primarily oriented toward facilitating the macroeconomic performance of agricultural production rather than assisting the needs of the hungry. Not until 1970 was a Congressional hunger lobby successful in beginning fundamental changes in the program (Berry 1984). Within just a few years, however, the nation's first comprehensive mechanism to attack hunger would be undermined, as the economic system on which it was based began to falter.

The Current Crisis

Limits to the continued development of the postwar production system began to appear in the early 1970s. The United States began to lose its position in the global economy as foreign corporations challenged U.S. control of international markets for manufactured goods and at the same time cut into the domestic market. These ascendant corporate powers were located in industrialized nations, such as Japan and West Germany, that had completed their recovery from the destruction of World War II, and were increasingly searching for outlets abroad. In addition, Third World nations began to assert control over their resources, many of which had been cheaply exploited for U.S. production. This increased the costs of manufacturing and drove down profits.

Inflation accelerated, even as unemployment grew and real wages fell. In contrast to the 1930s crisis, when large segments of the labor force were suddenly thrown into destitution, the majority of Americans in recent years have remained employed and above the poverty level. But when the assumptions of continued national growth and individual affluence were challenged, the possibility of financial failure became a very real prospect in the public conscience (Harrington 1984).

The affluence generated during the postwar had been sufficient to underwrite at least some degree of state assistance, but as the economic system began to weaken, a fiscal crisis ensued and the welfare state lost its ability to address needs. Rather than viewing failures in the welfare system as a consequence of problems in the larger economy, however, the argument was reversed. Indeed, Ronald Reagan rode to victory in the 1980 Presidential election, identifying profligate state spending, particularly in social assistance, as a source of economic problems. It was argued that national affluence had been squandered on social programs for the indolent, and only the elimination of such programs would restore the United States to its previous position of global prominence.

Thus began the dismantling of the welfare state. For example, the food stamp program lost $11 billion from budget cuts in the first two years of the Reagan Administration (Berry 1984). The national government attempted to pass responsibility for social assistance back to the states and localities, arguing that the postwar expansion of temporary federal programs during the Depression had been a mistake all along. Private philanthropies and voluntary organizations were encouraged to assume the assistance role.

The old philosophy that deprivation results from individual failure, rather than shortcomings of the market system, was forcefully reasserted as justification for the federal retraction of responsibility. Reagan's public remarks were exemplary of this individualist, blame-the-poor position. Asked by a schoolboy from the Midwest why America fails to feed many of its people, President Reagan replied that there was an abundance of food in the nation, but "the hungry are too ignorant to know where to get it" (Brown and Pizer 1987, 189). Given the recent increases in the nation's hungry population, one can only marvel as to the reasons for this sudden explosion of ignorance.

The incidence of hunger and deprivation has accelerated during this most recent period of crisis. The number of Americans living in poverty increased from 26 million in 1979 to 34 million in 1983 and 20 million Americans suffered from hunger at that time (Brown and Pizer 1987; Harvard University 1985). Reports from the Salvation Army, the National Council of Churches, the U.S. Conference of Mayors, and even the General Accounting Office and the U.S. Department of Agriculture have provided similar accounts of the rapidly rising incidence of hunger.

Although the food stamp program is designed to be countercyclical, such that program use increases as the economy declines, participation declined from 70 percent of the poor in 1979 to 59 percent in 1985, the result of budget cuts (Berry 1984). That proportion who do obtain food stamps receive an average of 45 cents per meal, which is clearly insufficient; Congressional testimony has shown that increasing numbers of food stamp participants have begun to appear in the nation's breadlines, after monthly benefits are exhausted (Berry 1984; Maney 1989). A Department of Agriculture study has found that fully 80 percent of the households using the basic food stamp expenditure consume a diet seriously lacking in primary nutrients (Brown and Pizer 1987). Although there exists no comprehensive count of the number of persons who regularly form breadlines, the Salvation Army alone provides 10 to 12 million meals annually (Glassner 1988).

Summary

During the early 20th century, mass production manufacturing was developing as the future base of the economy, but the types of social assistance available to help in times of need were derivative of a previous era, dominated by agricultural and crafts production. When the economic system failed in 1929, these forms of social regulation were insufficient, and inappropriate, to address the crisis formed from failures in the emergent production system, and breadlines lengthened as an indicator of increased hunger. The economy eventually recovered, based upon a coordinated regime of mass-automated manufacturing production and regulatory mechanisms, referred to in the aggregate as the welfare state. Hunger programs were late in coming and focused more on facilitating the production system than on feeding the hungry. Nevertheless, breadlines shortened due to a relatively strong economy and the assistance of social programs.

Since the early 1970s, mass-automated manufacturing has declined and new production systems, often based upon flexible arrangements, have emerged, but once again, the types of social assistance in existence are predicated upon problems derivative of a postwar economy and are poorly articulated to address new sources of deprivation. The fact that historical structures of social regulation have been designed to facilitate the dominant mode of economic production, rather than to confront problems accruing to the margins of the labor force so engaged, leads to scepticism that state policy will be used as a magnanimous gesture toward the nation's hungry, as the emergent regime takes shape. The breadlines will continue to lengthen, and the paradox of want in the midst of plenty will continue to deepen, unless or until mechanisms are devised to address the fundamental causes of hunger.

Acknowledgments

I thank Mark Ellis and Jill Quadagno, both at Florida State University, for their insightful comments on an earlier draft of this paper. Any errors of fact or interpretation are my responsibility.

References

Aglietta, M. 1979. *A Theory of Capitalist Regulation.* London: New Left.

Bernstein, I. 1970. *The Lean Years: A History of the American Worker, 1920-1933.* Baltimore: Penguin.

Berry, J. M. 1984. *Feeding Hungry People: Rulemaking in the Food Stamp Program.* New Brunswick, NJ: Rutgers University Press.

Brown, J. L., and Pizer, H. F. 1987. *Living Hungry in America.* New York: Macmillan.

Cox, K. R. 1989. The politics of turf and the question of class. In *The Power of Geography: How Territory Shapes Social Life,* eds. J. Wolch and M. Dear, pp. 61-90. Boston: Unwin Hyman.

Glassner, I. 1988. *More than Bread: Ethnography of a Soup Kitchen.* Tuscaloosa: University of Alabama Press.

Hallgren, M. 1933. *Seeds of Revolt: A Study of American Life and the Temper of the American People during the Depression.* New York: Knopf.

Harrington, M. 1984. *The New American Poverty.* New York: Penguin.

Harvard University. 1985. *Physicians' Task Force on Hunger in America.* Cambridge: Harvard University School of Public Health.

Kodras, J. E., and Jones, J. P. III. 1990. Academic research and social policy. In *Geographic Dimensions of U.S. Social Policy*, eds. J. E. Kodras and J. P. Jones III, pp. 237-248. London: Edward Arnold.

Leab, D. 1967. "United We Eat": The creation and organization of the unemployed councils in 1930. *Labor History* 8:300-315.

Maney, A. L. 1989. *Still Hungry after All These Years: Food Assistance Policy from Kennedy to Reagan*. New York: Greenwood.

Piore, M. J. 1987. Historical perspectives and the interpretation of unemployment. *Journal of Economic Literature* 25:1834-1850.

Piven, F. F, and Cloward, R. 1979. *Poor People's Movements: Why They Succeed, How They Fail*. New York: Vintage.

Poppendieck, J. 1986. *Breadlines Knee-Deep in Wheat: Food Assistance in the Great Depression*. New Brunswick, NJ: Rutgers University Press.

Steinbeck, J. 1939. *The Grapes of Wrath*. New York: Penguin.

U.S. Bureau of the Census. 1970. *Statistical Abstract of the United States*. Washington, DC: U.S. Government Printing Office.

Skid Row, U.S.A.: Place and Community

Jennifer R. Wolch
University of Southern California

"Skid Row" is a generic place-name commonly used to denote a specific zone in the American metropolis. Typically on the fringe of the central business district, Skid Row zones are places of economic and social marginalization. Over the past decade, Skid Rows in America experienced a dramatic increase in homeless population. As a result, images of Skid Row have become symbols of societal dysfunction, for both domestic and international observers alike. In this paper I briefly outline the history of American Skid Row districts, and describe one in particular—Skid Row, Los Angeles. My focus is on Skid Row as a place with specific land-use patterns and environmental qualities, and as a community composed of both residents and workers.

From Hoboes to Homeless

Although the term Skid Row had not yet been coined, precursors to Skid Row zones emerged in the early 19th century. These districts were typically anchored by early forms of human service provision such as workhouses, hospitals, and local philanthropic orders. Those dependent on charitable or state provisions due to poverty or illness were drawn to such "zones of dependence." Over time, marginalized groups and services designed to support them became concentrated, as those in need sought assistance and as service institutions sought locations accessible to clients (Dear and Wolch 1987). The mix of residents changed over the 19th century, for at least two reasons. First, innovations in medical and social service delivery, such as specialized asylums for the mad and orphans, removed segments of the dependent population from the urban community. Second, as the century wore on, increasing numbers of displaced agricultural workers, casualized

industrial workers, and immigrants flowed into and out of American cities in search of work. After the Civil War, the exodus of blacks from the south to northern cities led to the development of Skid Row areas strictly segregated by race (Hoch and Slayton 1989).

From the 1870s to the 1920s, migrant workers, variously called hoboes, tramps, and bums, crowded into what had become "zones of transiency." This was the heyday of the Main Stem in Chicago, the Bowery in New York, and the Tenderloin in San Francisco. Entrepeneurs responded to the demand for inexpensive living by providing cheap housing, services, and casual/day labor employment. The result was a vital community with its own argot and system of values, where transient men found support in one another's company. Their way of life was aided by municipal and charitable agencies, and the availability of single room occupancy hotels, lodging houses and hostels, inexpensive beaneries, stale beer parlors, pawn and thrift shops, and billiard halls. Despite these resources, life in such neighborhoods was difficult, and "zones of transiency" were widely viewed as unsavory environments.

By the 1940s, these communities were in serious decline. They had also acquired a label: Skid Row, named after Seattle's "Skid Road," a street used to skid logs to a waterfront and home to transient loggers. Although their populations fluctuated according to seasonal rhythms, variations in the business cycle, and the onset or end of wartime, the overall trend was one of shrinkage. After the second World War, demand for seasonal agricultural workers shrank because of farm mechanization, and the industrial economy entered the period of mass production for mass markets, necessitating a stable and skilled workforce. Those left in Skid Rows were the last of the hoboes, tramps, and bums, typically unable to work because of age, incapacity, or lack of job opportunities. Residence in Skid Row acquired a pejorative, double meaning: of being downwardly mobile ("on the skids") and being constrained to a marginalized environment (a "row"; Hoch and Slayton 1989, 88-89). After the Great Depression, the public sector took a greater role in regulating Skid Row life, through policing and the intervention of formal human services such as municipal shelters, public soup kitchens, and other relief agencies.

Skid Row was first studied in the late 19th century, when large-scale surveys of the poor were first conducted. During the 1920s and 1930s, Chicago School urban sociologists undertook richly detailed studies of "Hobohemia," its residents, and its social structures and problems (Zorbaugh 1929). During the post-World War II period, social scientists again assessed Skid Row areas. Some were impressed by the cameraderie and conviviality of Skid Row life, romanticizing its atmosphere and minimizing the severe hardships on residents (Wallace 1965). Other investigators pointed to high frequencies of social pathology and deviancy among the largely white, male Skid Row population, for example, depression, alcohol addiction, and social isolation, and connected these problems to the rundown housing stock and environs (Bogue 1963). These features became defining characteristics of what was now termed the Skid Row "homeless" population. Ultimately the social scientific diagnoses of Skid Row as a zone of dilapidation, deviance, and dependency prompted the demolition of Skid Row districts across the country, as urban politicians sought to "clean up" their cities, and business interests anticipated opportunities to expand the central business district.

Today, Skid Row districts still exist but retain only a small share of their former geographic area and population size. In most American cities, they have been transformed in two fundamental ways. First, their physical fabric has been drastically altered. This has occurred via demolition of single room occupancy housing and closure of inexpensive retail and personal services. Often in the path of urban renewal and speculative real estate investment, Skid Rows have increasingly been replaced by higher-value land uses. What remain rooted in Skid Rows (and unwanted by more affluent neighborhoods) are the human service agencies: missions, soup kitchens, drug and alcohol treatment centers, and shelters for the traditional Skid Row population.

Second, the population of Skid Rows has shifted from "old" homeless to the so-called "new" homeless, who are the latest wave of destitute in-migrants to these districts. In contrast to older white, male alcoholics typical of post-war Skid Rows, the new homeless population is characterized by its diversity. It includes young people and people of color, deinstitutionalized mentally disabled, women, families with children, and economically dislocated persons. Unable to obtain needed support services in their own neighborhoods, indigent people drift to Skid Row drawn by its historic role as provider of last-rung housing and emergency food and clothing. Skid Row human services have grown in number and scope over the past decade, to meet the needs of the "new" Skid Row homeless.

Skid Row as a Place

Skid Row, Los Angeles is east of the central business district, in the shadow of sleek postmodern office towers, glitzy retail complexes, a teeming Latino shopping district, and City Hall itself (Figure 1). It consists of 50 city blocks of mixed land uses: residential, commercial, institutional, and industrial. In this constrained area are infamous streets and corners ("The Nickel," "Thieves' Corner"), alleys, loading bays, and interior parking lots. As a physical environment, Skid Row

Figure 1. Makeshift shelter of a homeless person in downtown Los Angeles, with City Hall in background. Photograph by Stacy Rowe.

is a landscape of despair, hard-edged and inhospitable (Figure 2). The district is almost devoid of trees or other landscaping. The sidewalks are stained and dirty, and in some areas lined with old trash cans used as firepits. Parking lots are barren expanses surrounded by cyclone fencing or barbed wire. Many buildings are in disrepair, and some structures are not seismically sound (which often justifies demolitions). Surveillance activity is omnipresent; police vehicles cruise the streets, emanating from a police substation located on the Row, and designed according to principles of defensible space (blank walls, security entrances in smokey glass). Other social control mechanisms also operate, for instance overhead sprinklers outside many service agencies wet the sidewalk periodically and thus prevent loitering and curbside encampments. And, because poverty in Skid Row cannot be politely tucked away behind closed doors, the signs and symbols of despair and deprivation are unavoidable: used needles and syringes, cocaine pipes, and liquor bottles along with cast-off clothes, broken down cookstoves, and cardboard box-shelters.

Despite widespread demolition, Skid Row remains the region's largest concentration of single room occupancy hotels, with over 30 hotels and an estimated 6,700 hotel rooms (Hamilton *et al.* 1987). Increasingly, such hotels are operated by nonprofit organizations, rehabilitated, and let at subsidized rates to recipients of county-subsidized hotel vouchers or public welfare. In addition to hotels there are low-rent apartments, and nine missions and emergency shelters with a total of 2,000 beds, and social service agencies (nine day centers; nine food or clothing outlets; one legal aid center; nine employment and day labor centers; seven substance abuse agencies; and three mental health agencies). Two parks, once meeting grounds for drug dealers and users, have been redesigned and landscaped; they are now intensively managed by personnel from adjacent refurbished single room occupancy hotels, and heavily used by homeless residents. Skid Row is

also a vital business and industrial district. The area is home to the region's fish-processing plants, a rapidly growing wholesale trade in imported toys and electronics, and garment wholesaling and manufacturing.

The Row is surrounded by diverse central city zones, each growing and encroaching upon Skid Row. To the north is Little Tokyo, original locus of Japanese immigrant settlement now transformed into a zone of tourism and luxury hotels, largely (and ironically) as a result of offshore Japanese property investment. To the west is the Broadway Street corridor, the busiest Latino retail district in the region. And, to the south, lies a diverse area of industry and wholesaling activities, including clothing jobbers, wholesale produce and flower markets, printing and publishing houses, and (in greatest numbers) garment sweatshops. The result of this many-pronged land-use pressure has been a loss of low-cost apartments and single room occupancy hotels, and conversion of sites into parking lots for speculation. This growth has led to conflict over the siting and expansion of social services for the homeless.

Skid Row has also been powerfully shaped by local policy. The Community Redevelopment Agency of the City of Los Angeles has long imposed a policy of "containment" on Skid Row. Thinly disguised beneath a rhetoric of community development, this policy aimed to segregate and isolate the most marginalized urban residents and thus prevent a spread of blight to other neighborhoods. Until the explosion of homelessness during the 1980s, this kept other parts of Los Angeles relatively free of indigent people and supportive human services. But containment is unraveling as homelessness spills out across the city, to affect a broad mix of neighborhoods. Nonetheless, partly because of a moratorium on single room occupancy hotel demolitions in Skid Row, the district hosts the largest concentration of homeless people in the metropolitan region.

Skid Row as Community

The Skid Row community has several components, some of them antagonistic, others mutually supportive. There are two residential groups. One consists of homed (or domiciled) residents, who totaled almost 9,000 in 1980, and had grown to 11,000 by the late 1980s (Hamilton *et al.* 1987). Homed residents include low-skilled workers in downtown industries, for example, fish packing, garment manufacturing, services, retailing, and increasingly, artists living and working in the district's low-cost loft space. Since 1960, the proportion of young homed residents grew, as did the proportion of minorities. Blacks and hispanics constitute two thirds of the homed population in 1980. Although the Row

Figure 2. Towne Avenue, Skid Row. Photograph by Stacy Rowe.

population remains primarily male, there has been a slow increase in the number of women. Poverty is severe; more than half of all homed residents had incomes below the poverty line (U.S. Census 1980).

A second resident group is the homeless. Estimates of the homeless population vary widely, ranging from 500 to 4,000 judged to be without shelter on any given night, to several thousand who are living in temporary or unstable housing arrangements (Hamilton *et al.* 1987). Recent studies reveal a population profile not dissimilar from that of the homed group. The homeless are dominated by younger, minority males, although the number of families with children has increased (especially single women with children). Substance abuse is prevalent, and as many as one third suffer a chronic mental disability (Farr *et al.* 1986; Takahashi *et al.* 1989).

In addition to residents, there is a significant daytime population of human service providers and workers in the Skid Row business sector. Approximately 17,000 people worked in the area in 1980. Most worked in manufacturing industries, and the largest share was employed as operators, fabricators, and laborers (Wolch *et al.* 1989). Some workers are employed by local social services organizations and residential hotels.

Despite the common perception of Skid Row as a transient no-man's land, residents and workers together clearly constitute an urban community. As such, the area is characterized by social networks linking homed and homeless residents; connecting residents to service providers and the business sector; and tying service providers to the business community. These community networks can provide finances, food, transportation, and psychological support. Networks can also be damaging, for example, abusive spousal relationships, conflict between the homeless and hostile business owners, and victimization of residents and workers through crime.

Supportive Skid Row community networks are crucial to the welfare of residents and workers, for several reasons. First, to survive, many indigent people (especially the homeless) rely on reciprocity and sharing among friends. They also rely on employment provided by local businesses, and on the human service sector. These are not one-way relationships of dependency, however. Service providers "need" the homeless to rationalize their existence and expansion; some businesses rely on workers and residents as customers, and on residents for casual labor; and workers' ties with residents and police afford them a crucial margin of safety as they negotiate the journey from home to work through the Row. Second, the service-provider community of Skid Row is heavily interconnected and interdependent, although many differ in service phi-

losophy and compete for resources. Not only do members of this community rely on one another to coordinate and deliver services to clients, but their networks are critical to political efforts on behalf of clients, program initiatives, and personal agendas. Third, whether positive or negative, local social networks structure daily activity paths of Skid Row residents, creating routine in their everyday lives, and shaping their personal identities and self-esteem. Poverty and homelessness necessitate a daily path driven by reliance on formal and informal social supports, experienced predominantly in the deteriorated environment of Skid Row. In turn, the centrality of social support to survival, the dependency relations involved in obtaining support, and the dangers and stigma of living in Skid Row all work together to influence how residents feel about and define themselves as individuals.

The role of Skid Row community networks has been explored among the homeless (Rowe and Wolch 1990). The homeless typically have social supports that are predominantly community-based. These networks consist of homed and peer elements. The homed portion includes remnants of the social network from prior domiciled) periods; panhandling "clients" or donors; workmates in casual labor; social workers and other service providers. The peer component, in contrast, includes other homeless people: friends and family; homeless lovers or spouses; informal communities based in street encampments; and members of homeless political organizations (Figure 3). Community networks of the Skid Row homeless serve to bring continuity to their daily activity paths, as they seek out the same people at the same general places on an everyday basis. Moreover, social relationships can substitute for what would be fixed stations in the daily path of a domiciled person, such as home or work. For example, one homeless woman leaves her husband with all their belongings at their sleeping area, allowing her to depart

Figure 3. Members of homeless street encampment sharing cooking tasks. Photograph by Stacy Rowe.

and secure resources for survival. At day's end, she returns to the security and protection of her husband. Over time, their sleeping place varies but her day begins and ends wherever her husband is staying. Similarly, homeless people who panhandle often see this activity as analogous to a job, with set hours and habitual donors or "clients," although the panhandling spot may change periodically due to police interventions.

In general, community networks and daily paths in Skid Row affect the identities and self-esteem of the area's homeless population. Social ties may have both positive and negative effects on self-definition and morale. A "self-as-homeless" identity may be readily adopted, if the experience of homelessness creates a clearly defined role, recognition (as a leader or nurturing figure, for instance), notoriety, or other forms of attention previously unavailable to the person. Devastating effects on identity and self-esteem are perhaps more common, however. The immediacy of survival needs, precariousness of social resources, and threatening environment of Skid Row, can lead homeless people to postpone strategies to improve their life chances and to resign themselves to a negative "self-as-homeless" identity, deteriorating self-esteem, and hopelessness. Nevertheless, the support provided to the homeless by their community networks may parallel shelter itself in its effect on their quality of life.

The Future of Skid Row

What do the history and contemporary dynamics of Skid Row, U.S.A., tell us about its future? If trends of the postwar period continue, a safe prediction is that these districts will slowly vanish, to be replaced as office towers, condominiums, and retail services spill over from expanding central business districts. Such an obliteration of Skid Row would be a major loss, despite the harshness of the place. The demolition or conversion of Skid Row implies a loss of the lowest cost urban housing, single room occupancy hotels. It would force many homeless people to scatter throughout the region in search of shelter or sleeping places. Moreover, the fragile yet vital community resources and networks of Skid Row would be lost, leaving many homeless and indigent people without accessible human services. Nevertheless, the destruction of Skid Row has, in fact, already begun in many American cities.

Los Angeles' containment policy suggests an alternative to Skid Row demolition. Under such a policy, Skid Row might develop as a regional ghetto for the homeless and service-dependent. To prevent the real or perceived negative spillovers of Skid Row from affecting other neighborhoods, social service providers and shelters would remain in Skid Row, and new resources would be sited in Skid Row. Such a regional ghetto for the most economically and socially marginalized relieves more affluent neighborhoods from their obligations to help, while it imposes distinct costs on the Skid Row community. Not only does this approach create hardships for the business sector, but it warehouses the dependent and vulnerable. Their continued spatial segregation and isolation facilitates neither upward mobility nor an exit from homelessness or poverty; rather, it can reinforce coping problems and severely restrict opportunities for personal growth, education, and employment.

Harkening back to the heyday of Skid Row, a third alternative would enable these districts to regain their early 20th-century role as low-income but supportive urban neighborhoods. In this scenario, the district would boast refurbished single room occupancy hotels, low-rent apartments or tenant cooperatives; adequate parks and other public facilities; a wide variety of social services; and employment opportunities either in local businesses or sheltered workshop and training facilities. But unlike the ghettoization scenario described above, a revitalized Skid Row would preferably not be the sole locus of human services and shelter in the region. Rather, Skid Row would be only one node in a network of "shelter/service hubs" or places where newly homeless or service-dependent persons requiring longer-term support might choose to live, receive needed social welfare services, and establish supportive community networks.

In many cities, Skid Row districts show signs of all three scenarios. Skid Rows continue to shrink, as land-use pressures mount. While the homeless population has decentralized to some degree, Skid Rows retain their regional primacy as centers of homeless population and emergency shelter and services. And the potential value of Skid Rows as urban places can be discerned, as single room occupancy hotels and parks are slowly renovated, social service providers augment their support roles, and community networks expand. But ultimately, these futures conflict. Skid Row shrinkage precludes a role as regional homeless ghetto; ghettoization severely hinders chances for neighborhood revitalization. Since any one alternative imposes a different pattern of costs and benefits, proposed plans for Skid Row are hotly debated but seldom resolved.

Acknowledgments

I thank the National Science Foundation, Program in Geography and Regional Science, for research support. I also thank Michael Dear and Stacy Rowe for their comments on an initial version of this chapter and I gratefully acknowledge Stacy for the photographs.

References

Bogue, D. 1963. *Skid Row in American Cities.* Chicago: University of Chicago Press.

Dear, M. J., and Wolch, J. R. 1987. *Landscapes of Despair: From Deinstitutionalization to Homelessness.* Princeton, NJ: Princeton University Press.

Farr, R., Koegel, P., and Burnam, A. 1986. *A Study of Homelessness and Mental Illness in the Skid Row Area of Los Angeles.* Los Angeles: Los Angeles County Department of Mental Health.

Hamilton, Rabinovitz, and Alschuler, Inc. 1987. *The Changing Face of Misery: Los Angeles' Skid Row Area in Transition.* Los Angeles: Community Redevelopment Agency of the City of Los Angeles.

Hoch, C., and Slayton, R. 1989. *New Homeless and Old: Community and the Skid Row Hotel.* Philadelphia: Temple University Press.

Rowe, S., and Wolch, J. R. 1990. Social networks in time and space: Homeless women in Skid Row, Los Angeles. *Annals of the Association of American Geographers* 80(2):184-204.

Takahashi, L., Dear, M. J., and Neely, M. 1989. *Characteristics of the Homeless Population in Downtown Los Angeles, 1988-1989,* Working Paper # 23. Los Angeles: Los Angeles Homelessness Project, University of Southern California.

U.S. Census. 1980. *Census of Population.* Washington, DC: U.S. Government Printing Office.

Wallace, S. 1965. *Skid Row as a Way of Life.* Totowa, NJ: Bedminster Press.

Wolch, J. R., Law, R., and Wright, M. 1989. *The Employment and Residential Location of the Low-Skill Workforce in Downtown Los Angeles,* Working Paper # 18. Los Angeles: Los Angeles Homelessness Project, University of Southern California.

Zorbaugh, H. 1929. *The Gold Coast and the Slum.* Chicago: University of Chicago Press.

Strolling the Strip: Prostitution in a North American City

Rita Riccio
San Diego State University

Streetwalkers are no longer confined to "red-light districts" or to other specially designated zones in cities. Today women are free to choose their own solicitation sites. However, the determination of where to solicit is not random. There are still many social and spatial constraints upon the location of prostitution within cities.

As found in other geographic research on crime, the spatial distribution of visible prostitution (streetwalkers) is dynamic and is moving away from the downtown areas as a result of central city change (Jeffery 1977; Newman 1973). Other studies note the movement of criminal activities to the suburbs (Rengert and Wasilichick 1985; Reppetto 1974). For visible prostitution, many streetwalkers are relocating along major transportation arteries or suburban commercial strips. This paper focuses on the changing geography of visible prostitution and the role that modern transportation networks play in its distribution.

Prostitution is not ubiquitous throughout cities. Rather, it is usually found close to potential customers. Since many residents perceive prostitution to be a negative indicator of community status and associate its presence with other forms of crime, it is not tolerated in certain communities. This is particularly true of visible prostitution, the most ambulatory and obtrusive form of soliciting (Stopp 1978).

The persistency of prostitution in cities reflects the role of women in our society and gender inequalities (Rasheed 1985). Thus, a large disparity exists between the number of streetwalker arrests and the number of male patron arrests (San Diego Police Arrest Records 1986-1987). Most experts relate the existence of prostitution to poverty (Addams 1972; Sanger 1898). Yet, according to Symanski (1981), the social and economic

inequalities of women in a system controlled by men are the primary reasons for prostitution's existence. Prostitution, he says, is "deemed immoral by men who make it illegal but nevertheless patronize prostitutes" (Symanski 1981, 1).

Transportation and Visible Prostitution

Streetwalkers tend to work in three types of urban communities: old downtown areas, low-income minority neighborhoods, and commercial strips. These community types are considered de facto zones of prostitution, with women working and residing within (or close to) them (San Diego Police Arrest Records). But, these "zones" are not separate and distinct, rather they are linked by transportation arteries, enabling streetwalkers and customers to travel conveniently between "zones."

Streetwalkers use their mobility as a defense against arrests. As officials conduct "sweeps" or attempt to "clean-up" one part of the city, the women temporarily migrate to other areas. Multiple "zones" allow prostitutes to effectively time traffic flows in each neighborhood to maximize their visibility to patrons. Thus, due to the elusive and flexible distribution of streetwalkers, visible prostitution has spread throughout cities and is difficult to eradicate.

The automobile has also changed the way streetwalkers solicit in North American cities. In the past, prostitutes were called streetwalkers because they were pedestrians. Today they still solicit on the streets, but to customers in cars. Much of the soliciting occurs along commercial strips. The strip offers prostitutes a variety of advantages over other types of urban environments. It provides slow-moving, stop-and-go traffic and an ample supply of potential customers.

The commercial strip was designed to provide maximum accessibility and maneuverability for drivers. Many commercial strips were the main roads in and out of town before freeway development, and hosted small roadside motels. Some of these old motels still dot the landscape of commercial strips and help facilitate visible prostitution. Furthermore, the types of strip businesses and lack of residential land uses creates an anonymous landscape that invites deviant and criminal behavior (Ley and Cybriwsky 1974; Newman 1973).

Visible Prostitution in San Diego

Some North American cities are more attractive to streetwalkers than others. For instance, San Diego, with its strong military presence, and year-round tourists attracted by its Mediterranean climate, offers an ideal market for prostitution. In 1987, more than 350 women were arrested for soliciting, an increase over previous years (San Diego Police Arrest Records). Arrest records show three areas used by streetwalkers: downtown (the old "red-light district"), southeast San Diego (a low-income minority neighborhood), and El Cajon Boulevard (a commercial strip). Arrest records indicate a linear pattern that follows transportation routes connecting these three nodes (Figure 1).

In downtown San Diego, there were over 90 arrests in 1987, most clustered around Market Street, a four-lane thoroughfare of old apartment complexes, run-down hotels, and thrift shops. This street runs from

east to west and is home to transient workers, street people, and runaways. The majority of solicitations occurred in an area of downtown San Diego close to the former red-light district, the Stingaree. However, as urban revitalization changes the face of downtown San Diego, prostitution in the city is displaced.

This part of downtown is becoming the center for new investments and urban housing. Old waterfront establishments, such as topless bars and massage parlors, are being demolished and replaced with sleek office towers, new condominiums, and jazz clubs. Yet some of the old hotels and "adult" bookstores off Market Street remain despite nearby renovation. Eventually, as the economic effects from a recently opened convention center are felt, the traditional clientele of streetwalkers will move out of downtown and the streetwalkers will follow. Pushed out of the core, many streetwalkers migrate east, into Southeast San Diego. Most of the arrests in Southeast San Diego occurred on 32nd Street in a low-income minority community or along Main Street near the Naval Station (Figure 1). The arrests near 32nd Street and Imperial Avenue are unusual because they are in a residential neighborhood. Many of the women working in this part of the city are not full-time, professional prostitutes, rather they are "part-timers"—a local woman who occasionally solicits to supplement her income.

The professional streetwalkers arrested in this part of the city work south of 32nd Street on Main Street. Along Main Street are warehouses, auto-related businesses, and industrial complexes. More important to streetwalkers are the male employees at the National Steel and Shipping Yards, and the 32nd Street Naval Station. Together, Main Street and the minority neighborhood near 32nd and Imperial Avenue accounted for 63 arrests. Obviously, the majority of professional streetwalkers in San Diego spend most of their time working elsewhere.

More women (over 220) are arrested for prostitution along El Cajon Boulevard than anywhere else in San Diego. El Cajon Boulevard is a major traffic artery, extending 3.7 kilometers (6 miles) from downtown to the eastern suburbs. Many arrests occur along a few small sections of this strip. These "hot spots" along the boulevard are near streetwalkers' residences or landscape characteristics favorable for business (Riccio 1989; San Diego Police Arrest Records 1986-1987).

Like most commercial strips, El Cajon Boulevard has a drab commercial milieu, including bars, auto-shops, car stereo stores, pizza parlors, and old motels. The boulevard has elements that make it a potentially useful workplace for streetwalkers and their patrons: short blocks and numerous alleys for easy turns, left-hand turning lanes, and a wide straight street for

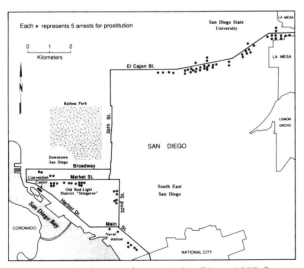

Figure 1. The distribution of arrests in San Diego, 1987. Source: San Diego Police Arrest Records.

unobstructed views of the sidewalk. Landscape features, such as empty lots, bus stops, and gas-stations, provide dual functions for streetwalkers. Customers can pull-over and pick-up prostitutes without interrupting the flow of traffic. Plus, the vacant lots associated with car washes and auto-stores are sometimes an alternative to motel rooms since some women prefer to work out of cars rather than rent motel rooms (Traitel 1986).

Conclusion

Central-city revitalization, the physical layout of transportation routes, and automobiles are major components in the changing distribution of visible prostitution in North American cities. As city residents move to the suburbs, prostitution follows. As urban cores become more expensive and fashionable, streetwalkers solicit elsewhere. The commercial strip is quickly replacing downtown as the focus of visible prostitution in cities. San Diego's El Cajon Boulevard is similar to commercial strips elsewhere in North American cities. Most show signs of deterioration as cities expand or revitalize (Kent and Dingemans 1977; McNee 1984). In Los Angeles, Hollywood Boulevard is a strip renowned for its illicit activities. Other examples are found in Phoenix, Las Vegas, New York City, and Denver.

Transportation corridors and automobiles connect the areas worked by streetwalkers, enabling them to easily relocate and migrate between neighborhoods. While in the past, prostitution was contained in one neighborhood, today streetwalkers may work in several communities throughout the city depending on the police enforcement patterns or the time of the day. Thus, prostitutes are better able to resist police efforts and to persist in North American cities.

References

Addams, J. 1972. *A New Conscience and an Ancient Evil.* New York: Arno Press and The New York Times.

Jeffery, C. 1977. *Crime Prevention through Environmental Design.* Beverly Hills, CA: Sage.

Kent, R., and Dingemans, D. 1977. Prostitution and the police: Patrolling the stroll in Sacramento. *Police Chief* 44(9):64-73.

Ley, D., and Cybriwsky, R. 1974. The spatial ecology of stripped cars. *Environment and Behavior* 6:53-68.

McNee, R. 1984. If you are squeamish. *The East Lake Geographer* 19:16-27.

Newman, O. 1973. *Defensible Space: Crime Prevention through Environmental Design.* New York: Macmillan.

Rasheed, J. 1985. Red-lights and the heavy shadows of hypocrisy. *The Far East Economic Review* 127:56-57.

Reppetto, T. 1974. *Residential Crime.* Cambridge, MA: Ballinger.

Rengert, G. F., and Wasilichick, J. 1985. *Suburban Burglary.* Springfield, IL: Charles C. Thomas.

Riccio, R. 1989. *The Landscape of Visible Prostitution.* M. A. Thesis, San Diego State University.

Sanger, W. 1898. *The History of Prostitution: Its Extent, Cause and Effects throughout the World.* New York: Medical Publishing.

Stopp, H. 1978. The distribution of massage parlors in the nation's capitol. *Journal of Popular Culture* 11:989-997.

Symanski, R. 1981. *The Immoral Landscape: Female Prostitution in Western Societies.* Toronto: Butterworths.

Traitel, D. 1986. Chula Vista Prostitutes using lots as motels businesses say. *San Diego Tribune* 21 August:B3.

Technological Failures and Toxic Monuments

Susan L. Cutter
Rutgers University

Toxic monuments evoke some past environmental desecration that lingers in our collective environmental memories. They are the environmental equivalent of historic sites or monuments. But, instead of commemorating a historic battle (Gettysburg), or national political tragedy (Wounded Knee), toxic monuments commemorate failures in technological systems. Their names conjure up popular images of vast environmental destruction, catastrophes, or human endangerment. While toxic monuments have not achieved the status, by and large, as popular tourist destinations or necromantic shrines such as Chernobyl (Clines 1991), their notoriety is very much a part of the contemporary landscape of North America.

Toxic monuments are specific locales (points) where some major environmental catastrophe occurred. The event rendered many of these places either temporarily or permanently unfit for habitation, or so fundamentally altered our perceptions of technological safety that major policy innovations were implemented. These watershed events, coupled with intense media and scientific attention to the catastrophe, produce toxic monument status. While pollution, waste disposal, and technological failures occur quite frequently, it is their uniqueness, the extent of degradation, and their politicization by the media, government, and the environmental community that elevates a normal accident to monument status.

Some of the most prominent technological failures that produced toxic monuments occurred outside North America. International examples arising from the civilian use of nuclear power abound. Chernobyl is the preeminent example. A 1957 fire in the reactor core made Windscale a British toxic monument. While there were no fatalities, radioactive iodine (^{131}I) was released within a 320-kilometer radius of the plant and detected in London, 480 kilometers away. Toxic monuments involving the military applications of radiation include Hiroshima and Nagasaki. In the hazardous waste class, a dioxin leak in Seveso, Italy, and the methyl isocyanate toxic cloud in Bhopal, India, provide good examples. Polluted wastelands such as the mercury contamination in Minimata, Japan, and the more recent oil spill in the Persian Gulf show that monuments can be found in most locations. Copsa Mica, Romania, provides evidence of chronic air pollution so pervasive that buildings are blackened from the soot and the natural landscape is virtually lifeless.

Classifying Hazards

Christoph Hohenemser and colleagues (1985) give a useful taxonomy of technological hazards using the concepts of hazard (threat to humans and what they value) and risk (quantitative measure of the consequences of hazards). Parameters that relate to the release of materials or energy such as its spatial extent, concentration, persistence, and recurrence are useful descriptors. We can also examine monuments based on the biological and human consequences of exposure, such as the population at risk, acute versus chronic effects, mortality, or transgenerational effects. In addition to examining monuments based on hazard and risk parameters, we can also describe their etiology, or agent of toxicity. Here we examine three different classes of monuments: radiation, hazardous waste, and pollution.

A discussion of hazards and the monuments they produce by category illustrates some of the commonalities in the failures of technological systems that result in the formation of hazardous landscapes. There may also be similarities in the public's underlying attitudes toward the technology or agent in question

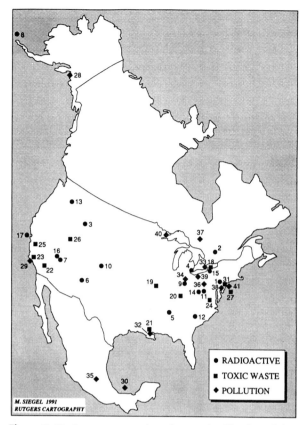

Figure 1. Toxic monuments based on a classification of their causal agent. Number codes locate sites discussed in this chapter.

thus affecting the level of concern and regulation (Zeigler *et al.* 1983). A brief spatial analysis of categories highlights the areal extent of hazardous landscapes and the concentration of the larger monuments in certain geographic regions. For example, North American monuments (Figure 1) are highly concentrated in the Great Lakes region, the mid-Atlantic region of the United States, and the American West.

The Making of Monuments

Radiation

Three Mile Island (Figure 1; site 1) is perhaps the best known North American toxic monument. The 28 March 1979 accident was noted more for its political impact on the nuclear industry than for the actual amount of radioactivity released, which was very small. Three Mile Island served as a watershed event in the development of commercial nuclear power in the United States by focusing on the potential consequences of a total core meltdown in the course of normal

accidents. But, Three Mile Island was not the first instance of a core meltdown on the continent.

On 12 December 1952, the NRX reactor in Chalk River, Ontario (site 2), exploded when uranium mixed with water. There were no fatalities but many people were exposed to radioactivity. Eight years later, another incident occurred in a different reactor at the same site. This time a fire in the fuel rods initiated a loss-of-coolant-water accident that spilled contaminated water in the reactor building. No one was killed, but it took three months to mop up the site.

Arco, Idaho, current home of the Idaho National Engineering Lab, is another toxic legacy (site 3). Human and mechanical errors resulted in a fuel-core meltdown in an experimental fast-breeder reactor in November 1955. A second fuel-core meltdown and explosion occurred in 1961, resulting in three fatalities. Farther east, another core melted down in 1966, this time at the Enrico Fermi reactor in Lagoona Beach, Michigan (site 4). By 1972, after operating only 30 days in its 6-year history, the Fermi plant was permanently encased, prompting a book and a song about the day "we almost lost Detroit" (Fuller 1975). Finally, a candle flame used to check the air flow in a cable room beneath the reactor control room started a loss-of-coolant accident at the Browns Ferry nuclear plant in Decatur, Alabama (site 5). At the time, this type of accident was thought to be a highly improbable event (1 chance in 1 billion reactor-years).

The military use of nuclear technology has created a number of monuments, including aboveground testing sites such as Los Alamos (site 6), the Nevada Test Range (site 7), and Amchitka on the Alaskan peninsula (site 8). These monuments illustrate the chronic and long-lasting effects of radioactivity. When compared with the sudden and shorter-lived radioactive releases often found in reactor failures, nuclear test sites cover greater expanses and the resultant nuclear landscapes (Goin 1991) become a more durable reminder of the generational consequences of nuclear proliferation. Weapons fuel (uranium) processing at the Fernald facility outside Cincinnati, Ohio (site 9), has a long history of environmental contamination of both land and air. Likewise, Rocky Flats, Colorado (site 10) (plutonium manufacturing); Oak Ridge, Tennessee (site 11) (enriched uranium production); Savannah River, South Carolina (site 12) (plutonium and weapons manufacturing); and Hanford, Washington (site 13) (plutonium production, processing, and high-level waste storage), are unequaled in the level of chronic contamination and spatial extent of degradation.

Low-level radioactive waste repositories—Maxey Flats, Kentucky (site 14); West Valley, New York (site 15)—are included, since they are now closed and in a caretaker status. Yucca Mountain (site 16), Nevada, the proposed, but now unlikely high-level waste repository, qualifies as a likely monument even before any waste is buried there because of the enormous controversy surrounding the project and the need to build containment structures that will last centuries. One interesting aspect of this project is an ongoing attempt to develop futuristic languages or universal symbolisms to warn of the danger 10,000 years from now. This is three times longer than the Sphinx has existed aboveground and we still do not know what it was for. After World War II, radioactive materials from Los Alamos and other labs were dumped near the Farallone Islands (site 17) off the northern California coast. Not until 1981, after the designation of the Gulf of the Farallones Marine Sanctuary, was this toxic monument rediscovered.

Hazardous Wastes

Chemophobia is rampant in North American society and is best illustrated by the public's concern over the uses of chemicals and the disposal of hazardous substances. As new dump sites were found, fears and anguish over chemical contamination increased. With respect to hazardous waste monuments, Love Canal in Niagara Falls, New York (site 18), and Times Beach just outside St. Louis, Missouri (site 19), are the best examples. The human abandonment of Love Canal with its highly toxic soil and water occurred as a result of decades of chemical waste dumping in the community. This chronic example is contrasted to Times Beach, which was created through the inadvertent use of dioxin-contaminated sediments as fill and by dioxin-contaminated oil on gravel roads to keep the dust down. Times Beach, long abandoned, is no longer shown on new maps of Missouri. Love Canal, on the other hand, persists as a community in Niagara Falls, as remediation efforts continue in parts of the community. Other notable toxic dump monuments include Kentucky's Valley of Drums (site 20), 48 kilometers south of Louisville, an illegal waste dump littered with thousands of barrels, and Devil's Swamp, near Baton Rouge, Louisiana (site 21), home to toxic sludge generated by the petrochemical facilities and their waste disposal contractors.

Rural areas have just as many monuments as the more industrialized regions. The Stringfellow Quarry Waste Pit (site 22) in the Jurupa Mountains just outside Riverside, California, is a good example. This site was well-known for its 121 million liters of assorted acids, heavy metals, organic solvents, and pesticides. "Acidgate" as it was locally known, was the preferred disposal

option for such regional manufacturing luminaries as General Electric, Hughes Aircraft, and Philco-Ford (Brown 1977). In a turn of events, a federally approved toxic waste site in Casmalia, California (site 23), just north of Santa Barbara, makes the list of monuments. Federal Superfund money pays Casmalia Resources to receive wastes from the Stringfellow Acid Pits site as part of the latter's clean-up, thus transferring the waste from one part of California to another and creating two monuments in the process.

Many other examples of chemical contamination help highlight its role as an initiating agent in creating monuments. For example, the James River in Virginia (site 24) was severely contaminated by the pesticide Kepone that poisoned workers and killed aquatic ecosystems in the river. Wildlife was seriously threatened, particularly birds, when the Kesterson National Wildlife Refuge in California (site 25) became contaminated with selenium leached from the runoff from the agricultural fields in the fertile San Joaquin Valley.

The Dugway Proving Ground near Toole, Utah (site 26), is another western monument noted for its role as a U.S. Army Chemical Corps testing station for chemical weapons (Lawless 1977). In 1968 nearly 6,000 sheep grazing in adjacent valleys were killed by nerve gas that drifted offsite. The 1970 disposal of World War II chemical and biological weapons into the Atlantic Ocean, 400 kilometers east of coastal New Jersey (site 27) provides yet another example. No specific name evokes an emotional response— perhaps the Garden State Trench might do.

Pollution

Pollution monuments are more ephemeral. They reflect major disasters at the time, but their effects are relatively short-lived when compared with the radioactive and hazardous waste events, yet their spatial extent is often greater. The *Exxon Valdez* spill in Prince William Sound, Alaska (site 28); the Santa Barbara Channel oil spill in 1969 (site 29); and the 1979 blow-out of the IXTOC-1 oil well in the Bay of Campeche, Mexico (site 30), illustrate major oil pollution episodes. IXTOC-1, the Mexican-owned well, released more than 2.7 million barrels of oil into the Gulf of Mexico. The spill caused an international dispute when oil washed ashore on Texas beaches, destroying fishing areas, wildlife, and recreational facilities.

Toxic clouds are another group of events that temporarily create toxic monuments (Brown 1987). A 1980 toxic cloud in Elizabeth, New Jersey, resulted from a fire at Chemical Control Corporation (site 31), a hazardous waste firm. After the fire, thousands of liters of hazardous waste remained on the land in addition to those flushed into the Arthur Kill from the fire-fighting efforts and subsequent rains. Other notable toxic clouds include Taft, Louisiana (site 32), a 1982 acrolein gas explosion that forced the evacuation of 17,000 people; Mississauga, Ontario (site 33), a 1979 train derailment that released chlorine and prompted an evacuation of 250,000 people for a number of days (Cutter 1987); and Miamisburg, Ohio (site 34), a 1986 white phosphorus toxic cloud forcing 40,000 to evacuate for days. While still etched in many people's minds, these toxic clouds were not nearly so lethal as the one that occurred at Bhopal, yet they are worrisome reminders of our technologically-advanced society.

Other incidents involving air pollution include the natural gas explosion and fire at St. Juan Ixhuatepec, Mexico (site 35), in 1984 that resulted in 500 deaths, 2,500 injuries, and total destruction within a seven-block area of the facility. Perhaps the most enduring air pollution monument is the first major air pollution episode in the United States in Donora, Pennsylvania (site 36), which killed 20 people. The continuing contribution of the smelters in Sudbury, Ontario (site 37), to regional and international problems with acid precipitation is one of Canada's entries as a pollution monument.

Water pollution monuments include the Arthur Kill (site 38), which connects Newark Bay to Raritan Bay and forms the border between New Jersey and Staten Island, New York. It is one of the nation's most polluted waterways, mostly from heavily industrialized runoff coupled with almost routine oil spills. The severe pollution of the Cuyahoga River, in Cleveland, Ohio (site 39), which periodically caught on fire during the 1950s and 1960s, is memorable as a monument, even though it is less polluted today. Pollution of Silver Bay, Minnesota (site 40), by the local taconite mines is another classic example. Finally, the "dead zone" in the New York Bight (site 41) provides the most poignant reminder of decades of wasteful practices. Years of ocean dumping by the New York metropolitan region plus waste carried by the Hudson River, rendered this part of the Atlantic Ocean devoid of all marine life.

Toxic Future

Based on the characteristics of hazards, we can distinguish among different types of monuments and their level of environmental impact. For example, radiation is clearly a feared technology with many aspects of the nuclear fuel cycle still unresolved. The myopia in the progression from a military to a civilian nuclear program without adequate safety and disposal programs in place, produced a series of toxic monuments. Only three of the monuments in this chapter involve failures in commercial reactors; the rest are by-products of waste disposal and the aging of facilities that produce weapons and nuclear fuel. The commonalities in this category include the duration of impacts, the fear that radiation evokes from the public, and the need for perpetual safeguards of both the technology and the monuments that are produced.

Chemical wastes are artifacts from industrial processes that viewed environmental contamination as a necessary by-product of progress. Until the health effects of chemical contamination became apparent, indiscriminate hazardous waste dumping prevailed. While these monuments have a temporal dimension as well, it is possible to rid most areas of contamination in years rather than centuries.

Pollution monuments, are the least pervasive temporally, but the areal extent of damage can be as large as the other monument categories. Once the source of the pollution is eliminated, the natural system will eventually recover. These types of monuments are also less deliberate; stemming from an accident rather than the willful dumping of hazardous waste or the testing of atomic weaponry.

Using the parameters of spatial extent, concentration, persistence, population at risk (human and non-human), and delayed or chronic consequences, my vote for the top five monuments (in no particular order) are Hanford (site 13), Savannah River (site 12), Times Beach (site 19), New York Bight (site 41), and the *Exxon Valdez* spill (site 28). These points remind us of our toxic past. But what about our toxic future?

We can only speculate on the creation of future monuments. Governmental policies, embodied in such legislation as the Superfund program, virtually ensure the consolidation of hazardous waste into "approved" locales, thereby intentionally creating monuments. Since the American military is largely exempt from environmental review, it is only a matter of time before we see more military toxic monuments, especially in light of the certain closure of many military bases and weapons-production facilities. Industrial policies, such as the creation of buffer zones around chemical facilities, and the physical removal of residents from existing neighborhoods, could result in monument formation. The coalescence of these points into regions beckons for the development of toxic parks. Places such as Canada's Chemical Valley (Sarnia, Ontario), its American counterpart, Chemical Alley (the Kanawha Valley, in West Virginia), the Golden Triangle in Texas (Port Arthur, Orange, Beaumont), the Toxic

Lowlands of Louisiana, and the Toxic Middens (Hemlock, Midland and Montague, Michigan) all vie for our attention.

References

Brown, Michael. 1977. *Laying Waste.* New York: Pantheon.
_____ . 1987. *The Toxic Cloud.* New York: Harper.
Clines, Francis X. 1991. Chernobyl beckons tourists to visit the radiation zone. *The New York Times* 4 Feb:A1.
Cutter, Susan L. 1987. Airborne toxic releases: Are communities prepared? *Environment* 29(6):12-17,28-31.

Fuller, John G. 1975. *We Almost Lost Detroit.* New York: Reader's Digest Press.
Goin, Peter. 1991. *Nuclear Landscapes.* Baltimore: Johns Hopkins University Press.
Hohenemser, Christoph, Kates, Robert W., and Slovic, Paul. 1985. A causal taxonomy. In *Perilous Progress: Managing the Hazards of Technology,* eds. R. W. Kates, C. Hohenemser, and J. X. Kasperson, pp. 67-89. Boulder, CO: Westview Press.
Lawless, Edward W. 1977. *Technology and Social Shock.* New York: Simon and Schuster.
Zeigler, Donald J., Johnson, James H. Jr., and Brunn, Stanley D. 1983. *Technological Hazards.* Washington, DC: Association of American Geographers Resource Publication.

Hydroclimate and Water Quality of Lake Tahoe

Marlyn L. Shelton
University of California, Davis

Lake Tahoe is a scenic asset of national significance. The spectacular geology and topography of the watershed and the extraordinary clarity and depth of the lake combine to create a landscape of unusual beauty. The natural purity and clarity of Lake Tahoe are rivaled only by Crater Lake, Oregon, and Lake Baikal in the Soviet Union. The exceptional clarity of Lake Tahoe is attributed to low erosion rates in the watershed; soils carried to the lake by runoff are the main source of nutrients for algae growth (Goldman 1989). However, the quality and clarity of Lake Tahoe are decreasing as the construction of summer homes and year-round resort hotels and casinos disturb the natural environment, increase runoff, and accelerate nutrient flows to the lake. Enhanced nutrient loading is deteriorating Lake Tahoe water, which is losing its famous transparency and gradually greening (Goldman 1989), threatening the environmental and economic values of the region (CTC 1987). This paper examines the link between hydroclimate and the transport of nutrients from the watershed to Lake Tahoe.

The Tahoe Basin

Lake Tahoe occupies a steep-sided graben fault basin formed during the uplift of the Sierra Nevada between 25 million and 40 million years ago (Crippen and Pavelka 1970). The basin is bounded on the west by the main crest of the Sierra Nevada and on the east by the Carson Range. The surface of the lake is at an altitude of 1,886 meters, but the surrounding peaks reach heights of 3,297 meters. The entire Lake Tahoe Basin is underlain by granitic rock and about half of the surface is exposed granite.

Pleistocene glaciers covered the crest of the Sierra Nevada and extended down the west slopes of the basin. Along the east side of the watershed, the drier Carson Range displays little evidence of glaciation. Consequently, the western Tahoe Basin has highly sculptured peaks, rounded valleys, and moraines. The eastern basin is characterized by more rolling peaks and subdued topography. The southern and western portions of the basin were especially productive sources of glacially derived sediments, which accumulated to depths of 0.5 kilometer on the floor of Lake Tahoe (Hyne 1969).

The watershed covers an area of 819 square kilometers exclusive of the lake, which has a surface area of 497 square kilometers and a depth of 505 meters (Figure 1). Approximately two thirds of the basin is in California and one third is in Nevada. In its current form, Lake Tahoe is about 11,000 years old, it contains 156 cubic kilometers of water, and it is the largest alpine lake in North America (Crippen and Pavelka 1970; Goldman 1989).

Soils in the Tahoe Basin are products of weathering of either granitic or volcanic parent material. They tend to be shallow and coarse, lacking humus and other material that hold moisture and bind the soil together (TRPA 1982). The growing season is relatively short due to the watershed's high elevation, and vegetation in the basin is mixed in response to differences in topography, soils, temperature, and moisture availability (Neilson 1973). However, coniferous forest covers about 85 percent of the watershed (TRPA 1982).

Hydroclimate of the Tahoe Basin

Runoff in the Tahoe Basin displays distinctive spatial and temporal variations in response to the coupling of the basin's hydroclimate and terrain. Precipitation and evapotranspiration are dominant hydroclimatic

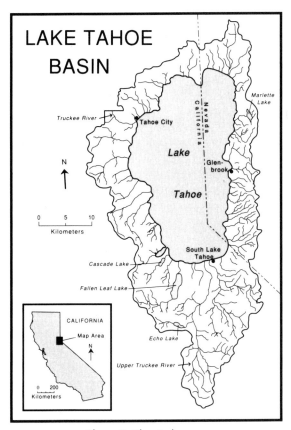

Figure 1. The study area.

temperature variations between lake level and the majority of the watershed are magnified by radiation-balance and energy-balance influences related to steep slopes and varying slope aspect. Energy loading and temperature drive the environmental demand for moisture. Potential evapotranspiration (Shelton 1978; Thornthwaite and Mather 1955) depicted in Figures 2 and 3 reflects the variation in energy loading and temperature for two stations at about the same elevation but on opposite sides of the Tahoe Basin.

Regional climatic controls impose a dry summer and a wet winter precipitation regime on the watershed. Frontal systems originating over the Pacific Ocean deliver more than 70 percent of the basin's precipitation between November and March. Topography plays an important role in the spatial distribution of precipitation and in determining whether the winter precipitation occurs as rain or snow. The western side of the basin has annual precipitation in excess of 200 centimeters, but the east side receives 65 centimeters or less. At lake level, 75 to 80 percent of the precipitation is normally snow; above 2,400 meters, 90 to 95 percent of annual precipitation is snow (TRPA 1982). The snowpack can be as deep as 550 centimeters with a

elements, and they quantitatively define natural moisture supply and the energy demand for moisture, respectively. However, they are caused by different physical processes and at any given place they are seldom the same in either amount or distribution through the year. In addition, evapotranspiration measurements are seldom available in a watershed, and temperature is the only readily available basis for estimating evapotranspiration in a watershed (Shelton 1985). The consequences of moisture supply and demand disparities on runoff are evident when monthly quantities of precipitation and evapotranspiration are compared. The climatic water balance (Thornthwaite and Mather 1955) is a proven technique for analyzing the hydroclimate of large watersheds (Gleick 1987; Shelton 1989). Figures 2 and 3 are climatic water balances for representative sites in the Tahoe Basin based on average monthly data for 1951 to 1980.

Spatial characteristics of temperature and precipitation are strongly influenced by topography in the Tahoe Basin. The high elevation of the watershed and local relief of more than 1,400 meters mask regional temperature controls that produce comparatively cool summer and cold winter temperatures. In addition,

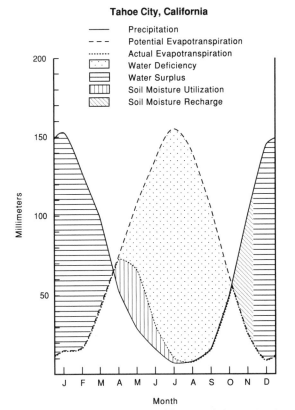

Figure 2. Tahoe City average monthly water balance, 1951 to 1980. Elevation 1,888 meters.

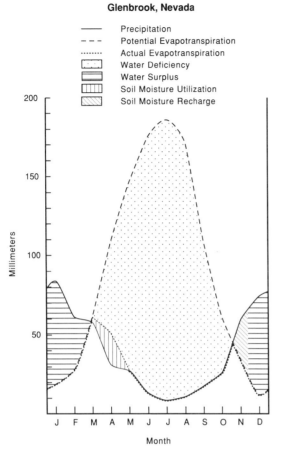

Figure 3. Glenbrook average monthly water balance, 1951 to 1980. Elevation 1,924 meters.

Seasonal Inflow to Lake Tahoe

Moisture inflow to Lake Tahoe consists of direct precipitation on the lake and groundwater and surface runoff from the watershed. These inflows occur at different times of the year: the precipitation on the lake is greatest from November to March (Figures 2 and 3), but runoff from the land area is related to snowmelt, concentrated in the late spring (Myrup et al. 1979). Crippen and Pavelka (1970) estimate that the long-term average inflow from direct precipitation on the lake is 262,000 cubic dekameters annually, which represents precipitation of about 53 centimeters on the lake surface. Annual runoff from the land area is 385,000 cubic dekameters, and 80 to 90 percent of this runoff comes from the California portion of the watershed. The contribution of groundwater flow into Lake Tahoe is poorly documented, but most groundwater likely discharges into stream channels before entering the lake due to the shallow depth of bedrock in the basin.

63 separate watersheds drain into Lake Tahoe (Figure 1). Most of the drainage areas are small and the tributary streams are relatively short, except for the Upper Truckee River, which is 34 kilometers long. The streamflow regime of all the tributary watersheds contains a strong seasonal component dictated by regional climatic controls and snowmelt. The Upper Truckee River hydrograph for 1972 to 1989 (Figure 4) shows that runoff is least in September and October and peaks

water content in excess of 100 centimeters and it is usually on the east side of the west rim of the basin (TRPA 1971).

Evapotranspiration consumes more than half of the moisture input to the watershed (Crippen and Pavelka 1970), and most of the moisture loss occurs on the west side of the basin where precipitation is abundant. Evapotranspiration would be even greater except that soil-moisture storage in the watershed is relatively low and most of the precipitation occurs during the cool months when the energy demand for moisture is low. Approximately 45 percent of the 805 millimeters of annual precipitation at Tahoe City is allocated to actual evapotranspiration (Figure 2) while 65 percent of its 473-millimeter annual precipitation is consumed by actual evapotranspiration at Glenbrook (Figure 3). Even so, the moisture deficit is nearly 1.5 and 2.6 times greater than actual evapotranspiration at Tahoe City and Glenbrook, respectively.

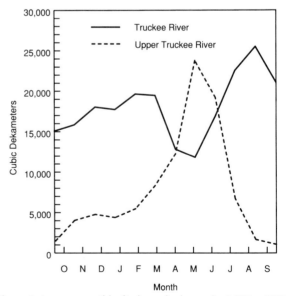

Figure 4. Average monthly discharge hydrographs, 1972 to 1989. Source: U.S. Geological Survey.

in May and June when 46 percent of the annual runoff occurs. Occasional summer thunderstorms in the watershed produce little runoff, and the peak runoff in May and June is due to melting snow that accumulates during the winter.

Fewer than 20 of the streams in the Tahoe Basin are gauged, but the Upper Truckee River and two adjacent tributary areas are estimated to yield over 40 percent of the annual runoff into Lake Tahoe (TRPA 1982). The Upper Truckee River alone contributes about 93,000 cubic dekameters or 24 percent of the annual inflow.

Water Outflow from Lake Tahoe

Water leaves Lake Tahoe by evaporation and by water discharged into the Truckee River. Evaporation from the lake surface accounts for 434,000 cubic dekameters annually, equivalent to a layer of water 88 centimeters deep over the lake surface (Crippen and Pavelka 1970). The evaporation maximum occurs in the fall months because evaporation from deep lakes, like Lake Tahoe, lags the peak occurrence of net radiation by several months (Myrup et al. 1979).

The Truckee River (Figure 1) is the lake's single natural outlet except for evaporation. Andesitic mudflows and lava form a natural dam at the outlet at Tahoe City, giving the lake a natural surface elevation of 1,886 meters (Strong 1984). The concrete outlet dam, completed in 1916, raises the surface level of the lake 1.8 meters when full (TRPA 1982). This regulation provides 888,000 cubic dekameters of additional water storage, about 0.5 percent of the total volume of Lake Tahoe. Average annual outflow through the dam into the Truckee River is 217,000 cubic dekameters.

The seasonal regime of the Truckee River for 1972 to 1989 (Figure 4) reflects the storage role of Lake Tahoe in delaying runoff and reducing the monthly variability of runoff. Peak discharge into the Truckee River occurs in August, which is one of the months of lowest discharge for the tributary watersheds like the Upper Truckee River. In contrast, May is the month of lowest discharge into the Truckee River, but May is the peak discharge month for the Upper Truckee River. The regulated storage of Lake Tahoe can delay runoff from the snowmelt season and hold this water for release during dry months when it is needed by riparian water-users downstream along the Truckee River. Annual flow into the Truckee River from Lake Tahoe represents about 26 percent of the regulated storage. The importance of this regulated

storage, equal to the average flow of the Truckee River for 4 years, has been realized during several droughts in this century; prolonged drought lowered the lake level below the outlet to the Truckee River in mid-September 1990, and Lake Tahoe reached a record minimum level of 1,885.36 meters early in February 1991.

The great volume of Lake Tahoe relative to the volumes of inflow and outflow causes it to have a relatively slow flushing action or self-cleaning time of 700 years or longer. The water residence time for most lakes is about one fourth this long. Sediments and nutrients do not remain in Lake Tahoe for such long periods, but the long residence time means that a large part of the sediments delivered to the lake settle on the lake floor (Crippen and Pavelka 1970).

Runoff and Water Quality

The hydroclimatology of the Tahoe Basin plays a significant role in the water quality of Lake Tahoe. The strong seasonal pulse of streamflow enhances the erosional ability of runoff and concentrates into a few months the capacity of streams to deliver sediments and associated nutrients to the lake. The sediments and nutrients removed from the watershed by runoff are particularly effective fertilizers for phytoplankton and algae in the lake (Goldman 1989; Leonard et al. 1979). In recent years, land development has accelerated erosion rates and reduced wetland areas that served as natural filters for sediments and nutrients (CTC 1987).

The potential for erosion in the Tahoe Basin is high because of the shallow and coarse soils. Under natural conditions, erosion is limited by the presence of mountain vegetation, which holds the soil in place and promotes infiltration of water. However, soils are easily eroded, especially on steep slopes, when vegetation is disturbed or destroyed because the climate and poor soils in the basin yield slow-growing vegetation that does not regenerate easily (Strong 1984; TRPA 1982). The natural sediment loading to Lake Tahoe is an estimated 3,100 metric tons per year (SWRCB 1980).

In the past 150 years, the Lake Tahoe Basin has been subject to two major human disturbances. Between 1870 and 1900, forests in the basin were heavily logged to provide mining timber for the Comstock Lode near Virginia City, Nevada. Beginning in the late 1950s, roads, houses, and casino-hotels were constructed to support year-round recreation in the basin. Evidence of the 1870s logging is not detectable in midlake sediments, suggesting that the watershed healed quickly from this disturbance. Great changes in recent sediments are clearly correlated with population growth and extensive construction activities of the last 30 years (Goldman 1989). Current sediment loading to Lake Tahoe is estimated to be 67,000 metric tons per year, an increase of more than 2,000 percent over natural sediment input (CTC 1987).

The increased nutrient loading associated with the large increase in sediment delivery to Lake Tahoe has resulted in steadily accelerating algal growth in the lake and a progressive loss of lake-water transparency. The growth rate of free-floating algae in Lake Tahoe has been increasing at a rate of 5.6 percent per year and has more than doubled since 1959 (CTC 1987). Prior to the increase in algal growth, it was possible to see depths farther than 36 meters in Lake Tahoe (SWRCB 1980). Transparency measurements indicate that Lake Tahoe has lost more than 10 meters of transparency over the last 30 years, and it will lose most of its famous transparency in about 40 more years if the present rate of loss of 0.4 meters per year continues (Goldman 1989).

The seasonal regime of flow into Lake Tahoe and the long residence time of water in the lake combine to exacerbate the influence of the large quantities of sediment and nutrients delivered to the lake. The concentration of runoff in 2 or 3 months dramatically increases erosion rates from disturbed surfaces lacking adequate land-management practices (CTC 1987). Changes in water quality occur slowly because of the long residence time of water in Lake Tahoe and because the elevated nutrient loading exerts its full effect through a gradual increase in nutrient concentration over a number of years (SWRCB 1980). Ultimately, extensive algal blooms occur when the seasonal runoff delivers large quantities of new nutrients to the lake, which already contains a high concentration of nutrients.

Prospects for the Future

Hydroclimate couples the terrestrial and aquatic ecosystems that are the key to the future health of Lake Tahoe (Goldman 1989). The structure, function, and linkage of these ecosystems must be understood to develop resource-management strategies to accommodate human use of the region without harmful degradation of the watershed and Lake Tahoe. The relationship between climate and runoff is particularly significant because of its role in soil erosion and the transport of nutrients from the watershed to Lake Tahoe.

The climate, fragile soils, and steep slopes in the Lake Tahoe Basin make this watershed acutely sensitive to human activities. Land-development practices that would have little impact in other regions cause severe erosion in this basin. Consequently, erosion control has become a major element in the program to slow the deterioration of water quality in Lake Tahoe. Various measures have been designed and implemented to prevent erosion at its source, to stabilize sediment-transport systems, and to treat sediment-laden waters. One particularly innovative component of this program involves the purchase and restoration of wetlands that filter both sediments and nutrients from the water before it enters Lake Tahoe. Implementation of a suite of proposed erosion-control projects could reduce the sediment loading of Lake Tahoe by nearly 50 percent or 30,000 metric tons annually by 1995 (CTC 1987).

The national and global significance of Lake Tahoe as one of the world's few large, high-altitude lakes known for the clarity and purity of its water was recognized in its selection for a joint U.S.-Soviet project to study Lake Baikal and Lake Tahoe. This project will begin during the summer of 1991 with an exchange of scientists and students who will jointly study at both sites. Such international collaboration may prove beneficial for preserving the environmental quality of both lakes.

References

California Tahoe Conservancy (CTC). 1987. *A Report on Soil Erosion Control Needs and Projects in the Lake Tahoe Basin.* Sacramento: Resources Agency, State of California.

Crippen, J. R., and Pavelka, B. R. 1970. *The Lake Tahoe Basin, California-Nevada,* Water Supply Paper 1972. Washington, DC: U.S. Geological Survey.

Goldman, C. R. 1989. Lake Tahoe: Preserving a fragile ecosystem. *Environment* 31:7-11,27-31.

Gleick, P. H. 1987. The development and testing of a water balance model for climate impact assessment: Modeling the Sacramento Basin. *Water Resources Research* 23:1049-1061.

Hyne, N. J. Jr. 1969. *Sedimentology and Pleistocene History of Lake Tahoe, California-Nevada.* Ph. D. dissertation, Department of Geology, University of Southern California.

Leonard, R. L., Kaplan, L. A., Elder, J. F., Coats, R. N., and Goldman, C. R. 1979. Nutrient transport in surface runoff from a subalpine watershed, Lake Tahoe Basin, California. *Ecological Monographs* 49:281-310.

Myrup, L. O., Powell, T. M., Godden, D. A., and Goldman, C. R. 1979. Climatological estimate of the average monthly energy and water budgets of Lake Tahoe, California-Nevada. *Water Resources Research* 15:1499-1508.

Neilson, J. A. 1973. *Lake Tahoe Vegetation II: Natural Vegetation Zones*, Environmental Quality Series No. 14. Davis, CA: Institute of Governmental Affairs.

Shelton, M. L. 1978. Calibrations for computing Thornthwaite's potential evapotranspiration in California. *Professional Geographer* 30:389-396.

———. 1985. Modeling hydroclimatic processes in large watersheds. *Annals of the Association of American Geographers* 75:185-202.

———. 1989. Spatial scale influences on modeled runoff for large watersheds. *Physical Geography* 10:368-383.

Strong, D. H. 1984. *Tahoe: An Environmental History.* Lincoln: University of Nebraska Press.

State Water Resources Control Board (SWRCB). 1980. *Water Quality Plan: Lake Tahoe Basin.* Sacramento: State of California.

Tahoe Regional Planning Association (TRPA). 1971. *Hydrology and Water Resources of the Lake Tahoe Region: A Guide for Planning.* South Lake Tahoe, CA: Tahoe Regional Planning Association.

———. 1982. *Environmental Impact Statement for the Establishment of Environmental Threshold Carrying Capacities.* South Lake Tahoe, CA: Tahoe Regional Planning Association.

Thornthwaite, C. W., and Mather, J. R. 1955. The water balance. *Publications in Climatology* 8:1-86.

Urban Sprawl and the Decline of the Niagara Fruit Belt

Hugh J. Gayler
Brock University

One of the abiding symbols of the North American human landscape, that causes much concern and bewilders people from other cultures, is the low-density nature of urban growth during the 20th century. Cities are no longer compact. Rather than orderly outward expansion, development is linear along major highways and sporadic developments are found around the countryside. Popularly known as urban sprawl, this form of urban growth, strikes many as ugly, inefficient, costly, wasteful of natural resources, and destructive of urban and rural lifestyles. Canada has been facing urban sprawl along the line of its Niagara fruit belt and its east-west freeway, the Queen Elizabeth Way (Figure 1), where for some time the points of urban expansion have threatened a unique agricultural resource. Here, traditional attitudes towards land ownership clash with new, more environmentally conscious concepts and land use.

The Problem

The horrors of the 19th-century industrial city fueled interest in a more humane, less dense, and better housed and serviced urban environment. The railway and the car allowed this housing development to occur at increasing distances from the central city, followed by the more recent decentralization of industry, offices, retailing, recreation, and public institutions (Muller 1981).

As the city expands, it impinges upon the agricultural resource base. Most cities in North America are set amid good agricultural land, a factor instrumental in their founding and early growth. But with expansion, particularly in the last 50 years, urbanization invariably consumes our best agricultural land and never gives it back (Beesley and Russwurm 1981; Furuseth and Pierce 1982).

This loss of good agricultural land on the Niagara fruit belt is a major concern to Canadians. This area, lying between Lake Ontario and the Niagara escarpment, has superior soil and microclimatic conditions for Canada's foremost tender fruit and grape growing area (Krueger 1965). However, the industry has encountered financial hardships over the years, resulting from small land-holdings, high urban-related taxes, labor difficulties, foreign competition, the vicissitudes of weather, changing consumer demands, the loss of the local canning industry, and a political and financial establishment that is perceived to offer little support. Many farmers view urbanization as a way out of the problem (Gayler 1982a; Krueger 1978). The postwar economic boom in southern Ontario, the lack of good land-use planning until the 1970s, and the opening of the Queen Elizabeth Way through this area in 1940 encouraged them to sell not only whole farms for the extension of an urban area but also plots of land anywhere there was a public right-of-way and a willing buyer. The improved accessibility resulting from the Queen Elizabeth Way led to some of Canada's worst urban sprawl; but the valuable nature of the agricultural land resource, the conflict between the ex-urban resident and agriculture, and the costs of providing services to these semi-rural areas have resulted in recent government policies aimed at stemming the tide of development. However, this results in conflict between two diametrically opposed groups, those for and those against preservation of agricultural land. The government compounds the problem by not enforcing its preservation policies.

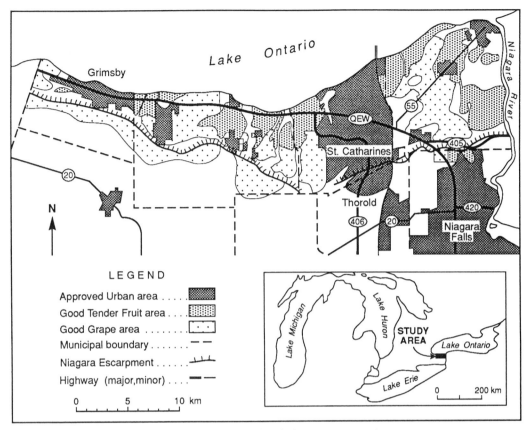

Figure 1. The Niagara fruit belt, Ontario, Canada

Attitudes toward Landownership and Land Use

North American landowners believe to varying degrees that they have a right to do what they like with their land, often with little consideration of the effects of such decisions on others; and government support for this right is judged by the strength or weakness of its laws relating to the rights of society over the individual. Compared with many other Western countries, land-use planning laws in the states and provinces of this continent do not greatly inhibit what landowners do. Some urban areas have no land-use plans and no zoning bylaws to organize uses and reduce conflicts. In rural areas, these have long been considered vacant holdings awaiting a higher and better economic (and presumably urban) use. Rarely are there plans that restrict urban uses and promote agriculture and the physical resource base; the protection of rural land uses is invariably through the initiatives of landowners, the physical unsuitability of the land for urban uses, and distance and lack of demand from urban areas.

The Niagara fruit belt is one of the few areas in North America where government has challenged in-dividual landowners' rights (Regional Municipality of Niagara 1988). The mechanism is the Official Plan process under the Ontario Planning Act, 1946 (as amended 1983), whereby in 1981 approval was given for urban areas that would be fully serviced and for rural areas where strict land-use policies would apply to agriculture, natural and scenic resources, and any further urban intrusions.

In Niagara the strictness of land-use policies outside urban-area boundaries relates to the quality of the land. On the Niagara fruit belt, on land with the greatest agricultural potential, only agriculturally related activities are allowed. However, this has a somewhat hollow ring when one considers that much of the urban sprawl occurred before 1981, and that there is nothing to prevent selling a whole farm to a non-farmer. Severing a residential lot for a farmer's re-tirement does not prevent selling that lot later to a non-farmer, and the regional government allows a number of urban-related activities on rural land under the Official Plan amendment. The exceptions to the rule and the observed attrition of farmland maintain the hopes of landowners that they can do what they want with their land. Nonetheless, there is little doubt that being on the wrong side of an urban-area boundary can result in farmers' dreams of selling out to developers and becoming instant millionaires turning into nightmares.

In spite of a Regional Niagara Policy Plan, landowners are not about to give up the fight for what they regard as a traditional right. A vocal minority, who either face financial hardships as farmers or resent being locked into farming, want planning policies eased for their benefit. Alternately, they want some kind of financial reward for serving as custodians of the land so that urbanites today and in the future can both enjoy the countryside and have cheap food. These issues have simmered among landowners since long before the Policy Plan was approved in 1981. However, the 1989 Canada-United States Free Trade Agreement and the present review of the Policy Plan focus attention, respectively, on the long-term decline of the fruit belt and the need to consider where future urban development is to go in the Niagara region: the one is forcing the farmer out of business, while the other could make the farmer an instant millionaire.

The Decline of the Niagara Fruit Belt

The Free Trade Agreement and the General Agreement on Trade and Tariffs created tremendous uncertainty for Niagara fruit farmers (InfoResults 1989). In spite of the various hardships noted earlier, at least until now, Niagara farmers have benefitted from tariffs on imports, an important factor when one realizes that in a similar area, immediately east of the Niagara River in the United States, there is virtually no tender fruit industry. With these tariffs gone, the prospects for grape growers are especially bleak. Even where they are growing the grapes that local wineries need, their future is uncertain because the wineries may buy cheaper foreign imports. This causes resentment since it was the wineries and government that had earlier encouraged grape farmers to switch varieties. The federal government recognizes these difficulties in its present Grape Acreage Reduction Program, whereby farmers are paid to reduce acreages by as much as 40 percent.

The present problem is part of a long-term decline in the industry. The area in tree-fruits declined by more than a third between 1951 and 1981; besides urbanization and other land conversions, improved varieties and management practices increase productivity per tree and trees per acre (Bond and Bruneau 1986). Similarly, the 40 percent increase in grape production for the region in the same period was achieved on only 12 percent more land. However,

the varieties of grapes introduced have allowed for expansion immediately south of the Niagara escarpment and a decline in acreage on the fruit belt.

Farmers can use the land for other forms of production. This is encouraged, since tender fruit and grape production yield poorer rates of return on investment than other agricultural activities. However, many farmers are financially unable to risk switching, may not have the skills necessary to do so, and may be too old to want to bother. Many are attracted by the even better returns that urban-related uses can bring, and there is considerable speculation in land in the event that this may happen, in particular along the Queen Elizabeth Way between Grimsby and St. Catharines (Krushelnicki and Bell 1989).

A vocal minority of farmers, many of them facing imminent financial hardship, recently canvassed the Ontario and Niagara regional governments to abandon their strict rural policies altogether and to allow urban-related activities anywhere. However, we fail to realize that there are many more problem farmers and more available land than required to meet the demands of developers. Moreover, the farmers wanting out are not necessarily in the areas where the next stages of urban expansion will take place. In effect, farmers are asking government to turn back the clock to the 1960s and to give them at least some financial reward by allowing small lots to be severed anywhere. This would continue the urban sprawl already seen on the fruit belt, unduly raising servicing costs (already high enough in urban areas, given the improved standards set by the Ontario government) and resulting in further conflicts with neighboring, more successful farmers. Government has so far not promoted wholesale abandonment, but local planners express concern at the recent increase in the number of exceptions that local politicians make when allowing residential lots in rural areas.

Urban Development in Niagara

In the past the local politician was very supportive of the landowner who wished to sell all or part of a holding for an urban purpose (Gayler 1979). In Niagara's small-town and rural setting, many of the politicians were either farmers themselves or sufficiently close to the business to appreciate the farmers' problems. Moreover, attracting urban functions improved the tax base in the many rural municipalities, bettered employment prospects for the whole area, and resulted in spin-off effects. The growth ethic is especially strong as politicians add up the benefits of urban development, although invariably giving little consideration to the economic and social costs and whether there is a net benefit.

During a long and bitter debate on the Regional Niagara Policy Plan in the 1970s, the costs to society of paving over the Niagara fruit belt came to the fore (Gayler 1982b). What had long been an academic discussion became an environmental and political issue as the forces for and against development began to challenge each other. The draft plan unduly threatened the fruit belt, for the amount of land set aside for future urban development was totally unrealistic. It was based more on 1950s thinking of the boom times than on the cold reality of a declining birth rate nationally and a problem local economy centered on heavy industry. However, this did not stop the environmental lobby from being swamped by local government, the development industry, and landowners, and it took the intervention of the Ontario government to introduce a sense of realism (Jackson 1982). In the end the Niagara Policy Plan was decided by the Ontario Municipal Board, a quasi-judicial body of the provincial government; besides setting the more realistic urban areas (Figure 1), it recommended the permanence of urban-area boundaries on the fruit belt, the redirection of urban development to the south of the Niagara escarpment, and strict policies on land use in rural areas.

During the 1980s negligible growth saved the Niagara fruit belt from further urban development, not the determination of politicians to abide by the policies of the Niagara Policy Plan (Gayler 1990). Elsewhere in southern Ontario, and certainly closer to metropolitan Toronto, there is continuing pressure on rural areas to accommodate the physical expansion of the metropolis, including the many people who prefer a rural or small-town lifestyle and the many urban-related functions, which out of necessity or preference seek rural locations.

Until recently Niagara has not been part of that growth scenario. The recession of the early 1980s resulted in considerable job losses in manufacturing, minimal population growth, a net emigration of people, and a reputation as a problem area. Moreover, since it is more than 80 kilometers (50 miles) from Toronto, it has shared less in the strong economic growth of the Toronto region than areas that are closer. The lands set aside for urban development in the Niagara Policy Plan were more than generous; and no consideration was given to either redirection to the south or any extensions to urban-area boundaries. However, this is changing. Various factors are now contributing to a quickening of growth, and there is the prospect of running out of developable land on the fruit belt within 10 years. This has led to the present review of the Niagara Policy Plan, but the very public nature of the review process is encouraging considerable lobbying from interest groups, especially problem farmers on the fruit belt.

One aspect of recent growth is the catch-up factor: changes that were postponed for economic reasons are now taking place, including new residential development, commercial, industrial and institutional expansions, and the various hard and soft services that the population needs. Also, St. Catharines (population 124,000), the largest city in Niagara, is emerging as the regional center. Certain environmental and physical attributes meanwhile contribute to Niagara's growing popularity as a retirement area and to further developments in its extensive tourist and leisure industry. Furthermore, the growing disparity in land and property prices between the Toronto and Niagara areas is encouraging both residential and non-residential development to seek locations close to the Queen Elizabeth Way on the fruit belt.

Urban growth has for a long while been greater in the fruit belt municipalities than in areas farther south. On the fruit belt, developable land is running out more quickly, and consideration of where new development is allowed is central to the Policy Plan review process (Regional Municipality of Niagara 1989). Meanwhile, the prospects of development or no development, the multi-million dollar outcomes that await such decisions, and the potential destruction of what remains of the fruit belt could lead to renewed conflicts between the forces for and against development.

Regional planners have proposed four options for future urban growth. Two support the 1981 Policy Plan and the redirection of growth to the south of the Niagara escarpment, either between Fort Erie and Port Colborne on the Lake Erie shoreline or in the area between Niagara Falls, Welland and Thorold. In both cases, ample land is already designated urban, although the latter area is closer to where development would most likely want to go. The third option is the status quo. Development until now has been accommodated in almost any municipality where it wishes to go; rather than redirection, urban-area boundaries would be extended where necessary. The final option recognizes that the greatest demands for land are close to the Queen Elizabeth Way, between Grimsby and St. Catharines, where extensive development should take place. Indeed, the current expansion of the highway from four to six lanes in this area will improve accessibility from metropolitan Toronto and encourage further urban development.

If allowed, the last two options would signal the end to the Niagara fruit belt; the speed would vary, but the result would be the same. The permanence of the urban-area boundaries would be lost and the strict rural land-use policies would be exchanged for landowners once again having the right to sell their land for urban-related purposes. Given the present state of agriculture on the fruit belt, many farmers favor these last two options, and speculators are already buying the land. On the other hand, more people probably favor some way of retaining Niagara's best land for agriculture. The key to the Policy Plan Review process in the not-too-distant future will be the stand taken by politicians at the local and provincial levels. Here lies the danger of fine words in favor of protecting the Niagara fruit belt but on another occasion allowing the hard-pressed farmer or the very convincing developer to convert land for urban purposes.

Conclusion

Canadian and American culture and technology have encouraged low density urban development and sprawl, contrasting markedly with other countries in the Western world; although lately the matter has been viewed in many areas as an unacceptable economic and social cost that must be remedied. This chapter has examined how urban growth is resulting in the loss for all time of our best agricultural land. Fundamental to the problem are traditional attitudes toward landownership and land use in rural areas. For the Niagara fruit belt, attempts to be environmentally responsible clash not only with farmers' rights but also with those of the development industry and growth-oriented politicians. Unfortunately, unless attended to, financial hardships and uncertainty in the agricultural sector will continue to promote land conversion for urban purposes.

References

Beesley, K. B., and Russwurm, L. H., eds. 1981. *The Rural-Urban Fringe: Canadian Perspectives.* North York, ON: York University, Department of Geography.

Bond, W. K., and Bruneau, H. C. 1986. *The Impact of Federal Activities on Fruitland Use: The Niagara Peninsula.* Ottawa: Lands Directorate, Environment Canada.

Furuseth, O. J., and Pierce, J. T. 1982. *Agricultural Land in an Urban Society.* Resource Publications in Geography. Washington, DC: Association of American Geographers.

Gayler, H. J. 1979. Political attitudes and urban expansion in the Niagara region. *Contact (Journal of Urban and Environmental Affairs)* 11:43-60.

————. 1982a. Conservation and development in urban growth: The preservation of agricultural land in the rural-urban fringe of Ontario. *Town Planning Review* 53:321-341.

————. 1982b. The problems of adjusting to slow growth in the Niagara region of Ontario. *The Canadian Geographer* 26:165-172.

————. 1990. Changing aspects of urban containment in Canada: The Niagara case in the 1980s and beyond. *Urban Geography* 11:373-393.

InfoResults. 1989. *Farming in the Niagara Region: Structure, Trends and Land-Use Policies,* Niagara Policy Plan Review No. 8. Toronto, ON.

Jackson, J. N. 1982. The Niagara fruit belt: The Ontario Municipal Board decision of 1981. *The Canadian Geographer* 26:172-176.

Krueger, R. R. 1965. The geography of the orchard industry of Canada. *Geographical Bulletin* 7:27-71

————. 1978. Urbanization of the Niagara fruit belt. *The Canadian Geographer* 22:179-194.

Krushelnicki, B. W., and Bell, S. J. 1989. Monitoring the loss of agricultural land: Identifying the urban price shadow in the Niagara region, Canada. *Land Use Policy* 6:141-150.

Muller, P. O. 1981. *Contemporary Suburban America.* Englewood Cliffs, NJ: Prentice-Hall.

Regional Municipality of Niagara. 1988. *Regional Niagara Policy Plan (Office Consolidation).* Thorold, ON.

————. 1989. *Where Next?: A Look to the Future,* Niagara Policy Plan Review No. 7. Thorold, ON.

Conflict in Yosemite Valley

Lary M. Dilsaver
University of South Alabama

Yosemite Valley has been called the incomparable valley by some, holy and eternal by others. Ralph Waldo Emerson simply noted that it was "the only place that comes up to the brag about it, and exceeds it" (*Yosemite Handbook* 1990). Since it came to public attention in the early 1850s, Yosemite has never failed to awe and inspire its visitors. The valley is but 1 kilometer in width and some 15 in length with sheer granite walls towering nearly 1,000 meters along most of its length. At various spots waterfalls seasonally mist or thunder over the walls and fall hundreds of meters toward the valley floor (Figure 1). Amazing vistas, towering trees, a rich fauna, and idyllic meadows and pools have combined to lend Yosemite a mystic quality.

These wonders have drawn increasing numbers of visitors over the years, which has spawned a negative side to the Yosemite experience. On Independence Day 1985, this problem exploded when the canyon's 1,500 parking spaces filled, leaving another thousand carloads of visitors in search of room to enjoy the park. Traffic accidents, stalled and overheated vehicles, petty crime, fights, and gridlock overtaxed park security forces and ruined the holiday of many visitors. That day the Park Service, which had expected such an eventuality, turned away cars at the gate (Frank Dean interview 1990). This was a logical solution both for safety and for protection of the park resources. Yet it marked a humiliating capitulation in a philosophical sense. Open park access for all who desire admission is a time-honored tradition in the national parks. That tradition stems in part from early Park Service efforts to attract attention and visitors, but beyond that it reflects democratic principles embodied in both government and culture. The alternate and upstart view to protect park resources at all costs, even if it means banning visitors, challenges this tradition. Yet that

view also is rooted in the process of park management (Dilsaver and Tweed 1990; Runte 1990).

The National Park Service faces many difficult problems in Yosemite today. In each case, these problems derive from the intersection of three incompatible conditions at the point called Yosemite Valley. First, pressure for recreation has grown continuously. The 1864 act setting aside Yosemite Valley to be administered by California, the 1890 act creating Yosemite National Park to which the valley was added in 1906, and the 1916 act establishing the National Park Service all stipulate the park to be a place for the "use and enjoyment of the people" (39 Stat. 535). But, increasing population, especially in California, a skyrocketing flow of international tourists, expanding demand for leisure time and activities, and a wealth of new recreational tools and toys escalate the strain annually.

Second, the acts also mandate preservation of park resources "unimpaired for future generations." As recreation pressure has increased and as the ability of scientists to detect symptomatic ecosystem damage has become more sophisticated, the ground rules for preservation have changed. Initially, park managers were content to engage in object preservation, carefully protecting important examples of flora, fauna, and scenery. Subsequently, as crowds trampled, defoliated, or simply packed into various parks, managers sought an immature systemic protection called atmosphere preservation; this was based largely on the advice of landscape architects. Finally, commencing in the late 1950s, academic scientists and the growing environmental movement forced the Park Service to adopt ecological preservation.

Each of these shifts was accompanied by considerable anguish and anger from some park users. The third factor, the inertia of tradition, had generated loyal constituencies for past policies suddenly found inap-

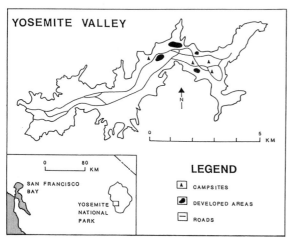

Figure 1. Yosemite Valley.

propriate either because of the precipitous increase in demand or altered preservation goals. In addition, American culture itself has repeatedly thrown emotional obstacles in the way of more restrictive protection measures (Dilsaver and Tweed 1990).

These three factors—recreation demand, preservation demand, and tradition—have always intersected at Yosemite Valley. But now the conflict grows more important and more intense as the resource becomes less capable of balancing use and preservation. Here we consider two specific resource problems in Yosemite Valley: the encroachment of forests onto valley meadows, as well as the congestion and concession housing. In each case, increasing demand, evolving preservation goals, and tradition have created today's debates. These debates ultimately challenge cultural values and the willingness to sacrifice for them.

Forest Encroachment on Meadows

When Europeans arrived in Yosemite Valley in 1851 they found a parklike setting with vast, open meadows and so little undergrowth that "a half day's work in lopping off branches enabled us to speed our horses through the groves" (Commissioners 1890). Frequent burning by the resident Ahwahnee Indians maintained the open character. As private recreation promoters and the State of California moved in, they uncharacteristically took a lesson from the Indians and continued occasional burns. Based on costly experience with forests in the eastern United States and the Old World, Europeans traditionally recoiled from forest fires. It took a powerful motive, the lucrative recreation business and protection of the scenic vistas that sup-

ported it, to engender deliberate burning of trees (Runte 1990; Yosemite 1987).

In the 1920s, however, two changes led to a new policy of complete fire suppression. First, rapidly increasing numbers of visitors dispersed around the valley in sprawling campgrounds and scattered concession houses. The visitors, their cars and property could not be endangered by the brush fires set in the past. In addition, the notion of atmosphere preservation gained strength, as did the traditional antipathy to forest fire. Blackened trees and meadows offended the sensibilities of visitors and the landscape architects who staffed the Park Service planning offices. Thus, in 1930, the Park Service implemented total fire suppression in Yosemite Valley. For the next four decades no fires consumed more than a few acres before increasingly sophisticated crews suppressed them (Heady and Zinke 1978).

The effects of fire suppression became evident immediately. Forest encroachment on meadows dramatically increased and botanical studies indicated that normal succession could eliminate nearly all the meadows in Yosemite Valley. For 40 years crews sporadically removed fallen timber and uprooted saplings manually, yet budget constraints and policy doubts among park managers made their efforts ineffectual. By 1960 meadow acreage on the valley floor had been reduced by 55 percent (Heady and Zinke 1978).

Then in 1963, years of analysis by university scientists culminated in a critical overview of Park Service preservation policy nicknamed the Leopold Report (after study team leader A. Starker Leopold). The report found pervasive problems with the Park Service's commitment to scientific management and with the recreation-oriented, atmosphere-management style of the previous four decades. Singled out for criticism was fire suppression and the problems of meadow encroachment and fuel buildup it engendered (Cain et al. 1963). The government responded by embracing the Leopold Report, calling for larger scientific staffs, and adopting a new philosophy of ecosystem management. From 1970 to 1988, the Yosemite staff set prescribed burns aimed at recreating the conditions extant upon arrival of Europeans in the valley (Yosemite 1987).

This latest shift in preservation policy has met resistance, however. Initial public confusion over deliberate burning of their parkland forests turned to anger when fire sear damage became noticeable, especially on the giant sequoias (*Sequoiadendron gigantea*) of the valley. And increased recreational use has made prescription conditions (low visitation and specific weather conditions) increasingly infrequent. The explosive Yellowstone fire of 1988 and the frightening

Yosemite fire of 1990 fanned hysteria about fire and forced the Park Service to temporarily reinstate total fire suppression. Finally, ecological research, a mainstay of the new preservation philosophy, has itself confused the issue by questioning whether the Yosemite Valley of 1851, burned for centuries by Indians, is the "natural" one to be recreated, particularly in view of evidence that succession to forest would be likely if human manipulation ceased (Graber interview 1990). Recreation, preservation policy, and tradition have become so entangled that a coherent vegetation management policy may never be fully accepted.

Concessions and Congestion

As delicate as the issue of forest and meadow management is, it pales by comparison with the emotion over the concession presence in Yosemite Valley. And to an even greater extent, the roots of this problem lie with the state and federal park managers. Here the intersection of rising recreation demand, shifts in preservation policy, and tradition creates a debate that borders on violence. Only months after its discovery by a band of gold miners chasing marauding Indians, the valley became a focus of recreation development. By the time Congress set it aside for state management in 1864, several entrepreneurs had squatted on the land and begun hotels and tourist services. During the remainder of the state's administration, park commissioners encouraged roads, tourist camps, and even a railroad in the valley. They only balked at assumption of the entire valley floor by farmers and the fencing that would accompany it (Runte 1990).

When the federal government took over operation of the valley in 1906, and particularly after the Park Service was formed in 1916, park managers and Washington administrators welcomed vast increases in visitors and the facilities to house and entertain them. By this means they sought to protect the fledgling Park Service and its parks from outright elimination or takeover by the larger and older Forest Service. Initial Park Service Director Stephen Mather further established a policy of one concession company per park, eliminating ruinous competition in what was then a speculative venture. In Yosemite, the Yosemite Park and Curry Company received approval to engage in massive expansion (Blodgett 1990).

The initial buildup of the concession in Yosemite Valley and the object preservation that allowed it faced its first challenge in the 1930s. Landscape architects had so far preserved the park atmosphere by insisting, with one exception, on small structures, scattered among the rapidly encroaching forest. By

this means, the visitor could enjoy a sense or atmosphere of wilderness while remaining close to the amenities. But by the late 1930s the largely auto-borne traffic to the valley approached a half million annually. Traffic jams, noise and chaos, and the frequent requests by the concessioner to build ever more housing frightened and saddened both veteran park enthusiasts and the park staff. In answer they established a line midway through the valley and stipulated that no structure would be allowed west of the line. Yet this left the eastern and most spectacular portion packed with as many as 4000 campers and an equal number of concession housing residents, as well as hundreds of employees and thousands of day visitors (Advisory Committee 1935-1953).

The arrival of ecosystem preservation had little direct effect on this argument because the botanical communities in the housing and camping areas had been obliterated and the cliffs and waterfalls seemed unaffected. Yet it arrived accompanied by a vigorous environmental movement among the public. The incessant clamor and stress of summer in Yosemite Valley, the unmistakeable human effect on meadows and streambanks, and the loss of the mystic quality of the canyon spurred a dramatic rise in the decades-old call for removal of cars and structures from the valley. Adding fuel to the fire was incessant probing for new areas of housing expansion and new recreational facilities by the concession company. When entertainment giant MCA took over the concession in 1973, a new aggressive developer faced off against an increasingly resolved and polarized environmental movement (Frank Dean interview 1990).

The apparent result of this polemic debate was the General Management Plan of 1980 drafted through a federally mandated program of public commentary. Members of the public who commented came overwhelmingly from the environmental movement, either actually or sympathetically. The General Management Plan called for the eventual removal of most valley structures to areas immediately outside the park and an increased commitment to free shuttlebuses rather than private automobiles. For a few years, the environmentalists exulted over their apparent victory.

During the 1980s, however, the rules seemed to change again. The Republican political administrations in Washington favored tourism and free (if monopolistic) enterprise at the expense of narrow preservation. Also, the aging of America diluted ranks of Americans who favored a wilderness experience and added to those who sought comfort with their scenery. Finally, tradition—the tradition of tens of thousands of families who had used concession facilities, and the traditional democratic notion that all the people deserved to experience Yosemite, not just those who could hike its trails—swelled among hitherto silent park users and the public at large.

In 1989, the Park Service released a draft reappraisal of the GMP that called for much further study before its implementation. Management Assistant Frank Dean (interview 1990) pointed out the need to study the areas that were to receive the evicted valley structures. He questioned whether those areas had the space, the water, the infrastructural capability to contain more than 1,000 housing units plus support structures. General Management Plan supporters now suggest that the Park Service is delaying, in hopes that the plan will be discarded eventually. They initially traced the cause to intense pressure from corporate giant MCA. However, now MCA has been sold to the Japanese corporation Matsushita Electrical, and the concession operation will apparently be sold to new American operators. The sale must conclude by 1993 when a new contract for the concessioner is to be signed. The valley removal issue is the pivotal factor influencing both the sale and the new contract negotiations (Dean interview 1990; Yosemite 1989).

Conclusion

In Yosemite Valley, the National Park Service faces many problems resulting from the collision of rising recreation demand, shifting preservation goals, and the onus of tradition. Will park managers resume burning to create a Yosemite similar to that of 1851, regardless of traditional resistance? Or will they allow forest encroachment to proceed either for traditional anti-burn reasons or because natural succession leads that way? Will they remove hundreds of structures enjoyed by tens of thousands of visitors in a spirit of democratic openness and capitalistic enterprise? Or will they wrench the concession from the valley in an effort to achieve the mystical quality described by early explorers? The only certainty for the future is that Yosemite Valley, the point of intersection for recreation, preservation, and tradition, will remain as controversial as it is beautiful.

References

Advisory Committee to Yosemite National Park. 1935-1953. *Reports.* Box 9, Yosemite Archives, Yosemite National Park.

Blodgett, Peter J. 1990. Striking a balance: Managing concessions in the national parks, 1916-33. *Forest and Conservation History* 34:60-67.

Cain, S., Gabrielson, I., Cottam, C., Kimball, T., and Leopold, A. S. 1963. A vignette of primitive America. *Sierra Club Bulletin* 48:2-11.

Commissioners to Manage Yosemite Valley and the Mariposa Big Tree Grove, Biennial Report. 1890. Sacramento: California State Printing Office.

Dilsaver, Lary M. and Tweed, William. 1990. *Challenge of the Big Trees.* Three Rivers, CA: Sequoia Natural History Association.

Heady, Harold, and Zinke, Paul. 1978. *Vegetational Changes in Yosemite Valley.* National Park Service Occasional Paper No. 5. Washington, DC: U.S. Government Printing Office.

Runte, Alfred. 1990. *Yosemite: The Embattled Wilderness.* Lincoln: University of Nebraska Press.

Yosemite Handbook. 1990. Washington, DC: Division of Publications, National Park Service.

Yosemite National Park. 1987. *Fire Management Plan, 1987 Revision.*

Yosemite National Park. 1989. *Draft Yosemite GMP Examination Report.*

The Grand Canyon Geographical Suite

William L. Graf
Arizona State University

In Ferde Grof's *Grand Canyon Suite* (1931), the musical notes convey a sense of grandeur, change, and complexity about the best-known landform on the North American continent. From an environmental perspective, the canyon is a geographical suite combining physical, cultural, and economic landscapes. The purpose of this paper is to introduce the canyon, and to trace its metamorphosis from a blank spot on the map to its present central location in the mental map of American environmental culture (Figures 1 and 2).

The geological origins of the canyon are obscure. The ancestral Colorado River probably flowed in a course entirely different from its present one. Streams issuing from the western slopes of the emergent Rocky Mountains probably drained northward until about 7 million years ago. Tectonic activity, including massive upwarping of the Colorado Plateau, resulted in drastic changes in regional drainage by reorienting flow in a southwesterly direction. The Colorado River etched its new course into the uplifting southern end of the Colorado Plateau to erode the Grand Canyon in a continuing process (Breed and Roat 1974).

Whatever its exact origins, the canyon now presents a stupendous appearance—a gorge excavated into nearly horizontal sedimentary and deformed metamorphic and volcanic rocks that span much of earth history. The Colorado River flows at an elevation about 1,100 meters below the northern rim of the canyon, which has a maximum width of more than 30 kilometers. Depending on the time of day and the lighting conditions, the geologic materials and complex slopes produce colors ranging from orange, red, and yellow to subdued blue and gray. Although there are wider canyons in the Colorado River system (Canyonlands, Utah), and deeper ones elsewhere in the American West (Hell's Canyon, Idaho), none are nearly so complex, colorful, and stimulating to the imagination.

A Geographic and Historical Perspective

The earliest human occupation of the Grand Canyon was by people related to the Desert Culture, a widespread group of Indians in the American Southwest. Their willow twig replicas of game animals, "split twig figurines," survived more than 3,000 years in caves of the Grand Canyon (Euler *et al.* 1979). Later, between A.D. 600 and about A.D. 1300, the Anasazi periodically farmed limited reaches of the canyon floor. Their modern descendants, the Hopi, occasionally used the canyon, but other modern Indians, the Hualapai, Paiute, and Dineh (or Navajo) used the rim areas rather than the canyon proper. Only the Havasupai live and farm below the rim (Schwartz 1983).

Like the Indians, the early European and American explorers saw the canyon as an impediment to travel. The first European to see the canyon was Captain Garcia Lopez de Cardena who encountered it in 1540 as part of Francisco Vasquez de Coronado's exploration of the northern frontier of New Spain. Non-Indians did not visit the canyon again until 1776 when Fathers Francisco Atanasio Dominguez and Silvestre Velez de Escalante encountered it while attempting to define trade routes between the missions of New Mexico and California.

The first Anglo-American to see the Grand Canyon may have been the fur trapper, adventurer, and frontier gadfly, James Ohio Pattie. His journey through the region in 1826 is poorly documented, as are the early explorations of the region by Mormon scout Jacob Hamblin who was active there in the 1850s. In 1857, the U.S. War Department authorized Lieutenant Joseph

Figure 1. A portion of the Grand Canyon seen from the North Rim in a sketch by W. H. Holmes, an artist who accompanied John Wesley Powell in his explorations of the region in the early 1870s. Courtesy of the U.S. Government Archives.

Christmas Ives to explore the lower Colorado River to determine its navigability. He reached Diamond Creek in the western Grand Canyon area, and rightly determined that steamboats were unlikely to ascend the canyon. James B. White claimed to have floated

Figure 2. General location of the Grand Canyon showing the extent of the Colorado Plateau and the regional river system.

through the canyon by log raft in 1867, but his story was later discredited.

John Wesley Powell, the eminent 20th-century geographer and geologist, named the canyon during his exploration of the Colorado River by boat in 1869. After traversing more than 1,000 kilometers of previously unknown rivers and passing through many spectacular canyons, Powell recognized this last one as awe-inspiring in its size and beauty and so named it "Grand." From that day to this, Powell's feat of geographic exploration and the Grand Canyon have been associated with each other in popular, historical, and scientific literature (Powell 1961).

Thus, by the time of Powell's first expedition in 1869, the Grand Canyon occupied the last great blank area on the map of the coterminous United States. The notoriety surrounding Powell's successful traverse of the river generated great interest in the canyon. By the early 1880s it began attracting tourists, and John Hance opened a log-cabin hotel on the south rim. Other entrepreneurs developed a camp and stage stop on the western end of the South Rim in 1883, and a rudimentary hotel at Diamond Creek in 1884. From that time on, the Grand Canyon began to make a transition from a peripheral blank spot to a central place on the American mental map.

The National Park

When the area that now is Grand Canyon National Park became part of the United States, it was, like most of the western region where the land was in public ownership, to be sold or given to citizens willing to invest in development. Congress set aside some lands for government administration as parklands, but western developers generally opposed designation of areas as national parks because park status excluded mineral exploration, grazing, and timber cutting. As a result, the Grand Canyon remained available for private development several decades after its inclusion in the United States.

The first attempt at environmental protection for the canyon was in 1893 when President Benjamin Harrison designated a portion of the canyon as a National Forest Preserve, excluding land sales for private development. By 1908 President Theodore Roosevelt, based on his hunting trips in the area, designated various parts of the canyon as a national game preserve, a national monument, and as parts of two national forest preserves. As the social value of the canyon became apparent, political pressure to preserve it increased, and in 1919 President Woodrow Wilson established Grand Canyon National Park in the central canyon area. Herbert Hoover designated the western canyon as a National Monument in 1932, and Lyndon Johnson did the same for the eastern canyon in 1969. Richard Nixon established the present boundaries of the park by combining the various administrative units together in a single national park in 1975.

During this transition from chaotic administrative units to a single national park, the Grand Canyon became one of the crown jewels of the American national park system. Along with Yellowstone and Yosemite, the Grand Canyon is considered by most Americans as a primary component of their natural heritage. The canyon is probably the single most recognizable natural landscape feature of the nation for many foreigners. From its inception as a national park, the number of annual visitors has steadily increased except during years of global wars or during global energy crises that restricted automobile travel (Figure 3). During this growth in public visibility, the canyon completed its transition from obscurity to national attention for preservation.

Development and Preservation

During the 20th century, the Grand Canyon attracted interest from developers and from preservationists. The growth of large-scale agricultural enterprises in southern California demanded water for irrigation, and protection from flooding by the quirky Colorado River. Passage of the Reclamation Act by Congress in 1902 and agreement on the initial legal division of water among the southwestern states in a compact signed in 1922

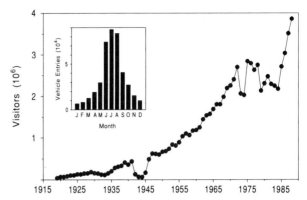

Figure 3. Number of annual visitors to Grand Canyon National Park and the monthly pattern for a typical year, 1965.

paved the way for the construction of the first high dam on the Colorado. Boulder Dam (later renamed Hoover Dam) was erected downstream from the Grand Canyon, and in 1935 its closure initiated Lake Mead. Further congressional acts and appropriations stimulated additional high dams on the river upstream from the canyon after 1948 (Graf 1985).

Water managers from California and Arizona viewed the Grand Canyon as the most logical site for the construction of additional dams in the 1950s. A long-running highly contentious public debate ensued between water developers and environmental preservationists that included predictions of the demise of western agriculture if the dams were not built and the national publication of retouched photographs showing the canyon inundated by an artificial reservoir. After a massive shift of public opinion favoring preservation, Congress decided against construction of the dams and the river remains free-flowing through the national park (Fradkin 1981).

Although dams were excluded from the Grand Canyon, a massive structure was authorized for the river immediately upstream from the park in Glen Canyon. Closed in 1962, Glen Canyon Dam satisfied the needs for flow regulation by structures that did not impinge upon the national park. The new dam also became a key part of the electrical power grid in the southwestern United States because of its efficient and relatively central location with respect to urban markets. The major contribution of the dam to the power grid is peaking power, that part of the total generated electricity that is subject to rapid increases. Unlike coal or nuclear-power generating stations, which provide the basic power capacity in the region, hydroelectric facilities such as Glen Canyon Dam can respond to major increases in power demands within a matter of minutes simply by releasing more water through the generating turbines.

After almost three decades of operating the dam daily to supply peaking power, it has become apparent that although the structure is outside the Grand Canyon, its operations directly affect the canyon environment. The variable releases of water from the dam in response to demands for peaking power produce 4-meter daily fluctuations in water levels of the river downstream in Grand Canyon. These artificial adjustments erode beaches, destroy riparian vegetation that is important habitat for birds, and change aquatic environments to conditions that are eliminating endangered native fish species (Stevens 1983). The major issue facing federal managers of the dam and the park is how to resolve the conflict between economic forces promoting power generation and social forces promoting preservation of the canyon with its diverse natural environment.

Economic development also threatens the dramatic vistas from the rims for which the canyon is famous. Too many visitors in limited spaces degrade the experience, and air pollutants from distant Los Angeles as well as nearby coal-fired electrical generating stations reduce visibility. New plans to control use of the park by enthusiastic visitors can address the first problem, but regional air pollution continues to defy solution.

Like Grofé's *Grand Canyon Suite*, the geographic symphony that is the Grand Canyon is complex in its majesty, power, and human value. To preserve the canyon for future visitors requires a clear understanding of the intricacies of the human and natural systems that caused its transition from blank spot on the map to central location in the American mental landscape. The compromises between preservation and development that have characterized the history of the canyon are likely to continue, with increasing emphasis on preservation. The remaining issue is how to geographically and socially distribute the cost.

References

Readers desiring detailed references for specific natural or cultural aspects of the canyon should consult the following general references as guides to the literature.

Breed, W., and Roat, E., eds. 1974. *Geology of the Grand Canyon.* Grand Canyon: Museum of Northern Arizona, Grand Canyon Natural History Association.

Euler, R. C., Gummerman, G. J., Karlstrom, T. N. V., Dean, J. S., and Helvy, R. H. 1979. The Colorado plateaus: Cultural dynamics and paleoenvironments. *Science* 205:1089-1101.

Fradkin, F. L. 1981. *A River No More: The Colorado River and the West.* New York: Alfred A. Knopf.

Graf, W. L. 1985. *The Colorado River: Instability and Basin Management,* Resource Publications in Geography. Washington, DC: Association of American Geographers.

Powell, J. W. 1961. *The Exploration of the Colorado River and Its Canyons.* New York: Dover Publications. (Reprinted version of *Canyons of the Colorado* [1895], New York: Flood and Vincent.)

Schwartz, D. W. 1963. An archeological survey of Nankoweap Canyon, Grand Canyon National Park. *American Antiquity* 28:289-302.

Stevens, L. 1983. *The Colorado River in Grand Canyon: A Comprehensive Guide to Its Natural and Human History.* Flagstaff, AZ: Red Lake Books.

P A R T   F O U R

DOUBLE
EXPOSURES

Second opinions and comparisons of two or more situations are useful strategies for revealing nuances of difference in geographical landscapes. Geographers often include comparative strategies in their research designs. Thus, the similarity of geometrical form of alpine cirques and the artificial amphitheaters of American football stadiums allow studies of surface temperatures and thermal radiation in relationship to landscape form. The Continental Divide provides a setting for comparing treeline responses on either side of the divide to differing combinations of microclimate, aspect, and elevation. Economic development strategies in downtown East Asian villages of Boston and Oakland, differences in the design of 19th-century Chicago suburbs, and historic preservation conflicts in urban and rural Florida offer other examples of a deliberate research approach to select counterpart settings for direct comparison. Knowledge of geographical environments is also enhanced through comparison of independent analyses of the same or similar situations by different researchers. Thus, we offer the views of physical geographers and human geographers in their interpretations of the San Andreas Fault and of Niagara Falls. Paired independent studies of hurricane origins and damage along North Carolina and Louisiana coastlines round out our double-exposure approach to geographical comparisons. We often see more of a composite picture when the frames have been double-exposed.

—Susan L. Cutter, Rutgers University

The San Andreas Fault

Antony R. Orme
University of California, Los Angeles

The San Andreas Fault is the most striking physical lineament along the western margin of the United States. It also poses a significant hazard to human life and livelihood, dramatized in October 1989 by the Loma Prieta earthquake. This earthquake was, however, but one of many that have resulted from episodic shifts of the fault over the past several million years. The dramatic landscape and seismic expressions of the San Andreas Fault are the focus of much research, while the human implications of fault movement attract studies directed toward earthquake prediction and land-use planning.

Origins and Tectonic Significance

The San Andreas Fault extends 1,200 kilometers northwestward across California, from the Salton Trough to Cape Mendocino, and dips at 70° to 90° into the earth's crust to depths of over 18 kilometers (Figure 1). It is a right-lateral strike-slip structure in that land on the far side of the fault is displaced laterally to the right, and there is also some vertical slip. In places the fault is a narrow zone of broken and pulverized rock; elsewhere it is a complex belt of parallel fractures and deformed rocks as wide as 10 kilometers.

In terms of plate tectonics, the San Andreas Fault represents a transform margin between the North American and Pacific plates. Both plates have a westward component but, because of its northward component, the Pacific plate is moving northwestward relative to North America. In reality, the San Andreas Fault proper is part of a broader system of near-parallel faults which reflect, to a greater or lesser extent, shearing between these plates over the past 15 Ma (million years ago). Whether active or not, these faults collectively form the San Andreas fault system, a zone of crustal wrenching as wide as 200 kilometers that now encompasses the populous Los Angeles and San Francisco metropolitan areas.

The origins of the San Andreas Fault lie within the margin that evolved during Cenozoic time between the North American continental plate and successive oceanic plates farther west (Figure 1 insets). In earlier Cenozoic time, this was a convergent margin wherein the westward-moving North American plate converged obliquely on and overrode the oceanic Farralon plate. As parts of the latter plate were subducted and the East Pacific Rise, an oceanic spreading center farther west, was in turn overridden, the North American plate began transferring onto the Pacific plate. The plate margin between this transferred mass and the main North American plate thus changed to a transform type, with the lateral shear being transmitted upward to the surface as strike-slip faults. Right-lateral movement along this margin probably began around 30 Ma and has dominated the past 15 Ma. Further, as the East Pacific Rise was progressively overridden, so lateral shear along this margin stepped farther eastward with time. Earlier strike-slip motion occurred along faults now off the California coast, then along the San Gabriel Fault, and now along the San Andreas Fault proper with some movement even farther east along the Hayward and Calaveras faults (Figure 1).

Farther south, initial shearing occurred along the Tosco-Abreojos-San Benito fault system off western Mexico. Later, as the North American plate overrode the East Pacific Rise, crustal extension formed the fledgling Gulf of California (Stock and Hodges 1989). About 5 Ma, the East Pacific Rise appeared beneath the mouth of the Gulf and the Baja California miniplate began accelerating away from mainland Mexico on the western limb of this spreading sea floor, namely the Pacific plate. Since then, as Baja California has

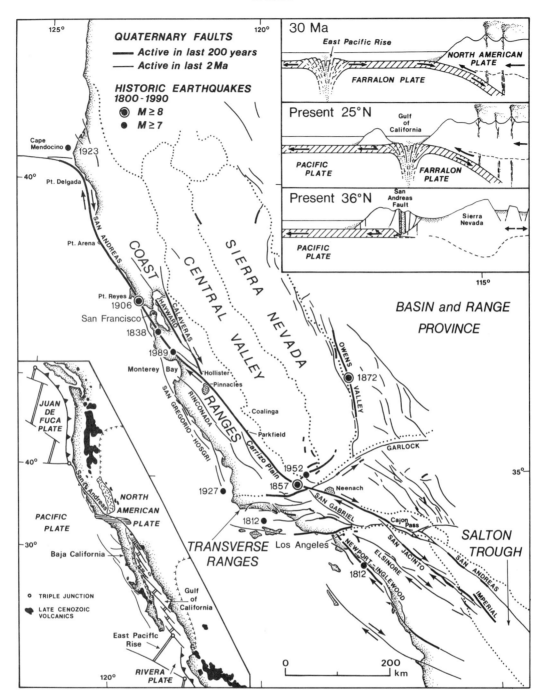

Figure 1. The San Andreas fault system, other Quaternary faults, and historic earthquakes in California. Insets show the evolution and present relations of the San Andreas transform margin.

been rafted northwestward at a mean rate of 56 to 60 millimeters per year (Larson 1972; Minster and Jordan 1987), transform faults offsetting the East Pacific Rise within the Gulf have propagated northward into the continental crust of the Salton Trough, forming sequential right-lateral strike-slip faults of which the San Andreas is currently the most prominent. Fur-

thermore, as the Baja California miniplate collided with the resistant roots of the Sierra Nevada, compression formed the Big Bend in the San Andreas Fault and severely buckled the Pacific Coast (Orme 1980). North of Cape Mendocino, these evolving plate relations remain at an earlier stage as fragments of the Farralon plate, namely the Gorda and Juan de Fuca plates, continue to be subducted beneath the North American plate, with associated volcanism in the Cascade Range farther east.

Surface Expression

The path of the San Andreas Fault across California may be divided into three segments (Figure 1). In the south, through the Salton Trough, a 50-kilometer-wide fault zone trends N45°W for 300 kilometers and comprises several active faults which converge on the Cajon Pass. For 180 kilometers from the Cajon Pass,

the Big Bend of the San Andreas Fault trends N75°W as a narrow lineament north of the Transverse Ranges. The fault then trends N35°W over the remaining 720 kilometers to Cape Mendocino, crosses the coast just south of San Francisco, and passes east of the Point Reyes Peninsula and Point Arena to Point Delgada before turning west into the Mendocino fracture zone.

Because of intense deformation and recurrent fracturing or persistent creep, rocks along the fault are relatively soft and easily eroded. Thus where the fault zone is narrow, its course is marked by a linear trough characterized by fault scarps, pressure ridges and sag ponds (Figure 2). Repeated displacement along the

Figure 2. The San Andreas Fault looking east-southeast along the northern edge of the Transverse Ranges, near Palmdale, California. The San Gabriel Mountains rise to the south, the Mojave Desert lies to the north. The California Aqueduct and the Palmdale Freeway now cross the fault in the foreground. Source: The Spence Collection, University of California, Los Angeles, 1954.

fault is well shown by offset streams and dislocated alluvial fans, notably in the Carrizo Plain and Transverse Ranges. Fault-controlled groundwater barriers commonly lead to springs and surface seepage. Local compression often causes uplift within the fault zone, as in the Ocotillo badlands in the Salton Trough. Mountain fronts alongside the fault are markedly linear with numerous truncated spurs. During earthquakes, liquefaction at depth may cause sand boils or "mud volcanoes" at the surface as water-saturated sediments flow upward along fractures. Continuing geothermal and volcanic activity within the fault zone, as in the Salton Trough where five rhyolite domes have extruded onto Quaternary sediments, have led to the plate margin's designation as a "leaky transform" boundary.

At a regional scale, compression of the Pacific plate against the roots of the North American plate has forced excess mass upward to form prominent mountains, notably the Transverse Ranges, which rose rapidly when the transform boundary collided with the Sierra Nevada and created the Big Bend in the San Andreas Fault. As these mountains rotated to their present east-west alignment and continuing compression triggered thrust faulting, marginal pull-apart structures such as the Los Angeles and Santa Maria basins also formed. The 260-kilometer left-lateral Garlock Fault may have been created by related displacement southeast of the Sierra Nevada. Farther north, the Coast Ranges are being compressed by active thrusting along their eastern margins (Wentworth and Zoback 1989). Lateral displacement of regional drainage is illustrated by the San Benito River which flows northwest for 100 kilometers along the San Andreas fault zone before breaking seawards to Monterey Bay.

Fault Movement and Plate Displacement

Because plate displacement along the San Andreas Fault is usually associated with earthquakes, an understanding of slip rates along the fault is important, particularly as recent movement may be the best clue to future fault behavior and earthquake prediction. Displacement along the fault may be deduced from a variety of temporal evidence. Long-term displacement is indicated, for example, by early Miocene volcanic rocks at the Pinnacles and Neenach, which originated from a common eruptive center around 23 Ma and have since been separated by 310 kilometers of movement along the San Andreas Fault (Matthews 1976). Marine sediments beneath the southern Central Valley and the Santa Cruz Mountains have been similarly offset by 325 kilometers since early Miocene times (Stanley 1987). Geologic discordances across the Gulf

of California indicate that Baja California has been rafted a similar distance northwest away from mainland Mexico during this interval. Although initial Miocene slip rates were as low as 3 millimeters per year, most evidence indicates a long-term slip rate of 13 to 14 millimeters per year over the past 23 million years. Such rates have not, however, been constant in time or space. For example, Colorado River deposits on the western margins of the Salton Trough have been displaced 130 kilometers from their source over the past 2.8 Ma, for a rate of 46 millimeters per year (Winker and Kidwell 1986).

Medium-term displacement is indicated at Wallace Creek, in the Carrizo Plain, which has been offset 128 meters along the San Andreas Fault over the past 3,680 years for a mean slip rate of 35 millimeters per year (Sieh and Jahns 1984). Nearby alluvial-fan deposits have been offset 475 meters over 13,250 years for a mean slip rate of 36 millimeters per year. Near the Cajon Pass, dislocated alluvial deposits indicate a slip rate of 25 millimeters per year over the past 14,400 years which, when added to slip of 10 millimeters per year on the San Jacinto Fault, is comparable to the Wallace Creek data (Weldon and Sieh 1985). Late Quaternary slippage in the Salton Trough approximates 23 to 35 millimeters per year (Keller et al. 1982).

For the past 100 years, short-term slip rates of 32 to 36 millimeters per year have been revealed between Parkfield and San Francisco Bay by periodic geodetic observations (for example, Prescott et al. 1981), augmented since 1985 by monthly observations of relative vectors using the Global Positioning System of survey satellites. The Farralon Islands on the Pacific plate 30 kilometers west of San Francisco are shifting northwards at this rate relative to "fixed" points east of the San Andreas Fault.

The medium- and short-term measurements together suggest a mean slip rate of 34 millimeters per year for Holocene times. This is an order of magnitude greater than the inferred Miocene rate of 3 millimeters per year and larger than the long-term rate of 13 to 14 millimeters per year. Movement along the San Andreas Fault has thus accelerated in the recent geologic past and may be greater now than at any time over the past 30 Ma.

However, magnetic anomalies and plate reconstructions across the Gulf of California indicate plate separation of 56 to 60 millimeters per year over the past 4 Ma (Larson 1972; Minster and Jordan 1987). To explain this discrepancy between plate displacement of 56 to 60 millimeters per year and a San Andreas slip rate of 34 millimeters per year, it is necessary to abandon rigid-plate concepts in favor of more deformable models and to invoke right-lateral motion of 8 to 27 millimeters per year on faults to the west and the extension of the Basin and Range Province to the east by 10 to 20 millimeters per year. These scenarios are well supported by evidence for slip along the 420-kilometer San Gregorio-Hosgri fault system offshore, which may have been displaced 100 kilometers since Miocene times, and by normal faulting east of the Sierra Nevada, which gave rise, among others, to the great M8 Owens Valley earthquake of 1872. In the Santa Cruz Mountains, the San Andreas Fault accounts for only 13 millimeters per year of relative plate motion; the remainder is distributed east and west, along the Hayward-Calaveras and San Gregorio fault systems respectively.

Earthquake History and Prediction

California has a historic record of earthquake activity extending back over 200 years and there is mounting evidence for datable prehistoric events. In 1769, the Portola expedition felt 24 earthquakes in one week as it crossed the Los Angeles basin. Franciscan missionaries, soldiers, and early newspapers later documented 19th century events but such records were inevitably incomplete, confined as they were to earthquakes actually felt in inhabited areas and uncertain as to which fault had moved. The first permanent earthquake laboratories were established by the University of California in 1887 and statewide seismographic coverage existed by 1932. Based in part on earlier systems, the modified Mercalli scale (MM) of earthquake intensities, which ranks shaking in ascending order from MMI to MMXII, was defined in 1931, and the open-ended logarithmic scale of earthquake magnitude (M) was introduced by Charles Richter of the California Institute of Technology in 1935. Since then, both scales have been used to describe earthquakes and retrofitted, as far as possible, to past events.

Since 1800, California has experienced more than 70 earthquakes of M6 or larger, including 10 over M7, of which 3 exceeded M8 (Figure 1). Of these, half have occurred along the San Andreas Fault proper, including the M8+ Fort Tejon earthquake of 1857 and the M8.3 San Francisco earthquake of 1906. The record of smaller but still troubling quakes is much larger: 188 events of M4 or greater occurred in California between 1975 and 1979, again about half along the San Andreas Fault. Most such earthquakes focus at shallow depths of less than 18 kilometers and the larger ones are thus particularly damaging at the surface.

In southern California, the 240-kilometer San Jacinto Fault is among the most seismically active strands of the San Andreas system, having experienced 9 events of M6 or larger since 1890. A complex freeway interchange between Interstates 10 and 15 actually straddles this fault! The nearby Imperial Fault has also been frequently active, with an M6.7 event in 1940 offsetting citrus groves while an M6.6 event in 1979 caused 30 kilometers of ground rupture with 55 centimeters of right-lateral and 19 centimeters of vertical displacement. Indeed, in the pervasively faulted Salton Trough, earthquake swarms of low to moderate intensity are quite frequent as strain is released upwards through thick, poorly compacted Quaternary sediments.

The largest historic earthquake along the Big Bend segment was the M8+ Fort Tejon earthquake of 1857 when the San Andreas Fault sustained as much as 9.5 meters of lateral slip over the 320 kilometers between the Cajon Pass and Parkfield. The fresh tectonic relief of this segment includes numerous sag ponds and fault scarps produced during this event. Intensities of MMVII or larger were felt from Los Angeles to Monterey but, owing to the then sparse population, damage was small (an MMVII event causes people to run outdoors, and damage to poorly built structures is considerable). Since 1857 this fault segment has remained locked, suggesting that stresses accumulate here over long periods until relieved during a great earthquake—the "big one" anticipated by southern California residents.

North of the Big Bend, tectonic creep increases to 32 to 36 millimeters per year along the San Andreas Fault as it passes through the Carrizo Plain to Parkfield, and then diminishes farther north toward Monterey Bay but is found along the Hayward-Calaveras system past Hollister. Creep may reflect the nature of basement rocks and structures. Because tectonic creep inhibits the accumulation of strain energy, which would otherwise be released by larger earthquakes, this fault segment has experienced mostly low to moderate events, notably the Parkfield earthquakes of 1932 (M6) and 1966 (M5.5). This creeping segment attracts scientific attention because periodic measurements of strain and creep may provide early detection of altered conditions preceding an earthquake.

The San Francisco region has suffered several moderate to large historic earthquakes within the San Andreas system. An M7 event caused 40 kilometers of surface ruptures along the main fault near San Francisco in 1838, while across San Francisco Bay two M6.8 events occurred on the Hayward Fault in 1836 and 1868. The great M8.3 earthquake of April 1906 ruptured along 450 kilometers of the San Andreas Fault from near Monterey Bay to beyond Point Delgada, with up to 5 meters of right-lateral slip in the trough east of the Point Reyes peninsula. This event raised studies of the San Andreas Fault to a higher plane (Lawson *et al.* 1908). Furthermore, realization that displacement died out within 30 kilometers east and west of the fault led to the theory of elastic rebound, which states that crustal rocks that are progressively strained by slow plate movement are returned to an unstrained state by the sudden fault displacements that cause earthquakes.

More recently, the M7.1 Loma Prieta earthquake of October 1989, whose epicenter lay in the Santa Cruz Mountains 100 kilometers south of San Francisco, was caused by compressive slip along a 40-kilometer segment of the San Andreas Fault that ruptured the crust from a depth of 18 kilometers to within 5 kilometers of the surface. Between 1906 and 1989, this fault segment had remained relatively quiet, or aseismic, and a 30 percent chance of an M7 earthquake had been predicted for the period 1988 to 2018! Displacement on the 70° fault plane amounted to 2 meters laterally and 1.3 meters vertically, but there was little primary surface faulting (Plafker and Galloway 1989). Instead, 10 seconds of seismic shaking at depth caused extensive surface cracking and many landslides, while liquefaction of water-saturated sediments and artificial fills caused much damage around San Francisco and Monterey Bays. Intensities of MMVII or larger were felt from Salinas in the south to Berkeley in the north. The earthquake caused 67 known deaths, more than $6 billion in property damage, and the collapse of several bridge and freeway structures, and left more than 12,000 people homeless. Some 51 aftershocks of M3 or larger occurred within 24 hours of the main shock. Despite this event, the probability of another M7 earthquake within the San Andreas fault system around San Francisco Bay remains high. There have also been frequent earthquakes at the far north end of the San Andreas Fault, notably the M7.2 event near Cape Mendocino in 1923. This area is a triple junction where the North American, Pacific, and Gorda plates meet.

Society is of course less concerned about past earthquakes than future events, and much scientific effort is now directed toward forecasting more precisely the time, place, and size of future shocks. Several avenues of inquiry are available for earthquake prediction but all have limitations. First, the historic record is useful in directing attention to active faults but is far too short for fine-tuned estimates of future earthquakes. The large 1838 and 1906 earthquakes on the San Andreas Fault near San Francisco occurred 68 years apart and, although strain was known to be accumulating, the next major event came in 1989, 83 years later, and focused on a different fault segment. Second, where the record can be extended back into prehistoric time, earthquake recurrence is better understood but still vague. For example, periodic fault

displacement and liquefaction of radiocarbon-dated marsh sediments at Pallett Creek, 50 kilometers northeast of Los Angeles, have been attributed to 9 large earthquakes between A.D. 575 and 1857 (Sieh 1978). Although an average recurrence interval of 160 years is suggested, the actual interval between events ranged from 55 to 275 years! Third, slip rates may be used to infer earthquake recurrence intervals. For example, the 9.5 meters of right-lateral slip sustained by the San Andreas Fault at Wallace Creek in the 1857 earthquake suggest, when compared with the medium-term slip rate of 34 millimeters per year, that the 1857 event was preceded by 280 years of strain accumulation (Sieh and Jahns 1984). Variations in fault offset and other assumptions, however, indicate that the latest three recurrence intervals range from 240 to 450 years! Fourth, geodetic measurements and strain-meters measure currently accumulating strain with much greater precision, but earthquake prediction remains confounded by variable crustal behavior, especially between locked and creeping fault segments.

Finally, even where the past role and future significance of the San Andreas Fault can be evaluated, seismic prediction in California is confounded because many earthquakes occur along other faults within the straining plate margin and some occur on previously unknown or buried faults. For example, wrenching along the Newport-Inglewood fault zone caused the M6.3 Long Beach earthquake of 1933, the damaging impact of which led to state regulation of public school construction (Field Act, 1933). An M7 event on this fault would have a stronger impact on the Los Angeles metropolitan area than an M8 event on the San Andreas Fault 50 kilometers to the northeast (Toppozada et al. 1989). The M6.4 San Fernando earthquake of 1971, which occurred on a largely ignored east-west thrust fault in the Transverse Ranges within that metropolitan area, caused sufficient destruction and death to promote statewide studies designed to restrict construction in active fault zones (Alquist-Priolo Special Studies Zones Act, 1972). The destructive M6.7 Coalinga quake of 1983 occurred close to the San Andreas Fault but on a buried thrust fault whose existence has since led to reappraisal of plate relations along the eastern margin of the Coast Ranges.

In short, California is earthquake country and the San Andreas Fault is the principal, but not the only cause of this activity. An earthquake of M7 or larger is likely to occur somewhere in California every 15 to 25 years, and the Loma Prieta event of 1989 emphasized the need for further diligence in developing and enforcing land-use policies designed to reduce the loss of life and property during future earthquakes. Scientific understanding of the physical dimensions of faults and earthquakes has increased rapidly in recent years and, whereas society may not yet be precisely forewarned, it must at least be prepared.

References

Keller, E. A., Bonkowski, M. S., Korsch, R. J., and Shlemon, R. J. 1982. Tectonic geomorphology of the San Andreas fault zone in the southern Indio Hills, Coachella Valley, California. *Geological Society of America Bulletin* 93:46-56.

Larson, R. L. 1972. Bathymetry, magnetic anomalies, and plate tectonics history of the mouth of the Gulf of California. *Geological Society of America Bulletin* 83:3345-3360.

Lawson, A. C., et al. 1908. *The California Earthquake of April 18, 1906; Report of the State Earthquake Investigation Commission.* Washington: Carnegie Institution.

Matthews, V. 1976. Correlation of Pinnacles and Neenach volcanic formations and their bearing on San Andreas fault problem. *American Association of Petroleum Geologists Bulletin* 60:2128-2141.

Minster, J. B., and Jordan, T. H. 1987. Vector constraints on western U.S. deformation from space geodesy, neotectonics, and plate motions. *Journal of Geophysical Research* 92:4298-4304.

Orme, A. R. 1980. Marine terraces and Quaternary tectonism, northwest Baja California, Mexico. *Physical Geography* 1:138-161.

Plafker, G., and Galloway J. P., eds. 1989. Lessons learned from the Loma Prieta, California, earthquake of October 17, 1989. *United States Geological Survey Circular* 1045.

Prescott, W. H., Lisowski, M., and Savage, J. C. 1981. Geodetic measurements of crustal deformation on the San Andreas, Hayward, and Calaveras faults near San Francisco, California. *Journal of Geophysical Research* 86:10853-10869.

Sieh, K. E. 1978. Prehistoric large earthquakes produced by slip on the San Andreas Fault at Pallett Creek, California. *Journal of Geophysical Research* 83:3907-3939.

———— , **and Jahns, R. H.** 1984. Holocene activity of the San Andreas Fault at Wallace Creek, California. *Geological Society of America Bulletin* 95:883-896.

Stanley, R. G. 1987. New estimates of displacement along the San Andreas Fault in central California based on paleobathymetry and paleogeography. *Geology* 15:171-174.

Stock, J. M., and Hodges, K. V. 1989. Pre-Pliocene extension around the Gulf of California and the transfer of Baja California to the Pacific plate. *Tectonics* 8:99-115.

Toppozada, T. R., Bennett, J. H., Borchardt, G., Saul, R., and Davis, J. F. 1989. Earthquake planning scenario for a major earthquake on the Newport-Inglewood fault zone. *California Geology* 42:75-84.

Weldon, R. J., and Sieh, K. E. 1985. Holocene rate of slip and tentative recurrence interval for large earthquakes on the San Andreas Fault, Cajon Pass, southern California. *Geological Society of America Bulletin* 96:793-812.

Wentworth, C. M., and Zoback, M. D. 1989. The style of late Cenozoic deformation at the eastern front of the California Coast Ranges. *Tectonics* 8:237-246.

Winker, C. D., and Kidwell, S. M. 1986. Paleocurrent evidence for lateral displacement of the Pliocene Colorado River delta by the San Andreas fault system, southeastern California. *Geology* 14:788-791.

The San Andreas Fault: Human Dimensions

Risa Palm
University of Oregon

The San Andreas Fault is an infamous and enigmatic line that is highly significant to California residents. This fault threatens life and property as it passes near or through such major population centers as San Francisco and Los Angeles. In some places it is demarcated by a major trench and long, narrow lakes (such as the San Andreas Lake). It also defines some of the most beautiful wetlands and seashore in the Point Reyes area on the northern California coast.

Because of earthquakes emanating from the San Andreas Fault zone, California is dubbed "earthquake country." Popular books discuss the earthquake problem matter-of-factly. A Sunset book sold in grocery stores, *Earthquake Country*, begins: "California is Earthquake Country. Earthquakes are a part of California's heritage and we all must learn to live with them" (Iacopi 1978, 4). *Life Along the San Andreas Fault* (Fried 1973) discusses the problems of living with the constant threat of earthquakes. Finally, the popular Thomas Brothers street atlases contain designations of "earthquake and flood zones." Yet, the earthquake hazard is not taken so lightly by California residents, particularly after the jarring experience of the 1989 Loma Prieta earthquake. It is clear that the San Andreas Fault is a line that continues to have a major impact on the lives of Californians.

The San Andreas Fault is not a single line at all, but rather a system of faults within California at the boundary between two relatively rigid crustal plates: the Pacific oceanic plate and the North American continental plate (for an illustration, refer to Figure 1 in the previous chapter, p. 144). This boundary area passes through several of the state's most densely populated urban regions (including the Los Angeles basin, the San Bernardino region, and the San Francisco Bay area). It is therefore "one of the most extensively urbanized tectonic plate boundaries on Earth" (Yerkes 1985, 25).

Considerable attention has focused on attempts to predict the location and magnitude of earthquakes. For example, the U.S. Geological Survey upgraded their prediction of a probable earthquake in the San Francisco Bay region in August of 1990, predicting a two in three chance of a magnitude 7 or higher earthquake along one of the major portions of the San Andreas system by the year 2020. Such an earthquake in this heavily urbanized region could cause loss of many lives, destruction of many homes, and disruptions in the lives of thousands of people.

The Federal Emergency Management Agency along with the U.S. Geological Survey estimate a high probability, exceeding 40 percent, that a large earthquake will occur within the next 30 years near Los Angeles (Lindh 1983; Wesson and Wallace 1985). Projected losses from such an earthquake include $25 billion (1980 dollars) in damages, with more than 50,000 persons made homeless and up to 12,500 deaths (Federal Emergency Management Agency 1980). Such a disaster would exceed any previous natural disaster in the nation's history. A small magnitude earthquake on the Newport-Inglewood Fault, which passes obliquely through the Los Angeles metropolitan area, is projected to have even more serious implications: as many as 21,000 deaths, 200,000 persons homeless, and $60 billion (1980 dollars) in damage. Clearly, the spectre of a disaster of major proportions caused by movement along the San Andreas Fault system hangs over the major metropolitan areas of California.

How Do Residents Respond to This Line?

General knowledge of the San Andreas fault and of earthquake hazards is widespread (Turner *et al.* 1979). But, for California residents to respond effectively to the earthquake hazard, they must be aware of its meaning, have resources to respond, and believe that they control their own destiny and the hazards affect them individually, and that the hazards are particularly salient in their daily lives. Let us deal with just one aspect of these necessary conditions: the availability of information.

Public Information

General information about earthquake hazards is available from an increasingly large number of sources. Teacher training programs introduce earthquake mitigation units into the public schools (Thier 1987). Scout organizations award earthquake mitigation badges. Local governments sponsor earthquake awareness days; the most ostentatious, sponsored by the city of Los Angeles, showed film footage of a simulated Los Angeles earthquake provided by Universal Studios (Mattingly 1987). Street atlases have overlays showing the surface fault rupture-zones. Telephone books contain information on emergency procedures to follow during an earthquake. Real estate agents must disclose location within a surface fault rupture-zone to prospective home buyers. Thus purchasers of residential property are specifically informed that their home lies within a "Special Studies Zone." The U.S. Geological Survey, the Office of the State Geologist, the Federal Emergency Management Agency, the Southern California Earthquake Preparedness Project (SCEPP), the Bay Area Region Earthquake Preparedness Project (BAREPP), the Business and Industry Council for Earthquake Preparedness (BICEPP), and other organizations provide brochures, maps, scientific papers, and other materials on the earthquake problem and on steps that can be taken to mitigate some of its worst effects. The *Los Angeles Times* and other California

newspapers have published maps and reports on earthquake hazards. In short, no one can claim that public information about the distribution of fault traces or recommended steps to prepare for a major earthquake are not in the public domain.

Public information about the availability of earthquake insurance is required by state law. Since 1985, every person subscribing to homeowners' insurance receives an offer of earthquake insurance, complete with costs and information about coverage and deductibles.

Risk Perception

To measure the receipt of this information, and its conversion into the risk perception of residents of the San Andreas Fault zone, Palm *et al.* (1990) posed four questions to probe perceived probabilities of a major (1906 San Francisco-type) earthquake causing serious damage to the resident's community and own home. They found that the general level of concern about a future earthquake causing damage to their own home was high among the 1786 respondents. In three of the four counties, more than 50 percent of respondents estimated that an earthquake of the size that occurred in San Francisco in 1906 had a one in ten chance of occurring in their community in the next 10 years (Table 1). This concern was highest in Los Angeles County where 69 percent of the residents thought the chance of such an earthquake was one in ten. More than 20 percent of respondents indicated that damage to their own homes would total over $200,000.

We must conclude that California residents have a great deal of information about earthquake hazards at their disposal and also that they are highly concerned with earthquake hazards. What then is the response of these residents? How do they protect themselves from the risk posed by the San Andreas Fault system?

The Adoption of Earthquake Mitigation Measures

Can residents respond effectively? Can anything be done by individuals and households to avoid at least some of the damage, destruction and devastation of a major earthquake? The answer is "yes." By preparing for earthquakes, states, regions, and individuals can lessen adverse effects and prevent deaths, injuries, and serious disruptions. Individuals and households can

Table 1. Levels of Concern in Relation to Perceived Risk of Earthquake Hazards (in Percentages)

	Contra Costa	Santa Clara	Los Angeles	San Bernardino
Estimated probability of a damaging earthquake affecting community				
>1 in 10	47.0	57.6	68.9	63.7
1 in 11 to 1 in 100	39.0	33.6	25.1	25.2
<1 in 100	14.0	8.8	5.9	11.1
Estimated probability of a damaging earthquake affecting home				
>1 in 10	45.1	49.5	61.0	61.4
1 in 11 to 1 in 100	39.3	38.7	31.7	32.8
<1 in 100	15.6	11.8	7.3	10.8
Estimated dollar damage from a major earthquake				
<$50,000	34.6	35.3	39.7	45.5
$50,001–$100,000	22.4	17.6	26.0	27.7
$100,001–$200,000	22.7	23.2	19.6	23.4
>$200,000	20.2	23.8	14.7	3.1
Likelihood of own home being seriously damaged by earthquake				
Very likely	3.6	5.2	12.0	15.6
Somewhat likely	27.8	33.5	44.6	40.4
Somewhat unlikely	31.1	31.0	21.1	20.2
Not very likely	37.5	30.3	22.3	23.8

take simple steps to make plans for family reunion after the earthquake; learn cardiopulmonary resuscitation (CPR) and first aid; learn how to shut off gas, water, and electricity; put latches on cabinets; store flammable and hazardous liquids on lower shelves, away from furnace and hot water heater; and maintain emergency supplies including flashlight, portable battery-operated radio, extra batteries, medicines, and emergency supplies of food and water.

Despite a long list of practical and inexpensive measures to prepare for the duration and immediate post-earthquake period, many households do not prepare for an earthquake. A 1977 survey of 1,450 Los Angeles residents (Turner et al. 1980) showed that a large percentage of people living in an earthquake-prone area believe they cannot prepare for an earthquake. In response to the survey statement, "There is nothing I can do about earthquakes, so I don't try to prepare for that kind of emergency," 41 percent agreed and, of these, 7 percent "agreed strongly." Almost one-third (32 percent) agreed with the even more fatalistic statement, "The way I look at it, nothing is going to

help if there were an earthquake" (Turner et al. 1980, 3). They continued: "most households are unprepared for an earthquake and that the prospect of an earthquake has stimulated relatively little preparatory action" (p. 101). Although more than 70 percent had a working flashlight, and more than 50 percent had a working battery radio and first aid kit, fewer than 30 percent stored food and fewer than 20 percent stored water or took any other precautions. Although 23 percent had inquired about earthquake insurance, fewer than 13 percent bought it. Only about 11 percent had structurally reinforced homes. About half of the families with young children had told them what to do in an earthquake, but fewer than 35 percent had set up emergency procedures in the residence, and fewer than 25 percent had plans for family reunion after an earthquake.

In 1979, I undertook a smaller survey of residents of Special Studies Zones (surface fault rupture-zones) in Berkeley and Contra Costa County. Its purpose was to ascertain whether residents who both received and recalled a disclosure that their property is within a

Special Studies Zone would be more likely to adopt the same set of mitigation measures as the general population of Los Angeles (Palm 1981). The study population included recent home buyers who understood the meaning of a special studies zone and were of higher average income and education than the general population. These residents were more likely to have inquired about earthquake insurance (41 percent), bought earthquake insurance (24 percent), and invested in structural reinforcements for their homes (9 percent). But they were generally less likely than the Angelenos to adopt such mitigation measures as instructing children what to do in an earthquake, establishing emergency procedures at the residence, making plans for reunion after an earthquake, having a working battery radio, rearranging cupboard contents, contacting neighbors for information, or storing either food or water. Thus, studies completed in the late 1970s showed an astounding lack of individual or household preparedness.

The adoption of earthquake insurance by homeowners as a measure to mitigate against some of the most severe financial effects of a major earthquake has also been studied. In a 1989 survey of 3,500 California owner-occupiers in Contra Costa, Santa Clara, Los Angeles, and San Bernardino counties, my colleagues and I (1990) found more homeowners who had purchased earthquake insurance than indicated by surveys in the 1970s. From approximately 5 percent coverage reported by Howard Kunreuther and colleagues (1979) in 1973 to 1974, the percentages of households with earthquake insurance rose to almost 40 percent in two of the four sample counties. In all four of the counties, those with insurance indicated that the most important factors in their decision to purchase insurance were related to possible damage to houses: "I worry that an earthquake may destroy my house or cause major damage in the future," "Most of our family wealth is tied up in the equity of our house, which might be lost if an earthquake destroyed or damaged it," "If a major earthquake occurs, the grants or loans available from the federal or state government will not be sufficient to rebuild my house," and to a lesser extent, "If a major earthquake occurs, the damage to my house will be very great, so insurance is a good buy."

Although the numbers and percentages of policyholders have increased, the majority of California owner-occupiers are still uninsured against the earthquake peril. A full 78 percent of Contra Costa County owner-occupiers do not have earthquake coverage, and in the counties with the highest insurance rates —Los Angeles and Santa Clara—60 percent remain uninsured.

Conclusion

The San Andreas region remains an enigma of human response to a dangerous and threatening environment. California residents are aware of earthquake risk. Many federal and state programs have been undertaken to prepare for a major earthquake and to minimize the loss of life and property. Yet, a vast majority of households do nothing to protect themselves from this environmental hazard. In this beautiful region—containing several of the world's most famous tourist destinations (the city of San Francisco, Disneyland, the Big Sur country)—danger lurks literally beneath the surface of the earth, marked by this infamous line: the San Andreas Fault.

References

Federal Emergency Management Agency (FEMA). 1980. *An Assessment of the Consequences and Preparations for a Catastrophic California Earthquake: Findings and Actions Taken.* Washington, DC: Federal Emergency Management Agency.

Fried, John J. 1973. *Life Along the San Andreas Fault.* New York: Saturday Review Press.

Iacopi, R. 1978. *Earthquake Country*, 3rd edition. Menlo Park, CA: Lane Books.

Kunreuther, Howard, Ginsberg, Ralph, Miller, Louis, Sagi, Philip, Slovic, Paul, Borkan, Bradley, and Katz, Norman. 1978. *Disaster Insurance Protection: Public Policy Lessons.* New York: John Wiley and Sons.

Lindh, A. G. 1983. *A Preliminary Assessment of Long-Term Probabilities for Large Earthquakes Along Selected Fault Segments of the San Andreas Fault System,* Open-File Report 83-63:15. Reston, VA: U.S. Geological Survey.

Mattingly, Shirley. 1987. Response and recovery planning with consideration of the scenario earthquakes developed by California Division of Mines and Geology. In *Proceedings of Conference 41: A Review of Earthquake Research Applications in the National Earthquake Hazards Reduction Program: 1977-1987,* Open File Report 88-13-A, ed. Walter W. Hays, pp. 550-554. Reston, VA: U.S. Geological Survey.

Palm, Risa. 1981. *Real Estate Agents and Special Studies Zones Disclosure: The Response of California Home Buyers to Earthquake Hazards Information,* Monograph No. 32. Boulder, CO: University of Colorado, Institute of Behavioral Science, Program on Technology, Environment and Man.

———, **Hodgson, Michael, Blanchard, R. Denise, and Lyons, Donald.** 1990. *Earthquake Insurance in California: Environmental Policy and Individual Decision-Making.* Boulder, CO: Westview Press.

Thier, Herbert D. 1987. The California earthquake education program. In *Proceedings of Conference 41: A Review of Earthquake Research Applications in the National Earthquake*

Hazards Reduction Program: 1977-1987, Open File Report 88-13-A, ed. Walter W. Hays, pp. 65-74. Reston, VA: U.S. Geological Survey.

Turner, Ralph, Nigg, Joanne, Paz, Denise Heller, and Young, Barbara Shaw. 1979. *Earthquake Threat: The Human Response in Southern California.* Los Angeles: University of California, Los Angeles, Institute for Social Science Research.

Wesson, R. L., and Wallace, R. E. 1985. Predicting the next great earthquake in California. *Scientific American* 252(2):35-43.

Yerkes, R. F. 1985. Geologic and seismologic setting. In *Evaluating Earthquake Hazards in the Los Angeles Region: An Earth-Science Perspective*, Professional Paper 1360, ed. J. I. Ziony, pp. 25-42. Washington, DC: U.S. Geological Survey.

The Meaning and Making of Niagara Falls

Patrick McGreevy
Clarion University of Pennsylvania

Only a handful of North American places are as well known around the world as Niagara Falls. When reports of the great cataract first reached Europe in the late 17th century, Niagara quickly became an emblem of the new continent itself. The completion of the Erie Canal in 1825 (Figure 1) made the falls accessible to travelers from the new middle class, whose ranks—on both sides of the Atlantic—swelled with the success of industrialization. By mid-century, Niagara Falls had become, in the words of one British visitor, "the goal and object of western travel" (Woods 1861, 244). Niagara's popularity waxed along with romanticism and the aesthetics of the natural sublime. A vast or overwhelming natural object was sublime if it produced a certain exhilarating experience of rapture commingled with terror. Many considered Niagara the ultimate embodiment of the natural sublime (McKinsey 1985).

Yet any late 20th-century visitor who arrives at Niagara Falls expecting to find a shrine of nature is likely to be puzzled by what people have made of this place. To begin with, the Falls is situated in a distinctly urban environment—a landscape of circuses, horror museums, and industry (Figure 2). Another curious feature of Niagara Falls is the tradition of honeymooning. The suicide tradition is a more closely guarded secret. Finally, what strikes many visitors as most curious of all is the tremendous contrast they see between the Canadian and American landscapes at Niagara. The former seems to have all the flowers, and the latter, all the pollution.

Much of what we see at Niagara Falls today seems at odds with the meanings people have attributed to the place. Yet the approach taken here is to turn to this realm of meanings—meanings not only of the Falls itself, but of the border as well—in an attempt to shed light on some of the puzzling features of Niagara Falls. Particularly in the 19th century, people expressed these meanings in a prodigious flood of words and pictures. Each of the following three sections focuses on a specific connection between the meanings attributed to Niagara Falls and its actual landscape.

A Place Apart

In the 19th century, traveling to Niagara was normally a once-in-a-lifetime event. Anna Jameson described the Falls in 1836 as "my childhood's thought, my youth's desire, since first my imagination was awakened to wonder and to wish" (Jameson 1838, 82). From afar, many travelers imagined a fantastic Niagara, a place so unlike their ordinary surroundings that it seemed otherworldly. It is not surprising, then, that many spoke of the journey to Niagara as a pilgrimage.

Medieval Christian pilgrims also idealized their goal from a distance. The pilgrim center, they believed, was different in nature from the towns and villages they had come from. The anthropologist Victor Turner argues that pilgrimage removes people from the ordinary world in the sense that social hierarchies, conventions, and routines are suspended. Turner likens pilgrims to initiates in a tribal rite of passage who are temporarily removed from familiar social structures before returning to tribal life in a new role (Turner 1973).

Some of Niagara's pilgrims believed that the Falls was a unique expression of God's power and majesty. This view achieved official recognition when Pope Pius IX, at the urging of Archbishop Lynch of Toronto, established a "pilgrim shrine" at Niagara Falls in 1861. A few years later, when Lynch founded a monastery at Niagara Falls, he commented: "God Himself has made the selection" (Lynch 1914, 99). Other travelers called themselves pilgrims more because of a reverence for nature than because of religious faith. "At Niagara,"

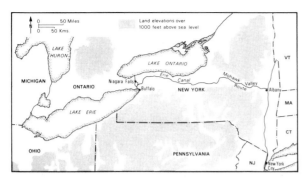

Figure 1. The completion of the Erie Canal in 1825 released Niagara Falls from its isolation deep in the North American interior.

wrote George Carlisle in 1850, 'you feel that you are not in the common world, but in its sublimest temple" (Carlisle 1850, 1073). Many who shared Carlisle's view considered the increasing intrusion of mills, hackmen, and hustlers a desecration. A movement to preserve the Falls eventually led to the creation of small reservations on each side of the border. James C. Carter, who dedicated the American reservation, said he had come representing the State of New York, to declare that Niagara was "not a property but a shrine." Therefore, he concluded, the state "marks out the boundaries of the sanctuary, expels from the interior all ordinary human pursuits and claims, so that visitors and pilgrims from near or far may come hither, and be permitted to behold, to love, to worship, to adore" (Carter 1903, 274).

On the face of it, the pilgrim's reverent vision seems incongruent with Niagara's bizarre side-show atmosphere. Since the early 19th century, visitors have been greeted with exotic markets, horror museums, circuses, and an endless series of stunts. Yet fairs and exotic markets always crowded around the great pilgrim churches of Europe. These activities, like pilgrimage itself, were outside the realm of ordinary life. They all belonged together, as Victor Turner notes, "in a place set apart" (Turner 1973, 208).

The honeymoon tradition came to Niagara, along with the railroad, in the mid-19th century. Marriage, particularly at this time, was a momentous transition. Like initiates in a rite of passage who are removed from their usual world to undergo a ritual transformation, the honeymoon couple entered a hiatus between two portions of life. Because of Niagara's image as a place apart from ordinary life, it was an appropriate site symbolically for the consummation of marriage. William Dean Howells wrote in his 1888 novel, *Their Wedding Journey*: "I think with tenderness of all the lives that have opened so fairly there. . . . Elsewhere there are the carking cares of business and fashion,

there are age and sorrow, and heartbreak; but here only youth, faith, rapture" (Howells 1888, 319).

Nature's Future

A number of 19th-century travelers likened the Niagara River to the stream of time. The Falls itself represented a radical break in that stream—a break separating the present from the unknowable future. A great outpouring of visions of the future appeared in the 1890s as hydroelectric development of Niagara Falls began. To many, the harnessing of Niagara seemed a capstone on the progress of industrial civilization. As one poet put it, "With power unrivaled thy proud flood shall speed / The New World's progress toward time's perfect day" (Copeland 1904, 12).

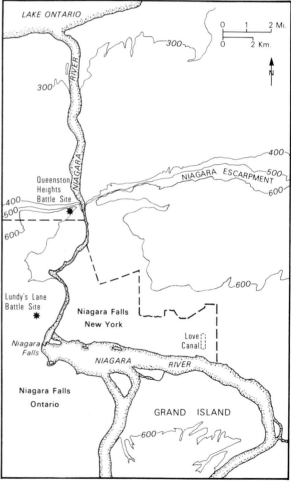

Figure 2. The two cities of Niagara Falls give this natural spectacle a very urban setting; today the Canadian city captures more of the tourist trade while the American city remains much more dependent on industry.

Most of these visions of the future were optimistic and took the form of massive utopian development schemes. Ironically these visions were based on the old sublime view of nature as unfathomable and exhaustless. Developers felt that if they could harness the power of Niagara, there was absolutely no limit to what they could do.

William T. Love, for instance, proposed in 1893 to build a navigable power channel around Niagara Falls to the site he called Model City. He promised "unlimited water power" that would lead to "rapid development such as no other city in the world has ever experienced." This was to be a carefully planned community, a Model City, "designed to be the most perfect city in existence" (Love 1895, 1). Although Love gained permission from local and state authorities to carry out his plan, an economic downturn bankrupted him after about 1.5 kilometers of the canal were completed.

Perhaps the most elaborate of these futuristic visions was King Camp Gillette's "Metropolis." Gillette argued that eventually there could be only one city in North America. Niagara Falls was the logical site because here was a power, he believed, capable of driving the concentrated industries of the continent as well as providing the energy needs for the 60 million inhabitants of Metropolis. The colossal city would stretch 230 kilometers (135 miles)—from Rochester to Hamilton—and the entire area, including Niagara Falls itself, was to be covered with a 33-meter (100-foot) platform (Gillette 1894).

A third scheme, proposed in 1896 by Leonard Henkle, called for a lavish palace, a kilometer in length and over 50 stories high, to span the Niagara River just above the crest of the falls. Henkle envisioned not only a great industrial and commercial center but a moral and political one, as well, for the palace was to serve also as a sort of United Nations Headquarters where each nation would send representatives to work together to eliminate war and poverty (*New York Herald* 1896).

One of the most daring proposals to surface in the 1890s was the idea of digging a 4-kilometer (2.5-mile) power tunnel around Niagara Falls. Vertical shafts capped with turbines would draw water from the upper river into the tunnel. E. D. Adams, a New York financier who led the effort, felt that past experience could shed little light on an enterprise this new and this massive. He enlisted the aid of a number of experts including Lord Kelvin, Thomas Edison, George Westinghouse, and Nikola Tesla. He even hired Stanford White to design a model worker's town. The plan created great excitement at Niagara. Tesla predicted that London and Paris would soon get their power

from Niagara. "Humanity will be like an ant heap stirred up with a stick," he exclaimed, "see the excitement coming" (Goldman 1971, 5). Unlike the other futuristic schemes, the power-tunnel project met tremendous success culminating in the first long-distance transmission of electricity in 1896.

The visions of the future presented in the utopian schemes of the 1890s provided a jump start for Niagara's industrial development. They made development alluring enough to justify greater and greater exploitation of the river. Today about two thirds of the annual flow is diverted for power generation. It was the old image of a stupendous, immeasurable waterfall that lured Gillette, Love, Henkle, and Adams. Niagara Falls was vigorously developed not because developers thought so little of the fall's natural splendor, but rather because they thought so much of it. Ironically, Love's uncompleted canal became filled with the byproducts of that development, and, instead of leading to the perfect future city, has become a place that, as far as human habitation is concerned, may have no future at all.

Front Door, Back Door

Since the 1830s visitors have noticed a distinct contrast between the two banks of the Niagara River. One mid-century traveler, for instance, found the American side "quite the reverse of attractive," but "the Canadian side of the Falls," he wrote, "boasts of charming scenery. Even in the snow, the neat cottages and houses—the plantations, gardens and shrubberies—evince a degree of taste and comfort which are not so observable on the American side" (Russell 1865, 49).

By the turn of the century, the accumulation of mills and power plants on the American shore struck many visitors as almost sacrilegious. More recently the Canadian landscape has been enhanced by a number of developments, notably the founding of a school of horticulture dedicated to beautifying the entire Niagara River area with flowers and gardens. The growth of the electrochemical industry in Niagara Falls, New York, created a gritty industrial landscape underlain at various points with toxic waste dumps; the Love Canal is only the most famous of these sites. Two hazardous waste disposal companies operate on the American side; they receive toxic waste from all over the northeastern United States. To complete the picture, a silo just north of Niagara Falls, New York, holds atomic waste from the Manhattan Project. The syndicated column "Dear Abby" received a letter from a 1985 visitor who commented that "the Falls on the American side were grossly neglected and looked terrible, but the Falls on the Canadian side were beautiful, bright with flowers and well-maintained." Abby received a flood of letters—1,900 that agreed with the original writer and fewer than 200 that defended the American side (Dear Abby 1985).

Certainly many factors have contributed to this contrast. Among them are the greater industrialization on the American side and the larger role of government in Canada. But the contrast is so pervasive and long-standing that it seems to require explanation of a different sort. I suggest that the way each side perceives the border itself may be crucial. It is as if the Niagara area is a front entrance for Canada, appropriately embellished and manicured, while for America it is a neglected back alley, a place perhaps for trash cans but not for flowers. Why should the Niagara border be particularly cherished by Canada but disturbing to the United States?

To begin with, the west bank of the Niagara River attracted one of the largest concentrations of loyalist settlers during the American Revolution, and the village known as Niagara or Newark became the first capital of Upper Canada. When the War of 1812 began, most Americans believed that Canada would be quickly conquered. The battles on the Niagara frontier were crucial, particularly those at Lundy's Lane, where almost 2,000 perished within earshot of Niagara Falls, and at Queenston Heights, where a huge federal monument now commemorates the war that preserved Canada and thereby demonstrated that the advance of America was not inevitable. Border tensions flared again during the Rebellion of 1837 when rebels set up camp on an island just above Niagara Falls where they received aid and supplies from American sympathizers. In 1866 a large group of Irish soldiers, just dismissed from the Union army, crossed the Niagara River in an attempt to force Britain to free Ireland by holding Canada hostage. Canadians perceived both of these events as American-inspired threats. More recently, Canada has faced invasions of American media, American popular culture, and American capital. Canadian nationalism has always been defined in relation to perceived American threats. A central problem for Canadian identity has always been to draw a line of demarcation between Canada and the United States. On the landscape, the border itself provides this line of separation. Because Niagara is the prime place where the integrity of the border has been tested, it has come to represent this separation, so vital to Canada, in a special way; and Canadians have expressed this in the attention they give to its landscape.

From the American perspective, the Niagara region appears to have a nearly opposite meaning. Americans have treated it as a place they would almost like to forget. Why? Sacvan Bercovitch and other Americanists have proposed that the roots of American nationalism lay in the religious fervor of the New England Puritans who, like the ancient Israelites, believed they must undertake an exodus from a corrupt Old World and embrace a new destiny, an "errand into the wilderness" (Bercovitch 1978). This myth proved adaptable and effective not only during the Revolution but also later when Americans began to see their destiny in the vast west, the frontier. The ritual of errand "implies a form of community without geographical boundaries," writes Bercovitch, "since the *wilderness* is by definition unbounded, the *terra profana* 'out there' yet to be conquered" (Bercovitch 1978, 26). Yet as Americans swarmed through the Mohawk Valley route via the Erie Canal and the railroads that followed, they encountered at Niagara their first border, a line that revealed a limit to the American wilderness. Canadian historian Ramsay Cook has observed that "the very existence of Canada, most 19th-century Canadians realized, was an anti-American fact" (Cook 1977, 20). This is perhaps why Americans have not showered Niagara Falls with attention. Indeed, western New York is still known as "The Niagara Frontier," a term which, even now, tacitly denies the finality of the border.

Conclusion

Images and perceptions, although they may seem at odds with the realities of places, can sometimes offer an avenue toward deeper understanding. In this essay, ideas associated with Niagara Falls helped to account for Niagara's honeymoon tradition, its carnival atmosphere, and its industrialized landscape. In the case of the contrasting banks of the Niagara River, insight flowed primarily in the other direction: the content of the landscape itself provided clues about Niagara's meaning to the two nations.

Acknowledgment

I thank Eliza McClennen for cartographic help.

References

Bercovitch, S. 1978. *The American Jeremiad.* Madison: University of Wisconsin Press.
Carter, J. C. 1903. Oration at the dedication of the State Reservation at Niagara, July 15, 1895. In *Nineteenth Annual*

Report of the Commissioners of the State Reservation at Niagara, pp. 263-277. Albany: State of New York.

Carlisle, G. 1850. Two lectures on the poetry of Pope, and on his own Travels in America. Delivered to the Leed's Mechanics' Institution and Literary Society, December 5 and 6. Reprinted in *Anthology and Bibliography of Niagara Falls*, ed. C. M. Dow, Vol. II, pp. 1072-1073. Albany: State of New York, 1921.

Cook, R. 1977. Cultural nationalism in Canada: An historical perspective. In *Canadian Cultural Nationalism*, ed. J. L. Murray, pp. 15-44. New York: New York University Press.

Copeland, G. 1904. Niagara. In *Niagara and Other Poems*, pp. 11-12. Buffalo: Matthews-Northrup.

Dear Abby. 1985. *Buffalo News* 11 September.

Gillette, K. C. 1894. *The Human Drift. Boston: New Era.*

Goldman, H. L. 1971. Nikola Tesla's bold adventure. *The American West* March:4-9.

Howells, W. D. 1888. *Their Wedding Journey.* Boston: Houghton Mifflin.

Jameson, A. B. M. 1838. *Winter Studies and Summer Rambles in Canada.* London: Saunders and Otley.

Love, W. T. 1895. *Model City Bulletin*, 10 August:(1). Model City, NY.

Lynch, J. 1914. Pastoral letter. In *Peace Episodes on the Niagara*, ed. F. H. Severance, pp. 102-110. Buffalo: Buffalo Historical Society.

McGreevy, P. 1985. Niagara as Jerusalem. *Landscape* 28(2):26-32.

———. 1987. Imagining the future at Niagara Falls. *Annals of the Association of American Geographers* 77(1):48-62.

———. 1988. The end of America: The beginning of Canada. *Canadian Geographer* 32(4):307-318.

McKinsey, E. 1985. *Niagara Falls: Icon of the American Sublime.* New York: Cambridge University Press.

New York World. 1896. Aladdin quite outdone: Giant palace to span the mighty Niagara cataract. 9 February.

Russell, W. H. 1865. *Canada: Its Defences, Condition, and Resources.* London: Bradbury and Evans.

Turner, V. 1973. The center out there: Pilgrim's goal. *History of Religions* 12:191-230.

Woods, N. A. 1861. *The Prince of Wales in Canada and the United States.* London: Bradbury and Evans.

Niagara Falls and Gorge

Keith J. Tinkler
Brock University

Niagara Falls, in the manner of an elusive and moving target, constitutes a focal point of unresolved debate in geological history right down to the present. The Falls were mentioned in the literature of travelers and placed on maps of North America long before they were officially recorded as having been seen (Sanson 1656). When Father Hennepin (1683) did officially record the sight of the Falls for Old World audiences, they became an instant spectacle along what was to become a primary route to the interior and the Upper Great Lakes.

Hennepin's highly colored account and exaggerated drawing (Figure 1), attracted the critical interest of the scientific community, for he estimated the height of the Falls at 600 feet (184 meters). But it gave no basis for estimating in later times whether the Falls were actively receding. By 1721 more sober accounts (Dudley 1722) accorded the Falls their present height of about 150 feet (46 meters), but their impact on the imagination remained. In the late 18th century the French military engineer P. Pouchot (1781) and the anonymous A. B. (1768), possibly a British intelligence agent, both deduced that there was active change at the Falls, and that the gorge, extending north 12 kilometers, had been eroded by the recession of the Falls. The matter of recession was actively discussed toward the end of the century with no definitive decision. Estimates that the Falls retreated by 6 meters (20 feet) in 30 years led to an inferred age for the gorge of 55,440 years, well in excess of contemporary geological estimates of 5,700 years (Tinkler 1987). Opposing factions denied that any active recession took place. Despite official surveys of the river at the end of the century to establish the national boundaries, no markers were left for the crestline. Even by 1843, when James Hall of the New York Geological Survey made an official trigonometrical survey of the Falls, there was still no reliable way to estimate the rate of retreat.

The inference that the Falls had retreated to their present position, leaving behind the gorge, was firmly established by the amateur geologist Robert Bakewell Jr. (1830). He demonstrated by basic geological mapping that the Niagara Escarpment was not a fault scarp, and that the Niagara Gorge was not created by an earthquake. From the evidence of old river banks above the gorge walls and downstream of the Falls he showed that they had receded gradually to their present position, and estimated 10,000 years for the age of the gorge. Charles Lyell used Bakewell's account until he visited the Falls for himself in 1841 in the company of James Hall of the New York Survey. Their professional positions and respective accounts (Hall 1843; Lyell 1845) established an "official" account of the Falls, which became the accepted view until the end of that century. Lyell conservatively estimated 35,000 years for the age of the gorge. Hall and Lyell added one significant element to Bakewell's account: Hall pointed out the peculiarity of the Whirlpool Gorge and suggested that it might connect with the break in the Escarpment at St. David; Lyell followed up the clue and established the existence of the buried gorge filled with drift. It followed that there could be no simple and catastrophic explanation for the Niagara Gorge, a complex history was clearly implied. About the same time the first measurements of Niagara River flow were made, and it was estimated that, in theory, Niagara Falls could power the entire industrial needs of the contemporary world.

Hall instituted an exact survey of the crestline, and Lyell entombed his version of the history of the gorge in successive editions of his *Principles of Geology*, and together they placed an effective stopper on further research. Not until the 1880s was the survey of the crest repeated and the recession then measured surprised observers, for a century devoted to the adoption of the principle of uniformity did not expect a retreat well over a yard a year. The measurement was repeated at short intervals for the next two decades until G. K. Gilbert made a definitive study in 1907. The matter was not just one of geological esoterica. The development of schemes to harness the power of the

Figure 1. A lithograph from Lyell (1845) of the engraving in Hennepin (1683). It was the basis of many derivative prints for the next century. There are anachronistic sailing boats on Lake Erie on the original print. The distant mountains may be interpreted to be the hills of southern New York State, easily seen from the north Lake Erie shoreline. The prominent "spout" of water on the right came from Chippawa Creek (now the Welland River), which made its separate way to the Falls until well into the 18th century.

Falls meant that precise information was needed so that power stations and water intakes could be located without fear of destruction from continued recession.

The turn of the 20th century saw a resurgence of geological, as well as the rise of industrial, interest in the Falls and the Gorge. Whereas earlier explanations were all based on local geology, by 1900 there was no doubt that a proper explanation of the Gorge depended on the position of Niagara within the late Quaternary evolution of the Great Lakes. The age of the gorge was still the focus of debate for there was no method of absolute dating. Estimates, argued in great detail, depended on several factors: the amount and duration of discharge, the adoption of short or long time scales for the earth and for the Quaternary, and assessments of how much of the present gorge was exhumed from an hypothesised extension of the buried gorge at St. David.

Powerful personalities and capable geologists were involved in this debate, but the relation of the gorge to the Great Lakes complicated the discussion. In 1907 J. W. W. Spencer published the *Falls of Niagara* under the auspices of the Canadian Geological Survey, and G. K. Gilbert (1907), for the United States Geological Survey, published a definitive statement on the rate of recession. In 1913 E. M. Kindle and F. B. Taylor published their Niagara Folio in the regional mapping program of the United States Geological Survey. This account was virtually reprinted as the Field

Guide for the International Geological Congress of 1913, which ensured that it became the established one. Two years later a massive memoir by F. Leverett and F. B. Taylor (1915) established a chronology of events for the evolution of the Great Lakes, into which the Niagara story fit (Figure 2). In both these joint publications Taylor wrote the sections on Niagara.

But the debate over Niagara was not finished. In Taylor's account, and in this respect Spencer (1907)

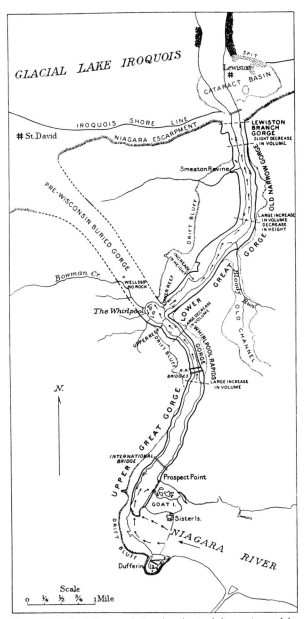

Figure 2. Taylor's figure relating the physical dimensions of the gorge to episodes in postglacial Great Lakes history (Kindle and Taylor 1913). The narrow Whirlpool Rapids Gorge, which is only 1 kilometer long, is attributed to the period when Upper Great Lakes drainage by-passed the Erie basin and flowed directly via the North Bay outlet, from Lake Huron to the Ottawa River.

agreed, the exhumed part of the buried gorge extended only as far as the north end of the narrow Whirlpool Gorge (Leverett and Taylor 1915). However, W. A. Johnston (1928) examined archived geological logs from borings made for the base of the Cantilever Bridge, at the south end of the Whirlpool Gorge. He concluded that they showed a large gorge infilled with drift, and only partly exhumed by the present river. Thus, on his reading, the buried gorge once extended to the present downstream end of the Upper Great Gorge. Spencer had only reported two of these logs and it was his view that they represented back fill from the collapse of the wall into the enlarging Upper Great Gorge. Taylor (1930) initially revised his interpretation of gorge history, but in his account for the International Geological Congress Field Guide (Taylor 1933) he offered both alternatives, his old view and the new one, and hinted that he did not think the matter was resolved.

The matter is still unresolved. P. E. Calkin and C. E. Brett (1978) retail the earlier Taylor version (Leverett and Taylor 1915), whereas in a recent account I. McKenzie (1990) offers the W. A. Johnston (1928) alternative. Since neither source debates the matter, and no resolution is attempted, Niagara remains ambiguous! A resolution would require further borings and careful logging, for although Johnston claimed the gorge infill was drift, the logs he published make no mention of the large igneous and metamorphic erratics normally found in Niagara Peninsula drift.

References

Anonymous A. B. 1768. The wonders of Canada. A letter from a gentleman to the *Antigua Gazette*. New York: 21 August 1768 (Reprinted in *Magazine of American History* 1:243-246.)

Bakewell, R. Jr. 1830. On the Falls of Niagara and on the physical structure of the adjacent country. *London's Magazine of Natural History* 3(January):117-130.

Calkin, P. E., and Brett, C. E. 1978. Ancestral Niagara River drainage: Stratigraphic and paleontologic setting. *Geological Society of America Bulletin* 89(8):1140-1154.

Dudley, P. 1722. An account of the falls of the river Niagara, taken at Albany October 10th, 1721, from Monsieur Borassaw, a French native of Canada. *Philosophical Transactions of the Royal Society* (April-May):69-72.

Gilbert, G. K. 1907. Rate of recession of Niagara Falls. *U.S. Geological Survey*, Bulletin 306. Washington, DC: U.S. Government Printing Office.

Hall, J. 1843. *Geology of New York (IVth District)*. Albany: State of New York.

Hennepin, L. 1683. *Description de la Lousiana, Nouvellement Decouvertée au Sud-Oüest de la Nouvelle France, Par Ordre du Roy*. Paris: Sébastien Huré.

Johnston, W. A. 1928. The Age of the Upper Great Gorge of the Niagara River. *Royal Society of Canada. Proceedings and Transactions*, 3rd series 22(4):13-29.

Kindle, E. M., and Taylor, F. B. 1913. Niagara Folio. *U.S. Geological Survey Atlas*, No. 190. (Note: the field [book] edition of this folio was published in 1914, and is identical except in size. Also note that Taylor wrote the sections on Niagara.)

Leverett, F., and Taylor, F. B. 1915. *The Pleistocene of Indiana and Michigan and the History of the Great Lakes*. U.S. Geological Survey, Monograph 53. (Note: the chapters are individually attributed, Taylor wrote on Niagara.)

Lyell, C. 1845. *Travels in North America* (2 vols.). London: John Murray.

McKenzie, I., ed. 1990. *Quaternary Environs of Lakes Erie and Ontario*. Waterloo, ON: Escart Press.

Pouchot, P. 1781. *Mémoire sur la Dernière Guerre de l'Amerique Septentrionale . . . 1755-1760* (4 vols.) Geneva: Yverdon.

Sanson, N. 1656. *Le Canada, ou Nouvelle France*. Paris: Pierre Mariette (loose sheet).

Spencer, J. W. W. 1907. *The Falls of Niagara; Their Evolution and Varying Relations to the Great Lakes; Characteristics of the Power, and Effects of its Diversion*. Ottawa: Geological Survey of Canada.

Taylor, F. B. 1930. New facts on the Niagara Gorge. *Michigan Academy of Sciences, Papers* 12:251-265.

———. 1933. Niagara Falls and Gorge. In *The Paleozoic Stratigraphy of New York*, International Geological Congress (XVI) Guidebook 4, Excursion A-4, ed. D. H. Newland, pp. 78-103. Washington, DC: U.S. Government Printing Office.

Tinkler, K. J. 1987. Niagara Falls: The idea of a history and the history of an idea 1750-1845. *Geomorphology* 1:69-85.

Cirque and City Form Effects on Energy Distribution

Ray Lougeay
State University of New York,
College at Geneseo

Anthony J. Brazel
Arizona State University

Melvin G. Marcus
Arizona State University

What do glacially formed cirques and American football stadiums have in common? Most obviously it is their amphitheater shapes. In turn, that similarity of form can cause similar diurnal trends of processes whether in the mountains or the desert or the city; for example, the march of radiation, reflectivity, temperature, and other energy-related fluxes. In this case, form is not only a product of processes, but the processes are also, in part, a product of form.

Generally speaking, physical geographers and other environmental scientists work from the *a priori* position that physical processes operate in the same manner regardless of their environmental setting. However, while processes such as heat and water fluxes may be ubiquitous, site conditions significantly affect magnitudes and rates at which the processes take place. The properties of earth surface materials, for example, control such factors of the energy distribution as emissivity, albedo, and heat capacity. On the other hand, once earth material properties are accounted for, similar terrain features produce similar energy patterns.

Radiative energy focused on a given point on the earth's surface is affected by terrain form. And, although these processes can directly change certain form parameters (height, width, depth) of the landscape (for example, glacier morphology), a significant consideration in climatology is how the reverse occurs—physical processes dynamically fashioned by landscape form.

This is illustrated for the natural setting of an alpine cirque and the constructed amphitheater of a football stadium. Here, the materials and settings are unlike, but the basic geometry is similar. The differences and similarities between these locations are identified by:

direct measurement of surface temperatures and surface emittance; temperature determinations from remotely-sensed imagery; and models that describe the thermal patterns.

Research Sites

The comparative sites are Fourth of July cirque in the Colorado Rockies and Sun Devil Stadium at Arizona State University, Tempe, Arizona. The cirque occupies a south-facing basin between 3,500 and 3,625 meters at 40°01′N latitude in the Indian Peaks of the Front Range. It is approximately 150 meters wide and 600 meters in length (Figure 1A). The original glacier has disappeared, but a summer neve field remains. The bottom of the cirque has mixed areas of snow, bedrock, periglacial deposits, and alpine tundra, depending on the season (Brazel and Marcus 1987; Marcus *et al.* 1981).

Sun Devil Stadium, which might be described as an "urban cirque," is about 220 meters long by 180 meters wide by 35 meters high and sits at an elevation of 362 meters and 33°25′N latitude (Figure 1B). Open to the south, it is constructed of concrete with anodized aluminum seats and a natural grass playing surface.

Radiation and Temperature Measurements

The measurements at Fourth of July cirque and the Arizona State University stadium were part of a larger investigation of long-wave wall emittance. It was postulated that heat enhancement from the walls would

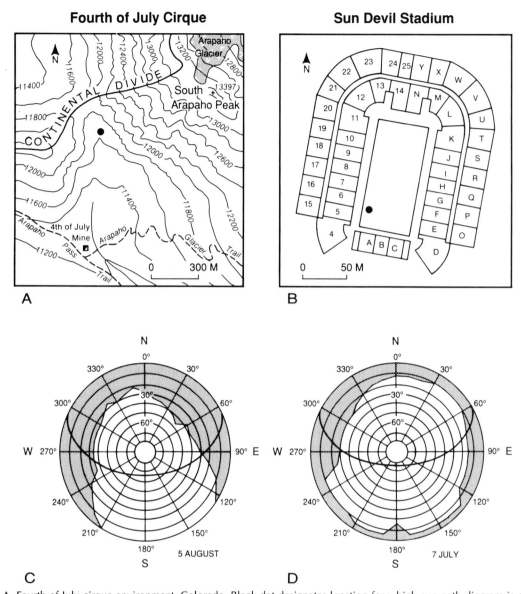

Figure 1. A. Fourth of July cirque environment, Colorado. Black dot designates location for which sun path diagram is constructed in Figure 1C. B. Sun Devil stadium, Tempe, Arizona. Black dot indicates location for construction of sun path in Figure 1D. C. Fourth of July cirque horizon chart. D. Stadium horizon chart. All figures were adapted from Brazel and Marcus (1987).

increase the thermal reservoir in the lower amphitheater (Brazel and Marcus 1987). The key measurements at each site consisted of hourly wall-surface temperatures. The terrain was divided into target sectors and measurements were taken with a Barnes Instatherm thermometer or an Everest Infrared thermometer. The sensors were at central positions that were also focus points for the construction of sky horizon charts (Figures 1C and 1D). Long distance observations were calibrated by close-up spot measurements of wall temperatures. The cirque observations (5 August 1980) and the stadium observations (7 July 1984) were taken under scattered and clear sky cover, respectively. Winds were essentially nil.

Figure 2 illustrates the temperatures for east- and west-facing walls as well as air temperatures 1 meter above the cirque and stadium floors. The wall temperatures are adjusted for emissivity (0.9 for Fourth of July; 0.7 for the stadium sections). The graphics are restricted to the period from noon onward, when measurements were made at both sites. However, the east- and west-facing stadium stands' surface temperatures are extrapolated back to 0930 hours—approximately the time of Landsat overpass. The morning build-up

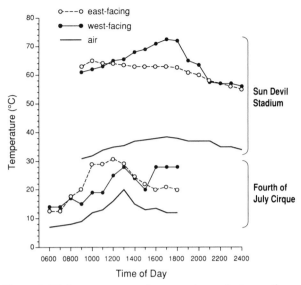

Figure 2. Wall temperatures of east- and west-facing walls on the cirque and stadium. Also shown are air temperatures recorded at dot positions shown in Figures 1A and 1B on the floor of the respective features. All wall temperatures corrected for emissivity.

of east-facing wall temperatures and subsequent drop in late afternoon is evident. Predictably, the west-facing wall heats later in the day and retains its temperatures later into the afternoon.

These are obviously two very different summer environments, one at more than 3,000 meters in an alpine landscape, the other a low elevation, urbanized, hot desert locale. Note, for example, the difference in ambient air temperature. Clearly, large temperature differences result from elevation and surface characteristics of the sites. Yet, because form is similar, the diurnal progress of the radiative processes also remains similar.

It follows that heat enhancement also is diurnally distributed. To test this distribution and its contribution to the thermal reservoir, we calculated long-wave wall emittance and sky long-wave radiation (after Idso 1981; Marks and Dozier 1979), incorporating sky view factors to partition sky and wall contributions. The impact of wall radiation in heating the valley floor is impressive—roughly 30 to 34 percent for the cirque bottom and 10 to 12 percent for the stadium playing field. The implications for snowmelt and urban cooling costs, respectively, are obvious. For example, in Fourth of July cirque about 0.65 meters of water equivalent snowmelt is attributable to wall radiation, a significant loss to an already marginal summer snowfield. While the absolute amounts of long-wave radiation emitted by the stadium walls were 10 percent higher than the cirque walls, the absolute amount of sky radiation from the desert atmosphere was almost twice as high as

from the mountain atmosphere above the cirque elevation. This is, in part, due to a combination of factors: sky view factor (Figures 1C and 1D); and increased air mass, and higher ambient temperatures at lower altitude.

Remote Sensing of the Stadium

Given good ground truth references, thermal imagery allows for more frequent and extended spatial coverage for surface heat analysis. One advantage of working with local thermal fields is that field measurements can be made at a scale easily adapted to current satellite imagery. For example, SPOT imagery brings pixel size down to about 10 by 10 meters, but not for thermal data. Landsat 5 data, with 120 by 120 meter pixels in the thermal band (band 6), can provide fairly fine resolution data of ground thermal conditions. The study of Sun Devil Stadium, with its detailed surface-energy data, provided an excellent test case of the effectiveness of imagery-derived measurements. Even at a relatively small scale, the stadium roughly occupies four pixel units, although edge effects are involved.

Surface-temperature values were determined from Landsat remotely sensed digital image data (Landsat Thematic Mapper band 6—10.4 to 12.5 microns for 0930 hours, 16 June 1989). Using microcomputer-based digital image processing software, the exact locations of target pixels were determined through visual interpretation of on-screen multispectral color composite imagery. Sample 120 by 120 meter pixels on the "walls" of the stadium were identified, and the digital number values for these pixels were retrieved from the band 6 data file. These digital remotely sensed values were then transformed to spectral radiance values (Table 1) and, in turn, changed to "at-satellite temperature" data (EOSAT 1986). Ground level observations enabled accurate assessment of the emissivity characteristics of the stadium surfaces, so that "at-satellite temperatures" could be adjusted to true surface temperatures. The Landsat Thematic Mapper thermal channel is relatively free of problems associated with atmospheric attenuation (Schott 1986).

Field results of 7 July were extrapolated back to the 0930 hours from the noon period (Figure 2). Table 1 lists comparative data from the field date and the Landsat image date. Although the imagery measurements were from 16 June and the field measurements for 7 July, both were taken on clear days during the high sun period. Morning air temperatures at the Phoenix airport station were within 2°C on the two days, although local dew points differ by 4°C. Taking the two temperatures 0930 hours and 1200 hours at face value, agreement is remarkable. For the imagery

Table 1. Sun Devil Stadium Remotely Sensed and Field Surface Temperatures (°C)

	16 June Emissivity corrected "At-satellite temperature" (0930 hours)	7 July Field temperature[a] (0930 hours)
West-facing	62	60
East-facing	66	64
Field surface	42	38
Air temperature[b]	34	32
Dew point[b]	19	23

[a]Field results extrapolated back to 0930 hours LST from Figure 2.
[b]Temperatures from Sky Harbor International Airport site some 7 kilometers to the west.
N.B.: From the EOSAT notes:
Spectral radiance = 0.1238 + (0.005632 × digital number)
"At-satellite temperature" (°C) = 273 + 1260.56/[ln (60.776/ spectral radiance) + 1]

and the field measurements, the temperatures were respectively 339°K (63°C) and 337°K (64°C) for east-facing stands, 335°K (62°C) and 338°K (65°C) for the west-facing stands, and 315°K (42°C) and 311°K (38°C) for the grass field. These data are adjusted for emissivity. Adjusting for time by extrapolating 7 July field measurements to 0930 hours, gives respective temperatures of 339°K (66°C) and 337°K (64°C) for the east-facing area and 335°K (62°C) and 333°K (60°C) for the west-facing stands. The difference of temperatures follow nicely in light of the 2°C higher ambient temperatures on 16 June. The cooler grass temperatures on 7 July may have resulted from less antecedent evapotranspiration on that day (note higher dew point temperature for 7 July, Table 1). Thus, for ground, cirque, and stadium measurements, Landsat remotely sensed thermal data truly respond to the morning heating regime as a function of stadium "topography."

Summary

Point and form thermal sources on the earth's surface illustrate the ubiquity of energy balance processes. At the same time, variations of earth surface materials, shown here for mountain and human-built desert sites alter the magnitudes and rates at which the processes operate. The nature of these alterations can have significant feedbacks in both natural and human-built settings. The impact on snow and ice features has been shown for Fourth of July cirque; comparable enhanced water losses have been identified for small glaciers and snowfields in the southern Rockies and San Juan Mountains of Colorado. In these marginal water-storage environments, which are threatened by ongoing climatic change, radiation heat enhancement processes may be the critical factor that "tips the balance" of the snowpack survival. In built environments, such as stadiums and urban canyons, wall enhancement leads to heat loading and consequent economic costs for air conditioning. Remote sensing, taken with good ground truth, provides opportunities to realistically extend measurements of the earth's surface.

References

Brazel, A. J., and Marcus, M. G. 1987. Heat enhancement by long-wave wall emittance. *Geographical Review* 77:440-455.

EOSAT. 1986. *EOSAT Landsat Data Users Notes*, Vol. 1, pp. 1-8. Lanham, MD: EOSAT Corporation.

Idso, S. B. 1981. A set of full equations for full spectrum and 8- to 12-m and 10.5- to 12.5-m thermal radiation from cloudless skies. *Water Resources Research* 17:295-304.

Marcus, M. G., Brazel, A. J., Lougeay, R., and Hyers, A. D. 1981. Long-wave radiation enhancement by cirque wall emittance, Front Range, Colorado. In *Research Papers in Climatology*, Geography Publication, No. 1, ed. A. Brazel, pp. 21-42. Tempe: Arizona State University, Department of Geography.

Marks, D., and Dozier, J. 1979. A clear sky long-wave radiation model for remote alpine areas. *Archiv für Meteorologie und Bioklimatologie*, Ser. B. 27:159-187.

Schott, J. R. 1986. *Radiometric Analysis of the Longwave Infrared Channel of the Thematic Mapper on Landsat 4 and 5*. Unpublished paper prepared for NASA/Goddard Space Flight Center, Greenbelt, MD.

Alpine Treeline in
Glacier National Park, Montana

Stephen J. Walsh
University of North Carolina

George P. Malanson
University of Iowa

David R. Butler
University of North Carolina

Projections from climate models indicate that anthropogenic emissions of carbon dioxide and other radiatively active trace gases will cause a rise in global temperature on the order of 3°C during the next century. This effect is likely to affect vegetation (cf. Malanson 1989), particularly at treeline, creating a problem for alpine tundra. Organisms isolated at the tops of mountains could be squeezed out of existence by an advance of organisms from lower elevations. Higher elevations in the mountains do not have the soil to support plant growth, and a forest advance may be more rapid than the potential development and colonization of new tundra sites. Alpine tundra is therefore threatened by subalpine forest in this scenario, and the sensitivity of treeline to climate warming is of central interest. The rates of climate change in response to increased concentrations of carbon dioxide will likely differ from rates of past climatic change, and thus geomorphic process-responses and ecotonal fluctuations will also change at different rates.

The response of treeline to climate change may be affected by the nature of the transition from trees through krummholz (stunted trees growing in patches as a result of environmental stress) to tundra. This ecotone, like others, is variable in its spatial pattern in three dimensions (Habeck 1969; Hansen-Bristow and Ives 1984). The linear pattern of alpine treeline can be spatially continuous or discontinuous around mountains depending upon environmental gradients and biophysical factors and processes operative at local to regional scales. Recent theoretical proposals in landscape ecology have argued that spatial pattern is critical to ecological processes because processes such as dispersion, nutrient flows, and energy and water movement in microclimates are spatially mediated. If we are to understand how treeline might respond to global climate change and to recognize areas of tundra that may be threatened, it is necessary to understand the relationships between pattern and process at existing sites. Therefore the objectives of this paper are to describe the biophysical factors and processes affecting the spatial pattern of alpine treeline and to suggest how alpine treeline may respond to global climate change as a consequence of its site and situation.

Study Area

Glacier National Park, Montana, is one area that shows striking evidence of variability in the position, condition, and character of alpine treeline (Figure 1). Glacier National Park is a United Nations-designated International Biosphere Reserve comprising approximately 0.4 million hectares (1 million acres) astride the Continental Divide in northwest Montana. The park is positioned at the eastern edge of the influence of Pacific maritime air in the northwestern United States and exhibits an ecotonal transition from mountains to the Great Plains. East of the Divide, the climate is much harsher, with lower temperatures, stronger winds, and drier conditions. Two mountain ranges occur within Glacier Park—the Livingston

Figure 1. The study area in Glacier National Park, Montana.

Range and the Lewis Range. Both sustain modern-day glaciers. The Continental Divide, itself a major boundary between the Pacific-influenced western part of the park and the drier eastern portion, follows the Livingston Range in the north and then shifts to the more easterly Lewis Range. The highest peaks in the park rise to about 3,100 meters. The park area was heavily glaciated during the Pleistocene, but a number of mountain crest and upland surfaces in the Lewis Range escaped glaciation.

Tree species comprising the upper treeline ecotone reflect the climatic differences induced by the Continental Divide. West of the Divide, upper treeline consists primarily of subalpine fir (*Abies lasiocarpa*), Engelmann spruce (*Picea engelmannii*), and subalpine larch (*Larix lyallii*), whereas east of the Divide the treeline is dominated by whitebark pine (*Pinus albicaulis*) and limber pine (*Pinus flexilis*). Because of the greater amounts of moisture west of the Divide, erosion from running water and past and present glaciers has produced particularly rugged topography. S. F. Arno and R. P. Hammerly (1984) describe this environment as a combination of extreme topographic and climatic stress. East of the Divide, the drier climate precluded glaciation during the late Pleistocene at some locations, leaving broad, gently sloping uplands at and above treeline (Butler and Malanson 1989). Upper treeline at these sites experiences less stressful growing conditions than do those on steeply sloping, soil-poor sites west of the Continental Divide; treeline at these eastern sites should therefore, as a consequence of improved soil and topographic conditions, be more rapidly able to respond

to climatic changes brought about by changes in levels of atmospheric carbon dioxide.

In Glacier National Park, treeline has responded to Holocene environmental changes, but well-dated chronologies are lacking. S. A. Elias (1988) showed by means of an examination of insect fossils at the southern edge of the park that treeline was at least 450 meters lower than present at about 11,400 BP (before present), a period marking the onset of Pleistocene deglaciation in the area (Carrara 1987). Although glacial evidence for Holocene climatic deterioration is largely restricted to the period of the Little Ice Age (roughly in this area 1600 to 1800 A.D.) (Butler 1989; Carrara 1987), D. R. Butler and G. P. Malanson (1989) showed that a mid-Holocene cold-climate episode induced widespread solifluction at relatively low elevations in the park and probably depressed treeline. Treeline was also lowered, but to unspecified levels, during the Little Ice Age. Since the late 1800s in Glacier Park, glaciers have retreated markedly, and forest has advanced only recently onto deglaciated terrain.

The concentration of atmospheric carbon dioxide has increased steadily since the onset of the Industrial Revolution in the western world. At present, atmospheric carbon dioxide is increasing and is expected to increase into the 21st century. Growth-chamber studies reveal that this increase in the concentration of carbon dioxide will result in increased growth rates of conifers in sensitive alpine locations at and near treeline. These increased growth rates, in concert with the expected global warming brought about by the "greenhouse effect," should advance the upper treeline into areas of alpine tundra. The areal extent of alpine tundra will correspondingly decline and could, in extreme cases, disappear completely as upper treeline elevations climb. Because alpine tundra is spatially isolated, the rise of treeline, at rates unprecedented in the Holocene, may eliminate tundra sites quickly, and lead to reduced biotic diversity through extinctions.

Remote Sensing and Alpine Treeline

S. J. Walsh and coworkers (1989a) used digital enhancements of Landsat Thematic Mapper digital data to understand alpine treeline conditions within a portion of Glacier National Park. They reported that two treelines are evident on the eastern side of the park: upper temperature-related treeline was complemented by a drought-related treeline at low elevations. Above the spruce/fir forest was mixed herbaceous alpine tundra, rock, and ice. Below the lodgepole pine was mixed brush and some herbaceous cover, made up of sagebrush and grass with scattered clumps of aspen. Elevation and slope both played a role in affecting the micro-climate that controlled the distribution of plant species. Their preliminary satellite analysis indicated that at high elevations the tundra does not form a continuous band around the mountains. In many of these areas the tundra was very patchy at a micro-scale in response to localized climatic conditions and geomorphic action, and some relatively important tundra was categorized as rock because a majority of the surface was barren. Even so, the macro-scale patchiness of the tundra was striking.

In this study, channel composites of Landsat Thematic Mapper data (6 August 1986 Thematic Mapper scene) were processed and analyzed to understand the general patterns and trends in alpine treeline conditions occurring throughout the park (Figure 2). The Thematic Mapper spectral data were initially preprocessed to correct for radiometric and geometric distortions of the data (Walsh *et al.* 1989b). Thematic Mapper channels 2, 3, and 4 were selected for image compositing. Channel 4 was assigned to the red color gun of the image processor, channel 2 to the green, and channel 3 to the blue. The derived composite was designed to highlight vegetation differences by differentiating treeline components based upon variations in plant pigmentation and leaf structure. In general, the black tones on Figure 2 represent water bodies; dark tones represent the spruce-fir forest of the subalpine and alpine environments; medium tones represent the brush, meadows, and herbaceous cover of alpine tundra; and the light tones represent the snow, ice, and rock surfaces dominating the upper elevations.

An interpretation of Figure 2 shows the northwest-southeast-trending expanse of subalpine forest dominating the western third of the Park. While approximately two thirds of the park is forested, the subalpine forest grades into the alpine environment characterized by diminished vegetation density representative of tundra and bare rock, snow, and ice surfaces. The spatial extent of this vegetation grading is related to elevation, topographic orientation, and local climatic controls. The eastern two thirds of the park is dominated by unvegetated and sparsely vegetated surfaces inter-

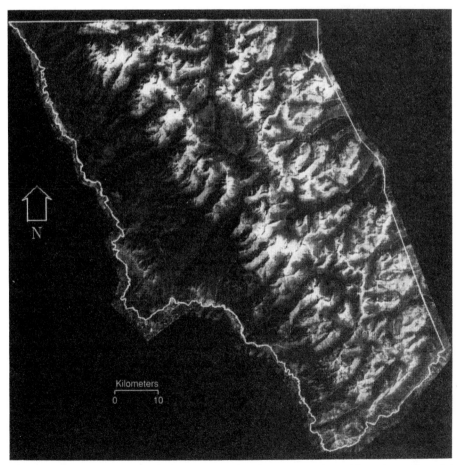

Figure 2. Landsat Thematic Mapper composite of Glacier National Park, Montana, derived from a sampling of the visible and near-infrared spectral regions of the Thematic Mapper sensor system.

rupted by valleys and hanging-valleys of subalpine forest, brush, and herbaceous cover. Because much of the alpine tundra is on the gently sloping uplands not far above treeline on the eastern side of the park (Butler and Malanson 1989), it may be particularly vulnerable to an advance of treeline.

Where vegetation has been disturbed, such as in snow avalanche paths and debris flows, the boundary between vegetated and unvegetated surfaces is distinct, and the transition zone from subalpine forest to alpine conditions is spatially irregular owing to the pattern and type of disturbance. Since topographic orientation exerts a substantial effect on vegetation, terrain conditions further served to modulate the pattern of alpine treeline. Extreme slopes exhibited a narrower transition zone from subalpine forest to tundra vegetation, while gentler slopes showed a more gradual and spatially broader progression to tundra. Steeper slopes further showed a less organized, sequential transition from subalpine forest to krummholz to tundra to bare surfaces. A high degree of variability was apparent in their sequence and in their areal extent.

The relationships between scale, process, and pattern are particularly important in alpine environments. The components of alpine treeline are unevenly expressed—their spatial pattern and areal extent varies in three dimensions. *In situ* observations of treeline components include measurements of individual and groups of trees (subalpine fir), individual components (krummholz), and individual community types (krummholz/tundra interface). Figure 2 shows redundant patterns of mottled tones occurring in the higher elevations, which indicate a spatially and spectrally mixed landscape. Because of the spatial resolution of the Thematic Mapper sensors (30 meters × 30 meters) and the spatial organization of the alpine environment, the tundra and krummholz components particularly reflect patterns characteristic of vegetated/unvegetated surfaces and patches of trees surrounded by tundra and alpine meadows, respectively. Variation in plant productivity is a key discriminator of alpine treeline components (Walsh and Kelly 1990). Measures of vegetation productivity vary as a consequence of the growth forms present in alpine environments and the geomorphic, climatic, and topographic variations of the landscape that are locally and regionally controlled.

Conclusions

The pattern of vegetation at an actual treeline is complex and variable. Its three-dimensional spatial pattern produces a band of vegetation that surrounds high mountain environments. In Glacier National Park, one seldom observes a continuous gradient along an elevational transect from tall trees, through shorter ones, to krummholz, and then to tundra. Instead, abrupt changes and reversal of trends are apparent. Local site factors, especially those related to the biological productivity and to disturbance of a site, are probably controlled in some degree by their biophysical and spatial interrelationships. That is, the position of the site in the landscape will affect the conditions for photosynthesis, especially available solar radiation, temperature, water, and nutrients and for events such as avalanches and fires. Topographic factors affecting productivity include convex versus concave slopes, which directs water either toward or away from a site as a consequence of its orientation and leaves it more or less exposed to desiccating winds; distance to lithological sources of nutrients, as when a geological contact upslope of a site may be its primary source of phosphorus, potassium, or calcium; and relationship to other mountain slopes that may affect its exposure and radiation balance. Factors affecting productivity that are related to disturbance include distance to geological contacts that produce rock debris and thus unstable substrates; position relative to slopes that act as avalanche source areas; and relationship to forest areas that, for reasons of fuel and topography, are more likely to burn, and specifically those that have burned in the past. In most mountainous environments the latter factors affect the pattern of treeline, and studies of simple transects in oversimplified situations may yield misleading interpretations of process. Such factors must be included in treeline studies to project the effects of carbon dioxide induced climate change in this landscape.

Alpine treeline, once considered a continuous line of vegetation surrounding mountains, is now recognized as a band of discontinuous vegetation and exposed surfaces representing a connection between the subalpine and the alpine environments. The nature of the biophysical factors and processes affecting alpine treeline influences the spatial and compositional pattern of the treeline. The site and situation of the alpine treeline are subject to change as a consequence of climate change. The spatial linkage between pattern and process will continually reshape the alpine treeline with those areas most sensitive to climate-induced changes responding most dynamically. The alpine treeline continues to represent a sensitive environment whose compositional and spatial patterns are interwoven and interlinked to biophysical factors and processes which are in themselves subject to change.

References

Arno, S. F., and Hammerly, R. P. 1984. *Timberline, Mountain, and Arctic Forest Frontiers.* Seattle, WA: The Mountaineers.

Butler, D. R. 1989. Glacial hazards in Glacier National Park, Montana. *Physical Geography* 10(1):53-71.

———, and Malanson, G. P. 1989. Periglacial patterned ground, Waterton-Glacier International Peace Park, Canada and U.S.A. *Zeitschrift für Geomorphologie* 33(1):43-57.

Carrara, P. E. 1987. Holocene and latest Pleistocene glacial chronology, Glacier National Park, Montana. *Canadian Journal of Earth Sciences* 24:387-395.

Elias, S. A. 1988. Climatic significance of late Pleistocene insect fossils from Marias Pass, Montana. *Canadian Journal of Earth Sciences* 25:922-926.

Habeck, J. R. 1969. A gradient analysis of a timberline zone at Logan Pass, Glacier Park, Montana. *Northwest Science* 43(2):65-73.

Hansen-Bristow, K. J., and Ives, J. D. 1984. Changes in the forest-alpine tundra ecotone, Colorado Front Range. *Physical Geography* 5:186-197.

Malanson, G. P., ed. 1989. *Natural Areas Facing Climate Change.* The Hague: SPB Academic.

Walsh, S. J., Bian, L., Brown, D. G., Butler, D. R., and Malanson, G. P. 1989a. Image enhancement of Landsat Thematic Mapper digital data for terrain evaluation, Glacier National Park, Montana. *GeoCarto International* 3:55-58.

———, Cooper, J. W., Von Essen, I. E., and Gallagher, K. R. 1989b. Image enhancement of Landsat Thematic Mapper data and GIS data integration for evaluation of resource characteristics. *Photogrammetric Engineering and Remote Sensing* 56(8):1135-1141.

———, and Kelly, N. M. 1990. Treeline migration and terrain variability: Integration of remote sensing digital enhancements and digital elevation models. *Proceedings, Applied Geography Conference* 13:24-32.

Storm Hazards along Louisiana Coastlines

Robert A. Muller
Louisiana State University

Bruce Fielding
Louisiana Department of Environmental Quality

Many of the coastlines of the United States are vulnerable to inundations by sea water during stormy weather. In the next century the potential greenhouse warming and associated sea-level rise, caused by melting of polar and alpine ice, may further increase the flood threat to coastal communities, metropolitan seaports, and coastal wetlands and ecosystems. The coastal regions of Louisiana represent a southern segment of the Atlantic and Gulf coasts with, in Köppen climatic classification terminology, a humid subtropical climate. Along the Atlantic and Gulf coasts inundations are normally associated with tropical storms and hurricanes in summer and fall, and to a lesser degree with mid-latitude cyclones during late fall, winter, and spring. Although mid-latitude cyclones occur much more frequently than tropical storms and hurricanes, vulnerability, fear, and destruction are normally reserved mostly for hurricanes, especially the rare great hurricanes. Here we focus on the frequencies of damaging winds, especially tropical storm and hurricane intensity winds, along the vulnerable Louisiana coastlines of the Gulf of Mexico (Figure 1).

Synoptic Weather Type Frequencies

The climate of coastal Louisiana has been organized into eight synoptic weather types (Figure 2) representative of lower atmospheric circulation patterns that generate local weather which affect environmental, biological, and economic systems (Muller 1977). Three of the eight synoptic types are stormy: frontal overrunning, with cooler temperatures and northerly to easterly winds and a cold or stationary front to the south over the northern Gulf of Mexico; frontal gulf return, with warmer temperatures, showers, thunderstorms, occasional severe weather, and southwesterly winds, all in association with cold or quasi-stationary

fronts over northern Louisiana and east Texas; and gulf tropical disturbance weather, associated with disturbed tropical weather systems over the gulf, ranging from weak easterly waves to great hurricanes such as Camille in 1969.

Routine hourly meteorological observations are not taken at any coastline locations in Louisiana, but hourly observations suitable for synoptic analyses are taken at the first-order station of the National Weather Service at New Orleans, in a coastal wetland environment about 80 kilometers (50 miles) from the open Gulf of Mexico. Synoptic weather-type calendars have been developed for New Orleans beginning with January 1961, and monthly frequency analyses have been completed for the 30-year "normal" period of the National Climatic Data Center for 1961-1990 (Muller and Willis 1983). On an average annual basis, frontal overrunning, frontal gulf return, and gulf tropical disturbance weather occurred 18, 13, and only 2 percent of the time, respectively. In winter, January for example, frontal weather occurs frequently, and for the 30-year period, frontal overrunning and frontal gulf return weather occurred 37 and 13 percent of the time, respectively. During July, on the other hand, the two frontal weather types were restricted to 11 percent of the time, and gulf tropical disturbance weather averaged 11 percent of the time. For the 30-year period gulf tropical disturbance weather peaked in September with a frequency of 12 percent. Gulf tropical disturbance weather varies greatly from one season to another, and during Septembers the frequencies range from 0 (for 6 years) to 43 percent in 1971.

Hourly observations of sustained wind speeds at New Orleans have also been evaluated by synoptic weather type to compare the potential for mid-latitude cyclones and tropical storms for generation of storm waves capable of erosion along the coastline of Louisiana. Wind speeds of 31.5 kilometers per hour (17 knots) or greater

Figure 1. Louisiana's Atlantic and Gulf Coast coastline towns for which tropical force and hurricane winds were estimated.

were observed for 129 hours on an average annual basis at New Orleans. These winds occurred on average between 16 and 22 hours per month from December through April, but less than two hours a month during June, July, and August. On an average annual basis, the two frontal weather types, frontal overruning and frontal gulf return accounted for 70 of the hours, about 54 percent of the annual total. The gulf tropical disturbance weather accounted for only about 11 hours per year on average, only about 8 percent of the annual total!

Although gulf tropical disturbance weather accounts for only a small proportion of the windy weather over southern Louisiana on average, wind speeds during tropical storms and hurricanes are much higher than during frontal weather events. At New Orleans gale-force winds greater than 52 kilometers per hour (28 knots) were observed during three hurricanes between 1961 and 1986—Hilda in 1964, Betsy in 1965, and Camille in 1969. These wind speeds are only an index of wind along the coastlines of Louisiana, because wind speeds at the coastline are normally higher than inland at New Orleans. In addition, swells from storms elsewhere over the Gulf of Mexico can generate destructive waves along the coastline.

Tropical Storm and Hurricane Frequencies

Tropical storms and especially great hurricanes— Camille in 1969 for example—are traditionally treated as the most dangerous natural hazards along the Gulf Coast, although mid-latitude cyclones account for many more frequent situations when winds, waves, and storm

tides can generate at least modest coastal flooding and erosion.

The National Weather Service classifies disturbed tropical weather systems into categories based on maximum sustained wind speeds. The weakest systems are termed tropical disturbances; systems are termed tropical depressions when sustained winds are between 37 and 64 kilometers per hour (20 to 34 knots); tropical storms have winds between 65 and 119 kilometers per hour (35 to 64 knots); and a disturbance reaches hurricane status when winds are greater than 119 kilometers per hour. Hurricanes, in turn, are divided into the five-level Saffir-Simpson scale on the basis of maximum sustained winds relative to disaster potential.

Hurricanes have enormous potential for disaster because of the destructive high winds, flooding from excessive rainfalls, and especially coastal flooding by storm surges. The 100-year rainfall record for southern Louisiana is 856 millimeters (33.7 inches) for four days during a hurricane in September 1940. Storm surges represent sea-level rises within the northeast quadrants of tropical storms and hurricanes approaching the Louisiana coastline from the south. National Weather Service computer models indicate that the storm surge in metropolitan New Orleans could reach 6.1 meters (20 feet) for one particular track of a great hurricane (Louisiana Office of Emergency Preparedness 1985). The destructive storm surge of Hurricane Camille reached 6.89 meters (22.6 feet) at Pass Christian, Mississippi, only 87 kilometers (54 miles) east of New Orleans.

The National Weather Service has no routine wind data for the coastline of Louisiana, but we can estimate the occurrence of sustained winds of tropical storm and hurricane force at locations along the coastline. The tracks, intensities, and dates of tropical storms and hurricanes over the North Atlantic including the Gulf of Mexico have been assembled in atlas form for the period 1871 to 1980 (Neumann *et al.* 1981), and annual summaries for subsequent years are published in the almanac issue of *Weatherwise* each year. We estimated the average radii of tropical storm- and hurricane-force winds about the center of storm tracks, and then for three selected towns—Boothville, Morgan City, and Cameron (Figure 1)—we estimated the maximum sustained wind class, tropical depression, storm, or hurricane at each site (Muller and Fielding 1988).

Table 1 summarizes the frequencies of sustained tropical storm and hurricane winds by decade at the three sites along the Louisiana coastline. Within the assumptions of the analysis, for 90 years, sustained winds of tropical storm force or greater occurred 22 times at Boothville, 30 times at Morgan City, and 19

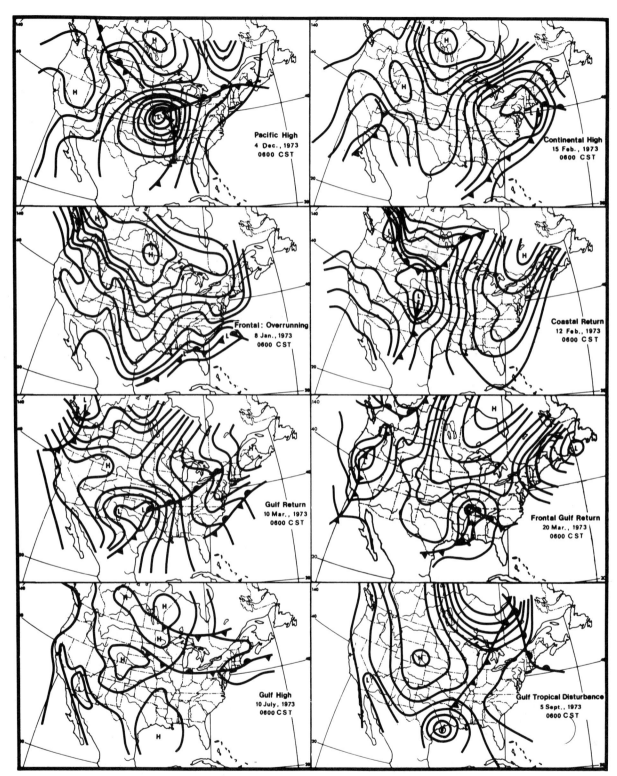

Figure 2. Synoptic weather types developed by R. A. Muller (1977) for southeastern Louisiana.

Table 1. Decade Counts of Tropical Storm (TS) and Hurricane (H) Force Events

Decade	Boothville TS	H	Morgan City TS	H	Cameron TS	H
1901-1910	4	0	2	1	1	0
1911-1920	2	1	2	1	1	1
1921-1930	1	0	1	2	0	0
1931-1940	0	0	3	0	2	2
1941-1950	2	1	6	0	3	0
1951-1960	4	0	5	0	2	1
1961-1970	0	2	0	2	0	0
1971-1980	2	0	3	1	1	1
1981-1990	2	1	1	0	4	0
Totals	17	5	23	7	14	5

times at Cameron. Hurricane force winds occurred 5 times at Boothville and Cameron, and 7 times at Morgan City. These data suggest that hurricane-force winds recur at a place once every 13 to 18 years, on average, along these coastlines, assuming a random storm-track pattern and no long-term climatic fluctuations or trends. Both of these assumptions may well not be correct. Similarly, for winds of tropical storm force or greater, the data suggest return periods of 3 to 5 years on average. We have not concluded that the greater frequencies at Morgan City are based on chance because the occurrence of preferred geographic tracks along this 418-kilometer (260-mile) segment of coastline are undetermined. Also undetermined at this time are the effects of greenhouse warming and sea-level rise on the frequencies and intensities of tropical storm activity along this vulnerable segment of coastline.

Acknowledgments

We thank John M. Grymes III, assistant state climatologist for Louisiana, for synoptic weather type frequency summaries and Barry Keim for assistance in the completion of this manuscript.

References

Louisiana Office of Emergency Preparedness. 1985. *Southeast Louisiana Storm Surge Atlas.* Baton Rouge, LA.
Muller, R. A. 1977. A synoptic climatology for environmental baseline analysis: New Orleans. *Journal of Applied Meteorology* 16(1):20-33.
———, **and Fielding, B. V.** 1988. Coastal climate of Louisiana. In *Causes of Wetland Loss in the Coastal Central Gulf of Mexico,* eds. R. E. Turner and D. R. Cahoon, Vol. II, pp. 13-29. New Orleans, LA: Minerals Management Service.
———, **and Willis, J. E.** 1983. New Orleans weather 1961-1980: A climatology by means of synoptic weather types. *Miscellaneous Publications* 83-1. Baton Rouge, LA: School of Geoscience, Louisiana State University.
Neumann, C. J., Cry, G. W., Casol, E. L., and Jarvinen, B. R. 1981. *Tropical Cyclones of the North Atlantic Ocean, 1871-1980.* Asheville, NC: National Climatic Data Center.

Coastal Storms in North Carolina

Simon Baker
East Carolina University

From the Virginia boundary in the north to the South Carolina line in the south, the Atlantic coast of North Carolina is made up of a 486-kilometer-long chain of barrier islands. They attain their greatest width from Cape Hatteras northward, reaching a maximum of 8 kilometers at Collington. At their narrowest, the islands may be a fraction of a kilometer wide. On average the chain of islands is about 2 kilometers in width (Figure 1). They range in elevation from sea level to about 42 meters at the top of Jockey's Ridge. This huge sand dune, just south of Collington, is the highest point on the coast. Some vegetated ridges get as high as 20 meters, but most of the elevations of the islands are well below this figure.

The total area of the barrier island system in North Carolina is 60,500 hectares (U.S. Department of Agriculture *et al.* 1977). This amount varies continually because of the constantly changing conditions of deposition and erosion occurring in the high-energy coastal environment. Four distinct zones of vegetation and topography may be observed where the islands are of sufficient width. On the ocean side are the beaches, common to all the islands and comprising 12 percent of the total area. Behind the beaches are two zones. The first consists of vegetated or unvegetated dunes. Further back is the mixed Maritime Forest. Together, these zones make up 51 percent of the barrier island area. Finally, the shoreline of the sounds is often lined with low marshes, which make up 37 percent of the total barrier island area. However, some of these zones may be missing on the narrower islands (Baker and Flint 1990; Graetz 1973; U.S. Department of Agriculture *et al.* 1977).

Island Origins

With rising sea level the entire barrier island system is migrating toward the mainland. The islands are thought to have originated 95 kilometers to the east of their present locations during the last period of maximum glaciation, 15,000 years ago. At that time the sea level was about 100 meters below its present level, and one theory is that the old coastal dune ridges became barrier islands. As sea level has risen with the melting ice, the islands have moved up the gently sloping coastal plain to their present locations. The movement has also been driven by dune migration, overwash processes, inlet dynamics, littoral drift, and storms. There is strong evidence that the islands roll over themselves; materials from the ocean side ending up on the mainland side as the islands retreat (Leatherman 1979; U.S. Department of Agriculture *et al.* 1977).

Hurricanes

As long as people have lived on the North Carolina coast, they have experienced a form of tropical cyclone known as a hurricane. There seems to be no regularity in the frequency of such storms. Their numbers vary from year to year and decade to decade; periods of intense activity follow longer periods of calm. The middle 1950s was a period of intense activity: three such storms making landfall in 1955 alone (Figure 2). There has been only one other like decade in the 20th century—1900 to 1909 (Table 1). Table 1 shows that the most active period annually occurs during July, August, September, and October. Storms reaching the North Carolina coast have their origins in the tropical waters to the south and east.

A mature hurricane has at its core a low pressure eye or center, which is relatively calm and sometimes cloud free, averaging 23 kilometers in diameter. Inwardly spiralling counter-clockwise winds often reach speeds of 160 kilometers per hour next to the eye. Any given location in the path of a hurricane may

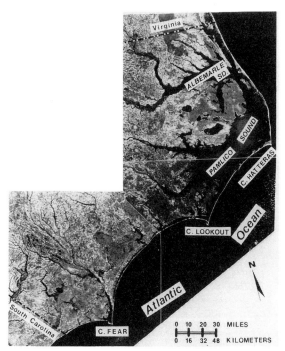

Figure 1. This mosaic of the North Carolina coast shows the barrier island system, the sounds, and the shoreline of the mainland. Three Landsat-3 multispectral scanner false-color composites were copied and fitted together. The satellite was 912 kilometers in space when these images were recorded.

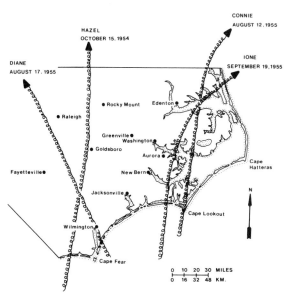

Figure 2. Within one year, October 1954 to September 1955, four hurricanes made landfall on the North Carolina coast, three of them in 1955 alone. The spiral paths are the storm centers. Hurricane-force winds of at least 119 kilometers per hour were felt 80 kilometers either side of each track. Gale-force winds of 64 kilometers per hour and higher were experienced as far as 322 kilometers on either side. Each storm traveled forward at about 20 kilometers per hour. Hazel was unusual in that her forward speed was about 50 kilometers per hour.

experience rainfall of 152 millimeters and more (Risnychok 1990).

Hurricane Damage

The most destructive force generated when a hurricane strikes land is usually the storm surge. Next in destructive power is flooding caused by the accompanying torrential rains, and least important is the damage caused by the direct action of high winds. The storm surge is a mound of water pushed up by winds ahead of a hurricane advancing landward from over a large body of water; it may be 80 or more kilometers wide as it crosses the coastline. Depending on the conformation of the shore and bottom, the storm surge may reach heights of 5 or more meters above the normal (astronomical) tide level. See Table 2 for the relationship between wind speed and surge height.

Hurricane Hazel

A direct hit by a hurricane on the barrier islands of North Carolina can destroy vegetated dunes and open new inlets. Where there are people and structures in the way, the consequences can be disastrous. Hurricane Hazel, a category 3 storm, made landfall on the South Carolina line on 15 October 1954 (Figure 2). The combination of storm surge and wind-driven waves resulted in severe damage to structures and the coastline itself for a distance of 130 kilometers to the east. The town of Long Beach, between the South Carolina state line and Cape Fear, was particularly hard hit. Grass-covered dunes 3 to 6 meters high in a continuous line 8 kilometers long simply disappeared along with 352 of 357 buildings. In addition to the spectacular property damage and dune destruction, 19 people were known to have died. Most of the victims lost their lives at or near the beach and the storm surge is believed to have been largely responsible. All parts of the North Carolina coast have always felt the direct impact of hurricanes making landfall or passing by close off-shore (Baker 1978).

Northeasters

Slowly moving northward along the coast, large extratropical cyclones—northeasters—are associated with frontal activity, precipitation, and winds from the northeast. In the northern hemisphere the winds move counterclockwise toward the centers of low pressure. Unlike a hurricane, which may dramatically pass

Table 1. Twentieth-Century Hurricanes on the North Carolina Coast by Saffir/ Simpson Categories

Decade	Category 3 and higher	Categories 1 and 2
1900 to 1909	None	11 July 1901 15 September 1903 14 September 1904 13 November 1904 17 September 1906 30 July 1908
1910 to 1919	None	3 September 1913
1920 to 1929	None	22 September 1920 25 August 1924
1930 to 1939	15 September 1933	12 September 1930 22 August 1933 18 September 1936
1940 to 1949	14 September 1944	1 August 1944 24 August 1949
1950 to 1959	Carol, 30 August 1954 Hazel, 15 October 1954 Connie, 12 August 1955 Ione, 19 September 1955	Barbara, 13 August 1953 Diane, 17 August 1955 Helene, 27 September 1958
1960 to 1969	Donna, 11 September 1960	Isbell, 16 October 1964
1970 to 1979	None	Ginger, 30 September 1971
1980 to 1989	Diana, 8 September 1984 Gloria, 27 September 1985	Charley, 17 July 1986

These storms either made landfall on the North Carolina coast or passed close enough offshore to cause damage. Based on information extracted from a variety of sources, mainly Baker (1978) and Hebert and Case (1990).

Table 2. Saffir/Simpson Hurricane Scale

Category	Central pressures (mbar)	Winds		Surge		Damage
		(km/h)	(mi./h)	(m)	(ft.)	
1	980 or more	119-153	74-95	1.2-1.5	4-5	Minimal
2	965-979	154-177	96-110	1.8-2.4	6-8	Moderate
3	945-964	178-209	111-130	2.7-3.7	9-12	Extensive
4	920-944	210-249	131-155	4.0-5.5	13-18	Extreme
5	<920	>249	>155	>5.5	>18	Catastrophic

Source: Risnychok (1990).

over a coastal location in a fraction of a day, a northeaster may affect the same location for several days. At the very least, a northeaster can generate visible beach erosion, while at its worst gale-force winds over several days may create a prolonged storm surge capable of great damage.

The passage of such low-pressure storms is usually followed by cells of higher atmospheric pressure with winds circulating outward from the center in a clockwise direction. Occasionally the location of a high is such that its winds may strike the coast from a northeasterly direction. No precipitation is associated with such an event.

Frequency of Northeasters

The winds of both types of northeasters are important mainly because they generate high waves and move them shoreward. It is the action of these waves on the shore and structures close to it that is the chief cause of concern. Studies of extratropical cyclones on the barrier islands north of Cape Hatteras identified 857 storms between 1942 and 1967 with winds strong enough to cause beach erosion. The minimum wave height in deep water capable of causing such erosion was 1.6 meters (Bosserman and Dolan 1968).

The most active months for northeasters were December through April. Each of these months had more than 90 erosion-causing storms for the 25-year period. Both high and low pressure, wind-generating systems were counted, with about 75 percent of the total identified as extratropical cyclones. Over the 25 years, winds generating wave heights of over 1.6 meters in deep water occurred on an average of every 10 days. A wave height of at least 3.4 meters occurred every three months and one of 5.2 meters every three years. Wave heights of over 7.0 meters occurred once in 25 years.

Ash Wednesday Storm

Probably the worst northeaster in recent history to affect the North Carolina coast was the so-called Ash Wednesday Storm, which blew during 6-9 March 1962. This was a classical extratropical cyclone and it changed the coast from Cape Hatteras northward by destroying many kilometers of grass-covered protective dunes. A new inlet 183 meters wide was cut just north of Cape Hatteras. Hundreds of beach homes were destroyed or damaged (Figure 3), and hundreds of automobiles were buried or submerged. Waves as high as 10 meters pounded the shoreline for four days (U.S. Department

Figure 3. Beach cottages at Kitty Hawk destroyed by wave action on 7 March 1962 during the Ash Wednesday Storm. Photograph courtesy of the Outer Banks History Center.

of Commerce 1962). Another northeaster, of shorter duration 11 years later in 1973, was noteworthy for the damage it did. The storm of 9-10 February 1973 was also responsible for much beach erosion. Again, dunes were destroyed and formerly protected structures were standing in the surf (U.S. Department of Commerce 1973) (Figure 4).

Figure 4. A. Two days after the northeast storm of 9-10 February 1973 the waters of the Atlantic are still turbulent. B. The frontal dune has been washed away and several motels are standing in the surf. C. Route 12 has been washed out and is not open to traffic, and D. there is evidence that the surf has penetrated farther inland. This location is 16 kilometers north of Cape Hatteras. Photograph by the North Carolina Department of Transportation.

The Future

Until the Second World War the islands were sparsely inhabited. Few roads were paved and just a few bridges connected them with the mainland. The inhabitants conducted life at a subsistence level and fished mainly for local consumption. During earlier times livestock raising and onshore whaling were important in different locations along the coast. There were a few Coast Guard stations with their personnel, and the beginnings of summer vacation colonies. Most of the permanent inhabitants lived on the soundsides of the islands and any storm that struck the coast expended its energy on empty beaches. The ocean side remained largely in a natural state until the period of prosperity following the war. Since then the sale of lots along the Atlantic shore has been brisk and people from outside coastal North Carolina have been building there (Dunbar 1958; Schoenbaum 1982; Stick 1958).

It is part of a national trend; property is being put in harm's way all along the U.S. coast from Texas to Maine. Figure 5 shows the national trend in death and property damage from hurricanes. Since 1900 deaths have been reduced dramatically and this can be attributed to improved evacuation procedures and communication such as radio, television, and satellites. However, the upward trend in property damage must only increase because beach cottages now line the ocean shores in many places in this state and along the Gulf and Atlantic coasts. They are sitting in the way of some future hurricane or northeaster. Furthermore, the trend is toward the construction of larger and more expensive condominiums on the coast. Inevitably, the storms of the future will endanger more lives and property than they do now. People will continue to live on the barrier islands or visit them because of the beauty and recreational opportunities found there. Does this mean that we are creating an impossible situation where death and destruction will be the price we pay for the use of the islands? This need not be so if we understand the nature of the barrier island storm environment and act on that understanding.

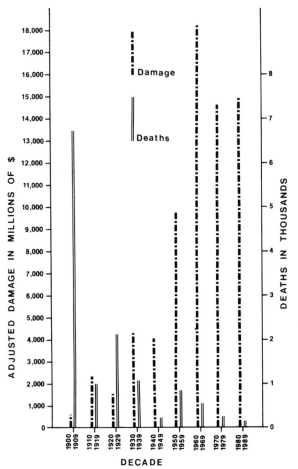

Figure 5. 20th-century deaths and damage (adjusted to 1989 dollars) caused by all hurricanes from Texas to Maine. Data source: Hebert and Case (1990).

References

Baker, S. 1978. *Storms, People, and Property in Coastal North Carolina.* Sea Grant Publication UNC-SG-78-15. Raleigh: University of North Carolina Sea Grant College Program.

———, **and Flint, K.** 1990. Development and change on the North Carolina coast: The case of Bogue Banks. In *Studies in Marine and Coastal Geography,* ed. P. J. Ricketts, pp. 1-16. Halifax: Saint Mary's University.

Bosserman, K., and Dolan, R. 1968. *The Frequency and Magnitude of Extratropical Storms along the Outer Banks of North Carolina.* Technical Report 68-4. Charlottesville: Coastal Research Associates.

Dunbar, G. S. 1958. *A Historical Geography of the North Carolina Outer Banks.* Coastal Studies Series No. 3. Baton Rouge: Louisiana State University Press.

Graetz, K. E. 1973. *Seacoast Plants of the Carolinas for Conservation and Beautification.* Sea Grant Publication UNC-SG-73-06. Raleigh: University of North Carolina, Sea Grant Program.

Hebert, P. J., and Case, R. A. 1990. *The Deadliest, Costliest, and Most Intense United States Hurricanes of This Century (and Other Frequently Requested Hurricane Facts).* NOAA Technical Memorandum NWS-NHC-31. Miami: National Oceanic and Atmospheric Administration.

Leatherman, S. P. 1979. *Barrier Island Handbook.* Amherst: The Environmental Institute, University of Massachusetts.

Risnychok, N. T. 1990. *"Hurricane!": A Familiarization Booklet.* Miami: National Hurricane Center.

Schoenbaum, T. J. 1982. *Islands, Capes, and Sounds: The North Carolina Coast.* Winston-Salem: John F. Blair.

Stick, D. 1958. *The Outer Banks of North Carolina, 1584-1958.* Chapel Hill: University of North Carolina Press.

U.S. Department of Agriculture, North Carolina Department of Natural and Economic Resources, and North Carolina State University Soil Science Department. 1977.

Soil Survey of the Outer Banks, North Carolina: Part I Text Material. Raleigh: North Carolina Department of Natural and Economic Resources.

U.S. Department of Commerce. 1962. Report for March 1962. *Storm Data* 4:137.

————. 1973. Report for February 1973. *Storm Data* 15:4.

East Asian Villages and Downtown Development

Harry L. Margulis
Cleveland State University

Life is made up of transient images of places and fleeting impressions of people. We buy Asian goods, eat exotic foods in Chinese and Japanese restaurants, and every once in a while exchange some words with a Korean "green grocer" or a Chinese "laundryman." Suburban Asian-American professional business persons are largely invisible, the contributions being made by Asian-Americans in all walks of life to our society hardly evident. Yet, as cities are transformed into advanced services centers, urban managers (corporate and public leaders) are finding that physical location and ethnicity are offering opportunities for rebuilding decaying neighborhoods while strengthening downtown development.

This study examines how urban managers are integrating East Asian communities in Boston, Massachusetts, and Oakland, California, into downtown development planning. Downtown East Asian villages constitute nodal settlements or as D. Lai (1988) notes "towns within cities." These settlements are increasingly performing culturally diverse regional and neighborhood functions, while at the same time serving as festival marketplaces for the larger metropolitan Asian and non-Asian population. Unlike early waves of immigrants who sometimes confronted hostility, overwhelmed social services, and were often ignored by urban managers, recent immigrants now face a less hostile environment, although an equally arduous journey to achieve economic and cultural assimilation. Urban managers are increasingly recognizing that the assimilative functions—the adaptation of non-urban institutions and cultures to the urban milieu (Gans 1962)—performed by East Asian villages can be harnessed to strengthen downtown cores. As mediators between the public and private sectors, urban managers are helping to reshape social development, land-use patterns, and the collective behaviors of these urban settlements.

The Growth of East Asian Population in the United States

In the last two decades, the source regions of immigration to the United States has changed substantially. American policy failures in Southeast Asia, the imminent return of Hong Kong to mainland China, changes in immigration policy, and Pacific Rim surplus capital investment in North American cities have increased immigration from Cambodia, the Philippines, Korea, Laos, Taiwan, the Peoples Republic of China, and Vietnam. Pre-1970 East Asian immigrants were largely poor, unskilled, and unemployed rural people driven from their native lands by hunger and poverty. Many post-1970 immigrants are refugees, former soldiers, agrarian farm workers, or housewives admitted on the basis of family reunification. However, a considerable number of Hong Kong, Japanese, and Taiwanese immigrants are skilled or semiskilled—doctors, engineers, architects, clerical workers, teachers, and technicians. Some are well-to-do or budding entrepreneurs operating small- to medium-sized businesses. In contrast, many unassimilated East Asians and many skilled non-English-speaking persons have found employment in factories and food-related establishments. For the latter, absorption into the social fabric of American life has been slower no matter how industrious or self-reliant. Still others bear the physical and emotional scars of displacement and resettlement.

Asian villages survive because, for many non-English-speaking elderly, low lifetime earnings and fixed incomes have drawn them back to the urban villages where friends and churches provide social support networks. For more recent East Asian immigrants, the urban village allows them to cling to their traditional lifestyles in familiar environments where generational assimilation takes place in a supportive multilingual setting (Chinatowns, Koreantowns, or Vietnamesetowns) where many middle-aged and elderly descendants of earlier immigrations still reside.

Ethnic Village Revitalization

Boston

Boston's Chinatown, a 28-block area astride the expanding midtown cultural district, is a mixed-use densely settled residential and commercial area. It is strategically located in the downtown area and close to major transportation arteries, such as the Massachusetts Turnpike and the Southwest Expressway (Figure 1). Residences are mostly found in three- or four-story apartment houses, mixed residential-commercial or residential-institutional structures, and a few larger apartment complexes (Tai-Tung Village, Castle Square, and Massachusetts Pike Towers). Most of the approximately 176 Chinatown businesses are new, the scale of operation is relatively small (Bourguignon 1988; Oriola and Perkins 1988) (Figure 2).

What makes Boston's Chinatown unique is the use of hinge block planning, tying the disposition of publicly owned parcels in the downtown with publicly owned sites in Boston's neighborhoods. A committee composed of the Boston Redevelopment Authority, Metropolitan Boston's Transit Authority's Real Property Department, the Public Facilities Department, the Governor's Office, the Mayor's Office of Neighborhood Services, and the Mayor's Office of Jobs and Community Services is working with the developer—Metropolitan Structures

Figure 2. A view of Beach Street, the Chinatown Eatery, in Boston, Massachusetts. Reprinted with the permission of Jenn Van Horne Photography.

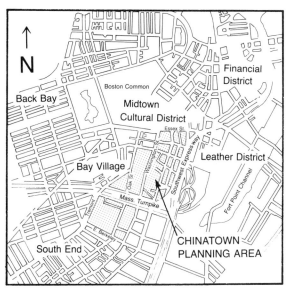

Figure 1. Boston's Chinatown is located in the downtown area, close to major transportation arterials, such as the Massachusetts Turnpike and the Southwest Expressway.

and Columbia Plaza Associates—to create a community benefits plan. The 50/50 partnership consists of a consortium of Boston-based Asian-American, black, and hispanic entrepreneurs and organizations. One Lincoln Street has been selected as a linkage redevelopment site (Metropolitan/Columbia Plaza Venture 1988).

One Lincoln Street is a 6,500-square-meter (70,000-square-foot) site located at the southern edge of the financial district adjacent to the northeast boundary of Chinatown. The project consists of a 35-story, 70,570-square-meter (760,000-square-foot) tower connected to a lower nine-story, 21,800-square-meter (235,000-square-foot) base containing 3,250 square meters (35,000 square feet) of retail space and a five-level, 900-space parking garage. The Boston Redevelopment Authority is to act as the development agency while the Chinatown-South Cove Neighborhood Council is to serve as the designated community review board.

Metropolitan Structures and Columbia Plaza Associates and Boston Redevelopment Authority, providing matching grants totaling more than $100,000, have contributed $48,000 for community planning. One requirement is that minorities and women receive job-training, and share development contract dollars and construction trade jobs. More than $1,500,000 in job-training linkage funds are available to train minority residents of Chinatown and Roxbury for jobs created by the project. Metropolitan Structures and Columbia Plaza Associates is also making a $400,000 challenge grant available on a two-for-one matching basis to create a minority on-the-job training-opportunities fund. Additional "incubator" facilities are to provide affordable office space and shared support services for both existing and starting minority businesses. Firms locating in these facilities can take advantage of arrangements whereby Metropolitan Structures and Columbia Plaza Associates provides minority businesses with technical assistance and timely access to project-development business opportunities.

Gentrification, commercial encroachment, and highway construction have reduced the availability of low- and moderate-income housing. Currently, 21 percent of the units in Chinatown are overcrowded and 42 percent of households are living below the poverty level. At least 2,000 new, affordable housing units are needed to meet current demand and, to meet this need, the process of rezoning the area has been initiated. A "Restricted Growth Subdistrict" and the "Chinatown-Bay Village Housing Priority Area" were created to encourage affordable housing production. Restrictive zoning limits building heights to six to eight stories and the floor area ratios to six or seven. In the Chinatown business district housing priority area, new construction or rehabilitation must have a minimum of 50 percent of the gross floor area set aside for residential use. In addition, all new construction in the Chinatown-Bay Village housing priority area must contain a minimum of 75 percent gross floor area in residential uses.

The Chinatown-South Cove Neighborhood Council is now positioned to influence planning for other redevelopment projects. For example, the Central Artery-Third Harbor Tunnel Project, proposed to replace the six-lane elevated section of the I-93 Central Artery with an eight- to ten-lane road, offers opportunities to reduce through-street traffic and to expand Pagoda Park. Approximately 33,070 square meters (356,000 square feet) of land should become available as the Central Artery is constructed (*The Sampan* 1989). Most of the land is owned by the Massachusetts Turnpike Authority, but the Chinatown-South Cove Neighborhood Council is seeking community control over this "Gateway" area. A number of development options have been recommended, for example, zoning

for large markets and businesses, a joint venture by community and private developers, or a joint venture between the community and Tufts University or the New England Medical Center. Whatever the outcome, the Chinatown-South Cove Neighborhood Council is likely to serve as the designated organization helping the community protect and preserve its heritage.

Oakland

Oakland's traditional Chinatown comprises a four-block area with approximately 18,580 square meters (200,000 square feet) of retail space (Figure 3). Scarcity of physical space, and rapidly increasing real estate sales prices and rents have forced Asian businesses across Broadway and west of Chinatown. Other businesses have moved up 13th, 14th, and 15th Streets and across Lake Merritt.

Some 40 restaurants, 40 sewing factories, 35 groceries and markets, eight financial institutions, eight bakeries or take-out food shops, 15 specialty stores, six jewelry stores, and numerous professional offices (dentists,

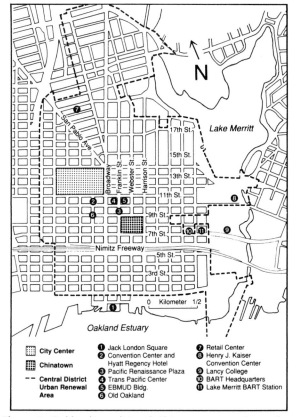

Figure 3. Oakland's traditional Chinatown comprises a four-block area surrounding the intersection of 8th and Webster streets. Reprinted with the permission of Jenn Van Horne Photography.

Figure 4. A typical Chinese green grocer in Oakland, California's, Chinatown.

physicians, optometrists, real estate brokers, attorneys, accountants, insurance agents) and services (hair-stylists, mechanics, travel agents) occupy commercial space (Figure 4). Approximately 20,000 shoppers visit the area each weekday. In addition, 11 family associations and tongs provide assistance to their members, and seven prominent service organizations and community and cultural facilities are also present (Dang 1988).

Traditional Chinatown abuts the southern edge of the Chinatown Project Action Area, a redevelopment area adjacent to Old Oakland, the Convention Center, and the Hyatt Regency Hotel complex. The redevelopment area links traditional Chinatown's commercial district to adjacent downtown development. Plans for the Pacific Renaissance Plaza, a $15,000,000 development to include office, retail, recreational, entertainment, residential, and open-space uses are now being finalized. Design standards and land-use controls are to ensure that Chinatown is physically integrated into the revitalized area.

The amended development agreement between the City of Oakland's Redevelopment Agency and Pacific Renaissance Associates ensures that the location, size, design, and site-operating characteristics are compatible with the successful development of Chinatown as an urban complex of residential, commercial, and cultural activities (City of Oakland 1987).

On parcels retained by the Oakland Redevelopment Agency, a 500-space public parking facility, a $2,500,000 Asian cultural center and library—a 2,230-square-meter (24,000-square-foot) facility—and subsidized affordable low- and moderate-income housing units are to be built. Moreover, the developer has consented to pay the agency an amount equal to 1 percent of the total development costs, excluding development costs related to the public improvements, residential improvements, and related residential parking, for the benefit of a minority non-profit com-

munity organization. The beneficiary should be the East Bay Asian Local Development Corporation, a broad-based community organization that performs multiple service functions for minorities.

Public actions, such as the Nimitz Freeway, BART Lake Merritt Station, Laney College, the Oakland Museum, the Oakland City Center, and the Chinatown Redevelopment Project, have removed 573 units from the local housing inventory. Conversion of privately owned residential hotels into offices, restaurants, and high-priced apartments have removed an additional 122 units. Consequently, affordable housing has become a serious problem. The East Bay Asian Local Development Corporation has thus become quite proficient in rehabilitating housing and equally influential in negotiating with local and private investors for funds to provide low- and moderate-income housing. For example, the City of Oakland through the Bay Region Housing Investment Development Group, a joint venture partnership with the East Bay Asian Local Development Corporation, has recently completed a $17,000,000 multi-unit project. Site acquisition and subsidies for 119 low- and moderate-income rental units were provided by the City of Oakland. Forty percent of the 119 units are low-income apartments; the rest rent at market rate. In addition to resident parking, the two organizations will build two levels of underground public parking to be operated by the City of Oakland.

The East Bay Asian Local Development Corporation has also been instrumental in renovating the four-story, wood-frame Madrone Hotel, a 32-room, single-room-occupancy hotel with two commercial storefronts, currently valued at $1,900,000. Another effort to preserve affordable housing involves the corporation's acquisition and rehabilitation of the Oakwood Apartments. The 56-unit, 4,505-square-meter (48,500-square-foot) wood-frame two-story apartment building now houses Cambodian and Laotian refugees, black, hispanic, and Fijian families. When acquisition, rehabilitation, and repairs are completed, the building should have a value of $2,400,000.

Conclusions

In cities such as Boston and Oakland, East Asian villages are being revived largely as a result of "the interpenetration of economies and societies at the international level, facilitated by the new communications technologies . . ." (Castells 1985, 32). Urban managers confronted by a resurgence and reconcentration of East Asian immigrants are striving to manage community change as well as to funnel some of this immigrant energy into downtown development. Working in conjunction with local development groups, urban managers are creating community benefit plans and physical linkages between these strategically located communities and downtown developments. As a consequence, these communities are serving as sanctuaries and economic launching platforms for immigrants and festival marketplaces for the larger metropolitan population. By guiding social processes and spatial development patterns, urban managers are helping to physically reshape these communities and, in turn, contributing to downtown regeneration.

References

Bourguignon, M. 1988. *Chinatown Survey Area Land Use Report*. Boston: Boston Redevelopment Authority, Policy Development and Research Department.

Castells, M. 1985. High technology, economic restructuring, and the urban-regional process in the United States. In *High Technology, Space and Society*, ed. Manuel Castells, pp. 11-39. Beverly Hills, CA: Sage.

City of Oakland. 1987. *Amended and Fully Restated Disposition and Development Agreement between the Redevelopment Agency of the City of Oakland and Pacific Renaissance Associates: II. A California Limited Partnership*. Oakland, CA: Oakland Redevelopment Agency.

Dang, T. 1988. *Community Economic Profile for the Oakland Chinatown Area*. Oakland, CA: Oakland Chinatown Chamber of Commerce.

Gans, H. 1962. *The Urban Villagers*. New York: The Free Press.

Lai, D. 1988. *Chinatowns: Towns within Cities in Canada*. Vancouver: University of British Columbia.

Metropolitan/Columbia Plaza Venture. 1988. *Parcel to Parcel Linkage I Update: One Lincoln Street and Ruggles Center*. Boston: M/CPV.

Oriola, D., and Perkins, G. 1988. *Chinatown Business Survey*. Boston: Boston Redevelopment Authority, Policy Development and Research Department.

The Sampan. 1989. 27(11):1, 8.

Riverside and Pullman, Illinois: 19th-Century Planned Towns

Eugene C. Kirchherr
Western Michigan University

By the 1880s, Riverside and Pullman were identifiable points or localities—literally free-standing satellite towns only 22.5 kilometers apart—within a nascent Chicago metropolitan region where scattered outlying settlements had not yet begun coalescing to form a conurbation. But these two towns would not remain inconspicuous points on the map. Despite their modest size, they were soon widely acclaimed as two of the finest *planned* communities in the nation: *Riverside*, an exemplary residential suburb (Figure 1), and *Pullman*, the quintessential company town (Figure 2). Admittedly differing in purpose and appearance, both towns nonetheless demonstrated creative design, innovative ensemble, and urban engineering that would be emulated in communities elsewhere.

This study, through a retrospective and comparative view of their early development, traces the events and processes that brought renown to these communities at a particular period. Accessible points, which once attracted numerous visitors, the towns were ultimately engulfed within the urbanizing region spreading outward from Chicago. Nevertheless, Riverside's remarkable design still stands in sharp contrast to that of neighboring western suburbs, while Pullman remains a distinctive, if aging, south side neighborhood of Chicago.

Foundations of Pullman and Riverside

The sites chosen for Riverside and Pullman were undeveloped tracts beyond the city of Chicago (as then delimited), though neither so distant nor remote as to be sequestered. Nevertheless, the development of the two towns was undertaken with markedly divergent objectives. Riverside was initiated in 1868 with Emery

Child's formation of the Riverside Improvement Company. Upon securing a 648-hectare tract along the Des Plaines River 14.5 kilometers west of Chicago's central area, the investors proposed to establish a dormitory or "bedroom" community where affluent families might enjoy pleasant surroundings without, of course, surrendering urban amenities to which they were accustomed (Riverside Improvement Company 1871) (Figure 3A). For their efforts, the developers naturally expected to realize a generous return. In need of an innovative and marketable community design, they engaged Frederick Law Olmsted Sr. and Calvert Vaux, the country's foremost landscape architects, to prepare a plan.

By 1880, George M. Pullman, prominent manufacturer of railway sleeping cars, had acquired 1,620 hectares in the Calumet district, about 21 kilometers south of downtown Chicago. Unlike the Riverside group, Pullman intended to establish a self-contained town comprising a manufacturing works of the Pullman Palace Car Company and a residential area where his employees and their families would be housed. He entrusted the town planning to two professionals—Solon Beman, an architect, and Nathan Barrett, a landscape designer.

While George Pullman maintained a mansion in Chicago, his industrial town was purposely distanced from the city he regarded "a repository of vice, disorder, and ugliness" (Buder 1967, 118). Practical considerations influenced his selection of the locale as well, especially the lower cost of land in the Calumet district compared with Chicago, and the access to iron and steel industries in nearby towns. And there was plentiful clay to be dredged from the bottom of Lake Calumet for making bricks to construct housing and industrial buildings in Pullman (Buder 1967).

Figure 1. Olmsted and Vaux's plan for Riverside.

View of Long Common and Junction of Roads.

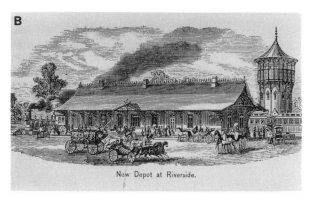

New Depot at Riverside.

Figure 3. A. Riverside's parklike setting. B. Rail depot and water tower in Riverside. C. Row housing in Pullman.

Both towns had access to water bodies and railroads. The water bodies were integrated into the storm-water drainage systems installed in each town. In other respects, the water features were perceived rather differently. In Riverside, areas along the Des Plaines River, with its graceful meander, were reserved by Olmsted for parks. By contrast, diminutive Lake Calumet, on the eastern edge of Pullman and connected to nearby Lake Michigan, was envisaged the site of an interior harbor (Figure 2). With the lake proving too shallow for larger vessels (Doty 1893), the Pullman Company relied on railroads to service its operations. The lake later became a center for recreational activities and events (Buder 1967).

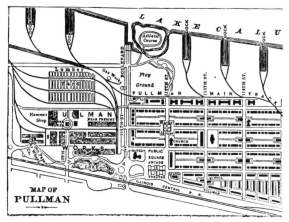

Figure 2. The original (or South) Pullman community in 1885. North is at the left on the map.

Trackage already running through the locales of Riverside and Pullman provided essential rail links with Chicago. Once town building was under way, depots were erected and a schedule of rail services set. The railroad was indispensable for Riverside's commuters who worked in Chicago (Figure 3B), though not so for the Pullman labor force who mostly lived within walking distance of their workplace. With the railroad, and later the streetcar, Pullman residents were never isolated, and could travel easily to Chicago for visits or other business.

The Town Plans

Olmsted and Vaux's plan for Riverside emphasized "gracefully curved lines, generous spaces, and the absence of sharp corners, the idea being to suggest and imply leisure, contemplativeness and happy tranquility" (Olmsted and Vaux 1868, 17). The pleasing physical plan, especially the curvilinear street pattern adjusted to local topography and drainage, deliberately attempted to preserve the natural landscape (Figure 1); indeed, Olmsted set aside more than 283 hectares for public use. Riverside was subdivided into large lots, generally 30.5 meters wide and 61 meters or more deep. The tree-lined streets, open spaces, and wooded plots were deftly blended into a setting described as "a village in a forest."

Viewed casually, the grid plan of Pullman, especially if compared with Riverside, appears quite ordinary (Figure 2). But on closer scrutiny, one discerns that Beman and Barrett skilfully executed an ingenious plan. Their plan separated the industrial and residential districts by a wide boulevard (Florence Boulevard, or 111th Street), a wall along the south end of the manufacturing zone, and a town center. Among the other attractive elements of Pullman were a landscaped pond (Vista Lake) fronting the company headquarters, and a grid with an interspersion of open spaces (Arcade Park) and other features (the Market Square, where fresh meat and produce were sold; the elementary school). Paved streets, sidewalks, fenced backyards, and paved alleys were ubiquitous (Ely 1885). With buildings set back from streets of 20 meters width, fronts of houses lining most streets were actually separated by 30.5 meters or more, resulting in untypically commodious surroundings for a working class district of that day.

Early Development

The pace of growth differed markedly in the two communities. In the early 1870s homes were under construction in Riverside along with a hotel, water tower, business block, and railroad depot. That initial spurt was not sustained, however, being deterred by a series of events including disruptions following the Chicago fire of 1871, the depression of 1873, and rumors of disease and an unhealthy environment in the vicinity of Riverside (Bassman 1936; Chamberlin 1874). Although the Riverside plan is ranked among Olmsted and Vaux's finest achievements, their actual involvement was brief as they terminated their association with the developers in early 1870 (Roper 1973). Then in 1873, the company declared bankruptcy. Fortunately,

others would guarantee that subsequent development did not deviate from Olmsted's design, most notably William LeBaron Jenney, a leading architect and Riverside resident, who is also credited with the design of the hotel, several homes, and the water tower that became the community's landmark. Despite reversals, Riverside gradually was taking form with a network of gas-lighted paved, streets, asphalt sidewalks, and service alleys. Through the 1870s, a number of homes were built and occupied (each connected to gas and sewer lines), private and public grounds dedicated, and schools and other facilities constructed (Collins 1985).

Riverside was already represented on maps of the Chicago area when ground was first broken for Pullman in the spring of 1880. In direct contrast to the sporadic growth of Riverside, Pullman would be completed and fully occupied within a few years. The population reached 8,600 in 1885 (Buder 1967). Not overlooking the higher density of housing, the working families of Pullman did benefit from amenities and services, including the provision of gas, water systems, paved streets, sufficient windows to provide cross ventilation and sunlight, and roomy basements. The company maintained the buildings, streets, parks, and front lawns, and arranged for the daily removal of garbage, ashes, and rubbish. Occupants had only to keep the inside of their homes clean and tidy.

Anticipating a diversified work force, a novel mix of housing types was introduced into the plan, predominantly row houses accommodating two to five families (Figure 3C), supplemented with tenements or "block houses," and detached or semi-detached single-family dwellings. The housing assigned an employee signified his status within the company (for example, unskilled single men in tenements, foremen in multi-level units, managers in detached residences). A tendency toward monotonous facades in row housing of that day was resisted by Beman who incorporated variations in the building designs (Ely 1885).

The plans for Riverside and Pullman each included a central area adjacent to the rail depot (Figures 1 and 2). A welcome openness characterized both town centers, in Riverside by preserving a commons area, and in Pullman by the spacious Arcade Park and the artificial Vista Lake. While the towns grew before zoning and related controls were being generally adopted, certain functions could be regulated (for example, industry was prohibited in Riverside, as were saloons in Pullman) and commercial activities were confined to designated buildings in the central areas. Riverside had its "Green Block," a structure housing a variety of commercial enterprises (Bassman 1936), though it was not comparable to the impressive, multifunctional Arcade Building and the nearby Market Square in Pullman (Doty 1893). A well-appointed hotel was opened in each town center to accommodate visitors.

The Political and Social Setting

Neither Riverside nor Pullman was constituted a political division at its founding. Riverside, however, incorporated as a village in 1875, and retained its status through later periods as newer suburban communities grew around it. Pullman, founded as a small industrial enclave within the jurisdiction of Hyde Park, never incorporated, and was administered by an agent appointed by and responsible to Mr. Pullman (Lindsey 1942). But in 1889 Pullman would grudgingly accede to the annexation of all of Hyde Park—including his model town—by the city of Chicago.

Riverside was a community of property *owners* of independent means. They were expected to fulfil certain obligations, for instance, to begin building a home within a year of purchasing the lot, ensuring a 9-meter setback of the house from the front property line, and planting two trees on the front part of the lot (Riverside Improvement Co. 1871). Such stipulations were no inconvenience to Riverside's landowners who took pride in their exclusive community of elegant, single-family homes designed by leading architects.

Pullman was totally under company management. Workers spent their days in Mr. Pullman's employ, returning afterwards to rented housing (no town property could be purchased). Residents shopped in company-managed facilities and enjoyed leisure time in company-maintained recreational facilities. Rents, reputedly higher than those paid elsewhere in the region, were not considered unreasonable because of the installed utilities and community services. Nevertheless, those who used the "public" library in the Arcade Building had to purchase memberships, and the Green Stone Church was available—at a charge—to any religious group for services and ceremonies. While the Riverside developers focused on attracting affluent families, Pullman was seeking to mold a contented, loyal work force, respectfully appreciative to be occupants of his model company town.

Retrospect

In the closing decades of the 19th century, Riverside and Pullman, two of several outlying points of new settlement near Chicago, achieved prominence as exceptional models of planned towns. Now subdued within the mosaic of an urbanized area, the communities are less easily identifiable as the focal points of attention they once were, though both continue to merit mention in works on planning history and are designated historic landmark districts.

Without diminishing the innovative aspects of the towns, their characterization as model communities does require qualification. Riverside, the commuter suburb, was undeniably attractive—Kenneth T. Jackson would say "blatantly elitist" (1985, 86)—yet such a community was certainly not an alternative for resolving the problems of low-income, poorly housed working people crowded in the cities. George Pullman seems to have been persuaded, as were others of that time, that "problems associated with slum housing and even those of morality could be controlled in an ideal community" (Pacyga and Skerrett 1986, 427-428). Critical observers, unimpressed by what was being hailed "the world's most perfect city" (Buder 1967, 73), characterized the town as a medieval barony "inhabited by serfs and chattel rather than free men" (Cohen 1961, 1). Sadly if inevitably, the Pullman experiment foundered. Townspeople grew resentful of an overbearing, paternalistic employer, culminating in the famous Pullman strike of 1894 (Lindsey 1942). Some years later, the courts brought an end to the company's domination of the community.

If Riverside and Pullman did not fully achieve their founders' expectations, neither did they totally fail. Neither town experienced uncontrolled incremental growth, and both had an impact on planning thought and practice in subsequent periods.

References

Bassman, Herbert J., ed. 1936. *Riverside Then and Now: A History of Riverside, Illinois*. Riverside, IL: Riverside News. (Reprinted, Chicago: University of Chicago Press, 1958.)

Buder, Stanley. 1967. *Pullman: An Experiment in Industrial Order and Community Planning, 1880-1930*. New York: Oxford University Press.

Chamberlin, Everett. 1874. *Chicago and Its Suburbs*. Chicago: T. A. Hungerford.

Cohen, Jerry. 1961. Timeless town: A restful oasis in wearying waste. *Chicago Sunday Sun-Times* 4 September(2):1-3.

Collins, Catherine. 1985. Riverside: A village where parks and open spaces are sacred. *Chicago Tribune* 9 August, pages unmarked in source copy.

Doty, Mrs. Duane. 1893. *The Town of Pullman: Its Growth with Brief Accounts of Its Industries*. Pullman, IL: T. P. Struhsacker. (Revised and reprinted, Pullman, IL: Pullman Civic Organization, 1974.)

Ely, Richard T. 1885. Pullman: A social study. *Harper's Monthly* 70:452-466.

Jackson, Kenneth T. 1985. *Crabgrass Frontier: The Suburbanization of the United States.* New York: Oxford University Press.

Lindsey, Almont. 1942. *The Pullman Strike: The Story of a Unique Experiment and of a Great Labor Upheaval.* Chicago: University of Chicago Press.

Olmsted, Vaux & Co. 1868. *Preliminary Report upon the Proposed Village at Riverside, near Chicago.* New York: Sutton Bowne.

Pacyga, Dominic A., and Skerrett, Ellen. 1986. *Chicago: City of Neighborhoods.* Chicago: Loyola University Press.

Riverside Improvement Company. 1871. *Riverside in 1871.* Chicago: C. & C. H. Blakely.

Roper, Laura Wood. 1973. *FLO: A Biography of Frederick Law Olmsted.* Baltimore: Johns Hopkins University Press.

Challenge Points over Historic Preservation in Florida

Ary J. Lamme III
University of Florida

Florida has always been a good place to get a tan or to be entertained at places like Disney World. Until recently, however, the state's cultural attractions lacked substance. Now the public is beginning to appreciate these features. In the case of historic sites, federal programs, such as the National Trust for Historic Preservation, and tax incentives for rehabilitation of historic structures have stimulated local efforts, while widespread recognition of a crucial connection to early Hispanic colonialism has prompted renewed interest in Florida's past.

In spite of this progress, contentious confrontation between historic preservation proponents and development interests resonated throughout the Florida peninsula during the 1980s. Because Florida is a key portion of the "Sunbelt," this confrontation has drawn nationwide attention. This paper considers two examples of the relationship between historic landscapes and development: the urban setting of St. Augustine and the rural setting at Cross Creek, in the interior karst landscape of North-central Florida. An examination of these two juxtaposed points of preservation concern is illustrative of challenges and opportunities facing historic sites in Florida and across many sections of the Sunbelt.

St. Augustine's Plaza

It is hard to believe that the fourth most populous state in the United States, a place that hosts 40 million tourists a year, was little more than an isolated, sparsely populated peninsula on the periphery of the nation less than 100 years ago. Although coastal beaches, interior springs, and the mild climate had been known and used by visitors from nearby states for some time, it was not until the 1890s that Florida began developing into a major tourist destination (Lamme 1989a).

Henry Flagler, who had earned a fortune working for John D. Rockefeller and the Standard Oil Company, first came to Florida in the late 1870s to provide his ailing wife with a gentler winter climate (Waterbury 1983). Flagler stayed in Jacksonville during his initial visit and then in 1883 at St. Augustine, both in the northeast corner of the state, because destinations farther south were too difficult to reach. A trip to Florida from the Northeast during the tourist season was a harrowing experience for most families. It could be accomplished by coastal steamer, which involved the discomfort of an Atlantic excursion in winter, or by a combination of rail lines and other modes of transportation.

The challenge of an attractive destination that was hard to reach from Northeast population centers stimulated Flagler's entrepreneurial spirit. He liked the historic atmosphere of St. Augustine, the oldest European city in North America, and thought it might become part of an "American Riviera" (Graham 1978). Eventually he owned three world-class hotels in that city as well as a string of similar establishments farther south. In addition, Flagler built the Florida East Coast Railroad, which ran 1,050 kilometers (650 miles) along the Florida coast and was completed to Key West in 1912. By that time, mass tourism had by-passed St. Augustine. Today, the city remains a relatively minor place of visitation in the galaxy of Florida tourist attractions.

However, St. Augustine cannot be ignored as a tourist destination. Although Florida was a Spanish colonial borderland, St. Augustine's heritage is receiving increased attention because of the growing Hispanic minority in the United States. The most prominent contemporary landscape artifact is the Castillo de San Marcos. Although St. Augustine was founded in 1565, this coquina fort was not completed until the 18th century, and is one of a long line of fortifications at St. Augustine. What about the historic town laid out earlier around a central plaza according to Spanish Laws of the Indies? It's there, but has been altered and obscured by accommodation to 20th-century commercial and transportation patterns.

St. Augustine's plaza, once the focal point of the Spanish colonial town, now exemplifies the tension between historicity and modernity in Florida (Figure 1). It is located along the waterfront of the Matanzas River and dates from the 16th century. Here the first church, government offices, shops, and homes of important officials were located. However, with the United States acquisition of Florida in 1821, the plaza began to change. St. Augustine slowly evolved into an Anglo town and the plaza was a logical place for a central business district, town park, and even the ubiquitous southern Confederate memorial. By 1895 a bridge spanned the river between Anastasia Island and St. Augustine, funneling traffic onto the plaza, from what has become Florida's route A1A. Today the traffic-clogged plaza is not reminiscent of a Spanish town. This unsatisfactory image is reinforced by an unimaginatively landscaped central section surrounded by a typical American commercial strip. Nothing remains to suggest the importance of the plaza to Spanish colonial life.

The plaza and route A1A literally intersect and are most certainly interrelated. They also have a strong figurative relationship that symbolizes one of the dilemmas of contemporary Florida. The plaza is a remnant

Figure 1. Traffic crosses the Matanzas River on the bridge in the background (route A1A) and empties onto the 16th-century St. Augustine Plaza. The monument in the center of the green to the right is a memorial to Confederate forces from the American Civil War.

historic feature; rich in meaning and association to those seeking a cultural experience or searching for historic roots. On the other hand, the road is a vital link in the tourist industry, providing access to recreational opportunities. Can these two features exist side by side without damage to the mission of either? In the case of St. Augustine the present answer is "no."

Cross Creek Village

"Waterfront property!" Such a real estate sign is bound to attract attention and the land will command higher prices because most people like to live close to water. For most people, a small, running stream creates a magical sound, a river or a lake can provide recreational opportunities, and wave action along the shore has a calming effect. The appeal of living near water is expressed in these lines from the poem "Simple Gifts of the Sea," by Kimberly Slevin (1986, 530):

> Watching,
> Gazing out to sea
> Emptying myself to feel the sea in me.

A unique North Florida place has abundant waterfront property, and all the challenges that go along with such land. Pulitzer Prize winning novelist and one-time resident Marjorie Kinnan Rawlings (1942, 1) described it:

> Cross Creek is a bend in a country road, by land, and the flowing of Lochloosa Lake into Orange Lake, by water. We are four miles west of the small village of Island Grove, nine miles east of a turpentine still, and on the other sides we do not count distance at all, for the two lakes and the broad marshes create an infinite space between us and the horizon.

Since Cross Creek is literally a short creek connecting two lakes there is plenty of shoreline in the vicinity (Figure 2). At the turn of the century, citrus agriculture surrounded the lakes, protected from chilly North Florida winters by their maritime influence. A series of freezes then began to push citrus production southward. The northern boundary of today's Florida citrus belt runs across the peninsula in the vicinity of Orlando. Cross Creek, 160 kilometers (100 miles) north of this border, saw its few remnant groves go out of production with hard freezes in the early 1980s. Nevertheless, a relatively prosperous future for the village of Cross Creek was ensured by two local resources. Marjorie Rawling's home, an authentic Southern cottage, is

Figure 2. Cross Creek looking toward Lake Lochloosa. The marshy shoreline along the creek and lakes is best suited for low-density recreational development.

maintained and operated as a historic site by the state of Florida. Visitation to that site surged with the release of a popular movie, "Cross Creek," based on her life. The second resource, lakes and associated shorelines, are potential areas of large-scale recreational development. However, such development would contrast sharply with the fish camp, small cabin recreational landscape of today.

Although Cross Creek has some unique historic credentials, it is really not much different from many other Florida locations. An extensive coastline combined with lake shorelines in the karst topography that dominates the interior of the state lure recreational development throughout the peninsula. Some places along the Atlantic, the Gulf of Mexico, and near major urban centers have been developed for decades, but many others remain—like Cross Creek—backwaters to large-scale development. The Florida legislature recognized this challenge by passing the Growth Management Act in the early 1980s. This Act required each county to develop a comprehensive land-use plan.

Because of its prominence, Alachua County adopted a restrictive land-use plan in 1985 that applies only to Cross Creek (Lamme 1989b). Although it allows some low-density development, that plan eliminated the possibility of massive recreational facilities in favor of the natural and cultural environments. The plan has since been challenged in Court and withstood those challenges. Like St. Augustine, Cross Creek's cultural resources are faced with development pressures. However, the Cross Creek outcome has been somewhat different. Here the relationship is less adversarial, for the public-planning process has derived a model prescribing a compatible relationship between the historic village and a recreational shoreline used within ecologically imposed limits.

Conclusion

In St. Augustine, the Hispanic-era town center is buried beneath a quasi New England green, encircled by commercial hubbub, and trampled by vehicular traffic. A major transportation artery intersects that historic plaza creating a mixture of material artifacts supposedly suggesting the more than 400 years of European occupation, but mostly representative of the modern era. This situation conveys a conflicting message for residents and visitors. In contrast, the historic landscape at Cross Creek retains a functional, appropriately scaled relationship between natural setting and human activity through conservation and preservation initiatives. Admittedly, Cross Creek development pressures are more recent and less aggressive than at St. Augustine. Nevertheless, the lesson of Cross Creek seems to be that development and preservation are compatible when the scale of development is limited through consideration of its effect on natural and cultural resources.

This essay on preservation challenge points in Florida does not seek to render a universal moral judgment concerning preservation of historic sites, for it is possible to justify some historic landscape alteration. A historic plaza or village can symbolize cultural heritage and stability as well as elitism and stagnation. A roadway or recreational facilities along a shoreline can be public necessities as well as harbingers of environmental degradation. Nevertheless, these two examples suggest one preservation truism. If they are typical of what is going on at other places in the Sunbelt, the relationship between preservation and development needs to be actively managed, as at Cross Creek, rather than left to an unguided evolutionary process, as at St. Augustine.

References

Graham, T. 1978. *The Awakening of St. Augustine.* St. Augustine, FL: St. Augustine Historical Society.

Lamme, A. 1989a. *America's Historic Landscapes: Community Power and the Preservation of Four National Historic Sites.* Knoxville: University of Tennessee Press.

————. 1989b. Preserving special places. *Geographical Review* 79:195-209.

Rawlings, M. K. 1942. *Cross Creek.* New York: Charles Scribner's Sons.

Slevin, K. 1986. Simple gifts of the sea. *The World and I* (July):530-541.

Waterbury, J., ed. 1983. *The Oldest City.* St. Augustine, FL: St. Augustine Historical Society.

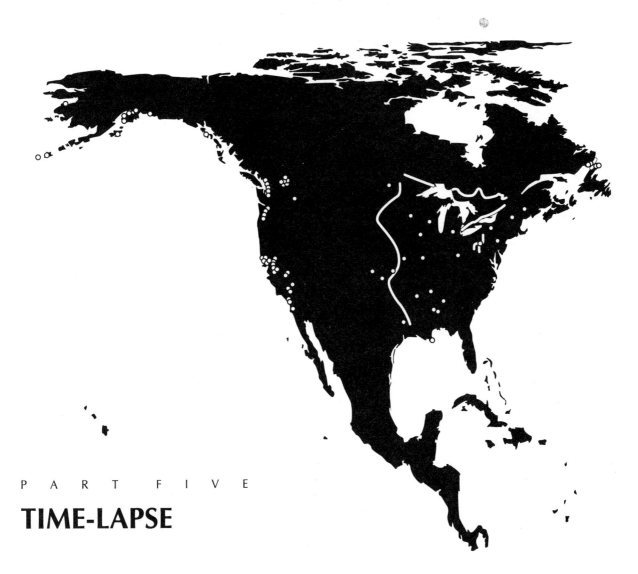

TIME-LAPSE

Spatial changes achieve a dynamism when considered through time, so if we are to understand patterns of change in geographic space we must also explore the time dimension. For example, sediment cores extracted from the earth may offer an unbroken record of environmental change over thousands of years. With them we reconstruct the sequence of biota and climate for specific point locations, which then can be averaged over space. In contrast, data from weather records and historical archives are often analyzed only for a sequence of dates or averages over short periods. Intervals and time spans circumvent technical problems in treating continuous data, but also reflect the paucity of information along the timeline. Thus, mudlump occurrence at the mouth of the Mississippi River is illustrated for broad time periods. Similarly, through careful archival work, interpretations of social and environmental change may be organized by broad eras. Examples include archaeological and archival documentation of early European contact points in Labrador and Newfoundland. Distinct periods in the migration of Russian religious groups characterize communities on North America's Pacific Rim. Specifically timed events and other circumstances help explain the diffusion of the "Moscow" place-name across the United States. Eras of settlement expansion along Ontario's Grand Trunk Corridor relate to transportation technology, while entrepreneurial prowess is the base for Pittsburgh's sequential development. And temporal categories of key events help expose why Birmingham, Alabama, was a critical point in America's struggle for civil rights. This section illustrates these approaches to assessing patterns of change over time through time-lapse methods.

—Laura E. Conkey, Dartmouth College

Since the Retreat of the Ice Sheet: A Wisconsin Wetland

Marjorie Green Winkler
University of Wisconsin–Madison

Natural events can be observed on many time scales. Aldo Leopold, in *A Sand County Almanac*, recorded the daily, monthly, and annual events taking place on his "poor farm" on the banks of the Wisconsin River in Sauk County, Wisconsin (Leopold 1949) (Figure 1). He also recognized the silent historians on the farm—the downed oak provided not only wood for fuel, but a tree-ring record of centuries of climatic and biologic change. Leopold (1949, 96) also wrote of other record-keepers on his farm:

> The peat layers that comprise the bog are laid down in the basin of an ancient lake. The cranes stand, as it were, upon the sodden pages of their own history. These peats are the compressed remains of the mosses that clogged the pools, of the tamaracks that spread over the moss, of the cranes that bugled over the tamaracks since the retreat of the ice sheet.

In this study I decipher the story recorded in the "sodden pages" of the peat from Leopold Marsh and describe the environmental changes since deglaciation (from 13,000 BP [years before present]) from the pollen and charcoal preserved in the peat. A single core, 7.75 meters deep, bored just south of Leopold's "shack," records the temporal pattern of environmental history, for one marsh site. However, it provides clues to regional climate and vegetation changes for a broader part of the northern Midwest, south from about latitudes 45°N or 42°N from the Dakotas in the Great Plains to at least northern Illinois (Winkler *et al.* 1986). North of latitude 45°N in the Midwest, the Great Lakes affect the regional climate.

Glacial History and Edaphic Setting

Leopold Marsh (43°33'N, 89°39'W) is part of a large floodplain wetland of about 1,600 square kilometers on the southern shore of the Wisconsin River in the northeastern corner of Sauk County, Wisconsin. It is within the Aldo Leopold Memorial Reserve (Figure 1). The site was within the proglacial lake basin of Glacial Lake Wisconsin adjacent to rolling topography of supraglacial till (Socha 1984). Since deglaciation and drainage of the proglacial lakes in the Baraboo area, peat has accumulated in parts of the ancient lake basins.

> The lake rose through the centuries, finally spilling over east of the Baraboo range. There it cut a new channel for the river, and thus drained itself. To the residual lagoons came the cranes, bugling the defeat of the retreating winter, summoning the on-creeping host of living things to their collective task of marsh-building (Leopold 1949, 98).

The present elevation of Leopold Marsh is about 244 meters. From the relationships of the proglacial lake shores (Glacial Lake Wisconsin and Glacial Lake Baraboo: elevation 291 meters; Glacial Lake Merrimac: elevation 260 meters [Socha 1984]), we can infer that Leopold Marsh was ice-covered when the Green Bay Lobe of the Laurentide ice sheet was at its westernmost extent. Then as the ice receded to the east, it was covered by Glacial Lake Wisconsin and Glacial Lake Baraboo. These relationships are recorded in the sediment stratigraphy of the Leopold Marsh core, which shows bottom-most sands and gravels (ice-deposited) overlain by clay. The marsh was uncovered for dep-

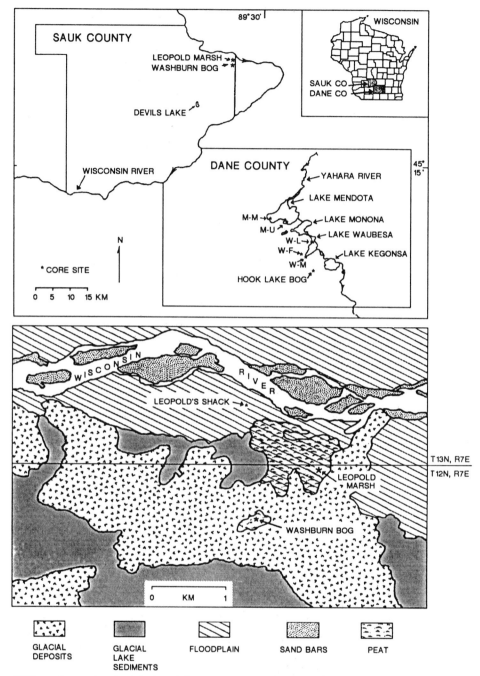

Figure 1. Maps of Wisconsin, southcentral Sauk and Dane Counties, and Leopold Marsh in Sauk County, showing the relationship of Leopold Marsh to glacial deposits, to the Wisconsin River, and to other paleoecological sites.

osition as more southerly Glacial Lake Merrimac formed and became the drainage route for the eastern extension of Glacial Lake Wisconsin (Socha 1984). A radiocarbon date of 13,130 ± 120 BP on basal organic sediments of the marsh indicates a minimum date for eastward routing of the Wisconsin River and a date for Glacial Lake Merrimac drainage. The large amounts of water that flowed to the east around the northern part of

the Baraboo Range probably incised the present channel of the Wisconsin River and sculpted the sandy delta ridges to form the contours of the river floodplain. Large sand plains and marshland in central Wisconsin north of the Leopold Reserve were derived from Glacial Lake Wisconsin deposits. These areas are now drained in part by the Wisconsin River as it meanders down from the north-central part of the state and cuts through the sandstone bluffs of the Wisconsin Dells before it reaches Sauk County. Leopold refers to Sauk County as Sand County because the region is underlain by sandplain deposits and also by sandy outwash. Attempts at farming the sands in the 1870s changed the landscape somewhat but the major land changes resulted from more recent machine-ditching. According to Leopold (1949), some of the ditching took place around 1910 to 1920, "the decade of the drainage dream," to create farms in the sand counties in the central part of Wisconsin. Peat fires were frequent after ditching, especially during drought years. They burned as much as 17,000 acres of marshland locally, and peat smoke "clouded the sun" for months and even years (Leopold 1949). At present, parts of the marsh have been ditched and drained, spoil piles of dredged peat remain, and peat has been removed to create ponds. Manipulations of land and water levels since European settlement have made determination of recent marsh history difficult.

Modern Vegetation

A sediment core was collected from the marsh on the Leopold Memorial Reserve about 1.3 kilometers southeast of Aldo Leopold's "shack" (Figure 1). Pollen, charcoal, and sediment analyses followed methods described in M. G. Winkler (1985). *Eleocharis acicularis* (spike-rush), *Dryopteris thelypteris* (marsh fern), *Cyperaceae* (sedges), *Juncus* (rushes), *Rosaceae* (Spirea), *Solidago* (goldenrod), *Habenaria psycodes* (purple fringed orchid), and *Lysimachia quadrifolia* (whorled loosestrife) surround the coring site. *Quercus* (oaks) and *Carya* (hickories) grow on ridges of low relief bordering the marsh, as do planted conifers. In the floodplain forests there are *Acer saccharinum* (silver maple), *A. rubrum* (red maple), *Betula nigra* (river birch), *Ulmus* (elm), *Salix* (willow), and *Populus* (poplar).

Sediment and Pollen Records

Sediment Stratigraphy

The basal 10 centimeters of the 7.75-meter sediment core (Figure 2) consisted of coarse reddish brown sand containing pieces of dark gravel. There was a transition at 7.65 meters to a 5-centimeter layer of blue-gray silts and clay, which in turn was overlain by 7.6 meters of dark, decomposed gyttja (lake sediment), peaty gyttja, and non-calcareous peat. A band of charcoal was evident from 20 to 23 centimeters from the top of the core. The top 20 centimeters of peat also were more fibrous.

Local Pollen Zones, Wetland, and Upland Pollen Changes

The Leopold pollen zones, the wetland evolutionary sequences, and the inferred climate from the pollen assemblages shown in Figure 2 are described in Table 1.

Local Vegetation versus Regional Vegetation Changes

Spring snowmelt upriver causes downriver flooding at floodplain sites. Because of large water-level fluctuations, pollen grains in floodplain sediments are more decomposed, and extra-regional pollen may be abundant. It is important, therefore, to compare floodplain stratigraphy to the well-preserved pollen records from nearby lakes or bogs so that regional and local changes can be identified. Regional changes were inferred from radiocarbon-dated pollen sequences from Devils Lake (Maher 1982) and Washburn Bog (Winkler 1988) in Sauk County and from Lake Mendota, Lake Waubesa, and Hook Lake Bog in Dane County (Winkler 1985; Winkler et al. 1986) at most, 80 kilometers southeast of the Leopold Reserve (Figure 1). The Leopold Marsh pollen sequence does not fit the regional chronology exactly, but similar interpretations of paleoclimate can be made.

Zone L2 pollen frequencies are similar to those for a broader region (Winkler 1988), but at Leopold Marsh the *Picea-Fraxinus nigra* vegetation assemblage lasted for about 1,500 years longer. Because of its location, Leopold Marsh was affected by glacial meltwater, the changing drainage patterns of the proglacial lakes, and the subsequent initiation of a modern Wisconsin River drainage pattern. While drier forests were growing on the better-drained uplands in the region, Leopold Marsh was in an area of extensive swamp forests. The relatively high percentage of organic matter (about 25 percent) in the sediments deposited from 13,000 to about 12,000 BP indicates that deposition took place under standing water. The sedimentation rate (about 0.8 millimeters per year) during the first 7,000 years of wetland development at Leopold Marsh was constant and the percentage of organic matter being deposited in the wetland from 12,000 until about 10,500 BP increased (from 25 to 80 percent). These two occurrences suggest that the Glacial Lake Wisconsin drainage channel stabilized rapidly before 13,000 years BP and Glacial Lake Merrimac drained at about 12,000 BP. At about this time, changes in the earth's orbit caused summers to become warmer, and winters colder, than at present (Kutzbach and Guetter 1984; Kutzbach and Otto-

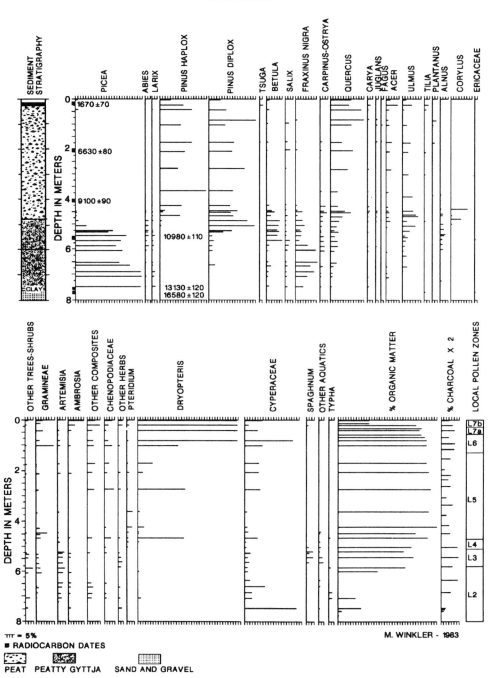

Figure 2. Percentage pollen diagram, radiocarbon dates, and sediment stratigraphy for Leopold Marsh, Sauk County, Wisconsin.

Bliesner 1982). This seasonality, coupled with abundant moisture and increased precipitation in the region, fed the ice sheet and contributed to the Greatlakean ice advance that drowned the Two Creeks forest about 180 kilometers northeast of the Reserve.

About 10,900 BP, when the ice advance ended and the ice sheet made its final retreat from the region (Maher 1982) water levels lowered and the black spruce-black ash swamp forests decreased. From 11,000 to about 10,500 BP (zone L3), a *P. mariana-Sphagnum* bog developed near the site. A mosaic on the former glacial lake bottom sediments of wet bogs and floodplain forests, as well as birch and fir in drier places, can be inferred from the pollen and charcoal results. The spruce-*Sphagnum* association lasted only about 500 years. After about 10,500 BP, because of increased warmth and a drier climate, shrubs, such as *Alnus rugosa* and *Corylus spp.* invaded the wetlands; grass and herbaceous plants (composites and chenopods) increased; and *Sphagnum*, spruce, and fir disappeared. The abundant charcoal during this time suggests that increased fires during drier seasons of the year contributed to the decline of spruce and to the concurrent growth of jack and red pine and birch forests in the region. A change in fire frequency in the region also mediated other Holocene plant communities (Winkler 1985). By the end of zone L3, at about 10,500 BP, and certainly,

by the end of zone L4, at about 9,500 BP, the Wisconsin River flowed through a channel similar to the one we know today and Leopold Marsh became a riverine wetland floristically similar to today's.

By 9,500 BP, white pine is a component of the upland forest at all sites in the region. After about 8,000 BP, however, the Leopold Marsh upland pollen sequence differs from regional vegetation changes by the persistence of abundant pine pollen and also by low abundance of oak pollen, indicating that a large component of upland pollen was washed into the Leo-

Table 1. Local Pollen Zones, Upland and Wetland Plant Assemblages, and Inferred Climate

Leopold pollen zone	Time period (years BP)	Vegetation (pollen)	Plant community	Inferred climate
L1	<13,000	No pollen		
L2	13,000–11,000	*Picea* greater than 30 percent, high *Fraxinus nigra*, Cyperaceae *Artemisia* (sage)	Black spruce-black and ash swamp	Cold, wet
L3	11,000–10,500	*Picea* less than 30 percent and decreasing, *Fraxinus* (ash) decreasing, relatively high *Sphagnum* spores, *Larix* (Larch), *Betula* (birch), and *Abies* (fir) pollen	Sphagnum-larch-spruce bog and black ash and willow floodplain forest	Cool, slightly drier
L4	10,500–9,500	High diploxylon *Pinus* (red/jack pine), high *Betula*, increased *Ulmus* (elm), *Ostrya-Carpinus* (iron-wood-hornbeam), *Quercus* (oak), *Corylus* (hazel), *Alnus* (alder)	Shrub carr and floodplain forest	Warmer, slightly drier
L5	9,500–6,000	High haploxylon *Pinus* (*Pinus strobus*, white pine) as well as diploxylon *Pinus*, high *Ulmus*, *Acer*, (maple), *Ostraya-Carpinus*, Cyperaceae, and *Dryopteris*	Sedge meadow and marsh and floodplain forest	Wet with drier intervals
L6	6,000–?3,000	Increased *Quercus*, Gramineae (grass), diploxylon *Pinus*, Cyperaceae pollen and *Dryopteris* spores	Marsh with frequent fires	Warmer, drier
L7a	?3,000–200	Increased *Betula*, *Acer*, *Tilia*, *Ulmus*, *Alnus*, and *Quercus*. Slight *Sphagnum* reappearance	Floodplains spreading	Cooler, wetter
L7b	?200–present	High *Ambrosia* (ragweed), decreased conifer-hardwood pollen	Sedge meadow-marsh	Sedge meadow-marsh, disturbed by drainage and frequent fires

pold site. Percentages of pine pollen remain high throughout the Holocene at Leopold Marsh, decreasing only in the topmost level of the core (Figure 2).

Organic matter in the Leopold Marsh sediment, above 80 percent until recently, indicates that river flooding brought a constant load of silt and clay to the marsh (but not an excessive amount) until European settlement. Charcoal is scarce throughout zone L5. Mesophytic trees such as *Ulmus, Ostrya-Carpinus, Tilia,* and *Acer spp.* are frequent in the pollen assemblage and remain dominant with pine until about 6,000 BP. After about 6,000 BP, there was less abundant and poorly preserved pollen in the sediments of Leopold Marsh. The sedimentation rate decreased from 0.8 millimeters per year to about 0.3 millimeters per year, indicating a decrease in water levels. *Dryopteris* spores become abundant, while increased grass, oak, and diploxylon pine pollen and high charcoal percentages indicate a decrease in moisture. *Salix* (willow) pollen increased, willow trees probably invaded parts of the wetland, and some floodplain forests (upstream as well as locally) were replaced by oak and pine sand barrens. The sedimentary evidence suggests that changes in Leopold Marsh after 6,000 BP were caused by summer drought in south-central Wisconsin in the middle Holocene and possibly by drier winters in northern Wisconsin as well. Elsewhere, as a result of a 12 to 19 percent decrease in annual precipitation in the region (Winkler *et al.* 1986), lakes lowered and upland vegetation changed from mesophytic forest to oak savannah.

An increase in *Ulmus, Acer, Tilia,* and *Alnus* pollen in zone L7a (after about 3,000 BP) indicates that precipitation subsequently increased and floodplain expanded. In the region, shallow lakes became *Sphagnum* bogs as precipitation increased after 3,000 BP (Winkler 1988). European settlement in this region is recorded by the pollen changes in the surficial sediments of the core (zone L7b). *Ambrosia* pollen increases to over 15 percent and pine and hardwood pollen percentages decrease (Figure 2), reflecting land clearance and logging by European settlers. The sharp decline in organic matter to about 30 percent was probably caused by soil erosion as land upriver was cleared. High charcoal in the top of the core and the date of 1670 ± 70 BP on this stratum (20 to 23 centimeters in the core), indicate that the peat from two millennia was burned off recently, possibly during the fires in the 1920s and 1930s when:

> Peat beds dried, shrank, caught fire. Sun-energy out of the Pleistocene shrouded the countryside in acrid smoke. No man raised his voice against the waste, only his nose against the smell. After a dry summer

not even the winter snows could extinguish the smoldering marsh (Leopold 1949, 100).

Conclusion

Changes preserved in the floodplain sediments of Leopold Marsh provide independent evidence of regional decrease in precipitation in the middle Holocene in the Midwest, previously inferred from lake-level and pollen changes from other sites in south-central Wisconsin. More than 13,000 years of climatically mediated fluctuating water levels have determined the development of Leopold Marsh. Only recently, since European settlement, has the river's dominance over the marsh been challenged.

> Great pockmarks were burned into field and meadow, the scars reaching down to the sands of the old lake, peat-covered these hundred centuries. Rank weeds sprang out of the ashes, to be followed after a year or two by aspen scrub. The cranes were hard put, their numbers shrinking with the remnants of unburned meadow. For them, the song of the power shovel came near being an elegy. The high priests of progress knew nothing of cranes, and cared less. What is a species more or less among engineers? What good is an undrained marsh anyhow? (Leopold 1949, 100).

Through the efforts of the Sand County Foundation to protect this land—The Aldo Leopold Memorial Reserve—from further development, Leopold Marsh will once more come under the river's domain and possibly, "The marsh might . . . keep on producing hay and prairie chickens, deer and muskrat, crane-music and cranberries forever" (Leopold 1949, 99) (Figure 3).

Figure 3. Sandhill cranes in the marshes of Aldo Leopold Memorial Reserve in Sauk County, Wisconsin. Photo courtesy of George Archibald, International Crane Foundation, Baraboo, Wisconsin.

Acknowledgments

I especially thank Nina Leopold Bradley and Charles Bradley for their interest and friendship. I also thank Al Swain, Brian Goodman, Liz Seehawer, and Frank Terbilcox for help with fieldwork and logistics; Sharon Barta for lab assistance; Bryan Richards and P. J. Behling for drafting. I also thank M. Kennedy and M. Woodworth. This research was funded by a fellowship from the Sand County Foundation and by NSF grants ATM 89-02849 and ATM 87-13227 to J. E. Kutzbach, Center for Climatic Research, University of Wisconsin, Madison.

References

Kutzbach, J. E., and Guetter, P. J. 1984. The sensitivity of monsoon climates to orbital parameter changes for 9,000 years BP: Experiments with the NCAR General Circulation Model. In *Milankovitch and Climate*, Part 2, eds. A. L. Berger, J. Imbrie, J. Hays, G. Kukla, and B. Saltzman, pp. 801-820. Dordrecht, The Netherlands: Reidel Publishers.

Kutzbach, J. E., and Otto-Bliesner, B. L. 1982. The sensitivity of the African-Asian monsoonal climate to orbital parameter changes for 9,000 years BP in a low-resolution general circulation model. *Journal of the Atmospheric Sciences* 39:1177-1188.

Leopold, A. 1949. *A Sand County Almanac and Sketches Here and There*. New York: Oxford University Press.

Maher, L. J. Jr. 1982. The palynology of Devils Lake, Sauk County, Wisconsin. In *Quaternary History of the Driftless Area. Wisconsin Geological and Natural History Survey*, Field Trip Guide Book No. 5, eds. J. C. Knox, L. Clayton, and D. M. Mickelson, pp. 119-135. Madison: University of Wisconsin Extension.

Socha, B. J. 1984. *The Glacial Geology of the Baraboo Area, Wisconsin and Application of Remote Sensing to Mapping Surficial Geology*. M. S. Thesis, University of Wisconsin, Madison.

Winkler, M. G. 1985. *Late-Glacial and Holocene Environmental History of South-Central Wisconsin: A Study of Upland and Wetland Ecosystems*. Ph. D. Dissertation, University of Wisconsin, Madison.

_____ . 1988. Effect of climate on development of two Sphagnum bogs in south-central Wisconsin. *Ecology* 69:1032-1043.

_____ , Swain, A. M., and Kutzbach, J. E. 1986. Middle Holocene dry period in the northern Midwestern United States: Lake levels and pollen stratigraphy. *Quaternary Research* 25:235-250.

The Shifting Forest Ecotone in Postglacial Northern Ontario

Kam-biu Liu
Louisiana State University

The ecotone between boreal forest and Great Lakes-St. Lawrence forest (Rowe 1972) in northern Ontario is one of the most widely recognized vegetation boundaries in North America (for example Braun 1950; Sjors 1963; Walter 1979). Bending southward to about 47°N, it marks the southernmost occurrence of the boreal forest biome on this continent (Figure 1). The position of the ecotone is controlled by a steep climatic gradient between Lake Huron and the James Bay lowland, along which mean annual temperature decreases from 5°C to -1°C over a distance of 500 kilometers (Chapman and Thomas 1968). It broadly coincides with the 2.2°C isotherm.

In this paper I highlight the modern phytogeography and vegetation characteristics of this ecotone; examine the changes in modern pollen assemblages along a meridional transect across the ecotone as a proxy for the vegetational gradient; and reconstruct the postglacial history of the ecotone based on radiocarbon-dated pollen-stratigraphic data from this region.

Phytogeography

The ecotone is a diffuse vegetation transition that does not follow any physiographic or edaphic boundary; it is hardly recognizable in the field. The Great Lakes-St. Lawrence forest region south of the ecotone is characterized by temperate tree taxa such as eastern white pine (*Pinus strobus*), red pine (*Pinus resinosa*), hemlock (*Tsuga canadensis*), beech (*Fagus grandifolia*), sugar maple (*Acer saccharum*), and yellow birch (*Betula alleghaniensis*), often in mixed associations with boreal tree taxa such as white spruce (*Picea glauca*), white birch (*Betula papyrifera*), and jack pine (*Pinus bank-*

siana). North of the ecotone, the boreal forest is dominated by boreal conifers and hardwoods such as white spruce, black spruce (*Picea mariana*), balsam fir (*Abies balsamea*), jack pine, white birch, and trembling aspen (*Populus tremuloides*) (Rowe 1972). Mosaics of boreal forest associations also occur within the Great Lakes-St. Lawrence forest, especially in cooler, wetter sites on north-facing slopes, valley bottoms, and swamps (Hills 1959). Braun (1950) described the broad ecotonal region of the Great Lakes-St. Lawrence forest as consisting of "inter-penetrating climaxes" between boreal and mixed forest associations.

Phytogeographically, northern Ontario is a tension zone where the temperate flora merges with the boreal and subarctic flora of North America. At least nine important tree species of the temperate forests reach their northern distribution limits in northern Ontario south of the ecotone, including eastern hop hornbeam (*Ostrya virginiana*), hemlock, beech, northern red oak (*Quercus rubra*), bur oak (*Q. macrocarpa*), basswood (*Tilia americana*), white ash (*Fraxinus americana*), green ash (*F. pennsylvanica*), and slippery elm (*Ulmus rubra*) (Little 1971). Several thermophilous tree taxa, notably white pine, red pine, yellow birch, sugar maple, red maple (*Acer rubrum*), big tooth aspen (*Populus grandidentata*), black ash (*Fraxinus nigra*), and white elm (*Ulmus americana*), venture into the southern part of the boreal forest region, mostly in scattered populations. A steep species diversity gradient thus exists in the southern part of northern Ontario between Lake Huron and the height of land (at about 48°N), where 12 tree species successively drop out of their distribution ranges within a distance of 200 kilometers, thereby leaving the boreal forest relatively impoverished in tree species. By contrast, only five tree species reach

204

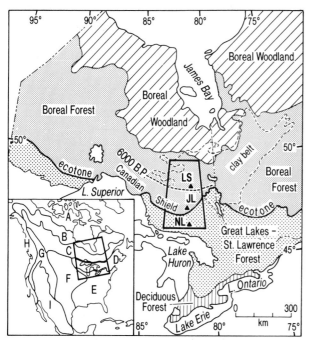

Figure 1. Vegetation regions of Ontario and adjacent Quebec (after Rowe 1972) showing the modern position of the boreal forest/Great Lakes-St. Lawrence forest ecotone (thick continuous line), and its inferred position at 6,000 BP during the Hypsithermal (thick dashed line). The thin dashed line delimits the Clay Belt. The thick-lined box delimits a broad transect consisting of 61 pollen surface samples. Radiocarbon-dated pollen-study sites (solid triangles) are Nina Lake (NL), Jack Lake (JL), and Lake Six (LS). The inset map shows the position of the ecotone (thick line) in the context of the major vegetation regions of North America. A = tundra, B = boreal woodland, C = boreal forest, D = Great Lakes-St. Lawrence forest, E = deciduous forest, F = prairie, G = mountain vegetation, H = Pacific coastal forest, I = desert shrubland and grassland, J = woodland and chaparral.

St. Lawrence forest, the upland boreal forest on the Canadian Shield, and the lowland boreal forest in the Clay Belt (Liu 1982).

The modern pollen rain of the Great Lakes-St. Lawrence forest is characterized by a preponderance of white pine, along with a slightly greater abundance of birch, maple, and hemlock. The pollen of jack or red pine (probably mostly from jack pine) are most frequent in the upland section of the boreal forest, between 47° and 49°N. The pollen of spruce, fir, and alder (*Alnus*) increase to maximum in the northern part of the boreal forest, especially in the Clay Belt lowland. The classification by several numerical techniques confirm that, despite the broadly transitional character of the vegetation, the Great Lakes-St. Lawrence forest and the upland and lowland sections of the boreal forest each has a distinctive palynological "signature" (Liu 1984).

Postglacial History

The part of northern Ontario straddled by the ecotone was covered by the Laurentide ice sheet until about 10,000 to 9,000 years ago. Soon after deglaciation the land was available for colonization by plants immigrating from the south. Paleobotanical evidence from adjacent regions suggests that a warmer climate occurred during the mid-Holocene, climaxing about 5,000 to 6,500 years ago (McAndrews 1981; Richard 1980; Terasmae and Anderson 1970). Radiocarbon-dated pollen records from three sites in northern Ontario document the postglacial history of the ecotone, with particular em-

their northern limits in the boreal forest region between the height of land and the James Bay lowland over a distance of 400 kilometers (Little 1971).

Modern Pollen Rain

Quantitative data on the abundance of tree species populations are not available for northern Ontario, but changes in modern pollen assemblages along a transect from the Great Lakes-St. Lawrence forest to boreal forest (see Figure 1) can be used as a proxy for the vegetation gradient across the ecotone. Figure 2 shows generalized modern pollen curves for major tree and shrub taxa along a broad transect between 46.4° and 50°N in northern Ontario. It is based on 61 pollen surface samples collected from different vegetational and physiographic-edaphic regions in the Great Lakes-

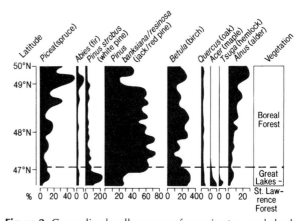

Figure 2. Generalized pollen curves for major tree and shrub taxa along a transect of 61 pollen surface samples from 46.4°N to 50°N in northern Ontario. Distance between latitudes along vertical axis is proportional to the number of surface samples within that latitudinal belt. Pollen percentages are based on an arboreal pollen sum.

phasis on its response to Hypsithermal climatic change (Liu 1990).

The three sites are aligned along a transect from about 46.5° to 48.5°N that parallels the steepest climatic gradient (Liu 1990) (Figure 1). Nina Lake, the southernmost site, is in the Great Lakes-St. Lawrence forest. Jack Lake is in the upland section of the boreal forest but only some 20 kilometers north of the ecotone. Lake Six is in the Clay Belt lowland section of the boreal forest, about 120 kilometers north of the ecotone and only about 18 kilometers north of the physiographic boundary between the Canadian Shield upland and the Clay Belt. The pollen records from these three small lake basins are expected to be sensitive to any north-south movements of the ecotone during the Holocene.

The basal organic sediments from Nina Lake, Jack Lake, and Lake Six yield radiocarbon dates of about 9,510 BP, 9,270 BP, and 6,970 BP, respectively. In all three pollen diagrams there is an early postglacial pollen assemblage zone dominated by spruce (*Picea*) pollen at the base. It is replaced by a zone dominated by the pollen of jack or red pine (*Pinus banksiana, P. resinosa*). The mid-Holocene sediments in all three sites contain maximum frequencies of white pine (*P. strobus*) pollen. In Lake Six the pollen and macrofossils of northern white cedar (*Thuja occidentalis*) co-dominate with white pine in this section of the core. The pollen of boreal conifers such as spruce and jack pine increase at the expense of white pine toward the top of all three pollen diagrams (Liu 1990). I used discriminant analysis (Liu and Lam 1985) to translate the pollen-stratigraphic changes in each pollen diagram into a curve of vegetation zonal indices, which summarizes the major trends of vegetational changes in terms of ecotonal movements (Figure 3).

Great Lakes-St. Lawrence forest (and as a corollary, the ecotone) did not exist in northern Ontario prior to 7,400 BP. At Nina Lake and Jack Lake the earliest postglacial vegetation was a spruce-dominated boreal forest or woodland without modern analog (Liu 1990). It changed to a boreal forest similar to the modern one after 9,000 BP when jack pine populations expanded and the alders (*Alnus crispa, A. rugosa*) immigrated from the south. White pine, beech, and hemlock immigrated to Nina Lake about 7,400 years ago, transforming the boreal forest to Great Lakes-St. Lawrence forest in the southern part of this region. The Great Lakes-St. Lawrence forest at Nina Lake, once established, has persisted there throughout the rest of the Holocene (Figure 3). For Jack Lake, the boreal forest was changed to Great Lakes-St. Lawrence forest about 7,300 years ago (Figure 3) as the ecotone continued its northward movement past this site fol-

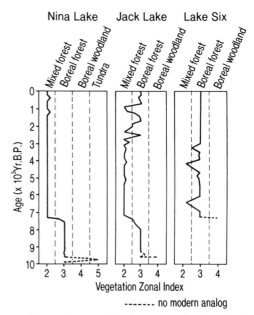

Figure 3. Vegetation zonal index curves for Nina Lake, Jack Lake, and Lake Six. Dotted line at the base of each curve indicates a vegetation without modern analog. Mixed forest is synonymous with Great Lakes-St. Lawrence forest.

lowing the rapid migration of white pine across northern Ontario. The vegetation zonal indices curve from Lake Six sheds light on how far north the ecotone reached during the Hypsithermal. The mid-Holocene vegetation around Lake Six was primarily a boreal forest, despite episodes of encroachment by Great Lakes-St. Lawrence forest (Figure 3). I therefore infer that during the Hypsithermal, approximately 7000 to 3000 years ago, the ecotone was situated along the physiographic boundary between the Canadian Shield upland and the Clay Belt, about 140 kilometers north of its present position (Figure 1). White pine was favored by the rolling topography and the well-drained, acidic soil derived from the sandy or bouldery till of the Canadian Shield upland, but was constrained by the lack of suitable habitats in the Clay Belt. By contrast, the abundance of calcareous soils and wetland habitats in the Clay Belt was favorable to the proliferation of northern white cedar, which is calciphilous and thrives in swamps and the margins of mesotrophic bogs. Thus the boundary between the Canadian Shield and the Clay Belt imposed a physiographic constraint to the northward advance of the ecotone during the Hypsithermal (Liu 1990).

The populations of white pine crashed at Lake Six after 4,000 BP due to Neoglacial cooling. By 3,100 BP boreal forest had been firmly established near Lake Six, signifying a southward retreat of the Great Lakes-St. Lawrence forest. At Jack Lake, Great Lakes-St.

Lawrence forest was reverted back to boreal forest after about 2,600 BP as the ecotone retreated to this site (Figure 3). After some fluctuations the ecotone has been established in its modern position only within the last millennium.

Summary and Conclusion

The diffuse ecotone between the boreal forest and Great Lakes-St. Lawrence forest in northern Ontario represents a tension zone where a number of thermophilous tree species successively drop out of their distribution ranges along a steep climatic gradient. Thus it is also defined by a floristic gradient along which the arboreal flora becomes progressively impoverished northward. For species population abundance, a notable change in the vegetation landscape in northern Ontario is the thinning of white pine populations northward—corroborated indirectly by the decline in white pine pollen frequencies in surface samples across the ecotone. The boreal forest and Great Lakes-St. Lawrence forest have distinctive palynological "signatures" identifiable from their modern and fossil pollen assemblages.

The Great Lakes-St. Lawrence forest entered northern Ontario only about 7,400 years ago and rapidly spread northward under a warmer climate. From 7,000 BP to 3,000 BP the ecotone advanced to the northern edge of the Canadian Shield upland in northern Ontario, about 140 kilometers north of its present position. It has retreated since then in response to Neoglacial cooling and has occupied its present position relatively recently during the late Holocene.

Acknowledgments

I thank M. L. Eggart for cartographic assistance, and M. Eldridge for help with word processing.

References

Braun, E. L. 1950. *Deciduous Forests of Eastern North America.* Philadelphia: Blakiston.

Chapman, L. J., and Thomas, M. K. 1968. *The Climate of Northern Ontario.* Climatological Studies, No. 6. Toronto: Department of Transport, Meteorological Branch.

Hills, G. A. 1959. *Soil-Forest Relationships in the Site Regions of Ontario.* Bulletin of Agricultural Experimental Station, pp. 190-212. East Lansing: Michigan State University.

Little, E. L. 1971. *Atlas of United States Trees. I. Conifers and Important Hardwoods.* Miscellaneous Publication 1146. Washington, DC: U.S. Department of Agriculture, Forestry Service.

Liu, Kam-biu. 1982. *Postglacial Vegetational History of Northern Ontario: A Palynological Study.* Ph. D. Dissertation, Department of Geography, University of Toronto.

————. 1984. Pollen-floristic zonation in northern Ontario. *AAG Program Abstracts 1984*, p. 97. Washington, DC: Association of American Geographers.

————. 1990. Holocene paleoecology of the boreal forest and Great Lakes-St. Lawrence forest in northern Ontario. *Ecological Monographs* 60:179-212.

————, and **Lam, N. S. N.** 1985. Paleovegetational reconstruction based on modern and fossil pollen data: An application of discriminant analysis. *Annals of the Association of American Geographers* 75:115-330.

McAndrews, J. H. 1981. Late Quaternary climate of Ontario: Temperature trends from the fossil pollen record. In *Quaternary Paleoclimate*, ed. W. C. Mahaney, pp. 319-333. Norwich, UK: GeoAbstracts.

Richard, P. 1980. Histoire postglaciaire de la végétation au sud du Lac Abitibi, Ontario et Québec. *Géographie Physique et Quaternaire* 34:77-94.

Rowe, J. S. 1972. *Forest Regions of Canada.* Publication 1300. Ottawa, Canada: Canada Forestry Service.

Sjors, H. 1963. Amphi-Atlantic zonation, Nemoral to Arctic. In *North Atlantic Biota and their History*, ed. A. Love and D. Love, pp. 109-125. Oxford: Pergamon Press.

Terasmae, J., and Anderson, T. W. 1970. Hypsithermal range extension of white pine (*Pinus strobus* L.) in Quebec, Canada. *Canadian Journal of Earth Sciences* 7:406-413.

Walter, H. 1979. *Vegetation of the Earth in Relation to Climate and the Eco-Physiological Conditions.* New York: Springer-Verlag.

The "Great American Desert" and the Arid/Humid Boundary

John E. Oliver
Indiana State University

From their sources to the foothills of the Rocky Mountains, the Missouri and Mississippi Rivers traverse a remarkable change in climate, from moist to subhumid and arid conditions. This gradient has been the focus of attention since the early days of westward expansion in the United States.

The Great American Desert

Until the Civil War (1861 to 1865), the public was convinced that a "Great American Desert" lay between the Missouri River and the Rocky Mountains. Between 1820 and 1850 maps of this Great American Desert were widely reproduced in atlases, histories, and textbooks (Billington 1949). Francisco de Coronado was the first to report the desert but its claim to existence rested more specifically upon reports of official expeditions. The Lewis and Clark Expedition, begun in 1803, made little mention of the desert, but these travellers had taken the northern passage along waterways, often lined by woods. Zebulon Pike reported of his 1806 expedition across the middle plains to the Rockies: "These vast plains of the western hemisphere may become in time as celebrated as the sandy deserts of Africa" (Webb 1931, 155). In 1819 to 1820 Major Stephen Long spoke of "the Great Desert at the Base of the Rocky Mountains." The idea of the Great American Desert was not put to rest until the settlement of the Great Plains and the development of agriculture there. The first seeds of doubt were expressed in the

railroad surveys of 1855 when ". . . the High Plains were compared to the Steppes of Russia. Authors were even venturing such expressions as 'subhumid' and 'semiarid'—common enough with us but new at that time" (Brown 1948, 370).

The idea of a great arid area on the eastern flanks of the Rockies coincided with the development of the discipline of climate and climatic classification. Boundaries assumed much importance, especially the humid/dry boundary.

Arid/Humid Indices and the Boundary

Most indices for differentiating humid from subhumid or dry regions relate precipitation to evaporation or temperature. In the *Encyclopedia of Climatology*, S. H. Stadler (1987) lists no fewer than 41 indices to define limits of dry climates, but that used in Köppen's climate classification is probably the best known.

H. M. Kendall (1935) chose the Köppen classification system to demonstrate variability of climate for the period 1915 to 1931 and to make suggestions concerning relocation of boundaries. Kendall produced a series of maps using Köppen's boundaries to plot annual rather than long-term values. Of particular note was the boundary between dry climate (B in the Köppen system) and humid climates (A, C, and D), a limit often referred to as the B/H boundary.

In deriving the boundary of the B climates, Kendall noted that Köppen's B/H boundary does not account

208

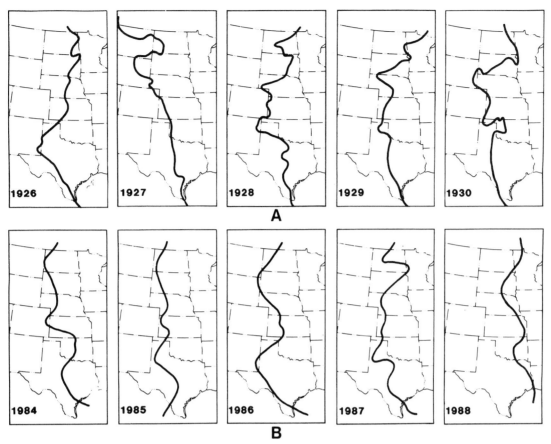

Figure 1. Boundaries separating the B/H climates for the years: A. 1926 through 1930 and B. 1984 through 1988. Dry climates are to the west of the identified boundaries.

for seasonal differences in rainfall concentration. Accordingly, he used a proposal by P. E. James (1922) to obtain summer and winter concentrations (Figure 1A).

Kendall's maps provide an image of the B/H variability in the first third of the 20th century. Given that the 1980s represent the warmest decade of the 20th century (Thompson 1989), it is instructive to investigate the variability of the B/H boundary in more recent years. Using a five-year period (1984 to 1988), data for 74 stations were analyzed by Kendall's method to produce individual-year boundaries (Figure 1B).

In the 1926 to 1930 series, the annual boundary in two of the three years varied considerably from the mean (Köppen-Geiger) boundary (Figures 2B and 2C). In 1927, especially in the northern plains, the boundary extended far to the west of the mean; by 1929, it was well to the east. This boundary varies somewhat less in the 1984-1988 series. With the exception of 1988, a notably dry year, the annual lines are closer to the Köppen-Geiger average than those of the early series.

Figures 2B and 2C show the extreme eastern and western limits of the annual B/H boundaries and relate them to the actual line identified on the Köppen-Geiger map of world climates. The broad region between these limits represents the transition between the arid and humid conditions. The width of the zone for 1926 to 1930 (Figure 2B) indicates considerable variation in the northern plain states, but much less south of

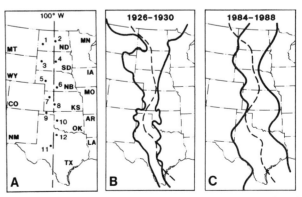

Figure 2. A. The study area showing location of stations listed in Table 1. B. The limits of the B/H boundaries for 1926 through 1930 and C. for 1984 through 1988. The Köppen-Geiger map boundary is shown by the broken line.

Table 1. Köppen-Geiger Climate-Years for Selected Stations

B climate-years	1926-1930	1984-1988	H climate-years	1926-1930	1984-1988
Dickinson, ND	5	2	Bismark, ND	2	4
Rapid City, SD	4	4	Pierre, SD	1	3
Ogallala, NB	2	5	Kearny, NB	3	4
Goodland, KS	2	3	Dodge City, KS	4	2
Hooker, OK	3	4	Elk City, OK	4	3
San Angelo, TX	4	3	Childress, TX	3	2

Nebraska. By contrast, except for North Dakota and Oklahoma, the limits are approximately equidistant from the mean Köppen-Geiger boundary for 1984 to 1988 (Figure 2C). The arbitrary nature of the defined B/H boundary is clearly evident in the maps.

Boundary Assessment and the Climate-Year

R. J. Russell's (1934) concept of the climatic-year helps evaluate how well the mean line represents the boundary (Gregory 1954). The climate-year is derived by applying the rules of a classification system to a single year rather than to long-term means. Russell noted that the annual recurrence of an identified climate type (for example, desert-year or steppe-year) is of greater climatic significance than a temperature/precipitation mean for long periods of observation. Thus, in a transition zone such as the B/H boundary region, analysis of climatic data permits comparison of annual frequency of dry B-years, with annual frequency of humid H-years.

Table 1 shows the B and H climate-years of selected transition zone stations (Figure 2A) for the two five-year periods. Stations classified as B or H climates in the Köppen-Geiger system are listed in separate columns. For the B climate stations, the number of B climate-years experienced is shown, while H climate-years are tabulated for Köppen-Geiger H climates. Only Dickinson, North Dakota, and Ogallala, Nebraska, experienced their classified types for an entire five-year span; many stations, while classified as an H or B, had more climate-years of the opposite type.

In the B climates, the frequency of dry climate-years is greater in stations representing the central tier of states for 1984 to 1988 while the humid stations experienced fewer H climate-years during the same period. From the station data, Nebraska, Kansas, and Oklahoma had an increased number of drier climate-years in the warm 1980s. The representative northern and southern stations indicate an increase in humid and a decrease in dry climate-years for 1984 to 1988.

Conclusions

The gradient from a humid to a subhumid climate in the traverse from the east to west, toward the foothills of the Rocky Mountains, has been recognized since the early exploration. Global and regional maps often show the dry/humid climate boundary, but this is a line that shifts appreciably year to year. Did historic perceptions of the Great American Desert relate to an eastward shift of the boundary at times of early explorations? Only continuing detailed research will tell us. Perhaps the most striking feature of recent years is the increase in the B climate-years in the central tier states during the 1980s.

The location of the B/H boundary and the frequency of B or H climate-years is more than of academic interest. While the analysis indicates the arbitrary nature of a line on a map, it also shows the great variability of climate within a region, a variability of important implications for moisture availability, agricultural productivity, and economic well-being.

References

Billington, R. A. 1949. *Westward Expansion: A History of the American Frontier.* New York: Macmillan.

Brown, R. H. 1948. *Historical Geography of the United States.* New York: Harcourt Brace.

Gregory, S. 1954. Climatic classification and climatic change. *Erdkunde* 8:246-252.

James, P. E. 1922. Köppen's classification of climates: A review. *Monthly Weather Review* 50:69-72.

Kendall, H. M. 1935. Notes on climatic boundaries in the Eastern United States. *Geographical Review* 25:117-124.

Russell, R. J. 1934. Climatic years. *Geographical Review* 24:92-103.

Stadler, S. H. 1987. Aridity indexes. In *The Encyclopedia of Climatology*, eds. J. E. Oliver and R. W. Fairbridge, pp. 102-106. New York: Van Nostrand Reinhold.

Thompson, R. D. 1989. Short-term climatic change: Evidence, causes, environmental consequences and strategies for action. *Progress in Physical Geography* 13:315-147.

Webb, W. P. 1931. *The Great Plains.* New York: Ginn.

Mudlumps

H. Jesse Walker
Louisiana State University

Warren E. Grabau
U.S. Army Corps of Engineers, Retired

Rivaled only by the bars, improvement of channels for navigation, or floods and their prevention as provocatives of debate and gratuitous theorizing are the curious mudlumps at the mouths of the Mississippi.

—RUSSELL (1936, 80-81)

These "curious" features (Figure 1)—mudlumps—have been recorded and discussed possibly since the time of Alvar Nuñez Cabeza de Vaca. He described them in the report that stemmed from his travels along the northern Gulf of Mexico. Because his description is accurate and because mudlumps are unknown from other river mouths,* his description has been taken as proof that he discovered the Mississippi River on 2 November 1528 (Fontaine 1872), some 13 years before Hernando de Soto crossed it hundreds of kilometers upstream.

However, not until 1699 did Pierre d'Iberville actually enter the river from the Gulf of Mexico. The long delay between learning that the large river existed and entering it from the sea is partly due to the presence of mudlumps (Figure 2). Had it not been for mudlumps, the lower valley might have become Spanish territory instead of French (McWilliams 1969).

Once the Spanish crown learned that La Salle had entered the Gulf of Mexico in 1684 after descending the river, it sent out at least 11 expeditions to search for the river's mouth. Some found it but turned away believing it to be unnavigable, a fact attested to by the name they gave it, "Río de la Palizada." Later d'Iberville described the Spanish palisades as "black rocks" and "trees petrified by mud" (McWilliams 1969). D'Iberville's attempt to enter the mouth of the Mississippi nearly met the same fate as that of the Spaniards. However, when he was looking for an entrance through the so-called palisades, darkness approached and the weather turned foul. In seeking protection he approached the shore and found that the perceived barrier was not continuous. By entering a channel between his "black rocks" he set the stage for the French occupation that led to the founding of New Orleans in 1717.

Once ships began to use the delta's passes, frequent strandings provided interested passengers opportunities to examine mudlumps. Even dredge boats were affected. For example, in August 1876, crews were forced to stop dredging on Pass à Loutre because they could not keep up with mudlump growth.

One of the first to write in some detail about these mudlumps was Sir Charles Lyell, who visited the Mississippi River in 1845 and again in 1846. He wrote

> In this region, where so rapid a conversion is going on from sea into land, a phenomenon occurs which is without parallel, so far as I am aware, in the delta of any other river. I often heard . . . of the swelling up of the muddy bottom of the gulf to a height of several feet, . . . and this in places where there had previously been a depth of several fathoms (Lyell 1889, 444-445).

Colorful descriptions were prevalent during sailing-ship days. However, once the passes into the delta were improved by dredging and the construction of jetties, ships were able to proceed without stranding. Not surprisingly, subsequent references to mudlumps were relatively few. With the exceptions of some reports, such as that by R. J. Russell (1936), mudlumps were generally ignored until the late 1940s when detailed field research was begun on their structure, nature, and origin (Morgan 1961).

Setting, Distribution, Form, and Longevity

South of New Orleans, the Mississippi River flows in one channel for some 150 kilometers before it reaches the Head of Passes where the flow diverges into several distributaries that carry it to the Gulf of Mexico (Figure

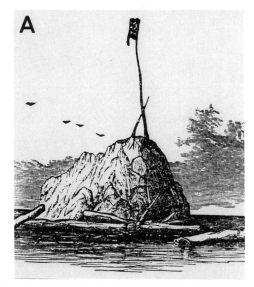

Figure 1. Mudlumps. A. Sketch of a 4.3-meter-high mudlump named The Wart at the mouth of Northeast Pass as drawn by C. Lyell (1889). B. Photograph from the mid-1960s, photographer unknown.

2). Near the mouths of these distributaries, conditions favor mudlump development. There are no confirmed reports of them in any other part of the delta.

Mudlumps occur as subaqueous and subaerial forms and are usually associated with distributary mouth bars. Their form and size are highly variable, and they continually change size and appearance. In their subaerial expression, they range from small pinnacles (Figure 1) to elongated, often S-shaped, islands as large as 8 hectares in area.

Subaerial mudlumps are exposed to processes that do not affect those beneath sea level. Wave action, especially from strong storms such as hurricanes, often drastically alters surface features, size, and shape.

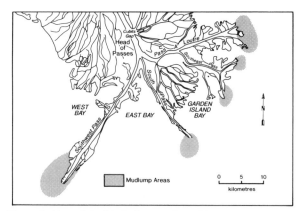

Figure 2. Map of the bird's foot portion of the Mississippi River Delta showing mudlump locations. From Morgan et al. (1968), reprinted with the permission of the American Association of Petroleum Geologists.

The life history of mudlumps can be complicated. Most apparently existed as submarine forms before emerging as islands. Once they appear as islands, some are destroyed in a very short time, others last many years. Some are eroded by waves only to reemerge. J. P. Morgan (1961) argued that the life span of mudlumps is related to their areal extent, their composition, their location with respect to wave action, the protection offered by other mudlumps or other relief forms, and the frequency of rejuvenation, among others.

Composition and Structure

As their name implies they are composed of fine-textured materials, especially clay. Near the surface, most are stratified layers of silt and clay reflecting the sediments of the bar or bottom that has been raised. In contrast, the core of mudlumps, which rises from subterranean layers, is fine-grained clay. Many mudlump cores extend to deeper than 150 meters with little variation in composition. Clay minerals are mainly montmorillonite, illite, kaolinite, and chlorite; some feldspars and carbonates may also be present. The amount of carbonate correlates with quantity of shell material (Morgan et al. 1963).

The surface features of mudlumps reflect their diapiric origin. The surface is usually irregular and displays upwarp and tilting with a seaward dip to the strata. The upward warp of the beds on one of the islands studied in detail ranges between 5° and 35°. Along its crest are normal faults within the zone of which are overturned beds (Morgan 1961).

In addition to such minor features, most mudlumps have gas (methane) and mud vents. These vents range to diameters of 50 centimeters and generally occur along fault lines. Gas vents are filled with water up through which bubbles inflammable gas. Mud vents, unlike gas vents, discharge fluid mud along with gas. The mud cones that result are circular and, because the mud has low viscosity, often broad. Because vents occur along a line of weakness, they tend to be replaced frequently by vents opening up elsewhere along the fault. Individual vents and cones are normally short-lived.

Mudlump Origin

Many theories have been postulated to explain the origin of mudlumps, including: invoking the processes that raised the Alps and Andes; gas pressure (a favorite during the 19th century); the beating of waves against river sediment in suspension; deposition by waves and tides; and the result of subterranean water courses (Russell 1936).

The currently accepted explanation involves the rapid deposition of dense sands over plastic prodelta and marine clays. Such a condition prevails near distributary mouths, the location of most mudlumps. The stress on underlying clays from heavy deposits is relieved by the intrusion of the clays into the overlying sand-silt bodies.

As described by J. M. Coleman and D. B. Prior (1982, 91-92) the features

> are thin spines of diapiric intruded mud, which forms islands or mudlumps. . . . These diapiric spines form rapidly, and vertical displacement of more than 100 m has been documented during a period of only 20 years.

Seasonal variations in the deposition rates affect the timing of mudlump formation. During flooding, when large amounts of sand are deposited at the delta front, mudlump activity increases. New lumps appear, old ones are revived, and gas and mud vents are reactivated.

Because mudlumps develop in response to distributary mouth deposition, they form in sequence as the mouth advances (Figure 3). They have been mapped off all passes and over the past 100 years numbered in the hundreds. During the period 1876 to 1973, 105 mudlumps were mapped in an area of 13 square kilometers off South Pass.

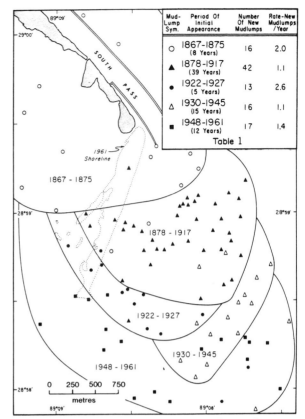

Mud-Lump Sym.	Period Of Initial Appearance	Number Of New Mudlumps	Rate-New Mudlumps /Year
○	1867-1875 (8 Years)	16	2.0
▲	1878-1917 (39 Years)	42	1.1
●	1922-1927 (5 Years)	13	2.6
△	1930-1945 (15 Years)	16	1.1
■	1948-1961 (12 Years)	17	1.4

Table 1

Figure 3. Map showing progression of mudlumps from 1867 to 1961. From Morgan *et al.* (1963), reprinted with the permission of Louisiana State University Press, © 1963 by Louisiana State University Press.

Conclusions

Mudlumps, which played such an important role in the early days of settlement, are still present but their effect is reduced. Prior to jetty construction, passes, such as South Pass and Southwest Pass, were broad and shallow and supported distributary mouth bars with mudlumps similar to those of unjettied mouths. By narrowing and lengthening outlets into deep water, jetties bypassed the bars and their associated mudlumps. They funneled sediments into deeper water and in general halted mudlump formation. For example, at Southwest Pass nearly 50 years passed from the time of jetty construction and new mudlump formation (Morgan 1961).

Note

* As this chapter was going to press, a letter from T. Tokuoka, Shimane University, Shimane, Japan, states that mudlumps are now forming at the mouth of a small river that flows into Lake Nakaumi. This river, recently diverted from its original course, is

now creating a delta. Because it is depositing coarse material on top of lake clays, forms similar (but on a much smaller scale) to those in the Mississippi River are developing, including mudlumps.

References

Coleman, J. M., and Prior, D. B. 1982. *Deltaic Sand Bodies.* Tulsa, OK: American Association of Petroleum Geologists.

Fontaine, E. 1872. Contributions to the physical geography of the Mississippi River and its delta. *Journal of the American Geographical Society* 3:343-378.

Lyell, C. 1889. *Principles of Geology,* Vol. 1, 11th edition. New York: D. Appleton.

McWilliams, R. G. 1969. Iberville at the birdfoot subdelta. In *Frenchmen and French Ways in the Mississippi Valley,* ed. J. F. McDermott, pp. 127-140. Urbana: University of Illinois Press.

Morgan, J. P. 1961. Genesis and paleontology of the Mississippi River mudlumps. In *Geological Bulletin,* No. 35. Baton Rouge, LA: Louisiana Department of Conservation.

Morgan, J. P., Coleman, J. M., and Gagliano, S. M. 1963. *Mudlumps at the Mouth of South Pass, Mississippi River: Sedimentology, Paleontology, Structure, Origin, and Relation to Deltaic Processes,* Coastal Studies Series No. 10. Baton Rouge, LA: American Association of Petroleum Geologists.

———, **Coleman, J. M., and Gagliano, S. M.** 1968. Mudlumps: Diapiric structures in Mississippi delta sediments. In *Diapirism and Diapirs,* Memoir No. 8., pp. 145-161. Tulsa, OK: American Association of Petroleum Geologists.

Russell, R. J. 1936. Lower Mississippi River delta. Reports on the geology of Plaquemines and St. Bernard Parishes, La. In *Geological Bulletin,* No. 8. Baton Rouge, LA: Louisiana Department of Conservation.

The Strait of Belle Isle: North American Nexus in Time and Space

Alan G. Macpherson
Memorial University of Newfoundland

The Strait of Belle Isle is a body of water 150 kilometers long and 14 to 20 kilometers wide, separating the island of Newfoundland from the North American mainland of Labrador and forming a passage from the Atlantic Ocean into the Gulf of St. Lawrence (Figure 1). The Labrador side of the strait carries a branch of the cold Labrador Current into the Gulf, while the Newfoundland side discharges Gulf water into the Atlantic. Floe ice is present from December to early July, and an occasional iceberg passes through into the Gulf in spring or early summer. The strait's subarctic climate is reflected in the scrub spruce forest that characterizes the Newfoundland shore and in the coastal tundra that dominates the Labrador side. Bleak and remote, the Strait of Belle Isle can nevertheless be viewed as a nexus in both time and space in the historical geography of North America. It is a nexus that dissolves into a number of instructive "point" events that occurred at particular locations within the strait between 7,500 BP and A.D. 1766, each with spatio-temporal "line" connections to "remote sources . . . and equally remote destinations" (Semple 1915, 27).

L'Anse Amour (7,530 ± 140 BP)

In 1974 archaeological excavation of a large burial mound at L'Anse Amour on the Labrador side of the strait revealed the skeleton of a child about 12 years old of undetermined sex, lying face down with the head to the west and a large slab of rock on the back. Accompanying grave goods included a walrus tusk, a decorated bone pendant, an antler harpoon head, a bird-bone whistle (still functional), two nodules of

graphite covered with red ochre, a small decorated antler pestle, a crescent-shaped ivory toggle, and several spear points in stone and caribou bone. Charcoal from ritual fires in the original excavation provided a radiocarbon date of 7,530 ± 140 BP for the occasion of this highly unusual Maritime Archaic Indian burial.

The L'Anse Amour burial mound is only one of a number of Maritime Archaic sites along the Labrador side of the strait and extending up the Labrador Coast as far as Saglek Bay north of Nain. Dated campsites in the strait indicate continuous occupance between 8,855 ± 100 BP and 2,410 ± 50 BP; in Northern Labrador occupance began later, around 5,995 ± 80 BP, on the island of Newfoundland around 4,990 ± 230 BP. The Maritime Archaic culture represents the easternmost extension of Palaeo-Indian colonization of North America between 11,500 and 9,000 BP, the L'Anse Amour site being one in a long line of campsites leading back to Clovis sites in southern Alberta at the southern end of the ice-free corridor through which the original sinodont settlers entered the continent from Alaska and northeastern Asia (Harris 1987; McGhee 1989).

L'Anse aux Meadows (A.D. 1003 to 1014)

Between 1960 and 1967 archaeological excavation revealed the site of a Norse settlement on a raised beach at the mouth of Black Duck Brook near the modern fishing outport of L'Anse aux Meadows. It consisted of three house complexes, each centered on a larger building, turf-walled according to the Icelandic "great hall" model, complete with floor hearths and cooking pits. It also included an open smithy with

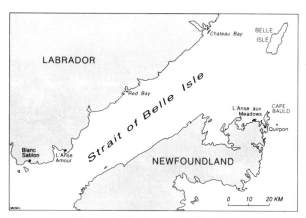

Figure 1. The Strait of Belle Isle.

clay-lined furnace (1200°C) and forge, and an adjacent charcoal kiln-pit. Artifacts were few, but included iron rivets, a ring-headed bronze pin of Celtic type, a fragment of smelted copper, a small-boat floor-board (spruce) with a trenail (pine—Scots?), a decorative finial (also pine), an Icelandic-style stone lamp, a soapstone spindle whorl, and a quartzite needle hone. Precise radiocarbon dating by atomic mass spectrometry gives A.D. 997 ± 8 (Kieser 1990), suggesting that the site may be identified with Leifsbúdir, established by Leifr Eiríksson of Brattahlíd, Greenland, around A.D. 1003 and re-occupied by over-wintering parties till A.D. 1014 (Harris 1987; Ingstad 1985).

The Norse settlement near L'Anse aux Meadows was a terminal point in the advance of the Norse pastoral frontier across the Atlantic between A.D. 786 and 1003. As such it connects with the Icelandic settlement of Greenland (A.D. 986), the Scandinavian and West-Viking settlement of Iceland (A.D. 874 to 930), and the earlier Norse invasions of the British Isles (A.D. 793 to 850). In terms of scholarship, it connects the modern archaeology of the strait with the annals and rich saga-literature of medieval Iceland.

Cape Bauld (A.D. 1497)

In the sparse second-hand reports of John Cabot's seminal voyage of discovery in 1497 the only reference that is geographically secure appears in John Day's statement that "the cape nearest to Ireland is 1800 miles west of Dursey Head" (51°35'N), indicating that Cabot was in the vicinity of Cape Bauld at the northern tip of Newfoundland and at the entrance to the Strait of Belle Isle. The most significant perception of resources was that "the sea there was swarming with fish . . . so many fish that this kingdom [England] would have no need of Iceland, from which comes a very great

quantity of fish called stockfish" (Williamson 1962, 209-214).

Cabot's discovery, which extended south and west as far as the southeastern shores of Cape Breton Island (45°30'N), pointed back in time to the century-old English fishery at Iceland, and—duplicated by the explorations of the Azorian Portuguese—pointed forward to the international fishery associated with the inshore waters and offshore banks of Newfoundland, the settlement of Newfoundland after 1610, and the fish trade into the Mediterranean, the Caribbean, Brazil and—in the 20th century—the North American market. Today, as a result of technological over-efficiency and political and business ineptitude, ecological crisis and serious economic decline in the fishery are jeopardizing the future of coastal Newfoundland, including the strait settlements.

Quirpon and Blanc Sablon (A.D. 1534, 1535, and 1541)

On 27 May 1534 Jacques Cartier of St. Malo in Brittany, Captain and Pilot to the King, reached the mouth of *La Baye des Chasteaulx*, the Bay of Castles—the French name for the Strait of Belle Isle—sailing in heavy ice:

> on account of the unfavourable weather and of the large number of ice floes [*glaces*] we deemed it advisable to enter a harbour in the neighbourhood of that entrance called Karpont, where we remained, without being able to leave, until June 9 (Biggar 1924, 9).

Once free of the ice blockade Cartier explored the Labrador shore of the strait, made the first ethnographic observations of native seal-hunters—Iroquois from the St. Lawrence—at Blanc Sablon, and began an extensive exploration of the Gulf of St. Lawrence, during which the official policy of native abduction was inaugurated. In 1535 Cartier's ships met at Blanc Sablon where he lingered 15-29 July before making his epic journey up the St. Lawrence to Stadacona (Québec) and Hochelaga (Montréal). In 1541, the rendez-vous was once more "Carpont" (Quirpon), where Cartier waited for some considerable time after 23 June before moving up to overwinter at Stadacona and revisit Hochelaga. Named after Le Kerpont, a harbor between the island of Bréhat and the Breton coast west of St. Malo, Quirpon was already a French fishing station and, after 1713, part of the French Shore of Newfoundland, which lasted officially until 1904. The early French migratory fishery was precursor to the French fur trade—Tadoussac (1550), Québec (1608), Montréal (1642)—and the settlement of New France. Blanc Sablon—also a pre-Cartier fishing station—is a toponymic reflection of the Blancs Sablons in the approaches to Brest in Brittany. It stands as a symbolic point of origin for official European impact upon native North American cultures (Barkham 1984; Biggar 1924; Harris 1987).

Red Bay (A.D. 1565)

In the autumn of 1565 a Spanish *galeon* was wrecked in the port of Butus or Buytres in the Granbaya—Red Bay in the Strait of Belle Isle:

> Having completed the whale fishing and being ready to return to Spain there came upon them in the said port a tempest which, after the bow moorings had broken, swung the ship around and blew her ashore where she remained aground, so that with all the cargo of whale oil that she had on board, a quantity of a thousand barrels, they left her in Terranueva (Story 1982, 73-74).

Owned and commanded by Ramos de Arrieta of Pasajes and outfitted by Joanes de Portu and Miguel de Beroiz, burgesses of San Sebastian in the Basque province of Guipúzcoa, she was probably the *San Juan*, one of a dozen Spanish Basque whaling galleons that frequented Red Bay, and one of a score that tied up in eight or nine harbors seasonally each spring along the Labrador side of the strait between 1543 and 1588, and took 14 to 18 thousand barrels of whale oil back to Europe each autumn (Barkham 1977; 1984).

Evidence for the wrecking of the *San Juan* and its location was found in 16th century insurance documents held in archives in Burgos and Valladolid in Old Castille. Since 1977, archaeologists from the Archaeological Survey of Canada have resurrected the wreck of the *San Juan* while others from the Memorial University of Newfoundland have uncovered tryworks and a large graveyard associated with Basque whaling activities at Red Bay. The contemporary lines lead back to a dozen ports along the southern coast of the Bay of Biscay, many of which were also major shipbuilding centers providing large ships for the Caribbean trade. The Basque whaling industry and Castillian sheep ranching were both backward linkages in the Flemish woollen industry; both were underwritten by the Old Castillian insurance companies. Basque ships carried Castillian wool and Terranovan whale oil to Flanders (Barkham 1973; 1980; Story 1982). Historically, the Basque whaling industry in the Grand Bay was the progenitor of all commercial whaling enterprises,

wherever and whenever they have occurred in the world.

Chateau Bay and Quirpon (A.D. 1763 to 1766)

In the summer of 1763, after a season of mapping on the east and south coasts of Newfoundland, James Cook sailed in the *Grenville* to survey the eastern entrance to the Strait of Belle Isle, and in particular to chart Château Bay on the Labrador side. In September 1764, in Cartier's old refuge at Carpoune [Quirpon] at the end of another survey season in the strait, Cook found himself presiding over one of the most poignant episodes in the cultural history of North America, when Jens Haven, a Moravian missionary with Greenland experience, met and spoke with a visiting party of Labrador Inuit in their own tongue. In the summer of 1765 the contact was reinforced when the Moravians John Hill and Christian Drachardt were landed in Château Bay from H. M. S. *Niger* to meet a large body of coastal Inuit and arrange a meeting with Hugh Palliser, the Governor of Newfoundland, to establish a "league of friendship"—and incidentally to dissuade the Inuit from crossing the strait to trade at Quirpon with migratory French sackships. Jens Haven and C. A. Schloezer, meanwhile, made an epic journey in the schooner *Hope* to survey the Labrador coast as far north as Davis Inlet (55°53′N) (Lysaght 1971).

The Moravian initiative, which effectively ended decades of conflict between Inuit and European fishermen, also resulted in the establishment of a Moravian mission field in northern Labrador with headquarters at Nain, which lasted from 1771 until after the Second World War. The antecedents reached back through mission fields in Greenland, Pennsylvania, Georgia, and the West Indies to origins in Germany where natural theology was introduced as an essential part of pietist missionary work. One result is instrumental weather records for northern Labrador that are among the earliest and finest in North America (Macpherson 1987).

In 1766 H. M. S. *Niger* returned to Château Bay, accompanied by Joseph Banks, a young gentleman-naturalist destined to meet Cook in the harbor of St. John's later that summer and to serve as his scientific chief on the famous voyage to the Pacific in 1768 to 1771 which mapped the coasts of New Zealand and eastern Australia. Banks, who had close connections with the Moravian community in London, made the first major collections of Labrador flora and fauna, and contributed specimens to the Moravian "Physick Garden" in Chelsea, the origin of the Royal Botanical

Gardens at Kew. He also visited the site of an ancient Basque whaling station in Chateau Bay which he attributed to "the Danes," that is, the Greenland Norse.

The presence of James Cook, Jens Haven, and Joseph Banks in the Strait of Belle Isle in the mid-1760s was a signal that the long period of resource exploitation from metropolitan bases in Western Europe was coming to an end. Survey, evangelization, and science, backed by political enlightenment, prepared the way for permanent settlement and civilised order. Between 1763 and 1830 a resident merchant fishery took shape in the strait, re-occupying long-abandoned Basque and Breton sites from bases in Devon, Bristol, and Jersey—later from Gaspé and Québec—entirely dependent upon migrant labour from those same sources. Harbors on both shores of the strait were first settled by independent fishermen between 1830 and 1850, predominantly from Britain, while Red Bay was occupied primarily by settlers from Carbonear in Conception Bay, one of the oldest settlements on the Old English Shore (Thornton 1977). Establishment of permanent communities led inevitably to political integration with a colony about to assume the responsibilities of self-government (1855).

Conclusion

In 1992 the past has finally converged with the present as recent research—archaeological, archival, historical, and geographical—has revealed the long story of human presence and marine resource exploitation in the strait. For the people of the strait the past itself has become a significant economic resource and a cross cultural experience, including mutual exchanges with the Basques and visits by latter-day Norsemen. The discovery of the geography of their past represents, for them, an intellectual and spiritual enrichment.

References

Barkham, S. 1973. Mercantile community in inland Burgos. *Geographical Magazine* 46:106-113.

————. 1977. The identification of Labrador ports in Spanish 16th century documents. *Canadian Cartographer* 14(1):1-9.

————. 1980. Finding sources of Canadian history in Spain. *Canadian Geographic* 100:66-73.

————. 1984. The Basque whaling establishments in Labrador, 1536-1632: A summary. *Arctic* 37:515-519, and note 3.

Biggar, H. P. 1924. *The Voyages of Jacques Cartier*, pp. 9-20, 94-95, 251. Ottawa: Publications of the Public Archives of Canada, No. 11.

Harris, R. C., ed. 1987. *Historical Atlas of Canada*, Vol. I: plates 2, 3, 5-7, 16, 22, 23, 25, 33. Toronto: University of Toronto Press.

Ingstad, A. S. 1985. *The Norse Discovery of America*, Vol. I. Oslo: Universitetsforlaget AS.

Kieser, W. 1990. Radiocarbon dating by accelerator mass spectrometry: Principles, techniques and applications. In *Examples and Critiques of Quaternary Dating Methods*, ed. Thomas W. D. Edwards, p. 182. Waterloo, ON: Canadian and American Quaternary Associations' Short Course 3, University of Waterloo.

Lysaght, A. M. 1971. *Joseph Banks in Newfoundland and Labrador*, pp. 47-49, 64-71, 129-132, 138-142, 181-183, 187-221, 293-421, and plates 5, 6, 16, 30, 40, and 91. London: Faber and Faber.

Macpherson, A. G., and Macpherson, J. B., eds. 1981. *The Natural Environment of Newfoundland, Past and Present*, pp. 61-76, 239-240. St John's: Department of Geography, Memorial University.

Macpherson, A. G. 1987. Early Moravian interest in Northern Labrador weather and climate: The beginning of instrumental recording in Newfoundland. In *Early Science in Newfoundland and Labrador*, ed. D. H. Steele, pp. 30-41. St John's: Avalon Chapter of Sigma Xi.

McGhee, R. 1989. *Ancient Canada*, pp. 47-54. Ottawa: Canadian Museum of Civilization.

Story, G. M., ed. 1982. *Early European Settlement and Exploitation in Atlantic America*, pp. 73-74, 41-52. St John's: Memorial University of Newfoundland.

Semple, E. C. 1915. The barrier boundary of the Mediterranean basin and its northern breaches as factors on history. *Annals of the Association of American Geographers* 5:27-59.

Thornton, P. A. 1977. The demographic and mercantile bases of initial permanent settlement in the Strait of Belle Isle. In *The Peopling of Newfoundland: Essays in Historical Geography*, Social and Economic Paper No. 8, ed. J. J. Mannion, p. 173. St. John's: Institute of Social and Economic Research, Memorial University of Newfoundland.

Williamson, J. A. 1962. *The Cabot Voyages and Bristol Discoveries under Henry VII*, Ser. II, No. CXX, pp. 209-214. Cambridge: Hakluyt Society.

The Grand Trunk Corridor in Ontario's Development, 1791-1986

Charles F. J. Whebell
University of Western Ontario

Ontario was created as the province of Upper Canada in 1791 and, ever since, it has depended on the lifeline of the St. Lawrence Valley. The valley's continental position—thrusting deep into the northern flank of the United States—has many times provided both security problems and economic opportunities for the province. Each time a major innovation in transport mode has been adopted, the line of communications reaching from Montreal (east of the provincial boundary) through to the Detroit River has been upgraded, with the consequence that this linear zone, called here the "Grand Trunk Corridor," contained in 1986 some 70 percent of the urban population of the province, which is itself 80 percent urbanized. Here, I encapsulate and update my work (Whebell 1969) and that of Maurice Yeates (1975, 1985).

Upper Canada faced serious problems of communications from the beginning. In 1797, the then Chief Justice observed that the settled zone approximated "a parallelogram of near 500 miles by not more than 20" and that it was "hardly possible to conceive anything more scanty, more irregular, or more uncertain than the means of communication" within that zone (Russell 1932, II:28). In response to these conditions, the government determined in 1799 to construct a "Great Highway throughout the province," between its eastern limit and the Detroit River (Russell 1936, III:46) (Figure 1). Such a road might well even then have been termed a "Grand Trunk" road, since the term was already in well established usage, for example in Bengal. This decision was instrumental in structuring the settlement and urban system of Ontario as it is today.

Decades passed before the road as a whole was tolerably useful. As late as 1832, surveyor Charles Rankin, was forced to walk from Hamilton westward over 60 kilometers, in bitter winter weather, to lay out the town plot of Woodstock, which lay along this route (Rankin 1833, No. 61). It was in winter that the roads of Upper Canada came into their own, as water transport ceased, and the snow erased the roads' ruts and mudholes. Perhaps the most spectacular journey along this road was the Governor General C. P. Thomson's dash from Toronto to Montréal in February 1840, reaching speeds of some 20 kilometers per hour. Lieutenant-Governor Arthur was moved to comment—most presciently—about winter's imposing its "periodic railway" upon the land (Arthur 1957, II:435).

The Railroad Era

Despite 20 years' scheming and promotion, railroads did not come into operation in Upper Canada until the 1850s, owing in part to its low levels of population and economic development. But the chief drawback to railroad-building in an era of U.S. protectionism was the fact that between tidewater on the St. Lawrence and the Detroit River, the line would be a dead end, a cul-de-sac, going nowhere very important and opening up only a modest incremental amount of settlement land.

During the 1830s and 1840s, however, the United States' Midwest was growing explosively, forming new and vigorous trading patterns within the Great Lakes Basin. Early in the 1850s, the United States agreed to a Treaty of Reciprocity (limited free trade) to take effect in 1854. Now the southwest of Ontario was no longer a cul-de-sac, but a channel directly into the United States' most expansive market. Improvements to year-round movement of commodities and people,

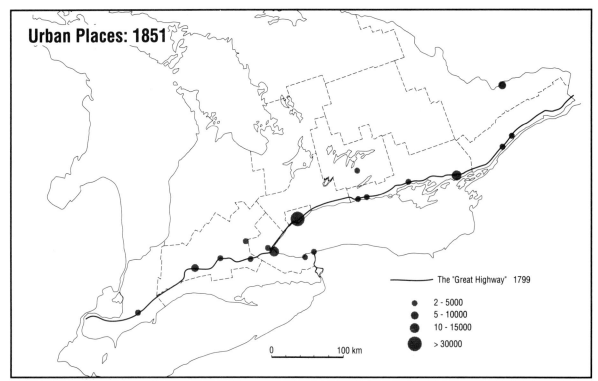

Figure 1. The need for a land route connecting the eastern and western limits was recognized early in the province's existence. Slow growth rates before the 1840s however precluded extensive urbanization; only a few places had significant industrial employment by 1851, notably Toronto, the provincial capital, and Hamilton, at the western end of Lake Ontario.

of which the railroads had already proved themselves capable elsewhere, became essential. In his position paper on the utility of railroads Francis Hincks, the finance minister, argued for government guarantees: "Canada is deeply interested in the success of Railway enterprize, as the productiveness of the public works must in great measure depend on the products of the West being carried to the Seaboard by the St. Lawrence route . . ." (Hincks 1849, 2).

The Grand Trunk Railroad was built a few years later to connect the ocean port of Montréal to the U.S. boundary at Lake Huron; it was later extended to Chicago. From the eastern limit of the province to Toronto it followed closely the route of the old Great Highway, but west of Toronto it took an alternative way through the Niagara Escarpment, some 40 kilometers north of that route, thence crossing the rivers of the middle Grand River system. At the same time another line (originally the Great Western, which later amalgamated with the Grand Trunk) had preempted the old highway route westward from Hamilton at the head of Lake Ontario. By the 1880s, a third main line through Toronto, part of the newly formed Canadian Pacific system, strengthened the

connectivity along the general east-west alignment of the old Great Highway. The Grand Trunk Corridor had assumed its permanent shape.

During the century or so that the railroad dominated land transport, the proportion of urban population in Ontario grew from about 15 percent in 1851 to more than 75 percent in 1951. Except for cities in the northlands regions, most of them founded in the 20th century, few important Ontario cities lie far from the Grand Trunk Corridor. Toronto, the provincial capital, was always the chief node in this network, growing from 31,000 in 1851 to just over 1 million in 1951. The cluster of cities in the middle Grand and middle Thames Valleys, the "Western Heartland" was particularly stimulated by the railroads, which gave scope to the development of local advantages of raw materials and waterpower.

The Motor Age

The towns along the Grand Trunk corridor generated great demand for motorable roads in the 20th century, with the result that the first of Ontario's numbered

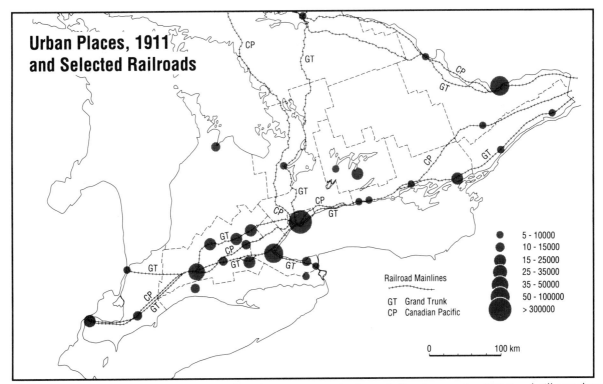

Figure 2. The east-west Grand Trunk lines passing through Toronto were the first main lines in operation (1854) and still carry heavy traffic. Railroads both stimulated and consolidated industrial urbanization, to produce the linear system of towns called here the Grand Trunk Corridor.

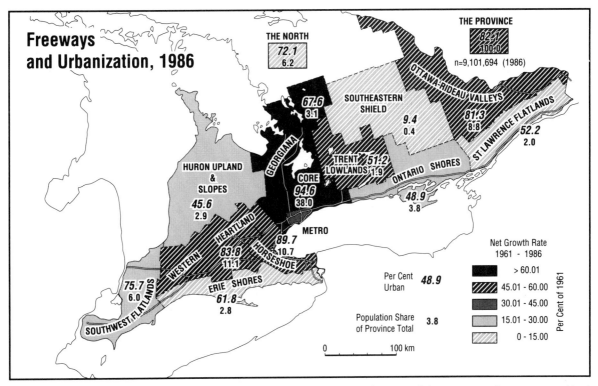

Figure 3. Since the beginning of the motorway (freeway) system in the 1960s, urban growth has increasingly concentrated in those parts of the corridor extending through the regions Core, Horseshoe, and Western Heartland. The shading indicates overall population growth by region during this time (1961 to 1986); data for the Core region include those for Metro Toronto.

provincial highways—King's Highway 2 (there is no No. 1)—approximates closely the old Great Highway. But this meant that traffic had to grind along the main streets of the towns, with much local complaining and loss of time. By the end of the Second World War, the need for specially built high-speed roads was obvious.

To by-pass such bottlenecks, the first true freeway in the province, Highway 401, was laid out in the 1950s with the same object as that of the Great Highway of 1799 and the Grand Trunk Railroad of 1854: to link the eastern and western borders. The towns of the Grand Trunk Corridor provided the traffic demand the planners had to accommodate; as a result this road, with its later branches (Nos. 402 and 403), approximates very closely the routes of the Grand Trunk mainlines of, say 1911 (Figure 2). The corridor now services well over two thirds of the total 1986 population of Ontario, and three quarters of its urban population; the regions through which it runs are themselves about 87 percent urbanized overall (Figure 3). The regions were devised to illustrate this point and show three important population parameters for the period 1961 (when Highway 401 came mostly into use) to 1986 (the latest census date): overall net growth from 1961 to 1986; each region's urban proportion, 1986; and each region's share of the total population of the province. The names chosen for these regions reflect in part their physical character.

The 1989 free-trade agreement with the United States, following on the special Auto Trade Pact of 1965, offers extended access to the U.S. market for a large number of Ontarians, most of whom are increasingly concentrated around the western end of Lake Ontario and in the Western Heartland, at the expense of many of the other regions. The Grand Trunk Corridor is clearly set to continue into the 21st century its historic role as the economic backbone of the province.

References

Arthur, Sir George. 1957. *The Arthur Papers*, 3 vols., ed. C. R. Sanderson. Toronto: University of Toronto Press.

Hincks, Francis. 1849. Memorandum suggesting a plan for aiding in the construction of railways. National Archives of Canada R.G.7 G 20,47, No. 5171.

Rankin, Charles. 1832. *The Rankin Papers*. Metropolitan Toronto Public Library.

Russell, Peter. 1932-1936. *The Correspondence of the Honourable Peter Russell*, 3 vols., ed. E. A. Cruikshank. Toronto: Ontario Historical Society.

Whebell, C. F. J. 1969. Corridors: A theory of urban systems. *Annals of the Association of American Geographers* 59:1-26.

Yeates, Maurice. 1975. *Main Street: Windsor to Québec City.* Toronto: Macmillan.

———. 1985. *Land in Canada's Urban Heartland.* Ottawa: Environment Canada, Lands Directorate.

Russian Religious Settlement along the Pacific Rim

Susan W. Hardwick
California State University, Chico

Since 1988, *glasnost* and *perestroika* have made it possible for more than 140,000 new Soviet emigrés to resettle in the United States. The majority left the Soviet Union for religious reasons and many have settled in Alaska, Washington, Oregon, California, and British Columbia. This North American Pacific Rim has served as an important migration destination for Russians for over two centuries. Of particular interest are the locations of six Russian religious groups that cluster in distinct enclaves dependent on religious affiliation. These groups include Orthodox, Old Believer, Doukhobor, Molokan, Baptist, and Pentecostal Russians. The term Russian refers to immigrants who perceive and identify themselves as "Russian" regardless of the specific Soviet province of their origin. Russian Jewish emigrés are thus not included because they are ethnically Jewish, not Russian. Figure 1 locates various nodes or points of Russian religious settlement along a north-south line extending from Alaska to San Diego, California.

Russians migrated to the Pacific Rim of North America in five waves: frontier North America, pre-World War I settlement, migration from 1917 to 1945, post-World War II to 1987 settlement, and glasnost-era migration that continues into the early 1990s. Each wave has left distinct and observable ethnic nodes based on religious affiliation and has created a unique cultural landscape.

Frontier America

Russian expansion eastward across Siberia in the 16th and 17th centuries ultimately carried fur hunters (*promyshlenniki*) to the west coast of North America. Most of these earliest Pacific Rim settlers came as employees of the Russian-American Company. James Gibson's comprehensive work on Russian expansion in frontier America defines the earliest migration into the region, from 1743 to 1799, as a phase of expansion distinguished by the establishment of permanent Russian settlements on top of high coastal promontories or at the mouths of rivers and bays (Gibson 1976, 5). Frontier settlement in Alaska found an early focus at St. Paul's Harbor on Kodiak Island and later at New Archangel (Sitka). These were centers of Orthodoxy as well as economic power. In gaining financial support from the government, the Russian-American Company also agreed to spread the teachings of the Russian Orthodox faith among the natives of the region. They brought icons, bibles, holy water, crosses, and their Eastern Christian values and culture with them to the New World. Evidences of the success of their mission on the contemporary Alaskan cultural landscape includes 78 Orthodox churches and chapels remaining in the region in 1991.

As Alaskan food supplies became scarce and furs depleted, a Russian colony was established on the coast of northern California at Fort Ross. An Orthodox church was constructed in the small settlement in the mid-1820s. It was the first Orthodox structure established in the Western Hemisphere outside of Alaska. Although no resident priest ever served at Ross, an active religious life was maintained at the colony with visiting priests conducting baptisms, weddings, confessions, confirmations, and funerals (Watrous 1991). When Ross was sold in 1841, most evidence of its church life was removed. However, today the chapel as part of a state park has been completely rebuilt and hosts regular Russian Orthodox religious services.

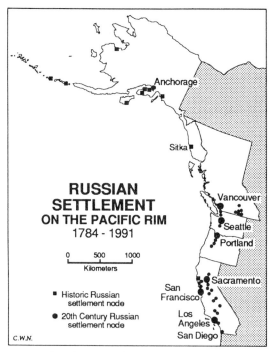

Figure 1. Nodes of Russian religious settlements along the North American Pacific Rim.

Pre-World War I

After the sale of Alaska to the United States in 1867, the prospect of new missionary work among European immigrants and non-Orthodox Americans, coupled with failing social and economic conditions in Alaska, encouraged the spread of Russian Orthodoxy southward. The Russian community in San Francisco, in particular, experienced a period of major growth as newcomers from Alaska arrived in the city during the early 1870s (Tripp 1980). Economic opportunities in California's then largest city encouraged Russian residences and businesses to cluster near the downtown. In 1872, the Russian Orthodox Church headquarters was transferred from Sitka to San Francisco, thus ensuring the longterm survival of Russian neighborhoods in the city (Tarasar 1975).

The increase in the number of Russians in the United States and Canada after 1900 was due in large part to conditions at home. Economic, social, and religious pressures developed during the Russian Civil War (1904 to 1905) and during the First World War. Religious expression was severely limited during this time. Because of religious persecution, Russian Molokans, a minority Protestant group, migrated to east Los Angeles between 1905 and 1907. By 1911, at least 1,000 Molokans lived in San Francisco and more than 5,000 lived in southern California (Young 1932). In the early years

of their settlement in urban California, this group of formerly rural Russian sectarians also established secondary settlement nodes in less urban parts of the American West. Small Molokan enclaves evolved in the San Joaquin Valley of central California, near Glendale, Arizona, in Oregon's Willamette Valley, and in the Guadalupe Valley of Baja California (Conovaloff 1990). Russian Molokans in Los Angeles and San Francisco continue to yearn for a rural lifestyle in the 1990s.

Other Russian religious groups also migrated to the west coast during this second wave, locating primarily in urban areas where economic opportunities were greatest. The first Orthodox parish in Portland, Oregon, was started in 1890. Seattle and Wilkeson, Washington, as well as Vancouver, British Columbia, also witnessed the beginnings of small Russian neighborhoods and the construction of several new Orthodox churches in the late 19th century (Tarasar 1975). Russian Baptists established a church in San Francisco and clustered in small settlement nodes in West Sacramento, Los Angeles, Seattle, and in California's rural Sonoma County (Hardwick 1986). Even more significant than Russian Baptists in total population were Russian Doukhobors, a religious group long persecuted by both the Orthodox church and the Russian government. In 1899, more than 7,000 Doukhobors settled on the Canadian prairie; however, Canadian quota restrictions, forced more than 12,000 to remain in Russia (Woodcock and Avakumovic 1968). Prior to and during World War I, the majority of Canada's Doukhobors resettled in the West Kootenay region of British Columbia (Hawthorn 1955). Today's largest populations are concentrated in Kootenay, Grand Forks, and Vancouver.

1917 to 1945

At the end of the Russian Civil War in 1922, new Russian emigrés, fleeing the Soviet regime, arrived in the United States and Canada in large numbers. Many had left their homeland to escape religious persecution. Russians came to the North American Pacific Rim during and after the revolution via three main routes. The great majority of them filtered across Siberia to Harbin, Manchuria, and into Sinjiang Province in China. Most eventually migrated here by way of Shanghai or Hong Kong, then San Francisco, Vancouver, Seattle, or Los Angeles. The second group left Odessa, Theodosia, and other ports on the Black Sea and took refuge temporarily in Constantinople. Many waited there for several years hoping for political change at home that would enable them to return.

The third group, arriving via the European route, settled initially in eastern and midwestern Canada and the United States, and eventually migrated to the west coast for permanent settlement (Day 1934).

As anti-Bolsheviks moved into parts of eastern Asia in the turmoil of the 1920s, the Asian routes in particular brought thousands of displaced Russians into Los Angeles, Portland, Seattle, and Vancouver. By 1928, the Russian population in San Francisco totaled 15,000 people (Tripp 1980). These new Orthodox and Baptist emigrés settled in well established Russian neighborhoods, a trend that was common throughout the region.

Los Angeles also had a large influx of new Russians immediately following the revolution. Most settled initially in Hollywood, just west of the earlier Molokan enclave. This diverse group of refugees also settled in other parts of the Los Angeles area, including Santa Monica, Venice, and Long Beach (Day 1934). Russian cultural centers, language schools, art clubs, restaurants, and new churches held the community together and maintained Russian culture in an increasingly multi-cultural city.

In West Sacramento, Russian Orthodox emigrés settled along the Sacramento River in an isolated neighborhood, constructing modest homes and small businesses on recently drained land. Another new parish of Orthodox emigrés was formed in Santa Rosa, California, in 1936 (Tarasar 1975).

1945 to 1987

A variety of Russian religious groups arrived on the North American Pacific Rim after World War II. Russian Old Believers, or *Staroveri*, migrated to the Willamette Valley in Oregon after several difficult years in Latin America. These strict Orthodox believers had resettled in Brazil in the early 1950s after more than three decades in Manchuria and Sinjiang. They, along with several thousand Russian Old Believers from Turkey, resettled in the town of Woodburn, Oregon, in 1964 with the financial support of the Tolstoy Foundation (Morris 1981). In 1968, frustrated with the perceived anti-religious influences of American culture on their children, five Old Believer families from Woodburn traveled to the remote Kenai Peninsula of Alaska and founded the Russian village of Nikolaevsk. More than 90 families currently live in Nikolaevsk and, since the early 1980s, 5 other Russian Old Believer settlements have located in Alaska. The area around Woodburn, Oregon, also remains an important center for Staroveri settlement. Earlier Molokan residents of Woodburn originally sponsored Old Believer families

in the 1960s establishing important ethnic linkages and adding vitality to the well-established Russian cultural landscape in the Willamette Valley of Oregon.

Also important in this era were the thousands of Russian Orthodox, Baptist, and Pentecostal immigrants who had fled Russia during its Civil War years. Settling primarily in Harbin, Manchuria, and the Sinjiang Province of China, these non-communist Russians had been forced out of China in the late 1940s because of the emerging dominance of the nation's new Marxist government. Most traveled to Hong Kong and were then taken to a refugee camp on a small island in the Philippines (Hardwick 1979). Many eventually resettled in South America and Australia. Some waited several years for permission to enter the United States, settling in older Russian neighborhoods in California, Oregon, and Washington in 1950. Of special note in this post-World War II migration was a tightly knit group of Russian Baptists from the Chinese village of Inning who added new energy and growth to the evolving Russian neighborhood in West Sacramento. This Protestant group laid the foundation for glasnost-era settlement in the area, and created a clustering of Russian Christians in California's capital city that has made Sacramento the preferred resettlement site for Russian Baptists and Pentecostals in the 1990s.

Post-Glasnost

The most recent wave of Russian migration to the North American Pacific Rim began in 1988 when Soviet President Gorbachev made the startling decision that victims of religious persecution could emigrate from the Soviet Union. Immediately, Pentecostal, Baptist, and Jewish Soviets began to plan for their departure. The majority of those who left the Soviet Union between 1988 and 1989 flew to American resettlement camps in Rome and Vienna. Most, including thousands of Baptists and Pentecostals, left the Soviet Union on Israeli exit visas. After four to seven months in the European camps, the migrants flew to various parts of the United States and Canada, their location patterns determined by the location of their sponsors. According to U.S. Immigration and Naturalization Service records, more than 80,000 had arrived in the United States by late 1989, approximately 55,000 more came in 1990, and at least 50,000 additional arrivals are predicted for 1991.

Emigrés now are processed directly by the American government in Moscow and all resettlement camps in Europe have been closed. After a brief period, many of these post-glasnost emigrés have resettled in California and Washington state, attracted by perceived economic opportunities, generous welfare benefits, and climatic amenities. My numerous ethnographic interviews indicate that the most important factor in their tendency to locate in places such as Sacramento, San Francisco, and Seattle is religious. Drawn to older Russian neighborhoods in these cities, Russian Pentecostals and Baptists find well-established religious and social networks awaiting them. A strong chain migration to places with antecedent Russian settlement nodes has intensified in recent years.

Conclusions

Ethnic relations and the patterns and processes of ethnic and religious settlement represent a complex set of issues in North America today. Although an understanding of the Russian experience has often been overlooked in the literature, recent changes in Soviet policy and the resultant significant influx of emigrés from the Soviet Union have made an understanding of their location patterns and adjustment processes vitally important. For almost a century, religion has served as a primary motive for emigration from the Soviet Union and has dominated migration decision making. Religious groups have tended to settle in tightly clustered Russian enclaves, both urban and rural. The Old Believers and Doukhobors have remained in primarily rural or quasi-urban settings, Molokans in both rural and urban nodes, and Orthodox, Baptist, and Pentecostal groups have settled in more urban locations. While a common cultural and ethnic background bind each group, religious differences have resulted in distinct nodes of settlement. Despite these religious differences, it is predicted that the Pacific Rim of North America will remain a persistent and preferred destination of Russian emigrés for many years to come.

References

Conovaloff, A. J. 1980. *The Molokan Directory.* Berkeley: Highgate Road Social Science Research Station.

Day, George Martin. 1934. *The Russians in Hollywood.* Los Angeles: University of California Press.

Gibson, James R. 1976. *Imperial Russia in Frontier America.* New York: Oxford University Press.

Hardwick, Susan Wiley. 1979. A geographical investigation of Russian settlement in an urban landscape. *The California Geographer* 19:87-104.

_____ . 1986. *Ethnic Residential and Commercial Patterns in Sacramento with Special Reference to the Russian-American Experience.* Ph. D. Thesis, Department of Geography, University of California, Davis.

Hawthorn, Harry B. 1955. *The Doukhobors of British Columbia.* Vancouver: University of British Columbia and J. M. Dent and Sons.

Morris, Richard Artells. 1981. *Three Russian Groups in Oregon: A Comparison of Boundaries in a Pluralistic Environment.* Ph. D. Thesis, Department of Anthropology, University of Oregon.

Tarasar, Constance J., ed. 1975. *Orthodox America: 1794-1976.* Syosset, NY: The Orthodox Church in America.

Tripp, Michael William. 1980. *Russian Routes: Origins and Development of an Ethnic Community in San Francisco.* Master's Thesis, Department of Geography, San Francisco State University.

Watrous, Stephen. 1991. *Religious Life among the Russians. Guide to Fort Ross.* Fort Ross, CA: Fort Ross Interpretive Association.

Woodcock, George, and Avakumovic, Ivan. 1968. *The Doukhobors.* Toronto: Oxford University Press.

Young, Pauline V. 1932. *The Pilgrims of Russian Town.* Chicago: University of Chicago Press.

The Moscow Connection

Irina Vasiliev
Syracuse University

Paris, Berlin, London, Madrid, Geneva, Athens, Petersburg, . . . Moscow. All are venerable cities of Europe. All have been used as place-names throughout North American. And not just once each, but many times over. A Paris in New York, Texas, and Ohio. An Athens in Georgia and in Ohio. Madrid, New Madrid; London, London Bridge, New London. All of them, singly and together, evoke a time when young American settlements delineated their new identities by looking to the old established European cities and took from them what most easily brought to mind images of sophistication and permanence—their names. Each name is scattered through nearly all of the states of the Union. Looking at any one name, a geographer is tempted to see patterns, like a child's "follow-the-dots" pictures. But in what order are the dots to be connected? Are all to be connected, or are some not related to the pattern? Is there only one pattern? What is the final picture, if there indeed is one? This paper examines naming patterns for the Moscows in the United States.

Since 1800, 49 populated places have been named Moscow in the United States. Many have disappeared or changed their names. Today, there are 27. Not all of the Moscows are connected to each other. Some belong to groups that relate to a central notion from which spokes radiate to each place. Others are linked by a form of place-name genealogy by which a linear structure describes the relationships among the Moscow points. For others, the links are unknown, but, even for these, the name Moscow summons the impression of a cultured, mature city, characteristics entirely lacking in the new Moscows of the 1800s in the interior of the United States.

Napoleon's Influence

The 1800s were a time of great growth and expansion in the United States as people moved west, settled into new villages and cities, and named or renamed their new settlements. One group of Moscows stretches from Maine to Alabama, and to Minnesota, and is linked by the single 1812 event halfway across the world—Napoleon's invasion of Russia and his subsequent defeat there. These Moscows are not connected to each other. Rather, they surround the idea of Moscow in 1812 and are linked to that central place and time by the impact of historic events there. Moscow was part of a global consciousness in the early 1800s that offered identity for places in need of names in a newly emerging nation.

In that year, the inhabitants of the town of Northfield in Somerset County, Maine, filed for incorporation. By the time the act of incorporation was passed, however, the name of the place had been changed to Moscow and the incorporation, in 1816, went under the new name. The news of the French advance onto Moscow, Russia, and Napoleon's consequent retreat in the year of the original incorporation filing had made a great impression on the residents of Northfield, enough so that they commemorated the old city by taking its name (Varney 1881).

Two states west, the Vermont village of East Calais (pronounced to rhyme with "palace"), in Washington County, is locally known as Moscow for a falling millstone's cracking sound that was equated with the ringing of church bells in Moscow, Russia, during Napoleon's campaign in 1812 (Swift 1977).

Meanwhile, a disagreement between a number of townspeople in Livingston County, New York, and one Samuel Miles Hopkins was responsible for his building the village of Moscow to thwart those others. He had spent two years in Europe in the 1790s, ". . . witnessing the peculiar social, political and scientific conditions of an eventful period—the time of Napoleon's splendid Italian campaign" (Smith 1881, 43). The mark left by his European travels was expressed in the name he chose for his village.

In 1917, however, the name was changed from Moscow to Leicester, to conform with the name of the township, post office, and train station. There is some indication that the Russian Revolution also played a part in this change (Mahoney 1976)—a simple example of how events in another part of the world again affected a tiny dot on the map of the United States.

During the years 1816 to 1817, after Napoleon's defeat, Louis Phillipe, the exiled king of France, lived in what is now Clermont County, Ohio. His presence there may have something to do with the naming of the village of Moscow by memorializing a major Napoleonic setback that, for Louis Phillipe, could have presaged his return home to France. Moscow, the village, was laid out in 1816 and became a prominent steamship center on the Ohio River (Crawford 1985).

A firsthand witness to the difficulties in Russia, Count Lefebre Denouettes had, before becoming one of a village's leaders in Sumter County, Alabama, ridden in Napoleon's coach during the 1812 retreat from Russia. This event led to his naming the village Moscow (Foscue 1978).

And finally, the result of a large forest fire in Freeborn County, Minnesota, in the 1850s, was nicknamed the Moscow Timber because it reminded people there of the accounts of the burning of Moscow, Russia, during Napoleon's campaign. The village was platted in 1857 and named after the nearby forest (Upham 1969).

Moscows of Russian Origins

Place-name linkage or lineage, in the form of a single-parent genealogy, can be found in a group of Moscows that transferred the name, as an offspring carries that of a parent, across the United States in a number of directions.

Before William Seward achieved notoriety for his folly of purchasing Alaska in 1867, Russian fur traders had made their way down the western American coast and established trading posts and small communities. Russian names, such as Russian River (originally Slavianka), Fort Ross, Sebastopol, among others, are still evident in northern California. One of the small communities on the Russian River, inland from Fort Ross, and across from Duncan Mills, was called Moscow (Josephine G. Nattkemper, letter to the author, 24 August 1987). It disappeared from gazetteers in the 1940s.

The 1830s were active years for naming Moscows by Russians or those with sort-of-Russian origins. The village of Moscow in Clay County, Missouri, 16 kilometers (10 miles) north of Kansas City, and long since absorbed by it, was named so because the first store was run by a Russian (Ramsay n.d.).

Early in that decade, Germans who had lived in Russia for generations began coming west to America. By 1894, a post office in Cavalier County, North Dakota, named Moscow was opened (Williams 1966). The community serviced by this rural post office consisted of Mennonites of German-Russian origin. It is curious to note that Moscow is in Waterloo township; immediately to the west is Moscow township. The Napoleon connection might have extended to these northern reaches, as well.

At the same time as German-Russian Mennonites were influencing naming patterns in North Dakota, a group of German-Russian Lutherans settled in Lackawanna County, Pennsylvania, an early haven for those persecuted for their religion. The community was started by the Reverend Rupert (Murphy 1928). It soon became known as Moscow and still carries the name.

Moscows of American Origin

The rest of the transfer Moscows have origins internal to the United States and have no Russians responsible for their names. In the 1870s, when Paradise Valley, Idaho, was petitioning for a post office, and needed a new name, the decision of what new name to choose fell to Samuel Miles Neff. This man had been born near Moscow, Pennsylvania. He traveled west with his family, first settling near what is now Moscow, Iowa, and finally arriving in Idaho. He chose Moscow (Otness 1980).

Moscow, Iowa, through which Sam Neff had passed, was laid out and named in 1836. Its early settlers are from Clermont County, Ohio—where there already was a Moscow (Figure 1)—and parts of Indiana and

Figure 1. Munchy Muskrat welcomes passersby to Moscow in Clermont County, Ohio, on the Ohio River, just east of Cincinnati.

popped up across the country. Post office regulations required that there be only one of each placename in a state (Stewart 1970).

Other times, a foreign name was assumed to bestow prestige on its community, thus attracting new settlers. The name Moscow served this purpose a number of times.

- In Kemper County, Mississippi, Moscow was settled in 1800. It is not far from Petersburg, which was also named for a city in Russia (Jackie Ratcliffe, letter to the author, 21 July 1987).
- The Jefferson County, Arkansas, Moscow was chosen either from a geography book or a newspaper (John L. Ferguson, letter to the author, 12 February 1986).
- Volume 12 of the *Dead Town List* of the Kansas Collection at the University of Kansas in Lawrence guesses that the Moscow in Cowley County was named for the city in Russia.
- Moscow Mills in Lincoln County, Missouri (Figure 2) was named in the 1870s for the foreign city (Ramsay n.d.).
- All that is left of Moscow in Woodward County, Oklahoma, is a Moscow Church, the remnant of a village named after the Moscow River (Shirk 1965).

As with the Napoleon group, all of these places are linked not to each other, but to the central idea of the original Moscow. The dots are again connected to a central dot, their order indicated by the date on which they were named, none related to any other.

Moscows of Mistaken Identity

A few other Moscows are connected to nothing, not to each other and not to any central notion. Like

Illinois—where there already were places named Moscow (McCoy and Witmer 1976).

At this same time, Alonzo Kies was moving to Hillsdale County, Michigan. He hailed from Moscow, in central New York, which had only recently, in 1824, changed its name from Moscow to McLean, in honor of a newly appointed Postmaster General (Norris 1984). Kies was instrumental in naming his new home in Michigan after the old (*Portrait and Biographical Album, Hillsdale County, Michigan* 1888).

Migration was not confined to movement north and west from the eastern culture hearth. Moscow, Tennessee—named so for no reason but the misunderstanding of an Indian word that means "between two rivers" (Siler 1985)—boasted a favorite son who, in 1846, moved himself and his family to Polk County in eastern Texas. David Griggs Green built a house and opened a blacksmith shop; soon others moved to the area, originally called Greenville. When the village became large enough to have a post office, the name had to be changed because another Greenville, in Texas, already had a post office. So, in memory of David Green's home in Tennessee, the post office was named Moscow (Heritage 1978).

The follow-the-dot pattern is fragmented for these transferred Moscows one or two—maybe three—places linked to each other at a time. Not a very cohesive group as a single unit but certainly a group of similar stories deserving of a classification category of their own.

Invoking the Image of Moscow

Of the rest of the Moscows, a large group was named simply for the Russian city. Often in the 19th century, exotic names were popular as many new communities

Figure 2. Moscow Mills in Lincoln County, Missouri, was named in the 1870s for the foreign city, as was the fashion of the time.

Moscow, Tennessee, they are mistakes; they were never meant to be the name of a Russian city. Moscow in Stevens County, Kansas, was meant to be named for an officer in Coronado's southwest expedition, a man named Moscoso. The residents, in their application for a post office, had shortened it to M-O-S-C-O. A postal clerk in Washington, upon receipt of their petition, being in a helpful spirit, and thinking the hayseeds in the west did not know their spelling, added a W to the name, changing its meaning completely (*The History of Stevens County and Its People* 1979).

Discussion

After this short survey, the question remains: Is there a comprehensive picture completed through the follow-the-Moscow-dots method with which this excursion began? On one hand, there is no picture containing all of the Moscows connected together in a single image, recognizable as belonging to one coherent design. On the other hand, it is possible to see them as together exhibiting a phase of the geography of movement and place-naming in the United States, something that reveals an underlying character of a young country's personality. The people involved in each place's name were integral to the development of the country's identity. Each "Moscow" links the new place to a historic event, acknowledges the prominence of an old idea, or, in the cases of the mistaken identity, engenders a story that identifies the new place and is remembered in the passing years. The name bestows a sense of place and time. Rather than being linked to each other, the Moscows are, more importantly, connected to the elements of geographic growth that made the Moscows possible in the first place—migration, frontier settlement, and information diffusion.

References

Crawford, Richard. 1985. Moscow early shipping center. In *The Clermont Courier*. Cincinnati, OH.

Ferguson, John L. 1986. Letter to the author, 12 February.

Foscue, Virginia O. 1978. The place names of Sumter County, Alabama. In *Publication of the American Dialect Society* No. 65. Tuscaloosa: University of Alabama Press.

Heritage Committee of the Polk County Bicentennial Committee and the Polk County Historical Commission. 1978. *A Pictorial History of Polk County, Texas (1846-1910)*. Polk County, TX.

The History of Stevens County and Its People. 1979. Hugoton, KS: Stevens County History Association.

Mahoney, Velma W. 1976. The Town of Leicester, New York, USA. Unpublished manuscript.

McCoy, George, and John J. Witmer Jr. 1976. *Wilton, Moscow, and Yesteryear*. Wilton, IA: Iowa Bicentennial Committee.

Murphy, Thomas, ed. 1928. *Jubilee History of Lackawanna County*. Indianapolis: Historical Publishing.

Norris, W. Glenn. 1984. *The Origin of Place Names in Tompkins County*. Ithaca, NY: Dewitt Historical Society of Tompkins County.

Otness, Lillian W. 1980. A great good country. Unpublished manuscript.

Portrait and Biographical Album, Hillsdale County, Michigan. 1888. Chicago: Chapman Bros.

Ramsay Place Name Files. In *University of Missouri, Joint Collection: Western Historical Manuscript Collection and State Historical Society of Missouri Manuscripts*. Columbia, MO: University of Missouri.

Shirk, George H. 1965. Oklahoma Place Names. Norman: University of Oklahoma Press.

Siler, Tom. 1985. *Tennessee Towns: From Adams to Yorkville*. Knoxville, TN: East Tennessee Historical Society.

Smith, James H. 1881. *History of Livingston County, New York 1687-1881*. Syracuse: D. Mason and Co.

Stewart, George R. 1970. *American Place-Names*. New York: Oxford University Press.

Swift, Esther M. 1977. *Vermont Place-Names*. Brattleboro, VT: Stephen Greene Press.

Upham, Warren. 1969. *Minnesota Geographic Names*. St. Paul: Minnesota Historical Society.

Varney, George J. 1881. *The Gazetteer of the State of Maine*. Boston: B. B. Russell.

Williams, Mary Ann Barnes. 1966. Origins of North Dakota place names. Unpublished manuscript.

The Point

Edward K. Muller
University of Pittsburgh

The Point is the small, triangular piece of land created by the convergence of the Allegheny and Monongahela Rivers, where they launch the Ohio River on a westward journey across the middle of the continent to its junction with the Mississippi River. At the headwaters of the Ohio, the Point is both a landmark of North America's physical and settlement geography and the functional and symbolic heart of one of the nation's great industrial regions—Pittsburgh and southwestern Pennsylvania. Southwestern Pennsylvanians, air travelers annually flying to or through Pittsburgh, and sports fans across the nation recognize the Point by the confluence of the three rivers and a fountain, which together mark the beginning of the Ohio River and the city of Pittsburgh.

The Allegheny River springs to life in southern New York state and journeys 523 kilometers (325 miles) southwest through rugged, timbered terrain to its rendezvous at Pittsburgh. The Monongahela River originates 206 kilometers (128 miles) south of Pittsburgh in the West Virginia mountains and meanders northward through coal-rich hill country. As the two rivers near each other, they turn westward and carve out a triangular piece of land, narrowing toward their junction. Like southwestern Pennsylvania in general, this land is complexly dissected into hills and valleys, until 1.6 kilometers (1 mile) from the rivers' convergence where it becomes a level floodplain (Cuff *et al.* 1989).

Native Americans, explorers, and fur traders recognized the strategic value of this flat parcel astride the principal river route to the continent's interior and the Mississippi River. Inevitably, the Point assumed military importance in the European struggle for political and cultural dominion over North America. Near the end of 1753, George Washington was sent by Virginia's governor to the Ohio region, where some Virginians had speculative land claims, to order the departure of an occupying French force from British territory. The French commander, charged with reasserting the French claim to the Belle Riviere, firmly refused. The encounter ignited the final contest between France and England for domination of the New World. During the 1750s the warring nations marched armies to the Point and built forts there in order to control the critical location. Fierce battles and smaller skirmishes bloodied the surrounding valleys. When the French finally burned their Fort Duquesne in the face of an overwhelming British army in 1758, they assured the Anglo character of the region, and ultimately the continent (O'Meara 1979).

The British erected Fort Pitt, named for British Prime Minister William Pitt, and a crude settlement called Pittsburgh arose outside its walls. Three decades of conflicts with native Americans in the Ohio valley replaced European embroilments and required a military presence at the Point until 1794, when hostilities finally ended in the Ohio region.

Peace on the frontier unleashed a rush of settlers to the Ohio Valley. Pittsburghers traded the military importance of the Point for the commercial prosperity of being situated at the gateway to the new West. In the opening decades of the 19th century, migrants plodded along the old British military roads over the Allegheny Mountains to Pittsburgh, where they purchased supplies, transferred to river crafts for the next leg of their journey, and gleaned information about the West in the young city's inns and taverns. The growing settlements of the Ohio Valley frontier became markets for Pittsburgh merchants, who benefitted from the break-in-bulk location between eastern overland routes and the western river network. At the mudflat of the Monongahela side of the Point, which constituted the city's wharf, merchants not only connected Philadelphia and Baltimore trading houses with frontier stores, but also invested in the manufacture of iron goods, glassware, and other items that enjoyed the protective barrier of the Allegheny Mountains (Baldwin 1937).

After 1820 new turnpikes, the cumbersome Pennsylvania Main Line Canal to Philadelphia, and steamboats enhanced Pittsburgh's commercial position. Hundreds of steamboats annually called at the Monongahela Wharf. Provisioners, merchant's countinghouses, and warehouses lined busy Water Street. Pittsburgh gained a reputation for its glass factories and iron-rolling mills, foundries and forges, which thrived on western markets (Reiser 1951). The city and its adjacent neighbors grew into an urban complex of 80,000 inhabitants by the 1850s.

Paradoxically, the Point's gateway role was being eclipsed at midcentury. As the frontier pushed relentlessly westward with each decade, it receded further and further from Pittsburgh's wharf and allowed downriver cities, like Cincinnati and Louisville, to compete for trade. The connection of eastern and western railroads in the 1850s eroded Pittsburgh's river-based middleman position because many commodities no longer needed to change transportation modes at the forks of the Ohio. Consequently, Pittsburgh experienced a slow truncation of its commercial horizons. The headwaters location no longer imparted a functional advantage to the city at the Point.

If the railroad crushed Pittsburgh's commercial aspirations, it fuelled an unprecedented expansion of the region's metal and coal industries during the second half of the 19th century. The burgeoning market for railroad rails attracted the attention of local ironmasters and capitalists, who initiated a long series of technological innovations, organizational changes, and investments, which catapulted Pittsburgh into the forefront of American steel-producing regions. Led by Andrew Carnegie, Henry Clay Frick, and their associates, Pittsburgh iron and steel manufacturers built integrated steel mills, metal-fabricating plants, and foundries along the three rivers, such that the Pittsburgh region produced one third of the nation's steel ingots by the early 20th century (Temin 1964). A bituminous coal bed 56 kilometers (35 miles) southeast of the Point proved ideal for making the vital coke fed to blast furnaces. Hundreds of mines, patch towns, and more than 30,000 beehive coke ovens for baking coal arose in the valleys, where the black gold was buried. Railroads knit this industrial region together, traversing short distances to river terminals or snaking through the hollows and river valleys to link the mines and ovens with iron-blast furnaces, steel furnaces, and steel finishers. Moreover, the railroads brought iron ore from the Great Lakes and reached out to the national markets that consumed the region's output.

Although mills surrounded it for miles in every direction, the Point lost its few early production facilities. Instead, it became the locus of capital, which financed and directed this industrial empire. Timber, coal, oil, natural gas, and railroad portfolios combined with pre-Civil War commercial fortunes to amass the financial resources capable of bankrolling enormous industrial ventures. Commercial banks, investment houses, and a stock exchange formed a local Wall Street in the 1880s. Manufacturers and inventors sought Pittsburgh investment houses, notably T. Mellon and Sons, for financial backing. Out of this cauldron of capital, natural resources, and industrial know-how emerged numerous manufacturing operations with headquarters offices at the Point, which became national leaders of their industries during the 20th century. Besides the large steel companies, Pittsburgh concerns specialized in the manufacture of steel, mining, and oil machinery; built railroad cars, brakes, switches and engines; launched barges and small ships; and engineered and erected industrial plants, bridges, and skyscrapers around the world (Muller 1989). George Westinghouse turned patents for railroading and electrical machinery into the Westinghouse Electric Company, the Westinghouse Air Brake Company, and Union Switch and Signal. Charles Martin Hall and Alfred E. Hunt produced the first aluminum ingots, leading to the Aluminum Company of America (ALCOA). Local oil-drilling experience and Mellon capital created the Gulf Oil Corporation out of the discovery of the great Spindletop field in Texas. Other companies established leading positions in industrial glass (Pittsburgh Plate Glass), chemicals and construction products (Koppers Company), and food processing (H. J. Heinz Company).

Like the coal and steel companies, these corporations built huge factories and adjacent towns throughout southwestern Pennsylvania. Both the city and regional milltowns attracted thousands of European immigrants. In 1920, Pittsburgh was the nation's seventh largest metropolitan area with 1,760,000 inhabitants. Work consumed the energies of managers and workers, but production and profits took precedence over social and civic responsibilities (Lubove 1969). Air and water pollution, land degradation, industrial dumping and landfill, and dismal human habitations produced a powerful, but grimy industrial landscape that was at once the wonder and disgrace of America's industrial power. The Point embodied these contrasts. The new skyscrapers of powerful corporations, banks, and capitalists towered over the fading warehouses, shipping concerns, and loft manufactories, which had lined the riverfronts since the early commercial era. The few newer warehouses and railroads failed to combat the spreading blight of the Monongahela Wharf and tip of the Point, where tracks and tramps were equally at

Figure 1. Railroads and warehouses on Pittsburgh's Point in the 1920s. Courtesy of the Archives of Industrial Society, University of Pittsburgh.

home. Recurrent floods and the region's air pollution painted the scene gray (Figure 1).

By 1945 Pittsburgh faced an uncertain future. Despite World War II's temporary economic boom, local industries showed their technological age, were poorly located for emerging Sunbelt markets, and were no longer part of the economy's growth sectors. The region's environmental and social problems had not been adequately addressed during its industrial heyday. The 15 years of depression and war since 1930 allowed conditions to deteriorate to the extent that the city repelled business prospects; and some major, home-grown corporations pondered relocation from the Point to New York City. The Point had become a symbol of industrial urban decay, and was recognized as such by national media (Teaford 1990).

At this darkest hour, the Point began a new career as a symbol of urban renewal in post-war America. Significantly, its original locational importance of the 18th century keyed this role in the redevelopment of downtown Pittsburgh. Under the leadership of Richard King Mellon, Pittsburgh's businessmen undertook an ambitious urban renaissance. Forging a crucial relationship with the powerful Democratic mayor, David L. Lawrence, corporate leaders focused their efforts on the Point where they not only worked and held major real estate investments, but also believed the nation took measure of the region. The partnership of public and private interests engineered legislation necessary to eliminate smoke, diminish floods, and exercise financial and planning authority across the traditional boundaries of local municipalities. Their actions led to extensive demolition and slum clearance, the organization of a public transit system, and construction of expressways, parking garages, low-income housing, industrial parks, a new sewage plant, an indoor civic arena, and an outdoor sports stadium (Lubove 1969).

Private development in downtown Pittsburgh accompanied this environmental clean up and infrastructural improvement, but only after the renewal of the tip of the Point was assured. For decades civic leaders worried about the blighted lands, where fabled forts had existed and romanticized historic figures battled for dominion of the continent. The physiographical, military, and frontier significance of this location inspired renaissance leaders in the 1940s to plan a symbolic park at the confluence of the rivers, which would remind Pittsburghers of their historic greatness and generate the spirit to rebuild for the future. On a pragmatic level, the governmental commitment to constructing Point State Park attracted outside private capital for the redevelopment of an adjacent 9.3 hectares (23 acres) into a modern office complex designed to keep corporate headquarters in Pittsburgh. In turn, the restoration of the Point and the office complex became the catalysts for Pittsburgh's downtown renaissance (Alberts 1980). Thus, symbolically, the park became the community's gateway to its future. When national media heralded the Pittsburgh renaissance in the 1950s as a model for renewal of urban America, Point State Park at the headwaters of the Ohio, framed by gleaming modern towers, became the hallmark photograph (Teaford 1990; Figure 2). The image of Pittsburgh's transition from industrial milltown to corporate office center signaled the economic transformation that was sweeping America. In its emphasis of physical redevelopment and its dislocation of poor residents, Pittsburgh's renaissance also tragically reflected the negative side of post-war urban renewal.

For two centuries, the Point has been instrumental in the development of southwestern Pennsylvania. It embodies the region's historical and geographical sense of place. But the Point's significance extends beyond southwestern Pennsylvania. Its history encompasses the broad patterns of North American urban geography from frontier settlement and commercial emporium to 19th century industrialization and immigration to 20th century decay, renewal, and economic restruc-

Figure 2. The park and redeveloped downtown of Pittsburgh's Point in the 1980s. Courtesy of the Archives of Industrial Society, University of Pittsburgh.

turing. To the nation, it signals the smoky steel city that renewed itself. Geographically, the Point is a national landmark, instantly recognizable by the fountain symbolizing the confluence of the three rivers, the headwaters of the Ohio River.

References

Alberts, R. C. 1980. *The Shaping of the Point: Pittsburgh's Renaissance Park.* Pittsburgh: University of Pittsburgh Press.

Baldwin, L. D. 1937. *Pittsburgh: The Story of a City, 1750-1865.* Pittsburgh: University of Pittsburgh Press.

Cuff, D., Young, W. L., Muller, E. K., Zelinsky, W., and Abler, R. A. 1989. *The Atlas of Pennsylvania.* Philadelphia: Temple University Press.

Lubove, R. 1969. *Twentieth-Century Pittsburgh: Government, Business, and Environmental Change.* New York: John Wiley and Sons.

Muller, E. K. 1989. Metropolis and region: A framework for enquiry into western Pennsylvania. In *City at the Point: Essays on the Social History of Pittsburgh,* ed. Samuel P. Hays, pp. 181-211. Pittsburgh: University of Pittsburgh Press.

O'Meara, W. 1979. *Guns at the Forks.* Pittsburgh: University of Pittsburgh Press.

Reiser, C. 1951. *Pittsburgh's Commercial Development, 1800-1850.* Harrisburg: Pennsylvania Historical and Museum Commission.

Teaford, J. C. 1990. *The Rough Road to Renaissance: Urban Revitalization in America, 1940-1985.* Baltimore: Johns Hopkins University Press.

Temin, P. 1964. *Iron and Steel in Nineteenth-Century America: An Economic Inquiry.* Cambridge, MA: MIT Press.

Birmingham, Alabama:
America's Civil Rights Struggle

Bobby M. Wilson
University of Alabama, Birmingham

In the first half of the 20th century, Birmingham was "the industrial city of the South" but relied upon a racially repressive labor system that contributed to a high level of underconsumption among blacks. By the early 1960s it was considered the most segregated city in America, the "Johannesburg" of America, making it the focal point of America's civil rights struggle. Fire hoses, dogs, and police, turned upon black school children in a Birmingham park in 1963, were part of an attempt by Eugene "Bull" Connor to preserve the racial status quo. Instead, they became vivid symbols of racial injustice in America. The events of the early 1960s were far-reaching; they sparked demonstrations in cities across America and provided a turning point in the national struggle for civil rights. President J. F. Kennedy spoke of the Birmingham movement as a base for introducing the Civil Rights Act and President Lyndon B. Johnson probably had in mind the events taking place in the city when he pushed through Congress the "Great Society" programs and the Voting Rights Act. This paper examines the political economy that made Birmingham the focus point of the civil rights struggle, including changes that led to a less repressive racial order following the civil rights movement.

Pre-Civil Rights Era

Birmingham depended upon and exploited for the production of coal and iron the large black labor pool of the Alabama "black belt" that became available after slavery. Although most blacks remained on the plantation as sharecroppers, many migrated to the more than 90 mine and mill communities throughout the Birmingham area in Jefferson County. At the time of Birmingham's incorporation in 1871, Jefferson County had a black population of only 2,506, but by 1880 thousands of blacks had migrated from the cotton farms in southern Alabama and surrounding states to mine coal and make iron. For each decade after 1890, the black population of the county increased between 50 and 75 percent. By 1920 the county's black population had reached 130,000, and by 1930 Birmingham had the largest black population of any city its size in the United States.

The 1890s were the beginning of rapid and widespread deterioration of blacks' political, economic, and social status in the South (Woodward 1951). By 1900 they were relegated to unskilled jobs, segmenting the labor force along racial lines (Marshall 1967). Black-white socioeconomic differentials were institutionalized in Birmingham. Nearly three fourths of employed black males in 1940 worked as laborers and operatives (largely repetitive, semiskilled work), but only one fourth of employed white males were in these occupations. Among employed black females, 85 percent worked in domestic and other services (for example, laundries). Service occupations and unskilled work accounted for 85 to 88 percent of black job placements during 1948 to 1951 (Hawley 1955). Most workers understood the differences between "white" jobs and "black" jobs. Segregation in the social and spatial fabric of Birmingham was "Caste in Steel" (Norrell 1986, 669).

Segregated housing reinforced the segmentation of labor along racial lines, contributing to the significance of the economic variable in Birmingham's housing pattern. Almost 60 percent of the actual housing segregation in Birmingham for 1940 was a product of black-white socioeconomic differences, compared with an average of 43.3 percent for 15 American cities studied by Karl E. Taeuber and Alma F. Taeuber (1965). For the few black households with high economic status, legal constraints eliminated any opportunity for residential mobility. Despite their socioeconomic status, blacks resided in racially zoned areas that allowed for no racial transition among neighborhoods.

The exploitation of laboring-class blacks because of skin color generated especially high levels of underconsumption in Birmingham. Before World War II Birmingham was so economically strapped that President Franklin D. Roosevelt called it the worst city hit by the Depression, the plight of blacks being especially bleak. At the height of the Depression nearly 75 percent of blacks in Birmingham were without work (Leighton 1937). They occupied 66 percent of the city's slum dwellings but made up only 40 percent of the population (Subcommittee of the Joint Committee on Housing 1948). For the most part, blacks were renters and whites homeowners, fragmenting the working classes not only according to race but class of housing.

Birmingham resisted the role of government as a redistributive agent and was slow to respond to the problems of underconsumption. Although changes occurred in the national political economy in the first half of the 20th century (for example, the New Deal and War on Poverty), the conservative political economy of Birmingham viewed economic growth as the primary cure for underconsumption. Birmingham's commercial and industrial elite had not inherited their social and economic positions but were self-made men who believed that others could be just as successful if they would only work (Lewis and *The New York Times* 1964).

This laissez-faire attitude, coupled with the industrialists' domination of the production process, provided little opportunity for social and economic equality. According to one local historian:

> Being predominately an industrial center, Birmingham found itself led in part by managers rather than owners. The majority of this managerial group, schooled in industrial techniques rather than humanistic studies, viewed Birmingham from an immediate rather than a long-range perspective. Primarily concerned with extracting from the city rather than with building it, they feared that racial change would prove politically and financially detrimental to the interests which they represented (Hamilton 1977, 143).

Civil rights leaders believed Birmingham to be the "toughest nut" in the South to crack. One leader noted, "You've got terribly powerful industrial forces there who have always been in favor of status quo. . . . You've got Negroes competing for jobs in the steel mills and iron mills. You've got a tradition of violence—notorious police brutality, the bombing of homes when Negroes move in. Birmingham is the worst city for race relations in the South" (Martin 1957, 115-116). Reverend Martin Luther King, Jr. was very much aware of this when he chose Birmingham as the focal point of the civil rights struggle in the South:

> In the entire country there was no place to compare with Birmingham. The largest industrial city in the South, Birmingham had become, in the thirties, a symbol for bloodshed when trade unions sought to organize. It was a community in which human rights had been trampled for so long that fear and oppression were as thick in its atmosphere as the smog from its factories. Its financial interests were interlocked with a power structure which spread throughout the South and radiated into the North. The challenge to nonviolent, direct action could not have been staged in a more appropriate arena (King 1963, 45).

The Civil Rights Era

The civil rights struggle of the 1960s was initiated when the Reverend Fred Shuttlesworth, taking cues from the sit-ins at Greensboro, North Carolina, organized sit-ins at stores in downtown Birmingham. This new round of bouts for civil rights placed Birmingham in the annals of civil rights history and changed the course of race relations in the United States. Local college students formed the Anti-Injustice Committee and demanded: desegregation of lunch counters, restrooms, and drinking fountains; the hiring of blacks as clerks and sales personnel; and general upgrading of black employees from solely menial jobs. *New York Times* reporter Harrison Salisbury (1960) described the racially repressive climate that he discovered in Birmingham in the spring of 1960, emphasizing the unwillingness of black and white leaders to speak out for fear of economic, social, and physical reprisals.

The successes of the civil rights movement were aided by significant changes in the city's political economy. The political-economic ties that depended on policies of strict racial segregation to make Birmingham "the industrial city of the South" were severed at the outset of the post-industrial era. The shift from managing race relations in the interest of industrialists toward racial justice was initiated, in part, as a reaction to the violent beating of the Freedom Riders on Mother's Day 1961. City newspapers spoke out against Connor and the City Commission for allowing the incidence to occur, accelerating the drive for governmental reform.

But a more significant reaction to the beating of the Freedom Riders occurred in Tokyo, where Birmingham business leaders were attending the International Rotary Convention (Vann 1981). The event, displayed prominently in Japanese newspapers, created tremendous negative publicity for Birmingham in Japan and throughout the world. In time, this unfavorable publicity made it difficult for the city to compete and Birmingham was unable to attract investments to replace the declining manufacturing sector. When the riot over the Freedom Riders erupted in 1961, one company that planned to build a new plant in the city decided not to and built its plant in Tennessee. Investment in new plants and expansion declined from $52 million in 1960 to $11 million in 1962 (*Wall Street Journal* 1961). While the number of jobs in the nation increased, the number in the Birmingham metropolitan area declined from a high of 246,000 in 1957 to 228,000 in 1963, contributing for the first time to a loss of population. The area ranked last among 13 southeastern metropolitan areas in both numerical and percentage gains in non-agricultural employment in the 1950s.

In the central business district, retail sales were off 40 to 50 percent in 1963. The Federal Reserve Bank reported in the four-week period ending 18 May 1963 that the city's retail sales were down 15 percent compared with the same period in 1962, while elsewhere sales were up 7 percent in Atlanta, 10 percent in New Orleans, and 15 percent in Jacksonville. There was also more commercial space for rent in the city than there had been during the Depression (Sutton 1963). The city's business and commercial elite, hit hard by these declines, organized a movement to change the form of government dominated by Connor.

Among the delegates at the Tokyo convention from Birmingham was Sidney Smyer, once spokesman for the American States Rights Association during the pre-civil rights era, but by 1960 a respected lawyer, real estate broker, and incoming president of the Birmingham Chamber of Commerce. Smyer with the Chamber of Commerce formed the Senior Citizens' Council, speaking for the city's commercial and corporate elite. The Senior Citizens' Council represented the employers of about 80 percent of Birmingham's labor force, including United States Steel. In December 1961 this group held its first meeting with black leaders, seeking to establish channels of communication between the races.

In one of its first actions the Senior Citizens' Council asked the Bar Association to study different forms of government and to select one that would best serve Birmingham and provide greater representation than the commission form of government (Cotman 1989). In August 1962 the Birmingham Citizens for Progress Committee organized a petition to request that the November election include a vote on whether to maintain Birmingham's Commission form of government or to switch to the Manager-Council or Mayor-Council form of government. The petition was successful. In the resulting election the Mayor-Council form won by a margin of only 700 votes. Connor's supporters filed a suit to invalidate the 1955 legislative act that allowed a Mayor-Council form of government, but the act was upheld by the court.

The first mayoral election ended in a runoff between Connor and Albert Boutwell. Boutwell, as member of the state legislature, chaired the Interim Committee on Segregation. Like Smyer, he was once a spokesman for the American States Rights Association, but by 1962 he was considered a moderate segregationist. Boutwell won the runoff election by 8,000 votes, but Connor and members of the city commission refused to vacate their positions, contending, "We were elected for four years and we're not leaving" (Vann 1981, 27). Thus, Birmingham had two governments with two mayors for almost 40 days, until the State Supreme Court upheld the validity of the new government.

Although the new city government never agreed with the Reverend King's method of nonviolent civil disobedience, it authorized the organization of the city's first bi-racial committee, the Community Affairs Committee, to bridge the gap between the white and black communities and repealed all segregation laws. Reverend King called the desegregation of Birmingham "the most significant victory for justice that we have seen in the Deep South" (Cotman 1989, 60).

Post-Civil Rights Era

Federal legislation opened doors of public facilities to blacks in the mid-1960s. John F. Kennedy's executive order, which mandated fair employment practices in companies holding federal contracts, forced United States Steel to use its authority to end discrimination. Title VII of the Civil Rights Act of 1964 made employment discrimination illegal and black workers became more aggressive at challenging discrimination in the workplace. Because of challenges, black steel workers in Birmingham received promotions and back pay as part of a 1973 consent decree that settled dozens of employment discrimination suits throughout the steel industry. In 1969 the first black was elected to the City Council, and in 1971, when the city was named an "All-American City" by *Look* magazine, a second black was elected to the council.

Taking advantage of the federal government's requirement in the 1970s for citizen participation in community development, blacks forced the city to go beyond the requirements to develop the most comprehensive neighborhood-based citizen participation program in the United States. The neighborhood program provided the political base to elect the city's first black mayor in 1979. While citizen participation and civil rights empowered blacks politically, there was no economic empowerment. The average unemployment rate among blacks in the 1980s was twice the rate for whites. Most jobs that blacks were now gaining access to because of civil rights legislation were disappearing as Birmingham moved from the industrial to the post-industrial era.

Conclusion

In the first half of the 20th century Birmingham was the industrial city of the South. It obtained this distinction by making use of a racially repressive labor system that made it the focal point of the civil rights struggle in the United States. Nowhere was the tension as great as in Birmingham, which set the stage for the fiercest struggle for civil rights in the United States. This struggle was aided by changes in the city's political economy which led to a less repressive racial order following the civil rights movement. Only when the city's elite no longer supported the old industrial order could civil rights leaders make gains in desegregating the city. But most jobs that blacks were gaining access to because of the civil rights movement were disappearing as Birmingham moved into the post-industrial era. High rates of black unemployment coupled with high levels of segregation persist.

References

Cotman, John Walton. 1989. *Birmingham, JFK, and the Civil Rights Act of 1963: Implications for Elite Theory.* American University Studies, Series X, Political Science, Vol. 17. New York: P. Lang.

Hamilton, Virginia. 1977. *Alabama: A Bicentennial History.* New York: Norton.

Hawley, Langston T. 1955. Negro employment in the Birmingham Metropolitan Area. In *Selected Studies of Negro Employment in the South,* ed. National Planning Association, pp. 213-328. Washington, DC: Committee of the South.

King, Martin Luther. 1963. *Why We Can't Wait.* New York: Mentor Books.

Leighton, George R. 1937. Birmingham, Alabama: The city of perceptual praise. *Harper's* 175:225-242.

Lewis, Anthony, and The New York Times. 1964. *Portrait of a Decade: The Second American Revolution.* New York: Random House.

Marshall, Ray. 1967. *Labor in the South.* Cambridge: Harvard University Press.

Martin, John Bartlow. 1957. *The Deep South Says Never.* New York: Ballantine Books.

Norrell, Robert J. 1986. Caste in steel: Jim Crow careers in Birmingham, Alabama. *Journal of American History* 73:669-694.

Salisbury, Harrison E. 1960. Fear and hatred grip Birmingham: Racial tension smoldering after belated sitdowns. *New York Times* 12 April:1, 28.

Subcommittee of the Joint Committee on Housing. 1948. *Slum Clearance Report.* 80th Congress of the United States, 2d Session. Washington, DC: U.S. Government Printing Office.

Sutton, Claude. 1963. Birmingham talks reach an accord on ending crisis—Dr. King accepts pledges from whites after cutting demands of negroes—four provisions in pact—store integration, job aid, biracial panel, and bid to free marchers promised. *New York Times* 10 May:1, 14.

Taeuber, Karl E., and Taeuber, Alma F. 1965. *Negroes in Cities: Residential Segregation and Neighborhood Change.* Chicago: Aldine.

Vann, David. 1981. *Events Leading to the 1963 Change from the Commission to the Mayor-Council Form of Government in Birmingham, Alabama.* Birmingham: Center for Urban Affairs, University of Alabama.

Wall Street Journal. 1961. Many southerners say racial tension slows area's economic gains—St Louis firm delays move to Alabama; industry hunt lags in Little Rock—race and recession tie in? 20 May:1, 21.

Woodward, C. Vann. 1951. *Origins of the New South, 1877-1913.* Baton Rouge: Louisiana State University Press.

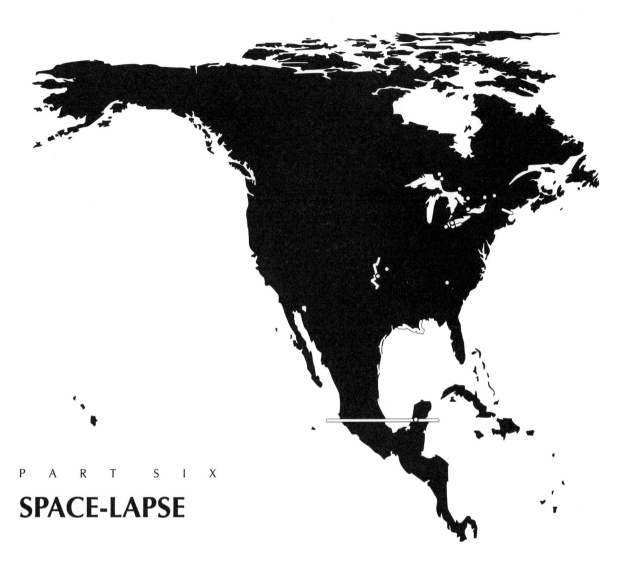

SPACE-LAPSE

The complexity and extent of geographical space often require abstraction of information in the spatial dimension. A traverse line or a selected sample of points may represent the information base for speculation and analysis about the character of geographical distributions. This is often true for studies on the geography of fauna and flora. The geometric arrangement of points or lines on a map inevitably attracts the curiosity of geographers. For example, the structural alignment of Maya ruins offers clues about the scientific accomplishments of past civilizations. The point pattern of towns on maps of coastal Maine sheds light on the development and areal organization of regional settlement systems. Boundaries of nature may coincide strikingly with the human organization of space for livelihood in Oklahoma. Current-day traffic patterns and peculiar bends in a road may reflect surveying practices of a previous century. And the locational origins of shoppers expose the influence of new marketing strategies. All these examples show breaches in the continuity of connection and gaps in information about intervening spaces. Special analytical tools, pioneered by geographers, help describe and analyze such patterns. Examples include centrographic analysis for generalizing about patterns of points (for instance, population distributions) and fractal analysis to discover the dimensional character of lines (for example, coastlines).
—*Robert A. Muller, Louisiana State University*

Pre-Columbian Alignments in the Yucatán: Edzná

Vincent H. Malmström
Dartmouth College

The Lines of Edzná

Edzná first came to my attention in early 1976 as the result of a deduction prompted by a computer analysis of the Maya calendar. My research revealed that the Maya seem to have reformed their calendar around A.D. 40 by moving their "New Year's Day" from 13 August to 26 July. Because they used the passage of the zenithal sun to calibrate the new date, they were obliged to have chosen a location along the parallel of 19.5°N—a line that intersects only the site of Edzná as it crosses the Yucatán Peninsula.

I found very little written about the site, for Andrews' book on Maya cities was just being published (Andrews 1975) and Matheny's excavations had yet to begin (Matheny 1976). The only two citations I found were an observation from J. Eric Sydney Thompson (1950) that a one-day correction in the calendar may have occurred there about the year 672—although he later reversed himself on this matter—and a pointed reminder on a map by the National Geographic Society (1972) that the site was of "Late Classic" (that is, A.D. 600 to 900) origin. Although the first citation seemed to support the notion that the place had some astronomical importance, the second definitely ruled out the calendar reform suggested by my computer study, because a "Late Classic" origin put the city's founding at least six centuries too late. Confronted with such contradictory evidence, I withheld all mention of Edzná and the supposed calendar reform from the article that I published in an astronomical journal (Malmström 1978).

Although Edzná was first reported to the scientific community in 1927, the site received little study until the late 1950s when a team from the National Institute of Archaeology and History began a process of partial excavation and restoration (Matheny et al. 1983). Since only one structure of any size—a five-story pyramid called Cinco Pisos—was in good enough condition to date for architectural style, Edzná was considered to be a relatively minor place. In 1968, however, after a University of Oregon architectural group led by George F. Andrews surveyed the site extensively, they concluded that Edzná was a "very large and important center with definite 'urban' characteristics" (Andrews 1975).

While continuing my research in Mexico in 1978, I chanced to meet Professor Ray T. Matheny and informed him of my dilemma regarding the age of Edzná, only to have him reply that there was no dilemma. His excavations revealed that Edzná dated back to about 150 B.C. and was perhaps the oldest major Maya site yet discovered, having a population approaching 20,000 at its peak. He encouraged me to investigate the site for astronomical alignments, but cautioned me that an astronomer he had consulted found none (Matheny 1976).

Within a week of my encounter with Matheny, a visit to Edzná confirmed two hypotheses and occasioned the discovery of another totally unexpected relationship in the spatial organization of the site. At the base of the Cinco Pisos pyramid I found an ingeniously designed gnomon (the index of a sundial) fashioned from a tapered shaft of stone surmounted by a disc having the same diameter as the base of the shaft. At noon on the day of the sun's zenithal passage (26 July), the disc at the top of the shaft enveloped the entire column in a cylinder of shadow, whereas on any other day of the year, a band of sunlight fell across it. Here, then, was strong circumstantial evidence that the astronomical exactitude Thompson postulated for Edzná was indeed possible, and that such a device allowed this ancient and important ceremonial center to serve much the same function in the calendrical year of the Maya as Greenwich does for diurnal time-keeping in our modern world.

The second confirmation came when I found that the meticulously laid-out ceremonial center, whose alignments Andrews (1975) noted but did not understand, was oriented to the sunset position on 13 August. This alignment I explained in 1975 at Teotihuacán, the great pre-Columbian metropolis about 50 kilometers to the northeast of Mexico City, and subsequently found repeated at dozens of other sites throughout the Mesoamerican realm (Malmström 1981). Because the 13 August alignment commemorated the "beginning of time" according to the Goodman-Martínez-Thompson correlation of the Maya calendar, it represented the most sacred day in the Mesoamerican year. To find the oldest major urban center of the Maya oriented to the same magico-religious point on the horizon as its larger contemporary on the Mexican plateau was striking proof of the communality of culture that already prevailed across the region by the time of the birth of Christ.

However, from examining the spatial relationships of the various structures at Edzná, it was apparent that the ruins of a large pyramid some 300 meters to the northwest of the Cinco Pisos (Matheny dubbed it *La Vieja* because of its antiquity) must have played a special astronomical role in the life of the city. It is the only structure which is high enough to intersect the otherwise-totally featureless horizon as seen from the top of the imposing five-story building. It appeared to have been constructed to fix some especially significant alignment as viewed from that structure; but, because the Northwest Pyramid is situated at an azimuth of 300° from Cinco Pisos, it could not have commemorated a solar position. This is because the maximum azimuth for the summer solstice sunset, as viewed from the latitude of Edzná, is 295°. But, if it was not a solar alignment, what could it be? The fact that the Northwest Pyramid lies exactly 5° beyond that point is, indeed, the clue. The orbit of the moon bears precisely this same relationship to the ecliptic of the sun. It would appear, therefore, that the Northwest Pyramid was constructed to fix the maximum northerly still-stand of the moon—the knowledge of which position was absolutely essential to predicting eclipses. Since the moon reaches this position only once every 18.61 years, a long period of observation was necessary to pin down this point with precision. This may be why the earliest lunar inscriptions of the Maya do not pre-date the middle of the fourth century A.D. (Lounsbury 1978).

Figure 1. Edzná, south of Mérida, on the Yucatán Peninsula.

Edzná appears to have functioned as the Maya's earliest astronomical center by virtue of their recognition and demarcation of three lines in the landscape. The first of these was the parallel of 19.5°N (Figure 1), even though the notion of a spherical earth with a coordinate grid-system was totally unknown to them. As observers of nature, what mattered to the Maya was that they could define the zenithal transit of the sun with precision. Because this happened to occur at the latitude of Edzná on 26 July, they adopted this date as the beginning of their new year—a fact recorded by a Spanish prelate who was in the region in the 1560s (de Landa 1983).

However, true to a tradition already well established within the so-called Olmec civilization, of which they were the principal heirs, the Maya continued to commemorate the date on which the present cycle of the world supposedly began. Likewise calibrated by the zenithal transit of the sun—this time at Izapa, the cradle of their calendar's origins near the Pacific coast of southern Mexico—this event explains the orientation of the entire city of Edzná to the sunset position on 13 August, that is, an azimuth of 285.5° (Malmström 1973; Figure 2).

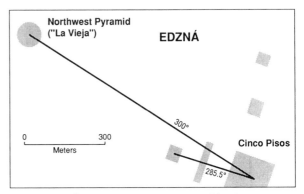

Figure 2. Key orientations at Edzná.

Finally, the precision with which the azimuth between the Cinco Pisos and the Northwest Pyramid was fixed suggests that Edzná likewise served as the earliest lunar observatory in the New World, and that it may have been here that the Maya finally established a means of predicting the eclipses that they, like superstitious peoples everywhere, always found so terrifying (Figure 2).

Edzná: A Point in Space

But why was Edzná located specifically where it was? What factors prompted the Maya to found what was to become their first major urban agglomeration at this particular location?

Unlike the Olmec, who appear to have consciously avoided the Yucatán Peninsula (Bernal 1969), Maya agriculturalists began pressing into the region from the southwest early in the second century B.C. They appreciated that the greatest challenge confronting the would-be farmer in this youthful karst area was the scarcity of surface water, a deficiency exacerbated by the monsoonal climate. Moreover, the sporadic downpours of the summer rains scoured the surface of the tabular limestone of most of its soil cover, washing it down into solution basins which the Spanish called *aguadas*, or washes. Edzná, as it turns out, was on the eastern edge of the largest aguada found anywhere in the Yucatán. A basin measuring almost 90 kilometers north-to-south by more than 20 kilometers east-to-west, the aguada of Edzná is the most extensive area of cultivable soils the Maya ever settled. Today it once again is the site of an ambitious agricultural development scheme.

Thanks to the fine-grained soils on the floor of the basin, the run-off from the summer rains tends to accumulate there, seasonally forming a sizable lake. This combination of productive soils and a more enduring water supply encouraged the Maya to develop an extensive network of radial canals reaching out across the aguada in all directions from Cinco Pisos (Matheny 1976). The earth excavated from the canals was mounded into low terraces on either side of the waterway to form the counterpart of the *chinampas* (floating gardens), which the Spanish found the Aztecs using to grow food for their metropolitan center a millennium and a half later. The canals themselves served both as arteries of movement for transporting foodstuffs to the city center but also, Matheny believes, for the production of fish as a source of protein. Thus, in many ways the premier Maya urban center was essentially a smaller version of both its contemporary, Teotihuacán, and the later Aztec capital, Tenochtitlán.

In theory, of course, Edzná could have been located anywhere around the perimeter of its aguada. So, what specific site factor was responsible for its having been positioned where it was? Without doubt, its Cinco Pisos pyramid constituted the very heart of its religious and astronomical precinct and, as such, must have served as its commercial center as well. The Andrews survey contended that Cinco Pisos was built atop a very large man-made platform (Andrews 1975), whereas Matheny states it was constructed on a natural rock outcrop (Matheny 1976). In either case, its location was chosen for the solid base it provided in the midst of a vast basin floored with productive but seasonally waterlogged clay soils. Once this node had been established, a network of converging canals was constructed to provide goods and services to this central place—an undertaking as momentous in its own right as the erection of the Pyramids of the Sun and Moon at Teotihuacán (Matheny 1976). But, before long, the Maya priests seeking to unravel the secrets of the heavens came to appreciate the site's wide-ranging and unencumbered view of the distant horizon likewise. Thus, to its role as the first major urban agglomeration of the Maya, Edzná could now add the distinction of becoming their earliest astronomical center as well—a fact that we have only belatedly come to recognize.

References

Andrews, George F. 1975. *Maya Cities: Placemaking and Urbanization.* Norman: University of Oklahoma Press.

Bernal, Ignacio. 1969. *The Olmec World.* Berkeley: University of California Press.

de Landa, Diego. 1983. *Relación de las Cosas de Yucatán.* Mérida: Ediciones Dante, S.A.

Lounsbury, Floyd G. 1978. Maya numeration, computation, and calendrical astronomy. *Dictionary of Scientific Biography* 15:759-818.

Malmström, Vincent H. 1973. Origin of the sacred 260-day Mesoamerican calendar. *Science* 181:939-941.

———. 1978. A reconstruction of the chronology of Mesoamerican calendrical systems. *Journal for the History of Astronomy* 9:105-116.

———. 1981. Architecture, astronomy, and calendrics in Pre-Columbian Mesoamerica. In *Archaeoastronomy in the Americas*, ed. R. W. Williamson, pp. 249-261. Los Altos, CA: Ballena Press, Center for Archaeoastronomy.

Matheny, Ray T. 1976. Maya lowland hydraulic systems. *Science* 193:639-646.

——— , **Gurr, Deanne L., Forsyth, Donald W., and Hauck, F. Richard.** 1983. Investigations at Edzná volume 1, part 1: The hydraulic system. *Papers of the New World Archaeological Foundation*, Vol. 1, No. 46. Provo, UT: Brigham Young University.

National Geographic Society. 1972. *Archaeological Map of Middle America: Land of the Feathered Serpent.* Washington, DC.

Thompson, J. Eric Sydney. 1950. *Maya Hieroglyphic Writing: An Introduction* (1st edition). Norman: University of Oklahoma Press.

Dimensional Regions of Coastal Maine

Allen K. Philbrick
University of Western Ontario

Dimensions do not exist separately in nature, but they certainly are useful in analyzing the basis for recognizing regional distinctions. Take the coast of Lincoln County, Maine, for example. The county presents an interface of sea and land like the fingers of two opposing hands. The fingers of land and opposite fingers of sea between them are 32 kilometers (20 miles) long, or more—rocky, forested peninsulas ruggedly alternating with equally long penetrations of the Atlantic Ocean intermingled with numerous islands (Figure 1A).

Known geologically as a drowned coastline caused by some combination of glacial erosion, post-glacial oceanic rise, or post-glacial reemergence (Bloom 1963), the land-sea contact zone is the context for three sets of human behavior. These are marine, landlocked, and a Janus-like interaction that looks both ways. This paper demonstrates how points, lines, and surface dimensions may be used to isolate patterns of human artifact, infrastructure, and perceived environment which define three dynamic regions of coastal Maine.

Surfaces, Lines, and Points

Start with the two dimensional surfaces of the water and land interface on the maps in Figures 1A and 1B. The two surfaces are joined by a truly intricate coastal line. Other lines representing human usage are the artifacts of transport shown on the maps in Figures 1C and 1D on land and beneath the water. Underwater estuarine sea- and riverbeds are incised by tidal channels followed on the surface by a wide range of marine craft from skiffs to ocean shipping. Channels 1 meter (3 feet) or more deep are shown. These comprise the tidewater routes from the open sea to harbors and landing sites for towns and villages. Over the land

surface of the mainland, islands, and peninsulas, the road network interconnects the settlement pattern of Lincoln and adjacent counties. The road net brings the land into contact with tidewater at local points that maximize access for both surfaces. At bridge points the two transport systems intersect three-dimensionally, land over water. This set of nodal points allows freedom of movement without serious interruption over both surfaces (Janelle 1964). Many other points represent the foci of activity in villages and small cities (Figures 2A and 2B).

Marine and Landlocked Orientations

The settlement pattern is subdivisible into two groups by position and orientation with respect to tidewater. The two orientations may be called marine and landlocked. In this section of the Maine coast it is, perhaps, not surprising how far inland access to tidewater reaches from the open sea.

Over the history of European penetration of eastern North America this orientation toward salt water has been important to the mentality and outlook of those so positioned. Similarly, those in a landlocked position tend to have settled there later and to have acquired a more landward mindset. Their orientation has become committed, while still aware of the nearness of the sea, by environmental perception of forest and farm, to products of the land.

Dimensions Combined into Regions

Dimensional descriptions combined in the action of the time dimension become real places. For, it is the activities of the occupants on the land and sea that

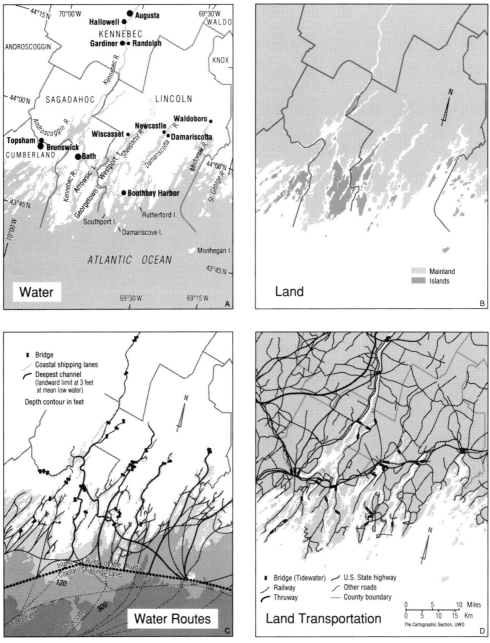

Figure 1. A. Identification and water surface. B. Land surface. C. Water routes—line pattern and bridge points. D. Land transportation—line pattern and bridge points.

turn the static patterns on the maps into three dynamic time-spatial behavioral regions of this section of coastal Maine. Connect the seaward limits of the points representing landlocked villages by an arbitrary line, and the landward limit of tidewater villages and cities by another. When this is done on the map in Figure 2C a transitional region that exhibits the characteristics of both appears between marine and landlocked regions.

It is precisely this transitional overlapping region that had the rail line and the historic first "post-route" of the new nation, now U.S. Highway 1, running through the middle of it. These major arteries connected the principal cities. They are the major market and shire-towns that serve the tidewater peninsular and island region seaward and the landlocked region landward. The linear and point nodalities of the transitional region made of it the core subregion of Lincoln County when it originally encompassed this entire section of coastal Maine.

Each island and peninsular-end must look to the highway-bridge connectivity of Route 1-oriented market places for goods and services. From west to east these places are: Brunswick at the falls of the Androscoggin River, home of Bowdoin College and still an industrial town; Bath, shire-town of Sagadahoc County, home of the Bath Iron Works famous for ship-building from the days of the tall-masted China clipper ships, on the Kennebec River; Wiscasset, shire-town of Lincoln County since 1760, formerly an international colonial port with its own customs house, on the Sheepscot River; the twin centers of Newcastle-Damariscotta, regional shopping centers at the falls of the Damariscotta River; and Waldoboro, an early German settlement on the Medomak.

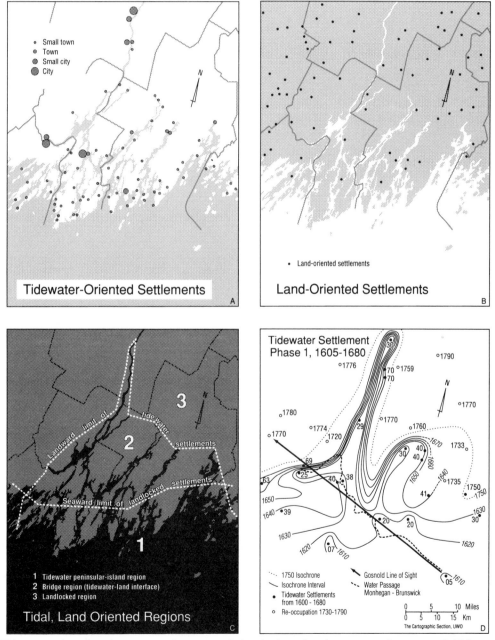

Figure 2. A. Tidewater-oriented settlements—point pattern. B. Land-oriented settlements—point pattern. C. Tidewater peninsular-island, bridge (tidewater-land interface), and landlocked region. D. First tidewater phase settlement, 1605 to 1680.

Upriver on the Kennebec, tidewater penetrates 64 kilometers (40 miles) to the state capital of Augusta, also shire-town of Kennebec County, past other market centers at Gardiner and Hallowell. Of the 35 bridges marking the intersection of tidewater-land transport nets, 26 are in this core transition region. It is fitting that this be so, for it emphasizes the bridge nature of the zone itself, Janus-like, serving both the landward and seaward regions it connects.

The circumstance of alternate fingers of peninsula and estuary means in practical terms that if you are at the tip of a peninsula in a car for transportation, you will have to travel 60 or 80 kilometers (40 or 50 miles) to reach the neighboring peninsula ends only 10 or 11 kilometers (6 or 7 miles) distant by water. Or by contrast from the landward end up the Sheepscot estuary at Wiscasset, for example, the distance by water to Newcastle is 60 or 80 kilometers (40 to 50 miles), which by land is a scant 10 kilometers (6 miles). This explains and enhances the focality of the market towns at the landward ends of the coastal estuaries, as a group connected by major highways. Accordingly, each such market town is in a position to compete for the business of two peninsulas and an inland service area, as well as the entire coastal region. Hence the larger size of those named centers at the strategic locations indicated.

In common sense terms the dimensional device has accurately described this section of coastal Maine regionally. For, it is true that landward from access to tidewater, the landlocked interior is mainly forested with small, cleared areas of farming, which is dependent upon the tidewater-core market centers for services and access for its products. It is equally true that the marine peninsular-ends and islands are isolated and special. They are the true ocean-recreational focus of marine-oriented tourism. The core's regional market centers thrive on the business generated by the wants of the marine vacationland of islands and peninsulas seaward of Route 1.

Historic Confirmation

The history of this region confirms this picture. A British Captain Gosnold's landing on Monhegan Island

Figure 3. Lupine on the banks of the Dyer River, at its confluence with the Sheepscot at Sheepscot Village. Sketch by Allen K. Philbrick.

Christmas day of 1602 is recorded. His sighting of snow-capped peaks (possibly Mount Washington) and his mission to explore "the main" led him by pinnace (Figure 2D) to the falls of the Androscoggin at the site of present-day Brunswick. Many European fishermen used the islands for fresh water and wood, and as sites for salting cod, secure from all but minimal contact with local Indians. Monhegan, Damariscove, and other islands are documented in this practice.

The isochrones of settlement on the map in Figure 2D reveal two phases, from 1605 to 1680, and after 1750. The struggle for possession of New England between France and England during the so-called French and Indian War saw the 1680 retreat of the settlers of the Massachusetts Bay Colony from this region to Boston for about 50 years. For example, Job Averill of Massachusetts started a mill at the reversing tidal falls of the Sheepscot River in 1730, which marked the return to that point to resume the development of the village of that name. He did this by arrangement with the proprietors of the Duke of York Patent of 1637 (Bradford 1952). The last isochrone on the map is for 1750, a full decade before the fall of Québec on the Plains of Abraham in 1760 led to the Treaty of Paris of 1763, ending the military conflict between the French and English in North America. By this time the marine character of the peninsular-end and island region of this part of coastal Maine had been well anchored in the previous century.

This dimensional portrayal illustrates how point (0), line (1), surface (2), volume (3), and action (4) dimensions can be the basis for delimiting significant regional analysis. As a method of thinking about and sorting the objects in the dynamics of the real world, dimensions help give shape to the behavioral regions of Coastal Maine (Figure 3). But, in the final analysis it is the people themselves over time, the fourth dimension, who have molded the land and watery substance of these three regions into one.

Acknowledgment

I thank Patricia Chalk, head of the Cartographic Section, University of Western Ontario, for transforming the original maps into their current form.

References

Bloom, A. L. 1963. Late Pleistocene fluctuations of sea level and postglacial rebound in coastal Maine. *American Journal of Science* 261:862-879.

Bradford, William. 1952. *Of Plymouth Plantation, 1620-1647* (copied into Church records in 1640s, found and published in England in 1856, returned to Boston 1897 by ecclesiastical decree). New York: Alfred A. Knopf.

Janelle, Donald G. 1964. *A Geographical Appraisal of the Bridge.* M.A. Thesis, Michigan State University, East Lansing.

The Meridian of the Plains: A Natural Boundary in Oklahoma

Stephen J. Stadler
George O. Carney
Mark Gregory
Oklahoma State University

Boundaries between adjoining natural regions are known as ecotones and can be distinct enough to approximate lines. The ecotone between the eastern deciduous forests and the Great Plains grasslands of the United States is a classic example of such a linear division. The marked east-to-west tonal differences in Figure 1 distinguish forests in the east from grasslands in the west. Within Oklahoma, the forest/grassland ecotone is a particularly abrupt transition in vegetation.

The Ecotone as a Natural Boundary

The boundary in Figure 1 marks several natural transitions. To the west are the Great Plains. Originally a mixture of tall and short prairie grasses consisting of big bluestem, sideoats grama, buffalo grass, Indian grass, and others, this landscape is now dominated by winter wheat in northern Oklahoma and rangeland and cotton in southern Oklahoma. Boundless grasslands seem to be prevalent in outsiders' perceptions of Oklahoma, but the natural mix of species is now all but gone; these lands are used today for crops or grazing.

Antithetical to stereotypes of Oklahoma, Figure 1 shows large portions of eastern Oklahoma covered by dark tones—indicative of forest. Directly to the east of the ecotone is a north-south trending belt known to American explorers as the Cross Timbers (Figure 2).

The native Cross Timbers species were post oak and blackjack oak interspersed with black hickory and some prairie tracts. These trees have been described as stunted because they rarely exceed 5 meters in height in the uplands. In addition, trees grew close together with branches near the ground. Armed with an undergrowth of vines and briars, the native Cross Timbers were difficult to traverse by foot or by horse. Immediately east of the Cross Timbers were substantial areas of tall-grass prairie whereas easternmost Oklahoma was covered with "full-sized" trees of the eastern deciduous forests.

The ecotone is approximately 200 kilometers to the east of the average position of Webb's (1931) famous humid/subhumid climatic boundary. Along the ecotone, precipitation averages 800 millimeters per year but has dramatic interannual swings. To the east are found greater precipitation averages and to the west are lesser precipitation averages. The ecotone is noted for frequent air mass conflict. Corcoran (1982) has suggested that its location is set by the plant stress caused by the striking east-to-west transition to greater summer vapor pressure deficits.

The underlying geology is mainly that of sedimentary beds with gentle westward dips. To the east of the ecotone, the surface formations are Pennsylvanian sandstones and shales that form a series of east-facing cuestas (Goodman 1977). Locally, the pattern is more convoluted and includes structural domes, the sites of major oil discoveries in the 1910s. To the west of the ecotone are Permian red beds which do not include cuestas. Far from being flat, the topography in the vicinity of the ecotone has as much as 40 meters of local relief and can be described as rolling to steeply rolling. In general, the topography to the east is rougher

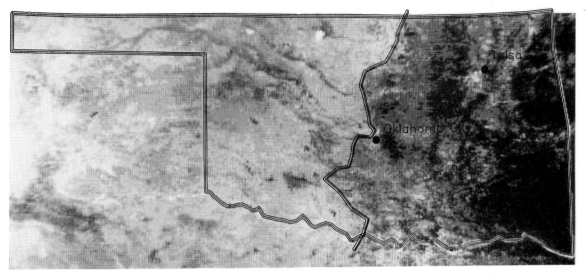

Figure 1. The ecotone. The forest/grassland boundary in Oklahoma is marked on springtime imagery from the visible channel of a National Oceanographic and Atmospheric Administration polar orbiting satellite.

than the topography to the west. Five major rivers cut west-to-east paths through the north to south grain of the topography (Figure 2).

In the Cross Timbers, sandstone soils were preferred sites for dense forests and shales created thinner soils associated with open forests and prairie enclaves. West of the ecotone, the Permian parent materials generated red, clayey soils, unsuited to sustaining a forest.

A Native American Boundary

The ecotone posed as a boundary to Native Americans for many centuries before it was seen by Europeans. The traditional range of the Comanches and Kiowas extended from the Rockies east to the Cross Timbers. These Plains Indians rarely entered the Cross Timbers because of the lack the buffalo on which they based

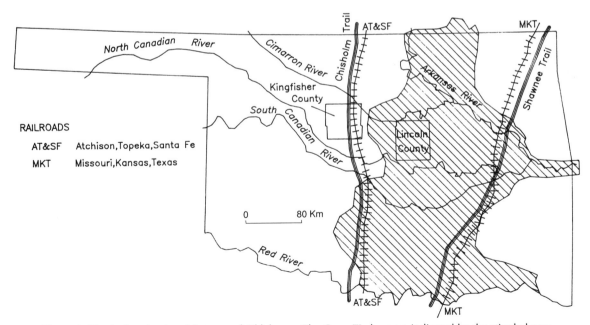

Figure 2. Physical and cultural features of Oklahoma. The Cross Timbers are indicated by the stippled area.

their existence (Foreman 1947). Inhabitants of the Cross Timbers had a village farming culture.

Starting in the 1830s, the United States government promulgated the relocation of Native American groups into Oklahoma. The infamous forced marches of the "Trail of Tears" brought the Five Civilized Tribes from the southeastern United States to eastern Oklahoma. The Cross Timbers were largely unoccupied and acted as a buffer between the sedentary Five Civilized Tribes and the nomadic "wild tribes" of the Great Plains. The government recognized the ecotone as an official boundary. For instance, one treaty proclaimed the signatories as having unlimited access "to hunt and trap in the Great Prairie west of the Cross Timbers" (U.S. 1846). However, the eastern Oklahoma tribes did not venture across the boundary (Hafen and Rister 1941).

A Boundary for Explorers

The term "Cross Timbers" emerged in the 1820s following the first explorations through the ecotone and possibly alludes to the north-to-south trend of the forested belt that crosses the east-to-west valleys of the major streams. Alternatively, it might refer to the experience of the westward traveler who encountered tall-grass prairie in eastern Oklahoma only to find it necessary to cross another forest before reaching the Great Plains (Costello 1969).

The Cross Timbers presented a considerable barrier to the explorers. Since no navigable rivers penetrated the Cross Timbers, overland travel was necessary. Recounting his 1832 Oklahoma travels, Washington Irving observed:

> The Cross Timbers is about forty miles in breadth and stretches over a rough country of rolling hills, covered with scattered tracts of post-oak and black-jack. . . . It was like struggling through forests of cast iron (Irving 1955, 95-96).

Likewise, the explorer Marcy commented:

> This forms a boundary-line, dividing the country suitable for agriculture from the great prairies. . . . It seems to have been designed as a natural barrier between civilized man and the savage (Webb 1931, 158).

On returning from an expedition to the Great Plains, Dodge grandiloquently commented:

> Oklahoma is an uncharted sea. The stars and a few physical features are the only guides. The Cross Timbers

is the Meridian of Greenwich to the navigator of the plains (Foreman 1947, 36).

A Boundary for Oklahoma's White Settlement

Oklahoma was a gigantic reserve for Native Americans, but that did not prevent increased white involvement in the post-Civil War era. Texas specialized in cattle production, but there were no railroads to carry the cattle to eastern markets. Consequently, trails through Oklahoma were used to drive Texas cattle to railheads in Kansas and Missouri. The Cross Timbers posed difficult travel and infinite possibilities for the loss of cattle into the thick undergrowth. Thus, the two most important trails, the Shawnee and the Chisholm, were routed to the east and west edges of this area (Figure 2).

The Cross Timbers also thwarted early attempts to build a railroad from east to west. In 1869, the north-south Missouri, Kansas, and Texas line was sited in the grasslands east of the Cross Timbers. In the late 1880s, the Atchison, Topeka, and Sante Fe ran a north-south line in the grasslands just west of the ecotone (Figure 2). Rail access became a great force in the public's demand to open lands as yet unassigned to any tribe.

The first legal settlers were allowed into the Unassigned Lands of central Oklahoma during the famous Land Run of 1889. This new Oklahoma Territory was enlarged from Indian lands in response to continuing pressure for white settlement. In the territorial era of 1889 to 1907 the greatest settlement pressure was on the grasslands west of the Cross Timbers (Debo 1949). The advent of the steel plow made it feasible for the white settlers to break the prairie sod and this was much easier than clearing the dense growth in the Cross Timbers. Thus, lands east of the boundary were developed into farms and ranches at a much slower rate than their western grassland counterparts. The settlement landscapes varied dramatically across the boundary. To the east, the original houses of Indians and whites were log cabins (Figure 3) and the occupants practised mixed farming. To the west, many houses were built of sod (Figure 4), and wheat and grazing dominated the landscape.

At the beginning of the 20th century, the Cross Timbers forests were partially cleared for cotton and tobacco agriculture. Debo (1949) termed this the "Plowman's Folly." The combination of thin soils with open-row crops soon took its toll in erosion. The amount of cropland has been declining for a half century in the Cross Timbers while just to the west of the ecotone it is little changed.

Figure 3. A home east of the ecotone. Logs were the primary building material in eastern Oklahoma. Reprinted with the permission of the Western History Collections, University of Oklahoma Library.

A Population Boundary

The Cross Timbers languished economically until the oil booms of the 1912-to-1930 era. Of Oklahoma's 23 "giant" oil fields (fields with an ultimate recovery of 100 million barrels or more), a dozen were in the Cross Timbers. Minor urban centers developed near these finds. But, unlike the slow but steady growth of population centers west of the boundary, centers in the Cross Timbers all stabilized or declined after the boom era.

Oklahoma's settlement pattern still reflects the presence of the forest/grassland boundary. The Cross Timbers remains a zone of little urbanization. Oklahoma City straddles the boundary, while Tulsa is east of the Cross Timbers (Figure 2). The settlement landscapes remain remarkably different. An example can be drawn between Lincoln County in the Cross Timbers and Kingfisher County to the west of the ecotone (Figure

2). The counties are comparable in size and overwhelmingly rural. Lincoln County's economy is based on mixed farming and ranching operations whereas Kingfisher County has an economy based on winter wheat. In 1987, Lincoln County farmers harvested 88,000 centares of wheat from 2,600 hectares while in Kingfisher County 2.27 million centares were harvested from 63,131 hectares (U.S. Bureau of the Census 1987). Winter wheat agriculture is more spatially extensive and employs fewer people per acre than mixed agricultural operations; accordingly, the population of Kingfisher County is six persons per square kilometer compared with 11 persons per square kilometer in Lincoln County. The difference in the agricultural setting plays into socioeconomic characteristics. Persons in Kingfisher County are more likely to have higher incomes, to have finished high school, and to vote for Republican candidates than their Lincoln County counterparts.

Conclusions

In linking a natural line to human endeavor we do not wish to engender the notion that the physical landscape determined the nature of life in Oklahoma. Yet, the present work demonstrates the interaction of a natural line with human society. The physical and cultural geographies are significantly different on either side of this natural meridian of the plains. In all realms—culture, transportation routes, climate, land use, soil types, agricultural economy, and vegetation—the two sides of the line show divergence rather than similarity. To the west, the former prairies have landscapes and cultures with traditional ties to the Midwest and the Great Plains. To the east, the Cross Timbers reflect the less prosperous landscapes and life-styles of the American rural South. The differences are profound in that no great distances, mountain ranges, water bodies, nor political boundaries separate them. The ecotone continues as a viable boundary for all manner of regional geographies. Thus, we present Figure 1 as a North American snapshot with connotations beyond that of the vegetative landscape.

Figure 4. A home west of the ecotone. Homesteaders frequently built with prairie sod where wood was unavailable. Reprinted with the permission of the Western History Collections, University of Oklahoma Library.

References

Corcoran, W. T. 1982. Moisture stress, mid-tropospheric pressure patterns, and the forest/grassland transition. *Physical Geography* 3:148-159.

Costello, D. F. 1969. *The Prairie World*. New York: Thomas Y.Crowell.

Debo, A. 1949. *Oklahoma: Foot-Loose and Fancy-Free*. Norman: University of Oklahoma Press.

Foreman, C. T. 1947. *The Cross Timbers*. Muskogee: The Star Printery.

Goodman, J. M. 1977. Physical environments of Oklahoma. In *Geography of Oklahoma*, ed. J. W. Morris, pp. 9-24. Oklahoma City: Oklahoma Historical Society.

Hafen, L. R., and Rister, C. C. 1941. *Western America: The Exploration, Settlement, and Development of the Region Beyond the Mississippi*. New York: Prentice-Hall.

Irving, W. 1955. A *Tour on the Prairies*, eds. J. B. Thoburn and G. C. Wells. Oklahoma City: Harlow Publishing.

U.S. 1846. *Public Statutes*, Vol. 7, pp. 474-475.

U.S. Bureau of the Census. 1987. *Agricultural Census, Part 38 Oklahoma*. Washington, DC: U.S. Government Printing Office.

Webb, W. P. 1931. *The Great Plains*. Boston: Ginn.

The Dogleg at Frank's Cutacross

David J. Nemeth
University of Toledo

Then it's not mere chance that you precisely are sitting opposite me? But what can be the idea behind it?
— ESTELLE, in *No Exit* (Sartre 1949, 19)

Improvements in urban automobile transportation systems should result in space-time convergence, bringing people and their exchanges of goods and services closer. Convergence did increase urban productivity and vitality within America's industrial cities through the first half of the 20th century. However, the decentralization of urban population, power, and wealth in many post-industrial cities like Toledo, Ohio, has resulted in transportation improvements, primarily at the loci of development boomlets along the urban fringes. Some urban transportation planners claim the commuting benefits of such urban decentralization (Gordon 1990). In fact, deteriorating transportation infrastructures linking suburbs through urban cores include many uncorrected design flaws that plague commuters. Grid-locks and traffic jams for cross-town travel are much more common than in the past. In Toledo, one long-uncorrected design flaw in its transportation grid has seriously impeded inter-suburban cross-town traffic flows—the "dogleg at Frank's cutacross."

Doglegs and Bottlenecks

Doglegs are short, crooked segments of linear traffic routes so called for their resemblance to a dog's hind leg (Figure 1A). Where doglegs occur in inner-cities within compact grid systems that have acute angles at road bends, commuters in transit through them may experience repeated stops at traffic lights. Commuters call such places "bottlenecks." Time wasted in transit means lost wages and lost opportunities to many commuters, so doglegs that create bottlenecks are avoided if at all possible. Avoidance strategies include shortcuts across an acute angle in the dogleg, maneuvers that are often illegal and dangerous to both motorists and pedestrians.

Frank's Cutacross

This case study introduces the notorious bottleneck (point of congestion) that has emerged in a dogleg (disrupted line) in the traffic pattern of mid-town Toledo, where north-south-bound Byrne and Secor Roads connect only through east-west-bound Dorr Street (Figure 1B). Average traffic flows through the dogleg are estimated at an extremely heavy 35,000 vehicles per day.

"Frank's cutacross" is a place-name that reveals some motorists' tentative solution to this traffic congestion problem. There is a huge sign that reads "Frank's Nursery" at the southeast corner of Byrne Road and Dorr Street, and there is a spacious street-corner parking lot beneath this sign. To avoid a long wait at the traffic light, impatient northbound drivers regularly use this lot as an illegal shortcut through part of the dogleg. As an opportunity to "beat the light" and perhaps save precious time, this place has gained landmark status among commuters. It is called "Frank's cutacross" primarily by some of the scofflaws who use it.

Toledo's commuters are understandably unhappy about the bottleneck at the dogleg at Frank's cutacross. They may also wonder while they queue up together at the stoplight there: "What can be the idea behind it?" The reasons this dogleg and bottleneck exist and persist have never been adequately explained to Toledo motorists.

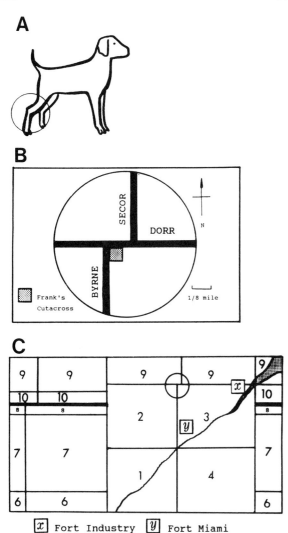

Figure 1. The dogleg at Frank's cutacross. A. A sharp bend in a road is metaphorically a "dogleg," that is, crooked as a dog's hind leg. B. Mismatched rectangular survey lines antecedent to Byrne, Secor, and Dorr roadways have created the dogleg at Frank's cutacross. C. The legacy of three distinct rectangular land surveys converging in the Toledo area is revealed here by a disarray of township numbers and mismatched boundary lines. Boundary lines encircled include interfacing section boundaries from the Michigan Survey (U.S. Public Land Survey of 1820) and the Twelve-Mile Reservation (1805). The heavy line interrupted by townships 2 and 3 of the Twelve-Mile Reservation rectangular survey is the Fulton Line separating the Michigan and Ohio rectangular surveys (U.S. Public Land Survey).

Survey Lines and Roadways

Though much obscured by time's passing, there is a good reason for the dogleg at Frank's cutacross—the original rectangular land survey systems that created the dogleg. These surveys resulted in land subdivisions within townships that were once essential for organizing orderly, stable, and productive settlements throughout rural America (Thrower 1966). Many surveyed sectional boundary lines became modified into rural transportation routes, as components of convenient farm-to-market transport systems. Most routes skirted private properties, bordering spacious squares and rectangles of cultivated fields without transecting them and thereby reducing their agricultural efficiencies and productivities. As automobile-oriented 20th-century industrial cities expanded into surrounding rural areas many of these old boundary-line corridors became integral parts of heavily traveled urban grid systems near the hearts of cities.

Rectangular Land Surveys

With the end of the Colonial War for Independence from Great Britain, initial legislation by the federal government promoted the creation of a westward-expanding United States of America. One ordinance (1785) provided for systematic surveys creating rectangular land divisions into 6-mile-square townships, each divided into 36 one-mile-square sections (Muehrcke 1986). Another ordinance (1787) provided for the creation of a state of Ohio among five states to be carved from the vast Northwest Territory wilderness located immediately westward of the original Thirteen Colonies. Ohio was destined to become the experimental ground for implementing a uniform national rectangular land-survey system, The United States Public Land Survey. This orderly system was used thereafter across the American frontier in advance of the sale and settlement of all public lands.

Because of the remoteness of the area, pacification problems, and frontier politics, Ohio's northern boundary remained uncertain until 1836—long after statehood in 1803 (Wendler 1977, 8). Persistent British-supported Native American resistance to America's westward expansion postponed experimentation with rectangular surveys in the Northwest Territory until 1805, a decade after General Anthony Wayne's 1795 victory at the Battle of Fallen Timbers. Wayne's military camp, on the lower Maumee River at Lake Erie, was called Fort Industry. It is significant as the settlement site about which the future city of Toledo would organize itself. A few miles upriver was a British stronghold, Fort Miami.

The Twelve-Mile Reservation

Wayne's victory included the capture of Fort Miami and led to the Treaty of Greenville (3 August 1795). This treaty brought on the rapid decline of Native

American populations in the Ohio Valley and initiated American surveys and settlements deep within the Northwest Territory.

The Greenville Treaty specifically provided for a 12-mile-square American military reservation. The geographer Elias Glover surveyed the reservation in 1805. He divided it into four 6-mile-square townships but did not center the 12-mile-square reservation on Fort Miami as intended in the Treaty of Greenville. Instead, he centered it on a trading post almost a mile south of the Fort. His choice is evidence that social affairs at community centers, and not a system of boundary lines in a peripheral wilderness, were foremost when rectangular surveys were first used in the United States (Sherman 1925).

The impact of Glover's decision on present traffic patterns is most significant. One of the major east-west arteries in Toledo, Dorr Street, now follows the northern boundary of the Twelve-Mile Reservation. A segment of Dorr Street now forms the transverse part of the dogleg at Frank's cutacross.

The National Land Survey

Glover's 12-mile rectangular survey preceded the full implementation of the U.S. Public Land Survey in northwest Ohio. The Reservation and some other early land divisions in and near present-day Toledo (for example, some French long-lots) had historic precedence over the U.S. Public Land Survey as it gradually became implemented. Their integrity is today preserved in cadastral maps and property transactions.

In the decades following Glover's survey, in the vicinity of old Fort Industry, confusion continued over the uncertainty of the northern boundary line for Ohio. Land in this region was eventually contested between the state of Ohio and the commonwealth of Michigan (formed 1805; statehood 1837), culminating in the bloodless "Toledo War" of 1835 to 1836. Their dispute was not finally resolved until 1836, upon federal order, in favor of Ohio.

Until resolved, the commonwealth of Michigan observed what was known as the Fulton survey line, established in 1818, as its southern boundary, claiming thereby the fast-growing settlements in and near old Fort Industry. These settlements would become incorporated as the city of Toledo in 1837. Meanwhile, the state of Ohio recognized the Harris survey line of 1817 as its northern boundary. The Harris line placed the Toledo site within Ohio's territory. The Harris and Fulton lines gradually diverged eastward from their mutual point of origin at southernmost Lake Michigan.

They were 8 miles apart where they intercepted Lake Erie at the Maumee River, with the contested settlements at old Fort Industry between them. Although Ohio finally gained the Toledo site, Michigan had by 1820 already surveyed and subdivided all the way south to the Fulton line using the U.S. Public Land Survey system. Similarly, Ohio had by 1821 also surveyed and subdivided all the way north to the Fulton line using the same national rectangular system—but using a different base line and principal meridian! Twelve-Mile Reservation survey lines are sandwiched between the two U.S. Public Land Survey systems (Figure 1C).

The U.S. Public Land Survey lines north of Dorr Street are those of the Michigan survey and not those of the Ohio Survey. Michigan survey lines intersect Dorr Street at right angles because Dorr follows the latitudinal northern boundary of the Twelve-Mile Reservation. Secor Road follows a meridional Michigan Survey section line, meeting Dorr Street to form the northern part of the dogleg near Frank's cutacross. The southern part of the dogleg is formed by Byrne Road, which follows a meridional township section survey line that radiates from the center of the Twelve-Mile Reservation and ceases at Dorr Street in the north.

In sum, it is an unfortunate mismatch of converging lines from the cardinal directions originating in two of these early 18th-century rectangular surveys, the Twelve-Mile Survey and the subsequent Michigan Survey U.S. Public Land Survey that explain the dogleg that now creates the bottleneck at Frank's cutacross.

Dying for a Hamburger

The region surrounding the dogleg can be divided into four quadrants for brief analysis of the traffic congestion problem. Diverse commercial land uses in all four quadrants facing the dogleg contribute to abnormally high daytime traffic congestion. University classrooms, student parking, and high-volume retail services dominate the northeast quadrant; student apartments and fast-food retailers dominate the northwest quadrant; fast-food outlets and other retail establishments fill the southwest quadrant; and, a diversified retail strip faces the roadway all along the southeast quadrant, concealing a quiet residential neighborhood.

Negotiating this dogleg may require long waits at several traffic lights. Cars, trucks, and buses are all part of the line-up. There are no crosswalks anywhere within the dogleg, despite the commercial activity and several bus stops there. Retailers bordering eastbound traffic on Dorr Street suffer most from the constraints that traffic engineers have placed on drivers passing through the dogleg. "Planned" circulation through the dogleg is designed to get commuters through it as safely and quickly as possible, but neglects the commercial interests of retailers and commuters en route. Traffic lights are timed to force commuters to join platoons while moving through the dogleg. These platoons move through quickly as drivers attempt to make up for time lost waiting at the traffic lights. Leaving a platoon anywhere within the dogleg gives added emphasis to the phrase "dying for a hamburger" and is not only considered discourteous by motorists, but requires slow-downs and lane crossings that endanger everyone. By sacrificing for safety's sake the utility of commercial establishments throughout the dogleg, Toledo traffic engineers have reduced trade and productivity and thus unintentionally subverted the basic purpose of this urban corridor at the heart of the city's market economy.

The abundance of traffic violations at Frank's cutacross is proof that timed delays at traffic lights are extremely frustrating for motorists; the large corner parking lot at Frank's Nursery invites an irresistible last-ditch escape from timed delays for some northbound commuters. Thus, frustration breeds anarchy, resulting in many accidents. Accidents, in turn, cause grid-lock throughout the dogleg.

For more reckless speeders there is a quiet middle-class, minority residential neighborhood behind Frank's that offers alternative short-cuts by connecting Byrne and Secor along several narrow, winding streets. Commuters may wonder why all north-south traffic has not already been funneled safely through this southeast quadrant. City officials discussed such a plan several decades ago, but failed to act (*The Blade* 1970). Clearly, a wide, intrusive traffic corridor would violate this neighborhood's integrity, replacing relative serenity with noise and noxious fumes. Political considerations alone eliminate this option at this time, and probably far into the future. Residents of this neighborhood are already irate about speeding northbound commuters using their residential streets as shortcuts onto Secor.

No Exit

This case study from Toledo reveals how 18th-century ordinances to create democratic order and fusion in the northwestern Ohio wilderness with the U.S. Public Land Survey have contributed instead to disorder, confusion, and even anarchy in 20th-century commuter flows. As time-space divergence increases in central Toledo, cross-town commuters at least need no longer wonder about the idea behind absurd delays and unsafe driving conditions they regularly endure within the dogleg at Frank's cutacross; this study has uncovered the geographical source for their predicament in some early American ideals about organizing orderly frontier settlement. The outlook for improvements in this congested traffic corridor is, however, grim. The noble ideas and institutions that created the U.S. Public Land Survey for rural agricultural settlement unfortunately have provided no reliable mechanisms for preventing or relieving subsequent disorders related to unforeseen economic, technologic, and social changes that accompanied rapid urbanization. Representative of these disorders is the bottleneck at the dogleg at Frank's cutacross, from which there now appears to be no exit.

References

Balsam, Carl W. 1935. *The Ohio-Michigan Boundary Settlement.* M.A. Thesis, University of Toledo.

The Blade. 1970. Residents protest Byrne-Secor traffic plan. 15 July 1979:1 (part 2).

Gordon, Peter. 1990. Suburban lives mean short drives. *CUPR Report* 1,3(Fall):3, 5.

Muehrcke, Phillip C. 1986. *Map Use* (2nd edition). Madison, WI: JP Publications.

Sartre, Jean-Paul. 1949. *No Exit and The Flies.* New York: Alfred A. Knopf.

Sherman, C. E. 1925. *Ohio Land Subdivisions,* Vol. 3 (of 4 vols.). Columbus: Ohio Cooperative Topographic Survey.

Thrower, Norman J. W. 1966. *Original Survey and Land Subdivision.* Chicago: Rand McNally.

Wendler, Marilyn V. 1977. *Maumee City: A Study of Urban Development in the Early Nineteenth Century.* M.A. Thesis, University of Toledo.

The Factory Outlet Mall

James O. Wheeler
University of Georgia

Emily E. Wheeler
Georgia College

The factory outlet mall represents a new retailing concept in the United States (Lord 1984). Beginning in the late 1970s and rapidly expanding during the 1980s, it has roots in the practice of some manufacturers to sell merchandise to customers who visit factory sites. The factory outlet mall is a concentration of stores each selling goods produced mostly by a single manufacturer. It is an excellent example of a retail magnet (a point) that draws customers over considerable distances (lines converging on the mall). The dominant attraction of these malls is lower prices and, like other malls, the opportunity to comparison shop. Most stores in factory outlet malls offer good quality merchandise, though some specialize in seconds and irregulars. Goods are shipped from the factories to the stores, including regular goods as well as closeouts and short lots. Most factory outlet stores do not compete with discount stores but rather with traditional retail chains; they do not sell the $5 shirt but the $30 shirt for $22 (Lord 1984). For example, in a survey of 14 outlet malls, shoppers whose income exceeded $75,000 were overrepresented by twice their presence in the U.S. population (Schwartz 1990).

In 1990, more than 5,000 factory outlet stores in the United States were served by more than 265 factory outlet chains; 44 outlet malls opened in 1989—the most in a single year (Schwartz 1990). In one so-called "off-price" or discount mall in Philadelphia, 30 percent of the stores are factory outlets, an unusual example of discount stores and factory outlets in the same mall (Sullivan 1990). Most factory outlet malls are east of the Mississippi River, perhaps because of ties to the location of the manufacturers (Lord 1984), though more such malls are opening in California (Klein 1990).

Factory outlet malls are built principally in two types of settings: near a major shopping center, or in a non-metropolitan area adjacent to an interstate highway. Which type of location is better is debatable within the industry. For example, Western Development Corporation, which developed a factory outlet mall in Gaithersburg, Maryland, announced the following rationale: "Off-price centers should be located in established, affluent trade areas, not in remote areas. Sites adjacent to or near existing regional malls often make ideal locations" (Black 1987, 33). By contrast, Liberty Village Associates, in Flemington, New Jersey, followed a decidedly different strategy:

> Location is crucial to the success of an outlet. While the center must be close enough to draw shoppers, it must be distant enough to avoid conflicts between the factory outlets and their major customers. As a rule of thumb, outlet centers should be located at least 50 miles from major retail stores in order to avoid such conflicts (Black 1987, 33).

Schwartz (1990, 45) concurs: "Location is everything. Malls must be far enough from the manufacturer's own clients." And yet some outlet malls made even different locational choices. Some of the first factory outlet malls located near major tourist attractions, just off interstate highways, as in Orlando, Florida. One factory outlet mall has recently opened in Pittsburgh's East Liberty section in hope of avoiding quiet weekdays and busy weekends—common in many outlet malls. The Commerce, Georgia, factory outlet mall adopted the non-metropolitan, interstate interchange location strategy, located approximately halfway between Atlanta and Greenville-Spartanburg on I-85 at the major north-south highway (441) crossing between these two metropolitan areas (Figure 1).

Commerce in Commerce, Georgia

In 1990, Commerce, Georgia, had a population of only 4,108, marking it as one of the more obscure central places in the U.S. rural South. When Interstate-85 was built through northeast Georgia in the 1960s, connecting the Carolina Piedmont cities with Atlanta—even then the foci of the Southeast's economy—the four-lane limited-access highway passed just 6.5 kilometers (4 miles) north of this small, sleepy, low-income settlement. Its primary industry in better times was textile manufacturing, making Commerce a representative rural Piedmont town.

Commerce is some 30 kilometers north of Athens, Georgia (1990 population of 156,000), via U.S. 441, and, more importantly, approximately 120 kilometers from Atlanta (1990 population of over 2.8 million). As a result of location, Commerce gradually lost many of its once higher-order central place functions to larger surrounding centers, including a great many jobs. By the late 1980s, its central business district was, visibly, in serious decline from its once historic and, indeed, ambitious past. The coinage of "Commerce" was, in fact, adopted in the spirit of boosterism in an attempt to foster a grand future for this isolated and remote settlement. What irony, then, that the major commerce in this area now lies 6.5 kilometers north of the old Commerce business district at a location of new-found accessibility.

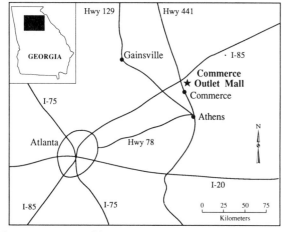

Figure 1. Location of the Commerce factory outlet mall in north Georgia.

Analysis

A license-plate survey conducted during the last three Saturdays in August and the first in September 1990, provided data for more than 2,200 vehicles during peak 1-hour periods. Since Georgia license plates display county names, we could measure the geographic reach of the factory outlet mall in attracting customers. The mall opened in 1989, but a few stores did not open their doors until early 1990. Thus, at the time the survey was conducted, little prior opportunity existed for advertising or for "word of mouth" to occur. The proximity of Athens, a college town with many out-of-state graduate students, may somewhat bias the results, although school was not in session while the survey was taken and though Georgia law requires that out-of-state vehicles be registered in Georgia within two months.

During the 4-hour study period, vehicles from 29 of the 50 states, including Georgia, were identified. Nine of the states were from west of the Mississippi River. All of the states east of the Mississippi were represented except for Indiana and Wisconsin in the Midwest and Connecticut, Vermont, Rhode Island, and Maine in New England. The four leading states outside Georgia were South Carolina (242 vehicles), Florida (89), North Carolina (85), and Alabama (37). The accessibility of the Northeast (excluding New England) and Middle Atlantic states no doubt relates to the role of I-85. In all, out-of-state vehicles represented almost 25 percent of those included in the survey.

The remaining 75 percent came from Georgia. Ninety of Georgia's 159 counties (56 percent) are represented (Figure 2). The 18-county Atlanta metropolitan area alone accounted for almost 45 percent of in-state visitors. More than 73 percent of total in-state vehicles represent metropolitan areas of Georgia.

The leading county is Jackson, although the factory outlet mall is just across the line in Banks County. One likely reason for the large number of vehicles from Jackson County—part of the Athens metropolitan area—is that many were owned by workers at the mall and we had no way to distinguish workers from visitors. We simply note that, if we could have eliminated workers' vehicles from our survey, out-of-state presence and the role of Atlanta would be even more impressive. The second leading county is Gwinnett, one of the fastest growing counties in the United States during the 1980s, and a county with a large, new regional shopping mall with four anchor department stores. The third and fourth leading counties are DeKalb and Fulton, in the heart of the Atlanta metropolitan area. Thus, despite the apparent re-

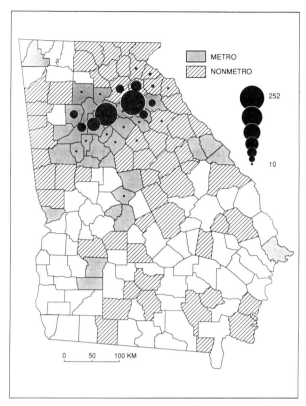

Figure 2. County origins of vehicles at the Commerce factory outlet mall within Georgia by metropolitan and non-metropolitan locations.

moteness of the Commerce factory outlet mall, the accessibility afforded it by I-85 means that a significant number of visitors choose the Commerce mall over several closer malls having much greater varieties of goods.

Most counties in northern Georgia above the Fall Line are represented in the survey. The factory outlet mall in south Chattanooga, Tennessee, probably explains the lack of representation by a cluster of counties in northwest Georgia. Although many counties in South Georgia are represented, most counties are not, especially in the southwestern and southeastern parts

of the state. The role of I-75 south of Macon may be related to the somewhat higher frequency of counties included in this part of Georgia.

Concluding Comments

Factory outlet malls are a new retailing concept, less than 15 years in existence. During the 1980s and early 1990s, their relative expansion was greater than that of traditional shopping centers, which are overbuilt in many areas. Factory outlet malls compete with traditional retail chains, such as Sears and J.C. Penney, by underpricing generally good-quality merchandise. Some outlet malls complete directly with national chains by locating near chain stores in regional shopping centers, especially those at interstate interchanges. Other factory outlet malls choose non-metropolitan locations 80 to 120 kilometers from major metropolitan locations via interstate highway. These malls attract customers from a surprisingly wide region. Unlike retailing under classical central place theory, with customers coming from a restricted surrounding hinterland, factory outlet malls capture traffic from interstates and attract customers away from locations near traditional regional shopping centers, as well as draw from a wide non-metropolitan market.

References

Black, Jane A. 1987. Off-price and factory outlet update. *National Mall Monitor* July-August:31-34.

Klein, Robert William. 1990. Outlet retailing, development taking off on the West coast. *Value Retail News* July:14-20, 26-31, 38.

Lord, J. Denis. 1984. The outlet/off-price shopping centre as a retailing innovation. *The Service Industries Journal* 4:9-18.

Schwartz, Ela. 1990. Factoring in factory outlets. *Discount Merchandiser* May:40-46.

Sullivan, R. Lee. 1990. Discounting at close range. *Discount Merchandiser* July:63-65.

Vegetation Change along a Bajada Gradient

Kathleen C. Parker
University of Georgia

The landscape throughout much of the Basin and Range Physiographic Province of the southwestern United States comprises fault-block and volcanic mountain ranges of diverse lithologies separated by broad alluvial basins. Geomorphic processes have resulted in a typical sequence of landforms from steep rocky uplands through pediments to alluvial fans or bajadas (Figure 1). Biogeographically, the Arizona Upland Section of the Sonoran Desert occurs within this physiographic province and is unique among the hot deserts of North America in its floristic and physiognomic diversity (Shreve 1964). When examined at a local scale, downslope changes in the physical environment along a bajada sequence and the associated distributional responses of this diverse collection of plant species constitute one of the more prominent lines in this arid landscape. These physical and biogeographic patterns, as well as some of the confounding factors that are apparent at a broader scale, are illustrated in Organ Pipe Cactus National Monument in southern Arizona.

Theoretical Environment/Vegetation Relationships

Alluvial fans are conically shaped landforms of fluvial and debris-flow sediment deposited where a stream emerges from mountainous uplands. In the case of water-laid deposits, aggradation occurs when the sediment load exceeds the capacity of the stream, which often results from a sudden decrease in velocity (Bull 1977). As the stream debouches from the uplands onto unconsolidated sediments, the less confined channel, the reduced stream gradient, the increased permeability of the substrate, or a combination of these changes decrease the water's velocity. Bajadas are adjacent alluvial fans that coalesce into aprons of sediment cloaking the base of the mountains.

Where bajada construction is primarily fluvial, sorting may produce a textural gradient from coarse soils in the steep, rocky uplands and on upper bajadas to progressively finer soils on the lower bajadas and alluvial flats. This topographic/soil texture gradient results in a gradient of moisture availability for plants (Phillips and MacMahon 1978; Yang and Lowe 1956). During heavy rainfall, coarse soils of rocky slopes and upper bajadas allow more rapid infiltration of water and hold it at higher water potentials than fine textured soils of lower bajadas and flats. Strong capillary action of fine textured soils draws subsurface water upward and promotes evaporative water losses. Therefore, lower bajadas are often relatively dry sites that only a few species can tolerate. High evaporation rates on alluvial flats may also concentrate salts, which further reduces the soil water potential on these sites and interferes with osmotically driven uptake of water by plants (Kramer 1983). This linear downslope moisture gradient may be apparent at a local scale on one part of the bajada but is often less obvious at a broader scale when the complex nature of fan construction complicates this simple gradient (see Bull 1977).

For many plant species in arid environments, water is the principal limiting factor; therefore, spatial variation in water availability has a profound influence on vegetation patterns. F. Shreve (1964) maintained that physical variables associated with the bajada gradient and their effect on soil moisture constitute the primary control of plant distributions in the Arizona Upland Section of the Sonoran Desert. Many scholars working at a local scale along bajada sequences have documented the influence of these variables on vegetation patterns (e.g., Bowers and Lowe 1986; Phillips and MacMahon 1978).

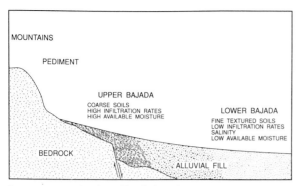

Figure 1. Generalized profile of a bajada sequence in the Sonoran Desert.

Vegetation Patterns in Organ Pipe Cactus National Monument

The Bajada Sequence

In Organ Pipe Cactus National Monument, compositionally distinct groups of species occupy different habitats along a line from alluvial flats to rocky uplands (Parker 1991). The vegetation of lower bajadas and alluvial flats is sparse and strongly dominated by *Larrea tridentata* (creosote bush) and *Ambrosia deltoidea* (triangle-leaf bursage) (Figure 2A). Where soils are particularly fine textured and somewhat saline, salt-tolerant *Atriplex* spp. (saltbush) assume dominance. Coarser alluvial soils closer to the uplands support more diverse open vegetation than the lower bajadas. The steep rocky uplands support the densest cover and greatest diversity of woody species. In addition to the more ubiquitous *Larrea tridentata* and *Ambrosia deltoidea*, a number of species are common: shrub (for example, *Simmondsia chinensis* [jojoba], *Jatropha cuneata* [sagre-de-drago]), cactus (for example, *Stenocereus thurberi* [organ pipe cactus], *Carnegiea gigantea* [saguaro]), and tree (for example, *Cercidium microphyllum* [foothill palo verde]).

Complex Patterns at a Broader Scale

When the scale is expanded beyond part of a single bajada, lithology, history of bajada construction, and slope aspect complicate the classical soil texture/moisture gradient associated with the bajada sequence. Organ Pipe Cactus National Monument, like much of the Basin and Range Physiographic Province, includes a variety of lithologic substrates. Soils in the monument area that have developed from granitic bedrock contain greater percentages of the coarse sand classes than volcanically derived soils (Parker 1991). Granite weathers more through mechanical shattering,

Figure 2. Plant communities of Organ Pipe Cactus National Monument. A. Fine textured soils of lower bajadas and alluvial flats are strongly dominated by *Larrea tridentata* (creosote bush). B. Common species on rocky granitic uplands include the columnar cacti *Carnegiea gigantea* (saguaro) and *Stenocereus thurberi* (organ pipe cactus), the trees *Cercidium microphyllum* (foothill palo verde) and *Olneya tesota* (ironwood), and the shrubs *Jatropha cuneata* (sangre-de-drago) and *Ambrosia deltoidea* (triangle-leaf bursage). C. Dominants on volcanic uplands include *Encelia farinosa* (incienso), shown in the lower left, *Simmondsia chinensis* (jojoba), in the lower right, and *Cercidium microphyllum*. The cactus in the lower center, *Opuntia phaeacantha* var. *discata* (prickly pear), is also common on volcanic slopes.

whereas basalt weathers slowly inward from the surface primarily through chemical processes (Birkeland 1984). Because of these and mineralogical contrasts, granitic soils are often coarser in texture than volcanic soils. These substrate-related textural contrasts, along with variation in insolation with slope exposure, influence moisture regimes independent of variation along the linear bajada sequence.

In Organ Pipe Cactus National Monument, distinct communities inhabit substrates of different lithologies; furthermore, particularly on granitic sites, aspect-related compositional differences are apparent. The vegetation on south-facing granitic hillslopes is dominated by *Encelia farinosa* (incienso), *Stenocereus thurberi*, *Jatropha cuneata*, and *Olneya tesota* (ironwood) (Figure 2B). *S. thurberi* and *O. tesota* are relatively frost-sensitive and therefore less abundant on colder north-facing slopes and basin floors where the more frost tolerant *Carnegiea gigantea* and *Cercidium microphyllum* are common. Volcanic sites are dominated by species common on other substrates as well (for example, *Encelia farinosa*, *Ambrosia deltoidea*, and *Cercidium microphyllum*). These sites, however, also include a number of species that are rare elsewhere in the monument (for example, *Simmondsia chinensis*, *Janusia gracilis* [janusia], *Ayenia microphylla*, and *Mimosa laxiflora*) (Figure 2C). The more marked aspect-related vegetation contrast on granitic than volcanic substrates observed in the monument area (Parker 1991) is representative of patterns reported elsewhere in the Sonoran Desert (Shreve 1964).

Even on a bajada of uniform bedrock, soil texture is seldom simply a function of slope angle or vertical position. Bajadas are complex depositional features often constructed over a long period (Bull 1977). Climatic conditions may vary from one depositional episode to another, perhaps with each episode constructing a different part of the bajada and involving detritus of different sources and clast sizes. Large floods may carry coarse material farther downslope than smaller magnitude events and bury finer sediment deposited previously. The constructional history of the bajada may dictate textural variation in the subsoil (that is, the zone occupied by many plant roots) to a greater degree than do pedogenic processes. Consequently, both surface and subsurface texture on bajadas and associated plant species distributions often vary in a complex manner rather than according to a simple linear gradient from top to bottom.

In arid landscapes like Organ Pipe Cactus National Monument, spatial variation in geologic substrate may also influence vegetation patterns by affecting the geomorphic disturbance regime. M. A. Melton (1965) found greater downslope mobility rates of surface boulders and cobbles on granitic than volcanic slopes in the monument, which he attributed to lithologic contrasts in weathering. Although the importance of geomorphic disturbances on regeneration dynamics and biogeographic patterns has been stressed theoretically and demonstrated empirically (Prez 1987; Swanson *et al.* 1988), such influences have received little attention in arid environments of North America.

Conclusions

When changes in the physical environment and the associated vegetation patterns are examined at a local scale along a simple bajada gradient, this prominent line in the landscape remains sharply in focus. But the expansion of scale to include different parts of a bajada, diverse lithologies, and constrasting slope exposures blurs this line. These other influences on soil texture and water availability interact to create a mosaic of moisture regimes, thus mitigating the dominant expression of the linear bajada sequence. At the subregional scale, vegetation patterns more clearly reflect this environmental mosaic.

References

Birkeland, P. W. 1984. *Soils and Geomorphology*. New York: Oxford University Press.

Bowers, M. A. and Lowe, C. H. 1986. Plant-form gradients on Sonoran Desert bajadas. *Oikos* 46:284-291.

Bull, W. B. 1977. The alluvial-fan environment. *Progress in Physical Geography* 1:222-270.

Kramer, P. J. 1983. *Water Relations of Plants*. New York: Academic Press.

Melton, M. A. 1965. Debris-covered hillslopes of the southern Arizona desert—consideration of their stability and sediment contribution. *Journal of Geology* 73:715-729.

Parker, K. C. 1991. Topography, substrate, and vegetation patterns in the northern Sonoran Desert. *Journal of Biogeography* 17:151-163.

Pérez, F. L. 1987. Soil moisture and the upper altitudinal limit of giant paramo rosettes. *Journal of Biogeography* 14:173-86.

Phillips, D. L. and MacMahon, J. A. 1978. Gradient analysis of a Sonoran Desert bajada. *Southwestern Naturalist* 23:669-680.

Shreve, F. 1964. Vegetation of the Sonoran desert. In *Vegetation and Flora of the Sonoran Desert*, eds. F. Shreve and I. L. Wiggins, Vol. 1, part 1. Stanford, CA: Stanford University Press.

Swanson, F. J., Kratz, T. K., Caine, N., and Woodmansee, R. G. 1988. Landform effects on ecosystem patterns and processes. *Bioscience* 38:92-98.

Yang, T. W., and Lowe, C. H., Jr. 1956. Correlation of major vegetation climaxes with soil characteristics in the Sonoran Desert. *Science* 123:542.

Biodiversity of the Lake Erie Archipelago

Brian Klinkenberg

University of British Columbia

S trewn between Port Clinton, Ohio, and Point Pelee, Ontario, in the western basin of Lake Erie, are a group of islands known collectively as the Erie Islands (Figure 1). The attraction islands have for geographers is felt no less by those who explore these islands than by those who explore exotic islands, for here are opportunities to study the effects of isolation on the resident flora and fauna, and to find species previously unknown in the area and, possibly, new to science. Many rare plants in Ontario and Ohio occur only on the islands, and herpetologists continue to debate the correct taxonomic status of the Erie Islands watersnake, unable to decide whether it represents a true endemic (Cook 1984).

The 21 islands that occupy the shallow western basin of Lake Erie represent the crests of two heavily eroded cuestas (Forsyth 1988). These cuestas were submerged only a few thousand years ago when the lake waters backed up as a result of the post-glacial rebound of the better known cuesta to the east—the Niagara escarpment. The islands now serve as stepping stones for migrating birds, butterflies, and plants. They are especially resplendent in the spring when warblers stop on their way north for the summer, the trees literally dripping with bright yellows and blues and greens.

Because of the ease of access to them, and the presence of Ohio State University's biological research station in their midst, the islands' flora and fauna are well known. Much of the knowledge accumulated over the years about the islands was presented at the Ninth Biosciences Colloquium of the College of Biological Sciences of the Ohio State University in May 1985 (Downhower 1988). It is from these proceedings that most of the data analyzed below are taken.

Species/Area Relationships

One of the first things biogeographers commonly do when exploring an island is to inventory the biota. The next thing is to determine whether any relationship exists between the total number of species found and the area of the island. This fascination with the relationship between the total numbers of species found and area has been going on for longer than 100 years (Kelly *et al.* 1989; MacArthur and Wilson 1967). In fact, well over 100 different species/area studies have been conducted (Connor and McCoy 1979) and most have found a relationship between the two.

Biogeographers have observed that island species/area relationships usually differ from mainland species/area relationships. One commonly noted difference is that islands tend to have fewer species than do equivalent sized mainland regions. This reduction in species diversity could be the result of many factors: a late frost could kill the only individuals of an annual plant species that flowered too early, and the seeds of the closest mainland population were unable to disperse to suitable habitat on the islands; or, an island's climate could favor one species over another, resulting in the eventual loss of habitat for the disfavored species, and so on.

Having discovered that there is a relationship between species numbers and area, the next challenge is to determine why. Several hypotheses have been put forth. We can summarize the more important of these as follows (Kelly *et al.* 1989):

- If individual species are randomly distributed about the landscape, larger islands will contain more species. This hypothesis—the *random placement*

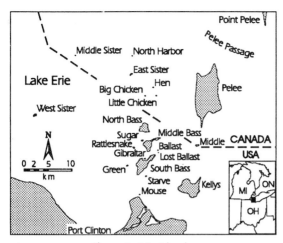

Figure 1. Erie Islands.

hypothesis—is a simple statistical explanation for the observed relationship.

- Larger islands usually contain a greater diversity of habitats; the greater the number of habitats the greater the number of species. This *habitat diversity hypothesis* is a simple biological explanation for the observed species/area relationship.

- A more complex biological explanation of the relationship considers the dynamics of species immigration and extinction, and how those events are affected by the island's isolation and size, respectively. That is, the more isolated the island the less frequently a new species "finds" it; the smaller the island, the fewer the individuals of any one species on it and the greater the likelihood that some event will extirpate that species from the island. This *equilibrium hypothesis* has received widespread attention over the last 30 years and is one that specifically addresses the differences observed between island species/area curves and mainland species/area curves.

- Another possible explanation for the observed species/area relationship is that larger species require larger territories, territories that can only exist on larger islands. However, this explanation, the *incidence function hypothesis*, can only be used to explain the relationship observed between larger faunal species and area.

Although species/area relationships have been observed on a variety of oceanic and freshwater islands and mainland areas, no clear-cut leader among the above hypotheses has emerged. Factors such as the amount of disturbance associated with each island generally confound the species/area relationship (Klinkenberg 1988; MacArthur and Wilson 1967), and make selection of any one hypothesis a complex

biological problem. Regardless of which hypothesis one believes is responsible for these relationships, determining its mathematical form is problematic.

If the number of species increases linearly with area, then large areas would possess improbably large species totals. Based on observations of species/area plots, many biogeographers have suggested that relationships be modeled or linearized by comparing the total number of species found on an island (S) with the logarithm of the island's area (A) in the semi-log model:

$$S = C \log A + D,$$

where C and D are numerical constants (Williamson 1981). However, certain biological models of species abundance lead to the conclusion that taking logarithms of both species numbers and areas should produce a linear relationship, as in the log-log model:

$$\log S = \log c + z \cdot \log A,$$

where c and z are constants (MacArthur and Wilson 1967).

Biogeographers who support the equilibrium theory generally favor the log-log model. With this model, more commonly known as the power model:

$$S = c \cdot A^z,$$

the value of the exponent z is said to have biological significance (Connor and McCoy 1979). That is, theoretical biological reasoning leads to the conclusion that species/area relationships observed on islands that are at equilibrium should exhibit a z value of 0.25 (with an expected range of values from 0.20 to 0.30). This value of z reflects the proportional increase in species expected to be associated with a proportional increase in area when the immigration of new species and the extinction of existing species are in a natural balance. Much higher values of z will be found where there are "islands within the islands," much lower values will be obtained when the islands are extremely homogeneous, or when the islands are not truly isolated from the neighboring mainland areas (Klinkenberg 1988). As with the various species/area hypotheses presented above, neither the semi-log model nor the power model has emerged as the mathematical model of choice among biogeographers.

The Analyses

A selection of papers in *The Biogeography of the Island Region of Western Lake Erie* (Downhower 1988) present species totals for five different groups on the Erie Islands: native plant species, alien plant species, terrestrial isopods, spiders, and reptiles. Although many species/area studies covering a wide variety of species assemblages have been reported in the literature, few examine the relationships among a variety of taxa across a single island chain (Schoener 1986).

Data on the native and alien plant species from 21 of the islands were obtained from B. Klinkenberg (1988). The species totals reported therein were derived from publications and collections dating since 1899, and represent a fairly comprehensive list. R. W. Dexter and colleagues (1988) presented information on the isopod distribution for 22 of the islands. Their species totals represent the results of 38 years of systematic collections. It must be noted that Pelee Island was sampled only in part, and because of the location of the Ohio State biological station on Gibraltar Island, the latter received the most intensive collection efforts. J. A. Beatty (1988) reported a similar Gibraltar Island collecting bias for his spider data. Nonetheless, Beatty's species totals represent the result of systematic collecting since 1958 on all of the islands. Reptile totals were reported for only nine of the Erie Islands (King 1988), but the species list was derived from published records since 1912 and extensive field-work, and therefore we assume that this would be a fairly complete record. It should be noted that species extinctions—which may have occurred—have not been taken into account in the totals used in these analyses.

The areas used in the following analyses were taken from B. Klinkenberg (1988). The mathematical technique used to determine the parameters of the species/area relationships (Table 1) was least median of squares regression (Rousseeuw 1984). This regression routine does not suffer from the outlier effect that traditional

least squares regression does. That is, with traditional least squares regression one anomalously high (or low) value can have an arbitrarily large effect on the derived parameters. By replacing the criterion that the regression minimize the *sum* of the squared residuals with the criterion that the regression minimize the *median* of the squared residuals, the effect of outliers is eliminated. Given the unequal collecting efforts reported by Dexter and colleagues (1988) and Beatty (1988), the use of a regression routine that minimizes the effect of unusual values (for example, higher species totals associated with Gibraltar Island because of the greater attention directed to that island), by giving them an effective weight of zero in the regression, was thought warranted.

Results and Discussion

While both species/area models (the semi-log model and the power model) provide reasonable fits to the five species/area relationships, the fit (r^2) associated with the power model was generally higher (Table 1). Since the exponent of the power model has been assigned biological relevance (MacArthur and Wilson 1967), the results of this model are considered in detail below (Figure 2).

The z values for the native plants, spiders, and reptiles are within the acceptable range of values associated with equilibrium conditions. Given the dramatic changes that have occurred on many of these islands over the last century (Downhower 1988), this suggests that these three groups have yet to respond to those changes and that their present equilibrium diversity reflects historic conditions. The differences in the c values—which should be compared only when

Table 1. Results of the Species/Area Tests

| Species group | Species totals | | Least median regression r^2 values† | | Power model parameters | |
	Islands	Adjacent mainland	Semi-log model	Power model	c	z
Native plants	704	1255	0.92	0.98	4.39	0.23
Alien plants	290	398	0.94	0.95	2.91	0.35
Isopoda	12	NA	0.70	0.48	1.47	0.10
Spiders	232	410*	0.81	0.91	2.71	0.28
Reptiles	17	25	0.98	0.96	0.64	0.23

* This represents the total for the Ohio mainland only.
† All of the r^2 values are statistically significant ($p < 0.05$).

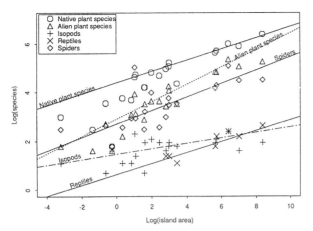

Figure 2. Power model species/area relationships.

the species/area curves have similar slopes—reflect differences in the sizes of the source pools of species able to colonize the islands (Table 1).

The z values for the alien plant species and the isopods, however, differ substantially from the other three groups. The z value produced by the alien species can be explained if one considers the nature of the distribution and dispersal of aliens. The agricultural fields and residential areas where most of the alien species are found exist as "islands" within the broader natural environment of each island, and it is predicted that the z value should be higher when islands exist within islands. Furthermore, on the larger islands a greater variety of agricultural products is grown, which would provide for a larger source of alien seeds and a higher than (biologically) predicted species increase with increasing area due, in part, to the increased immigration associated with the agricultural activities.

The z value associated with the isopods is, conversely, very low, as is the r^2 value. As noted above, low z values are associated with islands that have homogeneous taxa or that are not truly isolated from the surrounding mainland areas. Dexter *et al.* (1988, 107) note that "all of the species collected, with one possible exception, were old-world species introduced by agriculture, commerce, or driftwood. . . . Half of the species are cosmopolitan in distribution." Given the extent of human activity on the islands, it can be assumed that the lake does not represent a barrier to isopod distribution. The cosmopolitan nature of the distribution of many of the species also indicates that isopod habitat is relatively ubiquitous.

Conclusion

This brief overview of the biogeography of the Erie Islands shows that the relative increase in species diversity for three of those groups—native plants, spiders, and reptiles—is similar and falls within the limits associated with the equilibrium theory of island biogeography. The alien plant species exhibit a much greater increase in species diversity as the island area increases—there are proportionally more species on the larger, more disturbed islands than on the smaller, less disturbed islands. Conversely, the isopod distribution shows a relative lack of linkage with area—almost as many species are found on the smaller islands as are found on the larger islands.

The answer to the species/area question is potentially of great interest to many. For example, park planners need to know how many species are expected to survive in a remnant nature reserve over the long term, given that species immigration is likely to be severely curtailed, and species extinction is apt to increase. By studying these relationships on islands, biogeographers are attempting to understand what may occur as the natural landscape is increasingly fragmented and only isolated islands of the natural landscape remain. With today's increasing concern for preserving biodiversity, our ability to understand these relationships is ever more important.

References

Beatty, J. A. 1988. Spiders of the Lake Erie Islands. In *The Biography of the Island Region of Western Lake Erie*, ed. J. F. Downhower, pp. 111-121. Columbus: Ohio State University Press.

Cook, F. R. 1984. *Introduction to Canadian Amphibians and Reptiles.* Ottawa: National Museum of Natural Sciences.

Connor, E. F., and McCoy, E. D. 1979. The statistics and biology of the species/area relationship. *The American Naturalist* 113:791-833.

Dexter, R. W., Hahnert, W. F., and Beatty, J. A. 1988. Distribution of the terrestrial isopoda on islands of western Lake Erie. In *The Biography of the Island Region of Western Lake Erie*, ed. J. F. Downhower, pp. 13-23. Columbus: Ohio State University Press.

Downhower, J. F., ed. 1988. *The Biogeography of the Island Region of Western Lake Erie.* Columbus: Ohio State University Press.

Forsyth, J. L. 1988. The geologic setting of the Erie Islands. In *The Biography of the Island Region of Western Lake Erie*, ed. J. F. Downhower, pp. 13-23. Columbus: Ohio State University Press.

Kelly, B. J., Wilson, J. B, and Mark, A. F. 1989. Causes of the species/area relation: A study of islands in Lake Manaouri, New Zealand. *Journal of Ecology* 77:1021-1028.

King, R. B. 1988. Biogeography of reptiles on islands in Lake Erie. In *The Biography of the Island Region of Western Lake Erie*, ed. J. F. Downhower, pp. 125-133. Columbus: Ohio State University Press.

Klinkenberg, B. 1988. The theory of island biogeography applied to the vascular flora of the Erie Islands. In *The Biography of the Island Region of Western Lake Erie*, ed. J. F. Downhower, pp. 95-105. Columbus: Ohio State University Press.

MacArthur, R. H., and Wilson, E. O. 1967. *The Theory of Island Biogeography.* Princeton, NJ: Princeton University Press.

Rousseeuw, P. J. 1984. Least median of squares regression. *Journal of the American Statistical Association* 79:871-880.

Schoener, T. W. 1986. Patterns in terrestrial vertebrate versus arthropod communities: Do systematic differences in regularity exist? In *Community Ecology*, eds. J. Diamond and T. J. Case, pp. 556-586. New York: Harper & Row.

Williamson, M. 1981. *Island Populations.* Oxford: Oxford University Press.

The Fractal Nature of the Louisiana Coastline

Nina Siu-Ngan Lam
Hong-Lie Qiu
Louisiana State University

Geologic, geomorphic, and marine processes, as well as human occupation, have shaped the coast of Louisiana into one of the most distinctive coastlines in America. It is also the most rapidly changing coastline in the United States. Coastal retreat and the loss of land have attracted attention from both federal and state governments and have been targets for intense research. This paper illustrates a basic problem in studying coastlines—their measurement. Fractal measurement is used to demonstrate the intricacy and complexity of the Louisiana coastline and how such intricacy relates to coastal processes.

The Intricate Nature of the Louisiana Coastline

Forms and Processes

The Gulf of Mexico forms the southern boundary of Louisiana. This coastline is composed of two distinct parts, each subject to different processes. The eastern half (to the east of Marsh Island) is a deltaic plain formed by the seaward expansion of the Mississippi River and its distributaries, such as Atchafalaya River. The deltaic plain has gone through several stages, with varying river courses and dominant geomorphic processes in each. Deltas of younger ages superimposed on older ones result in a coastline of extreme irregularity. Based on physical, radiocarbon dating, and archaeological data, six stages of deltaic processes can be distinguished (Kolb and Van Lopik 1966) (Figure 1). The oldest delta, Sale-Cypremort, lies at the west of the plain and was formed approximately 5,000 years ago. Long periods of erosion by waves and currents

and downwarping has worn down much of its outer parts, leading to a less contorted coastline. The Balize delta at the southeastern tip of the deltaic plain is an active delta formed about 500 years ago. In contrast to the older deltas, it protrudes far into the Gulf with narrow ridges and passes that look like a bird's foot. No other delta in the world has such a unique and contorted shape as this one (Newton 1987).

The western part of the Louisiana coast is dominated by long-shore currents, resulting in a relatively straight coastline. Sediments carried to the Gulf by streams and rivers are blocked by long-shore currents, forming beach ridges. Live oaks, called "cheniers" locally, are often found growing along the crests of the ridges, thus earning the name chenier plain for this region. These ridges parallel each other and the shore, each chenier representing an earlier shoreline.

Boundary Controversy

The delineation and definition of the Louisiana coastline was not an issue until rich oil fields were discovered on the continental shelf. Legal battles were fought between the state of Louisiana and the federal government over how far Louisiana extends into the Gulf. A 5-kilometer zone from the shore was determined eventually by the court as Louisiana's offshore boundary (Kniffen and Hilliard 1988). A more fundamental question, however, remains: From what base line should the 5-kilometer boundary be drawn?

The answers are not straightforward. Some researchers argue that it is necessary to distinguish the difference between "shoreline" and "coastline" (Kniffen and Hilliard 1988). The shoreline is the actual line

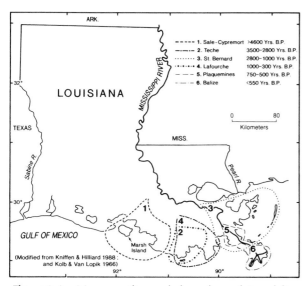

Figure 1. Louisiana coastline and chronology of river deltas.

between land and water and would include the numerous bays and points, whereas the coastline usually refers to the outer and general shape of the shore. Generally, it is much easier to measure the outer coastline than the inner shoreline, where bays, promontories, and sand bars are abundant. Although measurements of the Louisiana coastline vary, it is generally agreed that the outer coastline is about 600 kilometers, in contrast to a more than 1,600-kilometer shoreline. Since these measurements are often derived from maps and photos, they are subject to various errors of cartographic generalization.

Moreover, the coastline of Louisiana changes rapidly. While the shallow-water parts of the delta are advancing, coastal retreat is the dominant factor of change. Louisiana is the only state where the annual retreat rate exceeds 3 meters per year. Some parts of the coast have erosion rates exceeding 15 meters per year (Nummedal *et al.* 1984). Coastal retreat is due largely to sea level rise, documented at an average rate in excess of 1 centimeter per year. Other factors, such as reduced sediment supply, owing to human-made river structures and canal and pipeline dredging for petroleum exploitation, also increase the rates of retreat or subsidence. Since the boundary between Louisiana and the federal ownership of minerals shifts as the coastline retreats, the most immediate problem from coastal retreat is the loss of revenue from petroleum exploitation. A retreating coastline takes from Louisiana and gives to the federal government (Newton 1987). Locally, landowners lose income as the marsh breaks up to form bays and then lose their land if the new waterbodies prove to be navigable.

Fractal Analysis

How Long Is the Louisiana Coastline?

Measuring Louisiana's coastline length is the first and fundamental step toward understanding its forms, processes, and changes through time. Accurate line measurement is vital in deriving accurate rates of coastal retreat. Unfortunately such a seemingly simple and trivial question does not have a definite answer. There is no accurate value for the length of the coastline. The mysterious notion that the shorter the yardstick used to measure the length of a coastline on a map, the longer the resultant length has puzzled researchers for a long time. This so-called "Steinhaus paradox" (Steinhaus 1960) is further entangled by a number of questions or issues that have long plagued geography. What is the appropriate scale and generalization level of a map? How does it relate to the curvature and contortion of coastlines? And what sampling interval, resolution, or divider length should we use to measure the coastline?

French mathematician Benoit Mandelbrot (1967) experienced a similar problem when measuring the coast of Britain. He then explained the coastline measurement problem by means of fractals, a term he coined later in 1975 from the Latin *fractus*, meaning irregular (Mandelbrot 1977, 1982). Most natural curves, such as coastlines, are fractals whose dimensions can be fractional and lie between the conventional line dimension of 1 and plane dimension of 2. The more complex the curve, the higher the fractal dimension. Similarly, a surface may have a fractal dimension (D) ranging from 2 to 3. Mandelbrot asserts that fractal geometry is more appropriate in modeling natural phenomena than classical geometry. Since its inception, fractal theory has further been refined and applied in virtually every major discipline. Fractal analysis is widely used for measuring, as well as simulating natural forms and processes, and is becoming a major spatial analytical tool in geography (Goodchild and Mark 1987).

Properties of Fractal Curves

The main concept underlying fractals is self-similarity. A typical fractal curve is self-similar at any scale, meaning that any part of a curve, if appropriately enlarged, is indistinguishable from the curve as a whole. If you looked at the coastline in more and more detail, more wiggles would become visible and the shapes of these wiggles would look very similar to the coastline itself. Most geographic features, however, show this self-similar property only in a statistical sense over certain scale ranges. The range of self-similarity for coastlines could be used to determine the scale at which coastal processes operate.

The self-similarity concept forms the basis for the derivation of fractal dimensions for fractal curves. Suppose a coastline is measured by means of walking a pair of dividers along a map using two divider intervals, with the second interval only half the first one. For a straight line, it requires twice the number of steps to walk through the line using the second interval. For an irregular line, the number of steps required often increases more than double. The more irregular the line, the greater increase in the number of steps. Mandelbrot observed that when line length is plotted against divider length on logarithmic scales the points tend to follow a straight line for a self-similar curve. He then argued that the slope of a linear regression line (B) through these points is a function of fractal dimension and that $D = 1.0 - B$.

For a perfect fractal curve, all the points on the regression plot would follow a straight line, and B and hence D would be constant over all scales. Coastlines and many other geographic features, however, are not perfect fractals and self-similarity only exists in a statistical sense over certain scale ranges. It is important to capture the range over which a coastline shows self-similarity (that is linearity on the plot) and to estimate D within that range. Any interpretation of D without consideration of the self-similarity range for non-perfect fractal curves would be biased.

Data, Method, and Results

The Louisiana coastline (excluding islands) was extracted from five U.S. Geological Survey land-use data files. The total number of digital points for the entire Louisiana mainland coastline is 23,125 and the average distance between pairs of points is about 140 meters (Qiu 1988). Since different dominant geomorphic processes affect different parts of the coast, fractal dimensions were also calculated for the different parts to test whether D and self-similarity ranges change significantly. Six subdivisions were delineated based on their different geomorphic histories (Figure 2). The logarithmic relationships between line length and divider length for all coasts are shown in Figure 3. None of the coastlines tested is perfectly self-similar, therefore regression lines were fitted to only part of each curve and fractal dimensions were computed for those parts only. Table 1 summarizes the resultant D, number of points (that is, walks) used in the regressions, the corresponding self-similarity ranges, and the R^2 values indicating the regression fit. Within a scale range of 0.2 to 204.8 kilometers, the overall Louisiana coastline

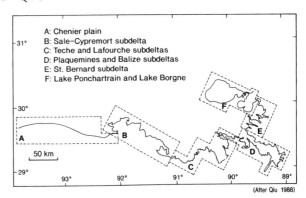

Figure 2. The Louisiana coastline and subdivisions used in this study.

has a relatively high fractal dimension ($D = 1.2728$) compared with the west coast of Britain ($D = 1.2671$), despite the latter being considered one of the most complex coasts of the world.

Interpretation

Of the six coasts, coast A has a fractal dimension close to 1 simply because it is almost a straight line. The others show self-similarity only over certain scale

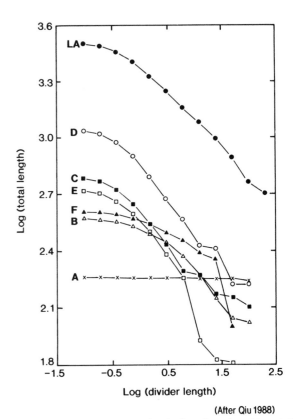

(After Qiu 1988)

Figure 3. Scatterplot between divider length and total length in logarithms for the overall coastline (LA) and the six subdivisions (coasts A to F).

Table 1. Ranges of Self-Similarity (ROSS) and Fractal Dimensions

Coast	ROSS (km)	Number of walks	Dimension	R^2	Years abandoned
A	0.2 to 51.2	9	1.0040	0.9852	NA
B	0.8 to 51.2	7	1.2755	0.9612	4,500
C	0.4 to 6.4	5	1.3604	0.9877	750
D	0.4 to 12.8	6	1.3652	0.9928	500
E	0.4 to 6.4	5	1.3370	0.9863	1,900
F	0.8 to 25.6	6	1.1484	0.9868	NA
Overall	0.2 to 204.8	11	1.2728	0.9870	NA

Source: Qiu (1988).

ranges and have dimensions over 1.1. Based on the ranges of self-similarity and the associated fractal dimension values, two groups of coasts can be identified. Group 1 consists of coasts B, C, D, and E and is characterized by higher dimensions. All coasts in group 1 are dominated by the deltaic processes of the Mississippi River and its distributaries. Group 2 includes coasts A and F, having lower fractal dimensions. They are dominated by coastal processes such as waves, tides, and currents.

Within group 1, the changes of fractal dimensions and the self-similarity ranges, to a certain extent, reflect the developing stages of the deltaic plain. The deltaic plain of southeastern Louisiana has been developed continuously, with the oldest (coast B) formed some 5,000 years ago. Whenever the Mississippi River advances a major lobe seaward, the overly extended river course is abandoned and favors the development of a shorter, more direct route to the Gulf. Once a lobe is abandoned, it is no longer replenished by river deposits and is subjected to marine processes such as tides, waves, and currents in the Gulf (Kolb and Van Lopik 1966). Such changes in dominant geomorphic processes result in progressive modifications in the geometry of the coast. Our results show that fractal dimensions of the coasts in group 1 tend to decrease with time, yielding a strikingly high R^2 value of 0.99, while the

ranges of self-similarity tend to increase with time, resulting in an R^2 of 0.82 (Figure 4). The youngest and most active delta, coast D, has the highest dimension ($D = 1.3652$) and a smaller self-similarity range (0.4 to 12.8 kilometers), while coast B, the oldest delta, abandoned about 5,000 years ago, has the lowest dimension ($D = 1.2755$) and the largest self-similarity range (0.8 to 51.2 kilometers) within the group.

Conclusions

Although none of the six coastlines examined is perfectly self-similar, their statistical self-similarity over certain scale ranges can be used, to a certain extent, to reflect the various coastal processes that have shaped the coastline of Louisiana. This paper demonstrates that in general coasts that are dominated by deltaic processes have higher fractal dimensions than coasts dominated by marine processes. Furthermore, deltaic coasts have decreasing fractal dimensions and increasing self-similarity ranges with increases in the number of years since delta abandonment. Larger self-similarity ranges could be linked to larger spatial scales at which coastal processes operate. This study also confirms that the marine processes of tides, waves, and currents, are operating at larger spatial scales than the deltaic processes of the Mississippi River and its distributaries. However, coastal data of different resolutions are needed to make more accurate assessments. Because of its uniqueness and importance, it is essential to examine the Louisiana coastline using different points of view. Through fractal analysis, the intricate nature of this line is shown to have no absolute length. Length, as well as many other cartographic measurements, is relative to scale. The concepts of scale, resolution, cartographic error, and hence, fractals, play important roles in refining our methods and in expanding the horizons of geographical interpretation.

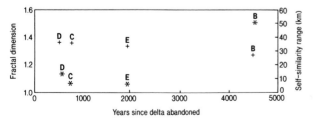

Figure 4. Scatterplot between fractal dimension (shown as +), self-similarity range (shown as *), and years since delta abandoned.

References

Goodchild, M. F., and Mark, D. M. 1987. The fractal nature of geographic phenomena. *Annals of the Association of American Geographers* 77(2):265-278.

Kniffen, F. B., and Hilliard, S. B. 1988. *Louisiana: Its Land and People.* Baton Rouge: Louisiana State University Press.

Kolb, C. R., and Van Lopik, J. R. 1966. Depositional environments of Mississippi River deltaic plain, southeastern Louisiana. In *Deltas in Their Geological Framework*, ed. M. L. Shirley, pp. 17-61. Houston: Houston Geological Society.

Mandelbrot, B. B. 1967. How long is the coast of Britain? Statistical self-similarity and fractional dimension. *Science* 156:636-638.

———. 1977. *Fractals, Form, Chance, and Dimension* . San Francisco: Freeman.

———. 1982. *The Fractal Geometry of Nature.* San Francisco: Freeman.

Newton, M. B., Jr. 1987. *Louisiana: A Geographical Portrait.* Baton Rouge: Geoforensics.

Nummedal, D., Cuomo, R. F., and Penland, S. 1984. Shoreline evolution along the northern coast of the Gulf of Mexico. *Shore and Beach* 52:11-17.

Qiu, H. L. 1988. *Measuring the Louisiana Coastline: An Application of Fractals.* Student Honors Competition Winning Papers, Cartography Specialty Group Occasional Paper No. 1. Washington, DC: Cartography Specialty Group, Association of American Geographers.

Steinhaus, H. 1960. *Mathematical Snapshots.* London: Oxford University Press.

The Population Center of Canada— Just North of Toronto?!?

Mark P. Kumler
Michael F. Goodchild
University of California, Santa Barbara

The concept of a center of population is often used in geographical studies. There are, however, several ways to determine a "center point" and at least as many ways to describe them. For very large areas the problem is complicated by the spherical shape of the earth. In this investigation three such centers are defined and determined for the population of Canada: the median center, the mean center (also known as the centroid, or center of gravity), and the point of minimum aggregate travel.

Background

In the first known publication of the center of a population (Hilgard 1872), the medial lines and the centers of gravity were determined for the population of the United States for the previous four censuses. Hilgard presented the medial lines and centers of gravity, and speculated that the ultimate population center of the United States would be "not far from St. Louis" (where, remarkably, it lies today—about 60 kilometers southwest of St. Louis). His medial lines appear to have been drawn parallel to the eastern coast; they divided the population in two—half living east of the line and half living west. Hilgard's centers of gravity were determined by "calculating the positions [for] the center of each state . . . with regard to the relative density of population in their different parts, and that all cities having over 50,000 inhabitants have been treated as separate centers . . ." (Hilgard 1872, 218). He ignored the spherical shape of the earth.

The popularity of center points flourished in the 1920s and 1930s. In the United States, the Bureau

of the Census (1923) determined and published the movement of the center of population for every census since 1790, and the movements of other agricultural and manufacturing centers since 1850. In Canada, H. E. M. Kensit published the movements of the centers of population, industry, and developed water, and predicted that in a few years the center of population would be, "paradoxically, in the middle of Lake Superior" (Kensit 1934, 269).

The study of center points was most fervent in the Soviet Union, where a group of geographers in Leningrad established the Mendeleev Centrographical Laboratory in 1925 (Neft 1966; Porter 1963; Sviatlovsky and Eells 1937). Named after the famous Russian chemist D. I. Mendeleev, who had discussed the concept of the center of Russia in one of his works in economic geography, the lab sought to establish centrography as a science by developing "laws of the distribution of phenomena based on the relationships and migrations of their 'centers of gravity'" (Poulsen 1959, 326). The lab met its demise in 1934, after recommending that commercial grain planting be limited to the traditional bread belt of European Russia; the recommendation clashed with a decision recently made by the Communist Party, and the laboratory was discredited and liquidated (Poulsen 1959).

It was also around this time that confusion first arose between two of the points considered in this study: the mean center and the point of minimum aggregate travel. In an attempt to popularize an understanding of the mean center of population, the U.S. Bureau of the Census added the following phrase to its otherwise correct description of the point: "If all the people in the United States were to be assembled at one place, the center of population would be the point which they could reach with minimum aggregate travel, assuming that they all traveled in direct lines from their residences to the meeting place" (Bureau of the Census 1923, 5; or see Eells 1930, for an exhaustive review of the confusion). The point just described is the point of minimum aggregate travel; this is *not* the same as the mean center.

Definitions

There have been many attempts to establish conventions for the terms and definitions of center points. Until such definitions are widely accepted it is necessary to carefully define what one means by the terms *median center*, *mean center*, and *point of minimum aggregate travel*.

Median Center

The median center is defined here to be the point that divides the set of points into two equal halves. In a one-dimensional distribution there will be an equal number of points on either side of the median point. In two dimensions the median center is the intersection of two perpendicular lines, each of which divides the distribution into two halves. The orientation of these lines is often arbitrary, and different orientations can yield different median points (Hayford 1902). On a sphere the lines become circles, either great or small. One could consider the intersection of two perpendicular great circles, in which case the solution is sensitive to the orientation of the circles, or one could use the established graticule of latitude (small circles) and longitude (great circles). The latitude-longitude median center is used here; it divides the population such that half lives east of the point and half lives west, while half lives north of the point and half lives south.

Mean Center

The mean center, also known as the centroid or the center of gravity, is the arithmetic mean, or average, of the locations of the points. It is the point on which the distribution would balance if represented by weighted points on a weightless line, plane, or sphere. Mathematically, the mean center is the point that minimizes the sum of the squared distances between it and all other points.

It should be noted that the mean center of points on a sphere is internal to the sphere. While this point

can be easily projected back onto the surface of the sphere, for the purpose of referencing the location of the point, the true mean center lies beneath the surface.

Point of Minimum Aggregate Travel

The point of minimum aggregate travel (or point of mat, or mat point) is the point that minimizes the sum of the distances (unsquared) between it and all other points. For a population, the mat point is the point that would result in the least total distance traveled if all people were to travel along straight lines to a single location. For spherical problems, distances are measured along great circle routes.

Several researchers have taken to calling this point the median (Gini and Galvani 1929; Neft 1966; Porter 1964; Scates 1933), on the basis that the median center described above is sensitive to the orientation of the axes, and that in one-dimensional problems the median center is also the point of minimum aggregate travel. While this argument has many merits, the more descriptive and unambiguous term—point of minimum aggregate travel—is used here, to avoid further confusion.

Data

The data for this study were provided by the Geocartographics Division of Statistics Canada. The population counts and approximate centroid locations for all census enumeration areas were extracted from the 1976, 1981, and 1986 Geography Tape Files The centroid locations were identified in degrees and minutes of latitude and longitude, and were converted to decimal degrees or radians for input to the computer programs.

Algorithms

Median Center

The simplest way to determine the median of a two-dimensional population distribution is to order the points by their coordinates in the *x*- and *y*-directions. The median in each dimension is then easily found by stepping through the ordered points and maintaining an accumulated population total. As soon as the accumulated total is greater than or equal to one half of the total population, the median in that dimension has been found. The process is repeated for the other dimension, and the intersection of the two medial lines is the median center. If the locations are specified in latitude and longitude, this algorithm generates the lat-long median center—the point north of which lives one half of the population and east of which lives half of the population.

If the number of observations is extremely large, and sorting or storing them is impossible or difficult, shortcuts can be taken. One could divide the range of coordinates in each dimension into a large number of small sub-ranges, or buckets, and then accumulate population counts only for each bucket. The bucket that contained the median value would be easily determined, and then the observations could be scanned again, storing and sorting only those that fell within the median bucket. This procedure requires two passes through the data values.

A different shortcut, used in this investigation, relies on the knowledge of the approximate location of the median center (perhaps from the previous census). It is essentially a "three-bucket" approach, in which a very thin bucket collects the observations near the median, and the populations at all other observations are simply accumulated into large buckets on either side. The locations and populations of the observations falling in the thin center bucket can be stored and sorted, and the median point easily determined.

Mean Center

Being the simple average location of the observations, the mean center is quite easy to compute. In one- or two-dimensional problems the coordinates (perhaps weighted by population) are simply summed and divided by the number of observations (or people). If the points are located on a sphere, and the locations are given in latitude and longitude, the locations must first be converted into their equivalent three-dimensional coordinates. This is easily done with the formulae:

$$x = \cos(\text{lat}) \cdot \cos(\text{lon})$$

$$y = \sin(\text{lat})$$

$$z = \cos(\text{lat}) \cdot \sin(\text{lon}).$$

The averages in the x-, y-, and z-dimensions are easily determined, and can then be converted back to latitude and longitude.

Point of Minimum Aggregate Travel

There are no known algebraic solutions to finding the point of minimum aggregate travel; the best solutions to date are iterative, and converge, often slowly, upon the final value. E. Weiszfeld (1937) developed the first heuristic for finding the mat point; many years later it was re-discovered independently by W. Meihle (1958), H. Kuhn and R. Kuenne (1962), and L. Cooper (1963). For an exhaustive review of this measure, including some interesting graphical and analog solutions, see Brian R. Rizzo (1982).

The Weiszfeld procedure begins with the selection of an arbitrary point as the initial seed location. The distances and directions from that seed to all other (possibly weighted) points are determined, and the trial point is displaced in the indicated direction. The procedure is then repeated until it stabilizes on a value, or some displacement threshold is reached. The procedure is not entirely robust; it will fail if the trial point ever coincides with one of the weighted points. This situation, known as the vertex iterate problem, results from an undefined division by zero. Although this problem did not arise in this investigation, perhaps because of the many digits of spurious precision that are automatically carried by most computers, it can usually be circumvented by either "nudging" the trial point away from the problem point, or, more simply, by re-starting the procedure with a different seed location. The algorithm is presented in numerical form in Figure 1.

Results

The mean and median center and the point of minimum aggregate travel were computed for the population of Canada for each of the past three censuses—1976, 1981, and 1986.

do

$$\hat{x} = \frac{\displaystyle\sum \frac{x_i}{[(\hat{x}-x_i)^2+(\hat{y}-y_i)^2]^{1/2}} \cdot \text{pop}_i}{\displaystyle\sum \frac{\text{pop}_i}{[(\hat{x}-x_i)^2+(\hat{y}-y_i)^2]^{1/2}}}$$

$$\hat{y} = \frac{\displaystyle\sum \frac{y_i}{[(\hat{x}-x_i)^2+(\hat{y}-y_i)^2]^{1/2}} \cdot \text{pop}_i}{\displaystyle\sum \frac{\text{pop}_i}{[(\hat{x}-x_i)^2+(\hat{y}-y_i)^2]^{1/2}}}$$

loop until stable

Figure 1. The Weiszfeld procedure.

Year	Median center	Mean center	Point of mat
1976	45°37'N	48°15'54"N	43°49'36"N
	79°24'W	84°21'25"W	79°23'29"W
1981	45°41'N	48°25'43"N	43°50'31"N
	79°26'W	85°00'30"W	79°25'46"W
1986	45°39'N	48°25'09"N	43°50'35"N
	79°26'W	85°10'32"W	79°26'23"W

The median center is near Burk's Falls, Ontario, just east of the Georgian Bay. The point moved north and west between 1976 and 1981, but only south between 1981 and 1986; it appears to be fairly well anchored by the two largest cities in Canada, at the approximate latitude of Montreal and the approximate longitude of Toronto (Figure 2). The mean center falls north of Lake Superior, approximately 55 kilometers northwest of Wawa, Ontario and 20 kilometers southeast of White River. The true mean center was 156.0 kilometers beneath the surface in 1976, 161.7 kilometers in 1981, and 162.3 kilometers in 1986. The point moved westward between the three censuses, but at a decreasing rate; it crossed the TransCanada Highway sometime in 1984. The point of minimum aggregate travel lies north of Lake Ontario, about 25 kilometers north of Toronto, and 5 kilometers south-southwest of Richmond Hill. It has also continued to move westward, at a decreasing rate; between 1981 and 1986 it moved just over two meters per day.

Different Center Points (Pros and Cons)

All of the points defined above have properties that make them useful for describing some aspect of the

distribution of a population. They also have their disadvantages.

The median center is easy to determine and is conceptually simple. It does, however, have two important flaws: it is sensitive to the orientation of the axes—different axes yield different points—and it is insensitive to distances from the median point. Large movements of population can occur within any quadrant without affecting the location of the median center, while the movement of a single person from one quadrant to an adjacent quadrant will change the location of the point.

The mean center is also easy to determine, and the weighted-points-on-a-weightless-plane concept is easy to understand, if somewhat abstract. It is not sensitive to the orientation of the axes and it is sensitive to distances. The mathematical property, however, that it minimizes the sum of the squared distances, is the source of the problem with this measure: the points, or people, are effectively weighted proportionally to their distance from the center—more distant people have greater influence on the location of the mean center than people nearby. This property is clearly unsatisfactory for describing the distribution of a population.

Conceptually, the point of minimum aggregate travel is the most appealing. The property of minimizing total distance is easy to understand and appreciate. Historically, the location of the point has been difficult to determine, but with modern computing power it is certainly feasible. Each iteration of the Weiszfeld procedure required approximately 6 minutes on an IBM PS/2, and the procedure always stabilized to one ten-thousandth of a degree (one third of a second of latitude or longitude) within 15 iterations. The mat point does have one flaw—it is insensitive to radial movement: If a person moves 1,000 kilometers directly toward or away from the mat point, the point will not move; if that same person, however, moves only a few kilometers in any other direction, the point will move accordingly.

Conclusions

Unfortunately, there is no ideal measure of the center of a population. The median center is insensitive to the distances of points from the center; the mean center puts undue weight on the distant points; and the point of minimum aggregate travel is insensitive to radial movements of the individual points. Of these shortcomings, the insensitivity to radial movement is the least severe, and the point of minimum aggregate travel is recommended as the best measure of the center of a population. The mat point can be deter-

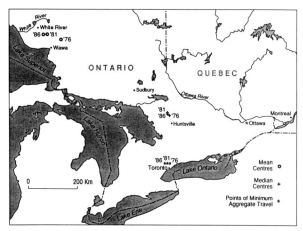

Figure 2. The population centers of Canada in 1976, 1981, and 1986.

mined for a large number of observations with a reasonable amount of computing time, and the characteristics of the point are easily understood. The point of minimum aggregate travel should be regarded as the center of a population.

Acknowledgments

The authors thank Statistics Canada for cooperation and financial support, and the University of California, Santa Barbara, Cartographic Laboratory for preparing the map.

References

Bureau of the Census. 1923. Center of population and median lines and centers of area, agriculture, manufactures, and cotton. In *Fourteenth Census of the United States, 1920.* Washington, DC: U.S. Government Printing Office.

Cooper, L. 1963. Location-allocation problems. *Operations Research* 11:331-343.

Eells, Walter C. 1930. A mistaken conception of the center of population. *Journal of the American Statistical Association,* New Series, 25(169):33-40.

Gini, C., and Galvani, L. 1929. Di talune estensioni del concetti di media a caratteri qualitativi [Of some extensions of the conceptions of average to qualitative characters]. *Metron* 8:3-209.

Hayford, John F. 1902. What is the center of an area, or the center of a population? *Journal of the American Statistical Association,* New Series, 58:47-48.

Hilgard, J. E. 1872. The advance of population in the United States. *Scribner's Monthly* 4:214-218.

Kensit, H. E. M. 1934. The centre of population moves west. *Canadian Geographic Journal* 9:262-269.

Kuhn, H., and Kuenne, R. 1962. An efficient algorithm for the numerical solution of the generalized Weber problem. *Journal of Regional Science* 4:21-34.

Meihle, W. 1958. Link length minimization in networks. *Operations Research* 6:232-243.

Neft, David S. 1966. *Statistical Analysis for Areal Distributions.* Philadelphia: Regional Science Research Institute.

Porter, Philip W. 1963. What is the point of minimum aggregate travel? *Annals of the Association of American Geographers* 53:224-232.

———. 1964. A comment on "The elusive point of minimum travel." *Annals of the Association of American Geographers* 54:403-406.

Poulsen, Thomas M. 1959. Centrography in Russian geography. *Annals of the Association of American Geographers* 49:326-327.

Rizzo, Brian R. 1982. *The Point of Minimum Aggregate Travel: Graphical, Analog, and Numerical Solutions.* M.A. Thesis, Department of Geography, University of Western Ontario.

Scates, Douglas E. 1933. Locating the median of the population in the United States. *Metron* 11:49-65.

Sviatlovsky, E. E., and Eells, Walter C. 1937. The centrographical method and regional analysis. *Geographical Review* 27:240-254.

Weiszfeld, E. 1937. Sur le point pour lequel le somme des distances de *n* points donnés est minimum. *Tohoku Mathematical Journal* 43:355-386.

ICONS

The urge to regard with awe certain landscapes and specific landscape features seems to be part of human nature. The term icon is frequently assigned to objects of great religious or cultural significance. And such phenomena fit admirably into the photographic metaphor around which we have organized this volume. What visitor can resist taking pictures at Niagara Falls, the Statue of Liberty, or the Empire State Building? The Shrine of Guadalupe in Mexico City falls into this class of objects as North America's most heavily visited religious site. Catholic pilgrimage centers in Quebec are also major points of convergence for the devout. But North American Indians, such as the Taos Pueblo in New Mexico, must contend with legal obstacles to gain and maintain access to their ancient sites of spiritual urgency. Aside from religion, national history, ecological ideals, commitment to style, and economics also validate sites and symbols of iconic significance. The historic core of Charleston, South Carolina, is the center of local and touristic homage to a tragic, but ultimately unifying, Civil War. Santa Fe's lifestyle and built landscape represent to some the ideal of an emerging postmodern ideology. Waldon Pond, Mount Monadnock, and the Appalachian Trail symbolize an environmental ethic, battles to preserve nature, and of simple, less ecologically damaging ways of living. While there is a serenity associated with most icons, particularly those of a religious nature, this is not so for one emblematic American landscape feature: the skyscraper. But is it an icon? As an embodiment of capital concentration and symbol of technological prowess and competitive zeal, some might say just that.

—Wilbur Zelinsky, Pennsylvania State University

281

Taos Pueblo's Struggle for Blue Lake

William J. Gribb
University of Wyoming

The Taos Pueblo, New Mexico, has been in existence since the mid-1300s. In more than 600 years of occupation they have challenged and endured attempts to control their sacred places. The most significant sacred site is Blue Lake (P'achale), the mythical beginning of the Taos people and everything on Earth (Figure 1). Their hopes, desires, and insights are exchanged with the Spirits through meditation and ceremony at Blue Lake. The loss or desecration of Blue Lake as a sacred place is likened to the loss or desecration of Mount Sinai, Mecca, or Mount Fuji. This paper examines the different periods of legal struggle over Blue Lake and the methods used by the Taos Pueblo to recapture jurisdiction over the site. It adopts the humanistic approach to interpreting the geography of place, following Y. F. Tuan (1976), and recognizes the mythologies of place, person, and tradition in adopting D. Harvey's (1990) approach to geopolitical history.

The legal jurisdiction over sacred places may not be held by those believing in its sacredness. The Middle Eastern wars since the Crusades are an example of the sorts of conflicts that occur to obtain a sacred site. In North America, almost every conflict between the American Indian and the advancing Europeans involved protection and preservation of traditional homelands and sacred places. With the time for armed conflict past, American Indians have turned to an often more successful battleground—the courts and legislative system.

Significance of Blue Lake

The Taos Pueblo is one of the most conservative of the Pueblos, a characteristic that has inhibited in-depth studies (Bodine 1988). But from the few studies available, the significance of Blue Lake (Figure 2) is apparent in at least three ways. The first is Blue Lake's relationship to the cosmology of the people, described by Nancy Wood (1989, 6):

According to one commonly accepted tribal legend, the Sun and Moon once mated above Blue Lake. While all the stars of the heavens watched, the People spilled out between the Sun and Moon, sliding down the handle of the Big Dipper into the water. There they lived for awhile in the Underworld, acquiring wisdom and knowledge. . . . The Feather People (Fiadaina) emerged first and went south, to a home near Ranchos de Taos, where they built a village of mud. Then the Shell People (Holdaina) emerged and made their way to Red River, twenty miles north, thence to Pot Creek, about fifteen miles south. The Water People (Badaine) came next and became fish who swam in the Ranchos de Taos Creek until observed by a girl of the Feather Clan. She ran to tell her people, and when they returned, the fish were standing up in the water. Two girls struck the fish with two bean plants and they became people. These three clans introduced themselves and together they went back to Taos Mountain to await the arrival of the other clans as they emerged from Blue Lake, one at a time. The Big Earring People (Fialusladaina), the Old Ax People (Fiadaikwaslauna), and the Knife People (Chiadaina) came out of the water and admired the beauty of the sky and trees, the waterfalls and the grass, the moon and stars. As the clans grew used to their new surroundings, the four-legged animals became brothers; streams became veins of life. . . .

The second way in which Blue Lake plays a key is in the social structure of the Taos Pueblo. As presented in the myth, all of the *kivas* (clans) originated in Blue Lake. Without membership in a clan, a Taos resident has no social or political presence in the pueblo. Ac-

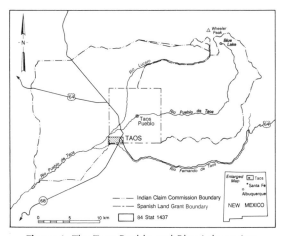

Figure 1. The Taos Pueblo and Blue Lake region.

cording to J. Bodine (1988) there are six active kivas at Taos, with several societies in each. Membership into a society is through an initiation ceremony. In late August a 32.2-kilometer pilgrimage is made to Blue Lake for the ceremonies. The whole process takes approximately eight days, beginning with the arduous walk to the Lake (3,505-meter elevation) and culminating in the ceremonies (Bodine 1988). Each year a different society will perform the new member initiation rights.

Third, Blue Lake is a religious retreat. According to Seferino Martinez, a Pueblo elder, "Blue Lake is the most important of all our shrines because it is a part of our life, it is our Indian Church. . . . We go there and talk . . . and for meditation" (Reno 1963, 21-22). Thus, Blue Lake's significance can be demonstrated by its place in the creation origins, its ceremonial role in Taosian social order, and its significance in Taosian religion and well-being.

Jurisdiction over Blue Lake

As with any sacred object or place, the inability to control or manage its use creates a tense situation for the believers. They do not know how those in control will use the place or how it may be desecrated by acts that may seem harmless to those in control. The Taos Pueblo wanted jurisdiction over Blue Lake so that it would not be desecrated or mismanaged and to ensure their access to the site for religious, spiritual, and ceremonial use. To understand how they eventually obtained control over Blue Lake, it is important to understand how they lost control.

There have been five periods of jurisdictional control over Blue Lake. During the first period, pre-contact,

Figure 2. Blue Lake, a sacred place of mystical serenity. Photograph courtesy of the Taos Pueblo.

the Taos Pueblo used the area. Archaeological findings confirm the pueblo's existence in the area since the early to mid-1300s (Ellis 1974). While ceremonial use of Blue Lake cannot be traced definitely to this period, early Spanish records indicate that the Pueblo used Blue Lake in ceremonies and pilgrimages (Ellis 1974).

The Spanish period technically begins with the earliest contacts by Francisco Coronado and then subsequent missionaries, in the early and mid-1540s. The legal status of the pueblos and their use areas was established in 1567 when a Royal Credula designated an area of 600 varas (one varas equals approximately 83.8 centimeters) in each cardinal direction "measured from the church" (Brayer 1939, 13). Unfortunately for the Taos Pueblo this did not include Blue Lake, making it part of the Spanish Crown. Definitional changes in land grants increased the pueblo grants to one league, 5,000 varas, in each direction (71.7 square kilometers), but this had no effect on the jurisdiction of Blue Lake. The royal decree of 15 October 1713 officially identified the Taos Pueblo as the four square leagues centered on the mission church (Dunham 1974), again excluding Blue Lake.

The Mexican period of jurisdictional control commenced with Mexico's declared independence from Spain. The Declaration of Independence, 28 September 1821, contained the principles that Indians were citizens of Mexico and that their land grants were private property (Brayer 1939, 17). This confirmed the Spanish land grants, but placed the jurisdiction of Blue Lake with the new Mexican government.

The fourth period consists of Blue Lake's jurisdictional control under the U.S. government, beginning with the signing of the Treaty of Guadalupe Hidalgo, 2 February 1848. This treaty ceded Mexican lands to the United States, and land grants became private property (Keleher 1929). But, because Blue Lake was not part of a land grant, it became part of the U.S. public domain. The grants were confirmed by the Surveyor-General of New Mexico in 1856 and approved by Congress on 22 December 1858 (11 Stat. 374). In the process of confirmation, the Taos Pueblo argued that their use area consisted of the region around Blue Lake and that the Spanish land grant was only that portion surrounding the pueblo. This point was agreed to by the Surveyor-General in 1864, but unfortunately the land patent identified only the square pueblo grant minus land claims within the grant area (Ellis 1974). Blue Lake was still under the jurisdiction of the federal government, but at least a recognition of the Taos claim was identified. Even though the jurisdiction changed from Spain to Mexico to the United States, the Taosians continued their religious and ceremonial uses of Blue Lake.

In 1906, President Theodore Roosevelt created the Taos National Forest (34 Stat. 3262), which included Blue Lake. Later the name was changed to the Carson National Forest (Executive Order No. 848 [1908]) but the effect was still the same, Blue Lake was now part of the National Forest system.

Three events provided legal foundation for the Taos Pueblo to build a defence for Blue Lake. First, in 1924 the Pueblo Land Act was passed (43 Stat. 636) to compensate the pueblos for non-Indian claims within their Spanish land grants. The Pueblo Board appraised the Taos Pueblo land losses at $458,520, but only awarded the Taos Pueblo $78,128.85. The pueblo waived their claim in return for the Blue Lake land patent (15 Indian Claims Commission 666). However, the Blue Lake land patent was not awarded.

The second event was the 1928 Executive Order No. 4929 by Calvin Coolidge. It ". . . withdrew from all forms of entry or appropriation . . . for the protection of the watershed," an area of 121.4 square kilometers, which included Blue Lake. Though the area was still under the jurisdiction of the Federal government, at least it was protected from all forms of entry. However, grazing and logging permits were eventually issued for the area, in direct contradiction to the order (Ellis 1974).

Finally, to provide retribution for the monies not received from the Pueblo Board of 1924, Congress passed a law in 1933 (48 Stat. 108) allocating monies to all pueblos unjustly compensated. The Taos Pueblo was allotted $297,684.67, a partial difference between the Pueblo Board's appraisal of the non-Indian lands and their awarded value. Again, the Taos Pueblo refused the money in return for title to Blue Lake. However, section four of the bill did identify the area from Executive Order No. 4929 to be segregated for "certain religious ceremonies." The federal government finally recognized Taosian use of the land, but ownership remained with the federal government. Though they had permission to use the land for ceremonial purposes, they had to pay a use permit. Over time, permits were issued to non-Indians for grazing, camping, and outdoor recreation (Ellis 1974), which the Taos Pueblo interpreted as desecrating their sacred lands.

In 1946 the Indian Claims Commission was created to settle land claims, treaty obligations, and inadequate compensation, and overall to be "fair and honorable" to the American Indian claimants (Barney 1974). The Taos Pueblo petitioned for a hearing, Docket No. 357, to receive their use lands and compensation for lands given to them by the Spanish but now part of the Carson National Forest. After more than 15 years of legal battle, the Indian Claims Commission decided in favor of the Taos Pueblo, 8 September 1965. The pueblo had petitioned that 1,213.97 square kilometers of land was taken without just compensation for the creation of the Taos-Carson National Forest, including the Blue Lake region. They also petitioned to receive their allocation of monies from the Pueblo Land Board. The Indian Claims Commission found that indeed this was their land but because of eight other Spanish and Mexican land grants that dissected the Taos area, they had claim to only 526.05 square kilometers, which included Blue Lake. The Indian Claims Commission also ruled that the Taos Pueblo should receive the $297,684.67 allotted to them in 1933.

Though the Indian Claims Commission agreed that the Taos Pueblo had use jurisdiction over a large tract of land, the commission did not have the power to grant the land. In an attempt to gain the patent to the land, the pueblo decided to petition for an amendment to the 1933 Act to give them title to the land. In 1969, Representative James Haley submitted House Bill 471 (S.750) amending the 1933 Act and giving the land to the pueblo (Hecht 1989). After many hearings and debates, the Senate passed the bill on 2 December 1970. The amendment was approved and signed into law on 15 December 1970 (84 Stat. 1437). The language of the law states

> . . . the following described lands and improvements thereon, upon which said Indians depend and have depended since time immemorial . . . as the scene of certain religious ceremonials, *are hereby declared to be held by the United States in trust for the Pueblo de Taos.* [emphasis added]

By placing the land in trust, the federal government holds the land patent, but the Taos Pueblo control its use and management. This was a landmark decision because it based land use on religious practice not economics. Since the return of Blue Lake to the Taos Pueblo, no non-Indian grazing permits have been renewed; all structures and fences have been dismantled and removed; and the frequency of Taosian spiritual visits to the area has increased.

Summary

The symbolic and religious significance of Blue Lake was a key factor in the persistence of the Taos Pueblo to refuse both the monetary compensation or the Act of 1933 to give them permit use but not jurisdiction over the area. The Taos Pueblo battled for almost 70 years to translate their symbolically significant site into a jurisdictional reality. More recently, the Northwest Indian Cemetery Protective Association filed suit to stop a U.S. Forest Service action that would have destroyed a traditional burial site in California (Akins 1989). A multi-tribal group received a permit by the U.S. Forest Service giving them access and use of the Medicine Wheel site in Wyoming. Thus, the federal government is becoming more aware and concerned with the preservation and use of Native American sacred sites.

Acknowledgments

I would like to thank Judy Jacobsen and the anonymous reviewers for their comments and suggestions.

References

Aikins, N. 1989. New direction in sacred lands claims: Lyng V. Northwest Indian Cemetery Protective Association. *Natural Resources Journal* 29:593-605.

Barney, R. 1974. The Indian Claims Commission. In *Pueblo Indians, I,* ed. D. A. Horr, pp. 13-16. New York: Garland Press.

Bodine, J. 1988. The Taos Blue Lake ceremony. *American Indian Quarterly* 12:91-104.

Brayer, H. O. 1939. *Pueblo Indian Land Grants of the "Rio Abajo," New Mexico.* Albuquerque: University of New Mexico Bulletin, No. 334.

Dunham, H. D. 1974. Spanish and Mexican land policies and grants in the Taos Pueblo region, New Mexico. In *Pueblo Indians, I,* ed. D. A. Horr, pp. 151-311. New York: Garland Press.

Ellis, F. H. 1974. Anthropological data pertaining to the Taos land claim. In *Pueblo Indians, I,* ed. D. A. Horr, pp. 29-150. New York: Garland Press.

Harvey, D. 1990. Between space and time: Reflections on the geographical imagination. *Annals of the Association of American Geographers* 80:418-434.

Hecht, R. A. 1989. Taos Pueblo and the struggle for Blue Lake. *American Indian Culture and Research Journal* 13:53-77.

Keleher, W. 1929. Law of the New Mexico land grant. *New Mexico Historical Review* 9:350-371.

Reno, P. 1963. *Taos Pueblo.* Denver, CO: Sage Books.

Tuan, Y. F. 1976. Humanistic geography. *Annals of the Association of American Geographers* 66:266-276.

Wood, N. 1989. *Taos Pueblo.* New York: Alfred A. Knopf.

The Shrine of Guadalupe

Charles O. Collins
University of Northern Colorado

Just 20 minutes from downtown Mexico City, lies the most visited pilgrimage site in the Americas (Figure 1). Among all Roman Catholic shrines, only the Vatican annually attracts more of the faithful. Yet as the bus pushes through the crowded streets there is little sense of a pilgrimage and the landscape offers no hint of holiness until suddenly a protected pedestrian walk appears, splitting the lines of cars and buses. Only then is one alerted to something extraordinary ahead.

Many of the devout arrive by bus or subway; some carry a candle or handful of flowers, but often they simply emerge from the horde of shoppers and commuters. The pilgrims on foot are a different matter. Frequently costumed, often in tribal garb, they move slowly up the middle of the street bearing banners or elaborate wreaths, their gaze transfixed. Bicycle pilgrims from a distant village, in an act of reverence, walk the final steps. The very pious kneel to pray amidst the traffic, oblivious to the noise and exhaust fumes. A young mother inches forward on hands and knees, a tiny sleeping infant slung beneath her in a bright *rebozo* (shawl).

On any weekend, 100,000 persons may pass the gauntlet of vendors at the main gate and pour into the Shrine of *Nuestra Señora de Guadalupe*. Ten times that number congregate on 12 December to celebrate "the Lady's" anniversary. In all, 12 million to 20 million believers visit each year (Carroll 1986; Johnston 1981), firm in their faith that the Virgin Mary appeared here to a humble Mexican peasant five centuries ago. They come to be blessed by the nation's patron saint, the "Little Brown Mother-of-God" whose image is preserved in a miraculous portrait, proof to these pilgrims of the Virgin's visitation. By the thousands and millions they come to gaze upon the Mother of Jesus, the dark Madonna wrought in their own likeness.

Whether Mexico's beloved "Brown Virgin" is the result of divine endowment or religious duplicity is still warmly debated. Even examination of the image by infrared photography and computer enhancement is inconclusive (Smith 1983). But for the Virgin's vast flock no scientific proof is required. To generations of Mexicans the compelling truth is that *La Morena* provides an identity, a "collective representation" (Wolf 1958, 38), of their worth and nationality. Through four-and-a-half centuries marked by foreign interference, cultural imperialism, rebellion, and a tortuous search for national identity, the icon of Guadalupe is the single element on which most Mexicans agree. Whatever her origin, she is certainly the most successful syncretic symbol in new world culture.

The Holy Place

Every visitor to the Shrine of Guadalupe is reminded that symbols "stand ambiguously for a multiplicity of disparate meanings" (Cohen 1974, ix). Colonial churches and chapels dot the hill of Tepeyac and the extensive courtyard, their ornate facades conveying a European message executed by Indian hands. Elaborate altars, fountains, gardens, and tombs add to the sense of accumulated time and cultural interplay. Sadly, many of the colonial structures twist and sink into the soft soil, threatening to disintegrate in the process. The old Basilica and adjacent Iglesia de Capuchinas are thus vacant, but owing to their historic and symbolic significance are being slowly restored.

Dominating the shrine today is a modernistic new Basilica. In purposeful contrast to the stolid, earthbound colonial churches, the new dwelling of Our Lady of Guadalupe soars skyward like a vast concrete cornucopia (Figure 2). Actually, the new Basilica is

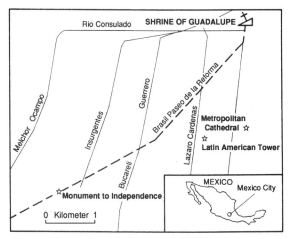

Figure 1. The Paseo de la Reforma linking downtown Mexico City and the Shrine of Guadalupe.

no taller than its predecessor but its rounded form and sweeping lines draw the eye upward and make it appear so. Symbolically, this new sanctuary lifts its revered mistress out of the past and thrusts her into the present and beyond.

The dissonance of traditional versus avant garde architecture is not lost on the Mexican populace. As construction proceeded toward a 1976 inauguration, the *strident modernity* of the new Basilica "raised some fears among the faithful that the Virgin might forsake her sanctuary" (*Time* 1976, 57). When new legislation simultaneously approved conversion of the old Basilica to a museum of religious art, public protest blocked the effort. The building was then locked to prevent a "popular clamour for return of the sacred image to its much loved, centuries old home" (Johnston 1981, 102). Such mass emotions underscore the Mexican

Figure 2. The new Basilica, completed in 1976, accommodates 20,000 worshippers. Built upon a thousand concrete pilings, it is designed to avoid the fate of its predecessor. Note the former Basilica at the right and its pronounced list.

identification with the image of Nuestra Señora, feelings beyond mere institutional iconography.

Modernization of Nuestra Señora's shrine, and the ensuing outcry, arose from political decisions, not ecclesiastical ones. Donations to finance the new Basilica were aggressively solicited on Mexican television and "Guadalupe bonds" sold in all the banks. Moreover, the public expected the government to cover any budget shortfall (*Commonweal* 1976). Several sources attribute the decision to build the new Basilica to Mexican President Echeverria (*Commonweal* 1976; Johnston 1981), a politician known for symbolic gestures, though seldom religious ones. Curiously, plans to modernize Mexico's singular national landmark and foremost religious shrine were revealed to the Abbot of Guadalupe, but not to the Cardinal of Mexico nor the Bishop's Conference.

For a contemporary Mexican president to take such a "religious" initiative is extraordinary; for the past 150 years church and state have been mutually hostile or at best concertedly indifferent. This strange marriage of politics and religion suggests that Graham Greene was correct—the Virgin is a ". . . patriotic symbol even to the faithless . . ." (Greene 1939, 98). It is more plausible to believe, however, that Echeverria sought a political sinecure by his symbolic gesture to Nuestra Señora de Guadalupe. Certainly there is precedent. Numerous aspiring leaders have embraced this "Mexican master symbol" (Wolf 1958, 34) as a political icon, notable examples being Father Hidalgo, instigator of Independence and the Revolutionary *cacique* (leader) Zapata.

The centerpiece of the Guadalupe Shrine is, without question, the "miraculous" centuries-old likeness of the Virgin preserved on a peasant's cloak. Without this tangible sign of divine visitation, without its miracle-working reputation, pilgrimage would falter and the Basilica would experience the underuse common to many Mexican churches. But the rapt devotion to the Madonna, an attachment that some ecclesiastics fear verges upon idolatry, fuels a continuing pilgrim zeal.

Incongruity persists inside the new Basilica. The venerated symbol, encased in gold, silver, and bulletproof glass, hangs to one side of the modern altar and massive pipe organ. In contrast to the smaller colonial Basilica, 20,000 souls possess a simultaneous, unobstructed view of their beloved *patrona*. Unavoidably, however, there is a loss of intimacy in the massive sanctuary. The "Little Virgin" is dwarfed and seems remote.

For the intensely devout, and the curious, a four-lane mechanical sidewalk provides an intimate audience with the Virgin. Out of the congregation's view, shawled

Mexican women kneel beside camera-laden tourists and silently they glide to and fro directly beneath the image.

The Apparitions

For 400 hundred years, within and beyond the Church of Rome, the authenticity of the "miraculous" origin of Our Lady of Guadalupe has been debated. In part, the controversy focuses upon the nature of an apparition experienced by the Aztec, Juan Diego. Analyzed still more vigorously is the apparition's physical manifestation that hangs in the Basilica, the reputed likeness of the mother of Jesus Christ.

The Virgin's story is known throughout Mexico. In 1531, 10 years after the Spanish toppled Moctezuma's Empire, a middle-aged Indian on his way to Mass was drawn to a former Aztec temple by beautiful music. Amidst its ruins the recently converted Juan Diego beheld the glowing image of a young woman. Speaking native Nahuatl, the apparition explained she was the Virgin Mary and urged that a temple be built on the site in her honor. Sent with that message to the Bishop in nearby Mexico City, Juan Diego was at first unable to convince the Bishop of his divine encounter. After several abortive trips between apparition and Bishop, the faithful Aztec received an extraordinary bouquet of flowers to authenticate his vision. Opening his *tilma* (cloak) to reveal the flowers to Bishop Zumarraga, all present beheld a likeness of the Virgin Mary on Juan's garment.

Juan Diego's apparition begets controversy. Some assert that Bishop Zumarraga immediately embraced the phenomenon and within weeks provided a small chapel to house its image (Johnston 1981). Others, however, suggests that the "miracle" was the consequence of a "... political fight ... among ecclesiastic and civil authorities" (Kurtz 1982, 194), and that the episode, including the personage of Juan Diego, was concocted after the fact by churchmen (Northrop 1946; Smith 1983). More recent conjecture even links the apparition to the Aztec practice of drug-induced hallucinations (Kurtz 1982).

Word of Juan Diego's visitor spread quickly, especially among the Indian masses, who were attracted by a propitious geography. Prior to Spanish rule the most popular of several Aztec shrines honoring Tonantzin, the "... preeminent female deity in the Aztec pantheon ..." (Carrol 1986, 184) was at Tepeyac, the site of the apparition's manifestation. This "topographic coincidence" (Lafaye 1974, 241) or "mystique of place" (Nolan 1953, 13) greatly renewed pilgrimage, which

was also stimulated by the related character of the two female deities.

Tonantzin and the Virgin symbolize fertility, rebirth, and the well-being of children (Kurtz 1982); both represent benevolent motherhood. Consequently, when the Spanish sanctioned the Virgin's manifestation under the title "Virgin of Tepeyac," many Indians found this a palatable transfiguration of Tonantzin. And because the Virgin appeared to an Indian instead of a Spaniard, "Our Lady" was further endeared to the masses, "... restoring in them the hopes of salvation" (Wolf 1958, 37), physical as well as spiritual. Indian conversion had "moved slowly" prior to the episode at Tepeyac, but was subsequently described as occurring in "... unprecedented, overwhelming numbers ..." (Carroll 1983, 108), with estimates of 9 million baptized between 1532 and 1548. To the devout, the "Mother of the True God" wilfully appeared at Tepeyac to symbolize Christianity's replacement of the Aztec religion (Johnston 1981); to the less credulous, the "... miracle of Guadalupe was not a mere replacement; rather, the two symbols were syncretized" (Kurtz 1982, 205).

Not all residents of New Spain were enthraled with this particular manifestation of Holy Mary in America. Ecclesiastical support was less than unanimous, often pitting one religious order against another. Concern was voiced about the genuineness of Indian conversions and their understanding of orthodoxy. Were pilgrims and converts drawn to Tepeyac by Mary, or by Tonantzin? Did they worship God or an image? If an image, was it European or Indian?

Spanish-born settlers in Latin America, including many clergymen and early secular leaders, supported numerous "daughter-shrines" of the Virgin of Los Remedios or the original Virgin of Guadalupe, both of which were European manifestations; they demonstrated little affection for a "painting on an Indian's cloak." But many New-World-born Creoles and disinherited Mestizos adopted the Virgin of Tepeyac as an indigenous icon, a counter-symbol to overweening European influence. Then, as the "Brown Virgin" endured and her popularity grew, Spaniards increasingly sought to "appropriate her for themselves" (Lafaye 1974, 230).

The Image

Apparitions are comparatively common in church history, most frequently linked with the "cult of Mary" (Carroll 1986). But seldom do these spiritual mysteries leave physical evidence. Consequently, artifacts of purportedly divine origin, such as the Shroud of Turin and the Image of Our Lady of Guadalupe, generate great hope, devotion, and invariably, debate.

Detractors of the Guadalupe apparition cite the absence of any contemporarily written historic references as evidence that Bishop Zumarraga (or someone) contrived or commissioned the image at a convenient moment to protect the Indians against secular abuse (Kurtz 1982) or to expedite their conversion (Smith 1983). It is also asserted that the image on the tilma replaced an earlier statue placed at Tepeyac by Spaniards to commemorate the original European shrine of Our Lady of Guadalupe (Carroll 1986). Allegedly, the European icon did not appeal to the Indians and the substitution was made. Some early church leaders also questioned the divine origin of the image, though their inspiration was often based on disputes between rival religious orders. Thus, in 1556, the Franciscan Provincial Bustamante criticized the cult of Guadalupe, favored by the Dominican Archbishop of Mexico City, and publicly charged that the image had been "painted by an Indian" (Johnston 1981, 69; Lafaye 1974, 238). Charges from secular sources pointedly refer to the "artistry of the Roman Catholic Church" (Northrop 1946, 26).

The subject of dispute is a seemingly simple painting on a crude Aztec cloak; the likeness is that of a young woman in flowing gown, standing with head slightly bowed and hands reverently clasped (Figure 3). Yet, this representation of Juan Diego's vision has become an exceedingly complex icon through both symbolic interpretation and physical manipulation.

The appeal of the Guadalupe Shrine, especially for the Mexican majority, is attributed to the "Indian" character of the image (Lafaye 1974). Skin tone, variously described as brown or olive in color, gives rise to the affectionate term *La Morena*, the dark Virgin. It is probable that at some point the "face and hands were painted darker" (Johnston 1981, 133) and the fingers shortened to make them appear "more Mexican" (Johnston 1981). Likewise, La Morena's garments are identified as the "gala dress of an Aztec princess" (Dooley 1962, 49). The absence of a child suggests additional intentional syncretism since ". . . churchmen in their artistry did not dare tamper with the spontaneous movement of the Indians' spirit by even attempting to insure orthodoxy of the Virgin by placing a Christ-child in her arms" (Northrop 1946, 27). Explicitly, if the indigenous population was drawn to Tepeyac by Tonantzin, who bore no child, neither should her successor.

Ironically, the image exhibits evidence of having been manipulated to be more "European." While authorship of portions of the image remains unexplained, other elements are clearly additions (Smith 1983).

Figure 3. The image responsible for much devotion and much controversy. Though purportedly more than 450 years old, the fabric and colors remain viable.

Tassels added to the gown suggest pregnancy, a gesture to orthodoxy; a brooch bearing a Christian cross is self-explanatory. The most symbolic additions, though, are the crescent moon and cherub that support the Virgin. These clearly suggest the Lady of the Apocalypse as described in orthodox scripture in Revelation 12:1-2 (Johnston 1981). Probably done between 1629 and 1634, these modifications are assumed by most devotees to be original.

Our Lady of Guadalupe evokes the essence of the syncretic symbol. She has linked Christian and pagan, simultaneously serves Indian, Mestizo, and European, and even provides limited common ground for Church and State. Attending the "birth" of New Spain, the colonial rebellion, and the revolutionary search for nationhood, the Virgin permeates the collective Mexican spirit. Throughout Mexico her image is more popular than that of Jesus Christ, and adorns homes, bull rings, restaurants, buses, and even houses of ill repute.

A Spiritual Oasis

Mexico is passing through troubled times. Unemployment, inflation, political dissent, and environmental decay beset the nation, their manifestation most acute in the burgeoning capital city. Yet here amidst a maelstrom of uncertainty the people find solace in a special place.

On 12 December 1990, the pilgrims converged upon the site of the Virgin's visitation. With virtually no hope of viewing the image, or of entering the Basilica, between 2 million and 3 million faithful came. It was enough to be near. A small army of police and medical personnel witnessed a human mosaic filling the streets for blocks beyond the shrine. The shuffling throng, largest in recent memory, raised hymns in various tongues—Spanish, Nahuatl, and Otomi—fervent in their faith that the Little Brown Mother of God heard and understood her children.

References

Carroll, Michael P. 1986. *The Cult of the Virgin Mary.* Princeton: Princeton University Press.

Carroll, Warren H. 1983. *Our Lady of Guadalupe and the Conquest of Darkness.* Front Royal, VA: Christendom Publications.

Cohen, A. 1974. *Two-Dimensional Man: An Essay on the Anthropology of Power and Symbolism in Complex Society.* Berkeley: University of California Press.

Commonweal. 1976. 26 March 26:194.

Dooley, L. M. 1962. *That Motherly Mother of Guadalupe.* Boston: St. Paul Editions.

Greene, Graham. 1939. *Another Mexico.* New York: Viking Press.

Johnston, Francis. 1981. *The Wonder of Guadalupe.* Devon, UK: Augustine Publishing Company.

Kurtz, Donald V. 1982. The Virgin of Guadalupe and the politics of becoming human. *Journal of Anthropological Research* 38:194-210.

Lafaye, Jacques. 1974. *Quetzalcoatl and Guadalupe.* Chicago: University of Chicago Press.

Nolan, Mary Lee. 1953. The Mexican pilgrimage tradition. *Pioneer America* 5:13-27.

Northrop, F. S. C. 1946. *The Meeting of East and West.* New York: Macmillan.

Smith, Jody Brant. 1983. *The Image of Guadalupe: Myth or Miracle?* Garden City, NY: Doubleday.

Time. 1976. 20 December 20:57.

Wolf, Eric R. 1958. The Virgin of Guadalupe: A Mexican national symbol. *Journal of American Folklore* 71:34-39.

Catholic Pilgrimage Centers in Québec, Canada

Gisbert Rinschede
Universität Regensburg

The Canadian province Québec has among all cultural regions of North America the largest concentration of Catholic pilgrimage places because of the strong cultural influence of French immigrants. Three pilgrimage centers are of significant international and national importance: Notre-Dame-du-Cap in Cap-de-la-Madeleine, Oratoire-St.-Joseph in Montréal, Ste.-Anne-de-Beaupré near Québec City, as well as eight other centers of national significance (Figure 1), and other sites are of regional and local importance.

This investigation of pilgrimage centers in Québec asks the question of how pilgrims affect pilgrimage sites and influence building pattern and functions by their religious and touristic activities. The description and analysis of site patterns is primarily focused on the three major pilgrimage places and especially on Ste.-Anne-de-Beaupré where the effects of pilgrim activities can best be distinguished.

Pilgrims as an Action Group

Approximately 5.5 million pilgrims annually (1988) visit the 11 largest pilgrimage centers in Québec, of whom 5 million visit the three largest centers: Oratoire-St.-Joseph, 2.5 million visitors; Ste.-Anne-de-Beaupré, 1.5 million visitors; and Notre-Dame-du-Cap, 1.0 million visitors. The remaining eight sites receive only 500,000 pilgrims, with 50,000 to 120,000 visitors each.

Pilgrims from overseas represent the French-speaking countries of France, Belgium, Haiti, and Martinique as well as European countries—Germany, Switzerland, Italy, Spain, Austria, Luxembourg, and Great Britain. Somewhat less frequently represented are Brazil, Australia, and New Zealand. The most important country of origin outside Canada is the United States (Figure 2). Most strongly represented regions are the Northeast, Mid-Atlantic, and Midwest. A few pilgrim groups also come from states in the Southwest where the concentration of Catholics is especially large. Besides Québec, the western and eastern bordering provinces of Ontario and New Brunswick form the largest contingent of pilgrims from within Canada. Aside from geographical proximity, a noteworthy share of French speakers characterizes Québec, its neighboring provinces, and the bordering U.S. states. Within Québec, the catchment areas of the three largest pilgrimage centers correspond to population distribution along the Saint Lawrence Seaway.

As a rule, the visitors of the pilgrimage sites from Québec are indeed pilgrims who come for religious reasons. Should the pilgrimage site also be of historical and cultural interest (Oratoire-St.-Joseph and Ste.-Anne-de-Beaupré) or should it be characterized by special festivals throughout the year (Notre-Dame-de-Lourdes), then tourists come as well. According to a survey in Ste.-Anne-de-Beaupré, the motive for 76.7 percent of the visitors was religious, for 23.7 percent touristic, and for 6 percent religious and touristic (Baillargeon 1988). A special type of pilgrim participates in the varied spiritual exercises, religious holidays, and conferences, that take place on pilgrimage grounds. These activities are most significant at Lac Bouchette and Sillery.

Most pilgrims remain only one day—eight hours or less—on the pilgrimage grounds. Because the distance between hometown and pilgrimage site is often considerable, the pilgrims visit other religious sites in the vicinity, if possible. Pilgrim groups from the northeastern United States and from the eastern and western provinces of Canada arrive after a several-day journey. Often they visit purely touristic attractions as well, which are accessible either along the way or in the vicinity of the pilgrimage site. For example, in the

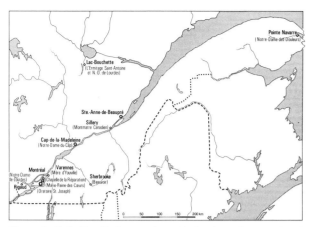

Figure 1. Catholic pilgrimage places in Québec of national and international importance, 1990.

urban surroundings of Oratoire-St.-Joseph, there are the sights of Montréal. The sights of Québec City are only some 20 kilometers from Ste.-Anne-de-Beaupré, which also gives access to natural wonders to the north.

School groups make up a considerable share of the total number of visitors during certain months of the year. They represent over 50 percent of all groups, that make a formal tour of the center at Oratoire-St.-Joseph during the months of minimal pilgrimage tourism (January to February and November to December). In the remaining months other groups predominate, such as senior citizen organizations, exclusive men's or women's clubs, dioceses, parishes, youth organizations, and travel agencies. Within organized groups, two thirds and sometimes as many as 80 percent of pilgrims are over 40. Near the end of the season (August to September), the over-50 age group is heavily represented.

As to socio-professional status, Canadian and American pilgrims differ (Doran-Jacques 1979). At Ste.-Anne-de-Beaupré approximately 64 percent of the Canadian pilgrims, compared with 73 percent of the U.S. pilgrims, were active in the service sector. We can partially explain this fact in that Canadian pilgrims tend to have rural origins, while their American counterparts are largely urban. Small numbers of sick and disabled pilgrims visit the three largest pilgrimage centers exclusively.

Not only English- and French-speaking Canadians take part on pilgrimage journeys to the religious centers of Québec, but other North American ethnic groups as well. The Indians pray to Saint Anne, whom they regard as their patron saint ever since their conversion to Catholicism. Annually, some 2,000 Indians of the tribes Huron, Mohawk, Micmac, Attikamègues, Montagnais, Abenaki, Algonquin, and Maliseet from Québec, Ontario, New Brunswick, Newfoundland, Labrador, and Maine take part in a common pilgrimage journey on the occasion of the birthday of Saint Anne and to the big Novena in late July. At this time, gypsies meet as well. Italian pilgrims come from the large metropolitain centers of Montréal, Ottawa, and New York. In recent years, first-generation immigrants have also made the pilgrimage—mostly in small groups by car, but also in larger groups. These include Haitians, Mexicans, Filipinos, and Vietnamese from the northeastern U.S. and Ontario.

The climate imparts a seasonality to pilgrims in Québec (Figure 3). Remote pilgrimage sites with predominantly regional catchment areas close during the winter season (September or October until Easter or late May). At most year-round sites the pilgrimage stream has its high season from May to September,

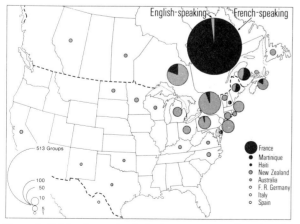

Figure 2. Origin of organized pilgrim groups (French and English speaking) to Ste.-Anne-de-Beaupré, 1988. Source: Information of the Shrine Office (1989).

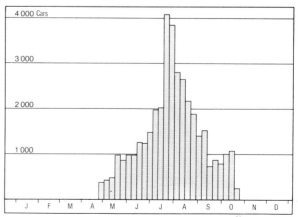

Figure 3. Number of cars on a parking lot at Ste.-Anne-de-Beaupré in the season of 1988. Source: Information of the Shrine Office (1989).

its peak in June and August, and its off-season from October to mid or late April. The vacation month of July is less significant for organized pilgrim groups, whereas the number of pilgrims who travel by private means increases at this time.

Pilgrimage journey on foot, which has a long tradition in the Old World, also developed in Québec. But today pilgrims journey on foot over shorter distances, such as from Beauvoir to Sherbrooke (15 kilometers) as well as over greater distances from St.-Theele (80 kilometers) or Montréal (150 kilometers) to Notre-Dame-du-Cap. Because of the pilgrimage centers' locations on the Saint Lawrence Seaway, the journey by boat played an essential role in their 300-year history. Today Ste.-Anne-de-Beaupré is still accessible by boat for pilgrims from neighboring Québec City. Near the end of the 19th century, the larger centers became accessible by train. For decades, train travel remained the most important means of transport. Today the pilgrimage journey by train has completely ended. Only after the turn of the century did the first automobiles transport pilgrims. Autos now transport most of the pilgrims (65 to 75 percent). Buses transport 25 to 35 percent. Few pilgrim groups, mostly ill pilgrims, arrive by plane.

Pilgrimage Sites

The pilgrimage sites of Québec originate from entirely different periods. The oldest pilgrimage site, Ste.-Anne-de-Beaupré, owes its founding to French settlers, who brought their devotion to Saint Anne to the region in the early 17th century. In the second half of the 19th century numerous new pilgrimage sites came into being, following the European model or imported from the religious orders. In Montréal, pilgrimage journeys to Frère André, a simple brother of the order who with the help of Saint Joseph performed numerous miracles, began in 1877. The founding of the pilgrimage site Notre-Dame-du-Cap (1878 to 1888) is also related to miracles. Other pilgrimage sites, founded in the beginning and middle of this century, owe their existence less to miraculous events, and more to the founding activities of individual orders.

The immediate location on the Saint Lawrence Seaway or the elevated location on a hill or mountainside is characteristic of Québec's large pilgrimage sites. The sanctuary is at the center of the religious sites (Figure 4). The shrine church usually constitutes this sanctuary. There are sometimes smaller old shrine churches, remnants of the original holy centers, and a series of chapels. In the mid-sized pilgrimage centers, there are outdoor churches with open-air altars. Replicas of the Lourdes-grottos constitute a part of the religious site. Almost all pilgrimage sites have Outdoor Ways of the Cross, Rosary Ways, Holy Stairs, Calvary Scenes, the Holy Sepulchre, or Gospel Walks. Monasteries are also located in the immediate neighborhood of the Shrine grounds, as religious orders founded many of the pilgrimage sites in Québec and administer them to this day. In the two largest pilgrimage centers,

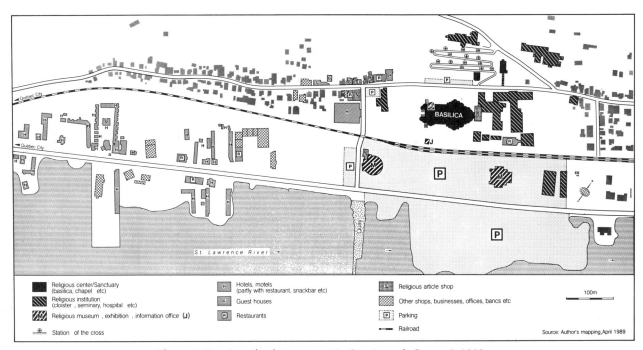

Figure 4. Functions (land-use pattern) in Ste.-Anne-de-Beaupré, 1989.

Oratoire-St.-Joseph and Ste.-Anne-de-Beaupré, various information facilities are available. At every pilgrimage site shops sell religious articles, religious literature, and souvenirs with symbols of the shrine. These shops represent an important source of income for the pilgrimage site. Almost every pilgrimage site has boarding and accommodation facilities. The religious orders also operate these. If there is enough space on the grounds, picnic tables are set up. Closely tied to the other facilities are the educational facilities on the grounds, such as the renewal, retreat, and spiritual centers at Lac Bouchette and Sillery. The large pilgrimage centers of Québec do not have the religious radio stations at their disposal as do corresponding sites in the United States (Rinschede 1990), but do publish their own magazines several times a year for intensifying their contact to the pilgrims.

The most important representatives and administrators of the Catholic pilgrimage sites in Québec are the religious orders, whose organizational skills appear especially suited to this task. However, approximately half of the pilgrimage sites were not, at their time of founding, under the administration of religious orders. The orders took them over much later in order to secure their continued existence. While 3 to 15 priests in addition to several mostly voluntary assistants run the mid-sized pilgrimage sites in Rigaud, Beauvoir, and Lac Bouchette, the three larger sites are according to seasonal demand under the auspices of 40 to 70 priests and members of the order as well as 100 to 400 secular assistants.

Pilgrimage sites emerged as an answer to the religious needs of the pilgrims. The shrine personnel have the task of fulfilling these needs, whether for personal prayer and meditation or for common participation in liturgical festivals. The pilgrimage journey has a therapeutic function, seen in the pilgrims' prayers of supplication. With their greater financial resources, the larger pilgrimage sites also have the task of supporting their respective orders in their social, pastoral, and missionary activities.

Pilgrimage sites and their surrounding areas are subject to the mutual influences of one another. In general, the pilgrimage site and its pilgrims affect the growth of population, settlement, and economic structures in its rural or urban surroundings. However, at the mid-

sized and smaller pilgrimage sites in Québec, as in the United States, such an influence is hardly noticeable. The oldest pilgrimage site, Ste.-Anne-de-Beaupré, has in this regard the greatest similarity with the larger European pilgrimage centers (Figure 4). The influence of the pilgrimage stream makes itself clearly visible outside the religious center, which consists of cathedral, monastery, way of the cross, scala santa, and other religious components. The settlements directly neighboring the religious center have several shops that see religious article, restaurants, and private guesthouses. Yet more obvious is the influence of the pilgrimage stream along the highway to Québec City, which is lined with numerous new motels.

Conclusion: Future Development

Pilgrims of various countries in the Old and New World visit pilgrimage places in Québec, and influence their economic structure, and the distribution of buildings such as places of worship, hotels, restaurants, and shops. Factors such as spare time and touristic activities, shift of religious activities from daily or weekly services to more periodical and episodical activities, forecast the further increase of pilgrim numbers. The development and expansion of pilgrimage centers will mainly occur in areas with favorable travel connections, near large agglomerations and major tourist centers.

References

Baillargeon, S. 1988, 1989. *Rapport sur les Activités Pastorales au Sanctuaire de Sainte-Anne-de-Beaupré pour la Saison 1987, 1988.* Ste.-Anne-de-Beaupré, PQ: Sanctuaire de Sainte-Anne-de Beaupré.

Doran-Jacques, A. 1979. *Le Pèlerinage à Sainte-Anne-de-Beaupré: l'Actuel 1958-1973.* Thèse de doctorat de 3ième cycle. Paris: École des Hautes Études en Sciences Sociales.

Rinschede, G. 1990. Catholic pilgrimage places in the United States. In *Pilgrimage in the United States*, Interdisziplinäre Schriftenreihe zur Religionsgeographie, Vol. 5, Geographia Religionum, eds. G. Rinschede and S. M. Bhardwaj, pp. 63-135. Berlin: Dietrich Reimer Verlag.

Walden

Michael Steinitz
Massachusetts Historical Commission

Joseph S. Wood
George Mason University

Twenty-four kilometers west-northwest of downtown Boston, just south of the heavily travelled Route 2 corridor, a scrubby oak-pine forest typical of a southern New England glacial outwash plain surrounds a modest kettle pond. The adjacent woodlands, wetlands, and stony, till-covered hills have preserved to a remarkable degree the appearance of natural character after 350 years of European occupancy of the region. The site's marginal agricultural attributes and its interstitial location at the boundary between two Massachusetts towns have protected this ordinary landscape. Near the townline and downwind of the center village, it has however proved useful for constructing a landfill. And, after the extraordinary real estate boom of late 1980s in eastern Massachusetts, the area also contains some of the last large local tracts available for development.

The pond by the dump and the roughly 10 square kilometers of woods and fields around it also form a place of rich and unusual historical association. They are part of an ensemble of natural and cultural landscape features that made up the village and countryside of 19th-century Concord, the setting for the musings of transcendental philosophers, poets, novelists, essayists, and reformers. The pond and the woods loom especially large in the American cultural imagination, for they are the place of the spiritual odyssey that gave title to Henry David Thoreau's masterpiece, *Walden; or, Life in the Woods* [1854] (1960). Here, Thoreau spent much of the years 1845 to 1847 trying to learn how simply life could be lived in material terms and how richly it could be lived in spiritual terms. At the urging of his friend and mentor, Ralph Waldo Emerson, who himself had purchased the woodlot at Walden Pond for its contemplative qualities, Thoreau wrote of his exploration and discovery. Visitors from around the world come to Walden Pond today to celebrate the historical and literary significance of Thoreau's work. Yet now Walden's sanctity is threatened, and efforts to preserve it have generated a controversy quite in the spirit of the place. Nearly a century and a half after Thoreau's experiment in living, the pond by the dump in the woods still symbolizes competing visions of the American landscape.

The Pond and the Woods

Antebellum Concord was the focus of a remarkable conjunction of human creativity and genteel place. English colonists, lured by fertile intervale meadows, first settled against a bluff on the edge of the Concord River in 1635, forming the farthest inland New England settlement of the time. Over the next 150 years, population dispersed in a pattern that reflected the variable fertility of the surrounding countryside, by-passing pockets of glacial till and scrub forest to be left as woodlot and waste. Concord grew in importance as the seat of Middlesex County and became the leading inland urban place in eastern Massachusetts, linked to Boston, a half-day ride away, and to the Connecticut Valley. To Concord came the British army in 1775, seeking the armory harbored there, and off went "the shot heard round the world."

Like center villages across New England, Concord underwent wrenching economic and social adjustments in the early 19th century: commercialization of agriculture, rise of a middle class, industrialization, and spread of the railroad. Places better situated for industrial development than Concord became more important economically, but Concord, close to Boston, came to serve as an appealing residential suburb. Moreover, for Romantic New Englanders, Concord became a central element in the literary landscape of 19th-century

New England. Transcendentalists and others found stimulation in antebellum Concord's 200-year-old cultural landscape, conceptualized the scene in sophisticated terms, and mirrored it in their writing. They contemplated woods and orchards, fields and pastures, and ponds and rivers of a settled countryside, which Thoreau distinguished from wilderness and from ruder rural regions. Emerson found Concord a serene yet urbane refuge from the commercialism of Boston. For the disaffected Romantic Thoreau, Concord satisfied a need for connection to a larger world while allowing a modicum of escape as well. As the Concord countryside evoked rich figurative language while socially and economically attached to Boston, so Walden Pond and its woods evoked rich figurative language while socially and economically attached to Concord Village (Wood 1991).

The woods about Walden Pond had long provided refuge for the town's outcasts, including freed slaves, drunkards, and the shanty Irish who had built the railroad. From this vantage point on the fringe, classical scholar Thoreau launched his critique of modernization and the aggressive, optimistic, profit-seeking society that surrounded him. The village, with its links to the urban-industrial world, represented the ambiguities of the modern 19th-century world he inhabited. The railroad, which cut through Walden Woods to Boston, became the central symbol of the technocracy that disfigured nature and alienated labor.

The pond and the woods came to embody an alternative set of values. Yet, Thoreau's theme was not simply withdrawal from society into an idealized landscape. In the humanist tradition, Thoreau meant metaphorically to establish a middle ground between and in transcendental relation to opposing forms of civilization and nature (Marx 1964; Nash 1973). While at Walden, Thoreau frequently visited friends in the village and stayed there overnight, and he entertained them at his cabin in the woods (Figure 1). Once he completed his experiment in simplicity, he returned to Concord, where he lived the remainder of his years, "a sojourner in civilized life again" (1960, 7).

Walden, Environmentalism, and Geography

Walden. Few secular place names evoke so clearly a moral philosophy and a world view. Walden was "the supreme idyll of Romanticism in America" (Kazin 1988, 57). It is an exemplar of nature writing, which reconciles nature and individual. Literary and historic significance have redounded upon the sites described by the poet-naturalist in the development of his ideas, ideas upon which are constructed all subsequent American inter-

Figure 1. Thoreau's cabin from a sketch by his sister Sophie in the 1854 edition of *Walden*. The cabin has become a material expression of Thoreau's concern for simplicity. A reproduction intrigues visitors to the pond and the woods today.

pretations of nature, succession, and renewal, the latter both in nature and in the spirit. Thoreau attempted to convince his contemporaries of their short-sightedness in prizing only the material and of treating wilderness as unattractive and unredemptive. His philosophy of continuity of humans with nature, that "We can never have enough of Nature" (1960, 267), became an intellectual foundation of the American conservation movement (Kazin 1988; Nash 1973). Thoreau's writings on "the tonic of wildness" (1960, 267) are seminal, the foundation for an intellectual tradition, a line of utopian and bioethical argumentation that directly influenced John Muir, Aldo Leopold, and countless other advocates of conservation philosophy. His writings found their mark as well in Olmsted's urban parks, development of the national park system, and wilderness preservation.

Walden is as well a lesson for geographers. As a naturalist and nature historian, Thoreau was a keen observer. He took measurements of the pond and the woods; he mapped them in detail; and he studied their ecology. He pursued archaeology to understand the relative value of the pond and the woods for human use in the terms of the native inhabitants, the farmers who had preceded him, and the industrialists who were close on his back. He recognized the commodification of nature and degradation of physical objects that resulted from commercialization. He understood the spatial organization of trade in the goods carried on the trains that passed by the pond and through the woods. He made a living, literally and figuratively, roaming the byways, fields, and woods of Concord as a surveyor of the land. And he had a most sophisticated sense of landscape in his head, noting, "I have frequently seen a poet withdraw, having enjoyed the most valuable part of a farm, while the crusty farmer supposed that he had got a few wild apples only" (1960, 73). The poet was Thoreau himself, for whom poetry and geography mingled constantly (Stowell 1970). His geographies were geographies of the mind as well as of the ground, and he knew well how to share them with his fellow Americans, how to make the most of the *genius loci*. Thoreau wrote in *Walden* perhaps the most important Romantic literary account of the New England landscape (Buell 1986).

Preservation of Walden Woods

If Thoreau is the patron saint of modern environmentalism, then Walden is his shrine, a high-order sacred place for the conservation movement (Kay 1988). The preservation of the holy water and sacred groves has therefore taken on special significance for conservationists. The Emerson family retained ownership of the 33-hectare woodlot where Thoreau built his cabin until the 1920s, when they transferred the land to the state with deed restrictions intended to preserve its historic character. The land became part of a 60-hectare Walden Pond State Reservation made up of the woodlands surrounding the pond. Although the reservation was ultimately expanded to 100 hectares, in the late 1980s less than a third of the area delineated as historic Walden Woods by local Thoreau scholars (Blanding 1988; Schofield 1989) had been designated as public land, and the fate of the unprotected, privately held land outside the conservation areas had become a source of controversy.

A long-term debate has continued over appropriate use of the pond and the woods. In 1866, the Fitchburg Railroad, which cut through the woods on the western edge of the pond, established a recreational park that proved to be a popular resort for weekend pleasure-seekers wishing to escape the Boston summer heat. Walden remains to this day the destination of choice for thousands of bathers and picnickers on a hot summer weekend. Park development and the sheer number of active recreational users lining Walden's shores have always been at odds with Thoreauvians' perceptions of Walden as a fragile historic landscape and nature

preserve. According to critics, the state fails to restrict access and visitation to stem the degradation of vegetation and the pond's shoreline. The Thoreau Society successfully sued the state in the 1950s over efforts to reconfigure the shoreline to improve bathing facilities. While the Massachusetts Department of Environmental Management, the present steward, has removed the most obvious modern intrusions on the site, one protectionist advocacy group, Walden Forever Wild, continues to call for a ban on all swimmers (Kay 1988).

Disputes over management of the public lands have recently been overshadowed by a highly publicized fight over the fate of two privately held parcels of land in the vicinity of the pond. Owners proposed large-scale construction projects overlooking the pond: an office-park complex 650 meters from the pond and a 139-unit condominium complex 1,300 meters from the pond. The office-park proposal drew local support from those seeking to expand Concord's tax base. The condominium proposal effectively by-passed local zoning restrictions by offering to subsidize a third of its housing units at affordable price levels, triggering the state's strict, anti-snob zoning law. This action gained the support of state housing authorities who had been pushing for moderate-level housing projects in exclusive suburban communities like Concord (Convey 1989). Ironically, the developer of the office park is Mortimer Zuckerman, publisher of the monthly journal *The Atlantic*, which in the 1860s published several of Thoreau's essays.

While public hearings, suits, and environmental reviews received widespread news coverage, preservationists were often portrayed as apologists for those who would exclude moderate-income residents from a wealthy suburban community (Convey 1989). With the state Executive Office of Communities and Development firmly behind the housing project, the prospects for successful blockage seemed remote, until rock star Don Henley, seeing news reports of the threat to the woods, joined forces with local activists in April 1990 to form the Walden Woods Project. The new coalition, studded with celebrities and politicians, convened high-powered news conferences, held benefit concerts, initiated a massive fund-raising effort, and began negotiations with property owners and public officials. Within a month Walden Woods was added to the National Trust for Historic Preservation's list of most endangered places. Henley countered housing activists by incorporating housing issues into his plan of action, promising to find alternative sites within the community (Canellos 1990). By early 1991, the coalition had secured the purchase of the condominium site, negotiations for an alternative affordable housing site were well under way, but confrontation continued with the intransigent Zuckerman.

The Meaning of Walden

The success of the Thoreauvians in preserving from condominium development Bear Garden Hill, where Thoreau picked huckleberries and found inspiration for many of his *Journal* passages, points to the symbolic power of Walden to inspire environmentalist action. Yet, the fight over the office park between preservationists and developers highlights the continued disjunction between ways landscape values are defined. Developers see a fringe landscape by the town dump that had been transformed into prime real estate by a booming regional economy. Preservationists see an intact ecological district, where Thoreau worked out his theories of forest succession, where he and others grounded their writings, and where modern concepts of wilderness conservation were formalized. In Walden Woods, the landscape is not the result of a dramatic historical event, such as nearby Battle Road in the Minuteman National Historic Park. Nor does the significance of Walden Woods as a cultural landscape find obvious expression in an ensemble of surviving historic buildings, in contrast to the homes of 19th-century authors in Concord Village. But then Thoreau did believe that a person is wealthy in proportion to the material things that he or she can afford to live without.

References

Blanding, Thomas. 1988. Historic Walden Woods. *The Concord Saunterer* 20(1-2):1-74.

Buell, Lawrence. 1986. *New England Literary Culture: From Revolution through Renaissance.* New York: Cambridge University Press.

Canellos, Peter S. 1990. Clash of liberal values over Walden. *Boston Globe* 5 April:25, 32.

Convey, Kevin R. 1989. Why Concord hates Walden. *Boston Magazine* July:81-83, 100-105.

Kay, Jane Holtz. 1988. Wall to wall at Walden. *The Nation* 246:867-872.

Kazin, Alfred. 1988. *A Writer's America: Landscape in Literature.* New York: Alfred A. Knopf.

Marx, Leo. 1964. *The Machine in the Garden: Technology and the Pastoral Ideal in America.* New York: Oxford University Press.

Nash, Roderick. 1973. *Wilderness and the American Mind* (rev. edition). New Haven: Yale University Press.

Schofield, Edmund A. 1989. The Walden Woods ecosystem. In *Walden Woods.* Concord, MA: Thoreau Country Conservation Alliance.

Stowell, Robert F. 1970. *A Thoreau Gazetteer.* Princeton, NJ: Princeton University Press.

Thoreau, Henry David. [1854] 1960. *Walden; or, Life in the Woods.* Boston: Ticknor & Fields [republished New York: Doubleday].

Wood, Joseph S. 1991. "Build, therefore, your own world": The New England village as settlement ideal. *Annals of the Association of American Geographers* 81:32-50.

The Grand (Mount) Monadnock

A. L. Rydant
Klaus J. Bayr
Keene State College

They who simply climb to the peak of Monadnock have seen but little of the mountain. I came not to look off from it, but to look at it.

—THOREAU (in Howarth 1982, 347)

A mountain is more than a physical feature to those who live in its shadow. It may represent the mundane, the necessity to survive, or loftier ideals, an expression of humanity's position in the cosmos. To New Englanders, Mount Monadnock represents both these perspectives, as well as a range of sentiments in between. "Monadnock" describes a classic geologic feature; it is a means of livelihood for farmers, loggers, and recreationists; it is a representation of Nature herself. This geographical description of the mountain discusses its "place" in New England from both the physical and cultural perspective. The physical overview provides a foundation for its site prominence, while the cultural history examines how Mount Monadnock has come to embody the soul of the region.

Physical Perspective

Mount Monadnock, elevation 965 meters (3,165 feet), is in the southwestern part of New Hampshire in the towns of Dublin and Jaffrey. The area dominated by the mountain carries the name Monadnock Region and stretches approximately from the southwestern corner of New Hampshire to the town of Milford, 69 kilometers (43 miles) east, and from the town of Marlow in the north to the Massachusetts state boundary, 53 kilometers (33 miles) south. The mountain towers 510 meters (1,686 feet) above Dublin Lake on its northern flank and approximately 605 meters (2,000 feet) above Thorndike Pond to its east (Figure 1A); it is a significant landmark that can be seen for some distance (Figure 1B). From the top of the mountain on a clear day one can see 104 kilometers (65 miles) eastward to the Atlantic Ocean and Boston; northward the White Mountains of New Hampshire are visible (152 kilometers [95 miles]), and westward the Green Mountains of Vermont (75 kilometers [47 miles]). On a clear day, the summit view encompasses 6 New England states. Generations of visitors have examined these rocks and imagined personality (the Imp, the Tooth) or antiquity (Sarcophagus, Doric Temple). Its singular form has also inspired a generic term in geology (Howarth 1982, 287).

Most of the mountain's surficial rocks date from the Lower Devonian and belong to the Littleton Formation (Fowler-Billings *et al.* 1949) (Figure 1C). The composition of the rock formation is quite varied and demonstrates a high degree of metamorphism. The southwestern part has Concord Granite from the New Hampshire magma series, which is Late Devonian.

The geology of Mount Monadnock reflects a turbulent development in conjunction with drifting continents. Once the floor of a great ocean, the sedimentary rocks were uplifted during the Late Devonian by the collision of the North American and African plates. Escaping magma cooled into granite and also rose, exerting pressure on the sediments. They metamorphosed into hard mica and schists (quartz mica schists, sillimanite schist, sillimanite garnet schist) and formed synclines that later fractured and jointed. These are best developed and visible today on the south side of the mountain directly beneath its summit, near the White Arrow Trail (Figure 1A).

Ages of erosion, occurring both prior to and following the uplifting, reduced the region's weaker rocks to a rolling plain (referred to as a peneplain by some authorities), leaving the more resistant rocks to stand above the peneplain surface. Such erosional remnants are dubbed monadnocks, a term coined by William Morris Davis in 1894 to describe a residual landform

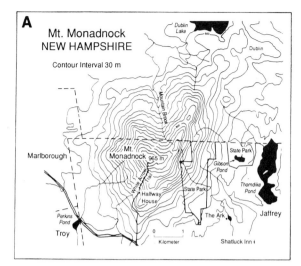

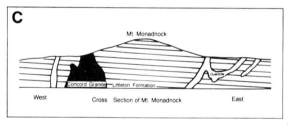

Figure 1. A. Mount Monadnock, location and land uses. B. Mount Monadnock, view from Dublin, New Hampshire, looking southeast. C. Cross-section of Mount Monadnock, west to east. After Fowler-Billings et al. (1949).

"that rises over the uplifted plateaus of denudation" (1894, 99).

Finally, great ice sheets more than 1.6 kilometers thick pushed across the mountain generally from north to south cutting deep ledges, smoothing, pulling, and plucking, creating "sheep backs" and deep angular grooves (Howarth 1982, 287). Many erratics and striations on the south slope demonstrate the great impact glaciation had on this feature, which is perhaps best characterized as a roche moutonnée. Its singularity in both form and space therefore gives it a unique place in history.

Although Monadnock is essentially a geographic point, elements of its position and development imbue it with a unique cultural and historic role in New England. Its place has evolved from that of a "no man's land," to a zone of marginal agriculture, toward a more symbolic representation exemplifying the emergence of an "American" culture via art and literature. Today this symbolic role is played out from a recreational perspective.

Symbolic Representation

"Monadnock" is an Algonquin name, essentially meaning "place of unsurpassed excellence" (Nutting 1925). While more than 22 spellings have been recorded, the image of the mountain has from the outset been consistently described by such terminology as "one going beyond all others in that vicinity for size," "at the honored or respected mountain," and "a mountain that stands alone" (Chamberlain 1968, 81).

A written account of the mountain first appeared in the chronicles of Governor Winthrop of Massachusetts in 1632. It was initially mapped in 1677 as part of the "White Hills" but not named until 1704 in an account of Indian killings in the region. During these pioneer years of settlement Monadnock was considered of little utility and was actually mapped as a "no man's land." In a citizens' petition, in 1787, Jaffrey residents requested 80 hectares (200 acres) of this "wasteland" for land to till and maintain their minister (Nutting 1925). Monadnock land was offered for sale in the mid-18th century for $0.02 per acre compared with the surrounding agricultural land's value of $3.33 per acre (Chamberlain 1968). Initially, it was at the fringe of a growing New England, and it remained so well into the early 19th century. Its main use was as pasture until about 1850, although several proximate base roads were opened as early as 1766, unknowingly setting the scene for later use.

Nature and landscape were common themes in the American art and literature that blossomed early in the 19th century. The pristine environs of southern New Hampshire thus took on special significance for a host of literary and artistic retreats. Intellectuals strove to create an American culture, unique and free from European dominance, traditions, and influence. Literature drew inspiration from the inherent conflicts between nature and an emerging urban society. New Hampshire's unspoiled environment and open countryside offered a sense of escape, as well as of horizon, perspective, and distance. It symbolized a place untouched by the harness of civilization. Thus, the landscape's most visible and prominent feature—

Monadnock—came to represent one's hoped-for condition in life. As early as 1785, Timothy Dwight, the President of Yale College, wrote that in Mount Monadnock "is seen the richest prospect in New England, and not improbably in the United States"; several years later he added that "a finer object can scarcely be conceived" (Nutting 1925, 56). It was grand, unique, solid, and exemplified permanence and stability. Artists and writers were drawn to the region and established "retreats" in Dublin, Harrisville, Peterborough (MacDowell Colony), and Claremont (Saint Gaudins) (Shonk 1974). Monadnock was transformed from a point to a symbol; it was the goal of many writers and artists to capture its essence (Brodie 1974). The list of notables is long and includes Rudyard Kipling, Allen Scott, Ralph Waldo Emerson, Henry David Thoreau, William Preston Phelps, Abbott Thayer, Luis Agassiz Fuertes, William Faulkner, William Ellery Channing, George Whitefield Chadwick, John Greenleaf Whittier, and Richard Burton.

Henry David Thoreau, Ralph Waldo Emerson, and Abbott Thayer were pre-eminent in setting Monadnock "above the others" and in elevating its position in the psyche of New England. Thoreau's stature in the emerging American literary base is well-known. He was the consummate ecologist and hiker. He visited Mount Monadnock repeatedly and wrote extensive diaries of his trips there. "You must ascend a mountain," he wrote, "to learn your relation to matter, and so to your own body" (Thoreau [1857] in Harding and Bode 1958, 497). Monadnock exemplified "hiking back into Nature." Thoreau deplored "tourists," those "false pretenders to fame" and the "defacers of mountain tops," and sought to set the Monadnock experience on a more spiritual, philosophical plane than was common in the region (1982, 305). He wrote more about Monadnock than perhaps anyone else. To truly understand the mountain you must understand Thoreau. In essence Monadnock best reflects the changes of mind and art that gave shape to his life. In his epic journals of the mountain he wrote, "They who simply climb to the peak of Monadnock have seen but little of the mountain. I came not to look off from it, but to look at it" (Howarth 1982, 347).

Similarly, Emerson, whose central concerns in literature were to impart life, strength, and above all self-sufficiency to American literature, also focused on Mount Monadnock as a key to understanding Nature and self (Shonk 1974). Such ideals are exemplified in numerous poems, Nature (1836) and The Sphinx (1847) and perhaps most explicitly in his 1847 work, Monadnoc, where he eloquently describes "Cheshire's [county] haughty hill" in such words as:

An eyemark and the country's core, Inspirer, prophet evermore; Pillar which God aloft had set so that men might it not forget (Emerson 1914, 45).

Abbott Thayer memorialized the mountain in two of his most famous paintings, Winter Sunrise, Monadnock and the Angel of Monadnock. The former, exhibited in New York's Metropolitan Museum, has been described by many as one of the supreme landscape pictures of the world. Thayer was a painter, a poet, and a naturalist. He wrote:

The outline of the mountain against the sky is as sharp as steel. Some painters soften such outlines, for the sake of atmosphere, but I can't make this one sharp enough. And those tiny spruces on the skyline—it's incredible how small they are, and yet their exact smallness is one of the things that gives scale to the mountain (Thayer 1956, 12).

Thayer etched the image of Monadnock in the minds of America and, in concert with the aforementioned literary figures, helped create Monadnock as a symbol beyond its mere physical presence. The mountain aura described by Emerson in Nature (1836) came finally to light. As the 20th century dawned, "Cheshire's haughty hill" ushered in a new phase or place for itself—recreation.

Mount Monadnock has always played a recreational role in the Monadnock Region, but only in the 20th century has this role come to dominate its place. During the region's early settlement history, the mountain was a favorite berry-picking and picnicking spot (Royce 1974) (a fact often lamented by Thoreau). The realm was expanded beyond the local scene by the introduction of railroad lines in the 1850s and the concomitant emergence of a tourist and hotel industry. Such landmarks as the Halfway House (1860), Shattuck Inn (1860), the Ark (1874), and the Monadnock Inn (1920) catered to a newly emerging and upwardly mobile middle class. The Woodshed Club and Monadnock Mountain Association were formed (circa 1910) to both enhance recreational opportunities and protect the area's natural beauty (Royce 1974).

Roads further opened the mountain in the 1920s, ushering in the automobile era, and transforming the mountain into a popular retreat for an increasingly urban society. The "Monadnock Experience" was conceived as a retreat from urban life, replete with clean fresh air, fine mountain brooks, and a view exceeding 160 kilometers (100 miles), the only place where all 6 New England states are visible (Royce 1974). The intoxicating aroma of Thoreau's "sun-baked spruce odor" assumed a new dimension and meaning.

The state of New Hampshire created a state park in 1930 and over the years has annexed land to produce a protected area of some 2,025 hectares (5,000 acres). Coupled with an adjoining 2,834 hectares (7,000 acres) of, as yet, relatively pristine lands, the region is clearly designed to embody the "Monadnock Experience."

Monadnock holds a special place in the mythology of the region and has done so for more than a century. Each year for the past six years a dance troop celebrates its "place" in a special summit performance. The park is one of the most heavily used recreational areas in New England. More than 135,000 visitors annually register to climb to its summit. It has long been known as the second most-climbed peak in the world, after Mount Fuji. In fact, with the recent introduction of motorized transit to Fuji's summit, Julia Older and Steve Sherman (1990) argue that Monadnock is now the most climbed mountain in the world. During the last 150 years use has grown so much that today's summit climate is, in fact, a man-made, arctic-alpine-like zone, reflecting a legacy of denudation, burning and overuse.

The pressures on Mount Monadnock's new "place" in the region are intense. As land values and property taxes rise, proximate large landowners are under increasing pressure to develop "underutilized" land. Growth in New Hampshire has been phenomenal in recent years; the state continues to lose prime agricultural and recreational land to housing tracts, malls, and a host of other urban developments. As such, Monadnock faces three key development issues: overuse, as well as conflicting use, the vulnerability of privately held parcels to development pressures, and a general lack of coordinated (regional) planning, which ultimately erodes the overall notion of the "Monadnock Experience" so carefully and affectionately crafted over the last three centuries.

Conclusions

Mount Monadnock has attained a place in both the New England and American culture. It is a distinct geophysical event, its proper name having been transformed into "a generic term for all such residual mountains that rise over the uplifted plateaus of denudation." From its physical presence it rose to become a symbol of human achievements and to represent an ideal in the cosmos of human—environment interaction. As an artistic and literary ideal, it came to represent man's hoped-for condition in life; as a modern recreational and aesthetic place it must face the increasing pressures of an encroaching urban society. Time will tell whether it remains, in Emerson's words, "Cheshire's haughty hill."

References

Brodie, Jocelyn. 1974. Mount Monadnock in American art. In *The Grand Monadnock*, pp. 19-22. Concord, NH: Society for the Protection of New Hampshire Forests.

Chamberlain, Allen. 1968. *The Annals of the Grand Monadnock*. Concord, NH: Society for the Protection of New Hampshire Forests.

Davis, William Morris. 1894. Mount Monadnock and Ben Nevis, I. *The Nation* 59:99-100.

Emerson, Ralph Waldo. 1836. *Nature*. Boston: James Munroe.

———. 1914. *The Works of Ralph Waldo Emerson*, Vol. V. *Poems*. London: G. Bell and Sons.

Fowler-Billings, Katharine, Nolan, L., Mork, D., and Chiasson, U. B. 1949. *Geologic Map and Structure Sections of the Monadnock Quadrangle, New Hampshire*. Concord, NH: State Planning and Development Commission and Highway Department. [Reprinted in 1979 by New Hampshire Forest and Lands and the Department of Resources and Economic Development.]

Harding, Walter, and Bode, Carl, eds. 1958. Letter to H. G. O. Blake. In *The Correspondence of Henry David Thoreau*, pp. 495-499. New York: New York University Press.

Howarth, William, ed. Monadnock journal. In *Thoreau in the Mountains*, pp. 286-365. New York: Farrar, Straus, Giroux.

Nutting, Helen Cushing. 1925. *To Monadnock*. New York: Stratford Press.

Older, Julia, and Sherman, Steve. 1990. *Grand Monadnock*. Hancock, NH: Appledore Books.

Royce, Charles H. Jr. 1974. Recreational history of Mount Monadnock. In *The Grand Monadnock*, pp. 23-27. Concord, NH: Society for the Protection of New Hampshire Forests.

Shonk, Bronson. 1974. Nineteenth century writers. In *The Grand Monadnock*, pp. 6-12. Concord, NH: Society for the Protection of New Hampshire Forests.

Thayer, Abbot. 1956. Letters. In *Barry Faulkner's Men of Monadnock*, ed. M. A. Dewolfe Howe, pp. 1-12. Keene, NH: Keene National Bank.

From Barrier to Corridor:
Transformation of the Appalachian Crestline

Douglas E. Heath
Northampton Community College

The Appalachian range in general, and its crestline in particular, was a formidable barrier that long constrained European settlement to the Atlantic seaboard. Americans eventually transformed the east-to-west barrier into a south-to-north corridor, the famous Appalachian Trail extending 3,435 kilometers (2,135 miles) from Springer Mountain, Georgia, to Mount Katahdin, Maine. Changes in attitudes about environment and technology catalyzed this transformation, the story of which has important links to the history of the American conservation movement.

The Appalachian crestline also illustrates a shift in disciplinary paradigms. Geographers of different periods have differed about its significance.

The Appalachian Barrier

There is no single Appalachian crest or skyline (Figure 1). Even employing the most restrictive of various regional delimitations, Appalachia has substantial width and contains at least three major crestlines: the Blue Ridge, the Allegheny-Cumberland Front, and the drainage divide above the headwaters of streams flowing to the Atlantic (Raitz and Ulack 1984).

Ellen Churchill Semple was the American geographer who placed the greatest emphasis on the national significance of Appalachia. Writing under the Davisian paradigm of "inorganic element-organic response," she perceived "the Appalachian Barrier" as a "geographic condition" that exerted great "influence" on American history. Strategically, the barrier consolidated British settlement and "took away the temptation to wide expansion that was defeating the aims of the Spanish

and French" (Semple 1903, 38). Culturally, the barrier produced contrasting results: on the east it "transformed the hunter into a farmer and the gentleman adventurer into a tobacco-grower," whereas on the west it "condemned . . . the Cumberland Plateau . . . to isolation, poverty, and a retarded civilization" (Semple 1903, 38, 72). Politically, the barrier fostered democracy in that it "did not permit the development of large estates . . . and aristocratic organization of society" but instead generated "a wide mingling of ethnic elements . . . [and] produced a race of men sturdy in their self-reliance . . . filled with the spirit of enterprise . . . [and inclined] to frame laws for themselves" (Semple 1903, 62).

Modern historical geographers have assiduously avoided deterministic conclusions such as Semple's, which leave no room for human agency and, though presented as scientific, are essentially untestable. Two examples of these more restrained conclusions are that "settlement and routes of transport conformed to the natural surface features" (Brown 1948, 93) and that the choice of the Atlantic divide for the Proclamation Line of 1763 complicated and undermined Britain's attempt to maintain colonial subordination (Meinig 1986).

In 1790, the center of the U.S. population still lay to the east of Baltimore, but by 1820 it had shifted westward all the way across the mountains as great processions of settlers moved through or around the Appalachian barrier (Watson 1967). Subsequent economic activity in the bypassed highlands focused on exploiting natural resources, particularly coal and timber, and often caused severe and enduring damage to the environment. Americans had turned the great

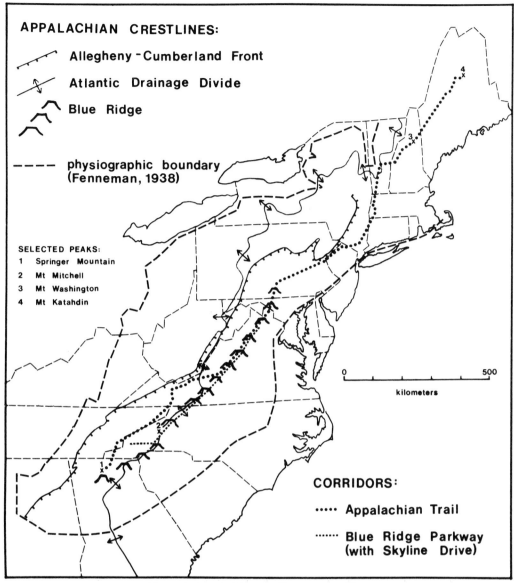

Figure 1. Appalachian crestlines and corridors.

barrier of the 17th and 18th centuries into a 19th-century place of plunder. To understand how they would make it a 20th-century hiking corridor requires a digression on American attitudes.

American Attitudes about Environment and Technology

A potent bias against cities and urbanism has persisted through American history, "voiced in unison by figures who represent major tendencies in American thought" (White and White 1962, 15) and echoed, if not amplified, by much of the general public. Before the 1890s, wilderness and primitivism were also widely

scorned, although the Romantic writers and artists of the early 19th century were a major exception (Nash 1973). In striking contrast, Americans have always held strongly positive attitudes about farms and villages and their accompanying way of life (Marx 1964). To Leo Marx "the pastoral ideal . . . is located in a middle ground somewhere between, and yet in a transcendent relation to, the opposing forces of civilization and nature" (Marx 1964, 23). Such "middle landscapes" have "metaphoric powers . . . [able to evoke] an image in the mind that represents aesthetic, moral, political, and even religious values" highly prized by most Americans (Marx 1964, 128).

Relying primarily on literary sources, Marx argues that the jarring intrusion of the railroads into the rural northeast in the 1840s brought a new way of life and an accompanying consciousness that clashed sharply with the pastoral ideal. In the face of accelerating technological advance, Americans have been attempting to accommodate "the machine in the garden" ever since (Marx 1964). To Lynn White (1967), the Judeo-Christian creation myth, rather than the literary pastoral, is our most potent cultural abstraction. With God-given "dominion" over nature and the charge to "multiply and subdue the earth," Judeo-Christian civilization has a unique license, if not a mandate, to conquer the wilderness and to exploit natural resources (White 1967). Although most Americans seem to welcome unbridled technological advance as the means of attaining the new Eden of effortless abundance, lingering doubts can erupt in powerful expressions of the opposite viewpoint—the anti-modernism of the radical amateur tradition in American conservation, epitomized by John Muir (Fox 1981).

Roderick Nash (1973) chronicles one such eruption, the development between 1890 and 1920 of a "wilderness cult," manifested in the origin and growth of the Boy Scout movement, the development of camping and mountain climbing as new forms of recreation, and the enormous popularity of *The Call of the Wild*, the Tarzan novels, and the "natural history" genre of nonfiction. Nash cites the popularization of Frederick Jackson Turner's frontier thesis as a critical stimulus; rather than a hostile enemy to be vanquished, Americans began to perceive wilderness as a "beneficent influence . . . linked with sacred American virtues" (Nash 1973, 146). Another critical stimulus was the pace and type of societal change. Many Americans were deeply troubled by the power of huge corporations, the ethnicity of recent immigrants, the expansion of cities, and the growth of scientism. For them "the familiar Romantic rhetoric took on new urgency . . . [and] the wilderness acquired special significance as a resuscitator of faith" (Nash 1973, 156-157).

The "wilderness cult" exerted a limited influence on American society. It significantly weakened the pervasive traditional bias against wilderness but failed to set the direction of the emerging conservation movement, which was dominated by the utilitarians under the leadership of Gifford Pinchot. By advocating the scientific management of natural resources in the public interest, the utilitarians clashed with both the corporate interests outside the conservation movement and the preservationists within it. The latter were led by John Muir, who argued that wilderness is the best use for large tracts of public land (Nash 1973).

The "wilderness cult" also appealed to a limited segment of society; it flourished among Americans of older stock, not among the new immigrants. The con-

servation movement, including both its utilitarian and preservationist factions, was an overwhelmingly Anglo-Saxon enterprise until its metamorphosis into the broader environmentalism of the 1960s (Fox 1981).

A Pastoral Vision of a New Appalachia

In 1921, the conservationist and social reformer Benton MacKaye proposed a continuous footpath "over the full length of the Appalachian skyline" (MacKaye 1921, 328). To the late-20th-century hiking enthusiast it appears self-evident that the wilderness ideal inspired the Appalachian Trail, but Benton MacKaye actually proposed to establish a new rural society, and thus the pastoral ideal was his primary inspiration. He envisioned the formation of "special communities in adjoining valleys" to provide lumber for trail shelters and community camps, food for vacationers and the local population, and employment for those who chose to remain in the region (MacKaye 1921, 328). Benton MacKaye saw natural barriers such as the Appalachian crest as "dams and levees impounding and channelling the floodwaters of the metropolitan culture" (MacKaye 1928, 170-171) and therefore as logical locales for building an "indigenous culture," a revived 20th-century version of the Jeffersonian ideal. Here people could experience genuine "living," as opposed to mere "existence" in the "metropolitan culture" (MacKaye 1928).

Benton MacKaye spent his formative professional years in the U.S. Forest Service, during the climax of the conservation movement, and served another two years as a community planner in the Department of Labor before taking an extended leave of absence to develop his plan. In this way he was responding to one of the basic dynamics of the conservation movement: the tension between the zealous amateurs who were its fountainhead and the bureaucrats who took it over and institutionalized it (Fox 1981). Only the former could have made such a bold and original proposal.

In many ways this vision of a future rural society along the Appalachian crestline was the antithesis of the region's past. Logging had been profitable to outsiders but destructive to any given highland locality, a sharp contrast with Benton MacKaye's proposal for sustained-yield forestry serving local interests. Upland farming had generally failed due to severe soil erosion accelerated by the practice of plowing up and down the slope (Watson 1967). Failed farmers formed the core of what became an enduring regional stereotype: the cantankerous moonshiner living in a crude shack tucked deep in a remote hollow. This historical experience and its associated negative images made it difficult to tap the wellspring of American love for "middle landscapes" that otherwise could have stimulated popular support for Benton MacKaye's regional plan.

The Making of the Appalachian Corridor

Benton MacKaye's small group of reformers made limited progress on the project until the Appalachian Trail Conference was established in 1925. Soon an urbane Yankee lawyer named Myron Avery had taken charge of the Appalachian Trail Conference and was energetically leading the effort to build the trail by coordinating local hiking clubs whose members performed the labor, obtaining valuable support from government agencies that managed land along the route, and securing rights-of-way from private landowners, often with a simple handshake. Charles H. W. Foster characterizes the relationship between Myron Avery "the doer" and Benton MacKaye "the dreamer" as "a remarkably successful, though not always agreeable alliance" and points out that after Myron Avery's arrival "MacKaye's regional planning concept was set aside" (Foster 1987, 12). In 1937, the Appalachian Trail Conference completed the trail, a task of unprecedented size and complexity for a volunteer conservation organization. The Appalachian Trail Conference has fought ever since to protect its slender and vulnerable creation with a buffer zone or "greenway" to maintain at least the sense of wild nature in the face of vigorous encroachment by ski resorts, trailer parks, cottages, condominiums, and houses (Foster 1987).

Ronald Foresta (1987) contends that establishing the Appalachian Trail Commission was decisive not only in the construction of the trail but also in its "transformation . . . from an instrument of social reform to a recreational facility" (Foresta 1987, 84). Like Avery, most members of the Appalachian Trail Commission were successful professionals who volunteered to build the imaginative hiking trail because it appealed to their sense of civic duty and avocational interest in nature. Their published writings extol the wilderness but are devoid of the socioeconomic issues that so concerned Benton MacKaye. By their behavior, attitudes, ethnicity, and social class they exemplified the "wilderness cult" that influenced their formative years. It is therefore not surprising that they desired merely to escape occasionally from the pressures of the "metropolitan culture" for an inspiring foray into the wilderness, rather than to discard their urban lifestyles in favor of the "indigenous culture" that MacKaye hoped to create.

Ronald Foresta also argues that the public-land managers in the 1920s lacked interest in Benton MacKaye's utopian ideas, in part because these seemed increasingly impractical as industrial capitalism became more entrenched, and in part because growing prosperity and the automobile made recreation their new institutional mission. Thus the complementary interests of volunteer trail builders from the Appalachian Trail Commission and U.S. public-land managers drastically narrowed the scope of the project. Because Benton MacKaye's proposal never attracted working-class support, this transformation was "smooth, uncontested, and forgotten" (Foresta 1987, 78). The "fatal flaw" of the proposal was its timing, for "circumstances no longer allowed such an ambitious or direct assault on modern industrial society" (Foresta 1987, 78). Benton MacKaye may have realized this by 1928 when he acknowledged that "maybe we are doomed to be engulfed [by the metropolitan culture]" (MacKaye 1928, 181).

Although Benton MacKaye acquiesced to this constriction of his project, he fought vigorously against the New Deal programs for building skyline drives, which posed a direct threat to the trail itself (Fox 1981). The Appalachian barrier was becoming a corridor for another machine in a different garden: the automobile invading the azalea, laurel, and rhododendron. Benton MacKaye, Bob Marshall, and six others responded in 1935 by founding the Wilderness Society but had little success with their first mission. The Skyline Drive and Blue Ridge Parkway now extend 924 kilometers (574 miles) and very closely parallel approximately 240 kilometers (150 miles) of the Appalachian Trail. In the long run, however, the Wilderness Society won many battles, most notably passage of the landmark 1964 Wilderness Act establishing a formal system of American wilderness preservation. In deciding in the 1930s to defend the Appalachian crestline against the automobile, Benton MacKaye and his allies added a critical middle chapter to the story of the long struggle within the conservation movement between the preservationists and the utilitarians.

Since 1937, the Appalachian Trail has been widely acknowledged as the inspiration for other long paths, most notably the Pacific Crest Trail. In 1968, Congress designated both footpaths as National Scenic Trails. Charles H. W. Foster, a former chairman of the Appalachian National Scenic Trail Advisory Council, has recently examined the complex management partnerships that have evolved since 1968. He touts these arrangements as an important model of "institutional bioregionalism," which is a new set of concepts concerning the conditions necessary for successful long-term management of an environmental system that stretches across the boundaries of many jurisdictions (Foster 1987).

Conclusion

The Appalachian Trail, the Blue Ridge Parkway, and Skyline Drive are major lines of recreational travel in the very locations where the land once acted as a formidable barrier to movement. In this way these lines symbolize the triumph of civilization over nature, but the Appalachian Trail also reflects the success of the conservation movement, or more specifically of its preservationist faction, in fostering respect for the value that remnants of wilderness have for people living in a modern society. A sad and familiar irony is that attempts to preserve wilderness often lead to destruction of its essential qualities. As a well-marked, well-trodden path with numerous shelters, the Appalachian Trail is at best a semi-wilderness, and hiking it with published guides and high-tech gear is hardly an exercise in primitivism. But the fact remains that many backpackers and day-hikers on the Appalachian Trail and elsewhere now seek physical and spiritual rewards in a way that their ancestors would find surprising and perhaps incomprehensible. The Appalachian crest has long been a major line on the cultural landscape, but Americans have transformed the premier wilderness barrier of the east into an intensively used recreational corridor as they have altered their relationship with nature and reconsidered the accompanying gains and losses.

Acknowledgments

The Northampton Community College Foundation provided a release time grant to write this essay. I thank Ellen Heath, the editor, and the referees for their constructive criticism of the manuscript.

References

Brown, R. H. 1948. *Historical Geography of the United States.* New York: Harcourt, Brace, and World.

Fenneman, Nevin M. 1938. *Physiography of Eastern United States.* New York: McGraw-Hill.

Foresta, R. 1987. The transformation of the Appalachian Trail. *Geographical Review* 77:76-85.

Foster, C. H. W. 1987. *The Appalachian National Scenic Trail: A Time to Be Bold.* Harpers Ferry, WV: Appalachian Trail Conference.

Fox, S. R. 1981. *John Muir and His Legacy: The American Conservation Movement.* Boston: Little, Brown.

MacKaye, B. 1921. An Appalachian Trail: A project in regional planning. *Journal of the American Institute of Architects* 9:325-330.

——— . 1928. *The New Exploration: A Philosophy of Regional Planning.* New York: Harcourt Brace.

Marx, L. 1964. *The Machine in the Garden: Technology and the Pastoral Ideal in America.* New York: Oxford University Press.

Meinig, D. W. 1986. *Atlantic America, 1492-1800.* New Haven: Yale University Press.

Nash, R. 1973. *Wilderness and the American Mind.* New Haven: Yale University Press.

Raitz, K. B., and Ulack, R. 1984. *Appalachia: A Regional Geography.* Boulder, CO: Westview Press.

Semple, E. C. 1903. *American History and Its Geographic Conditions.* Boston: Houghton Mifflin.

Watson, J. W. 1967. *North America: Its Countries and Regions.* New York: Frederick A. Praeger.

White, L. Jr. 1967. The historical roots of our ecological crisis. *Science* 155:1203-1207.

White, M., and White, L. 1962. *The Intellectual versus the City.* New York: New American Library.

The Symbolic Center of Charleston, South Carolina

John P. Radford
York University

The modern tourist in Charleston, South Carolina, might pass through the intersection of Broad and Meeting Streets without noticing anything particularly remarkable. Located near the tip of the peninsula, away from the major shopping plazas and industrial areas in a district that obviously qualifies as "historic," the intersection is dominated by the tall spire of an attractive white church. The other three corner sites are occupied by public buildings of differing architectural styles, scarcely taller, it seems, than the palmetto trees lining the sidewalk. Strings of small shops and offices adjoin these buildings, but a few steps beyond are residential areas—pleasant indeed, though modest (Figure 1).

Yet the intersection of Broad and Meeting is not only the heart of Charleston, but the centerpiece of a cityscape with enormous symbolic significance in the historical geography of the South, indeed of the United States as a whole. One of the oldest surviving groups of buildings in the United States, the nucleus was established when Charleston was the fourth largest city in colonial America. For more than two centuries this set of urban artifacts has embodied an elusive and partly invented continuity.

From the beginning of settlement, this position in the city's layout was conceived as its civic focus, set off from the commercial district behind the wharves on the Cooper River. The original plan, Anthony Ashley-Cooper's "Grande Model," which pre-dated settlement on this site in 1680, envisioned a 0.8-hectare (2-acre) open square in this location (Severens 1988). In the 1740s, the marshland site was cleared of the fortifications built by the earliest European settlers. The construction in the 1750s of St. Michael's church on the southeast corner and a State House

diagonally opposite established the basis of the intersection. A Guard House was then built on the southwest corner, so that by the 1760s this crossroads was firmly established as the center of the town (Fraser 1989). With the transfer of the state capital to Columbia in 1790, the State House became the County Court House. The northeast corner of the intersection was occupied by a retail market until the Branch Bank of the United States was built there in 1801. The bank building was taken over in 1818 and converted to a City Hall, thereby completing the institutional symmetry of the intersection.

During the decades before the Civil War this so-called Four Corners of Law reached functional maturity both as a symbolic center and as a locus of control. It was the major point of orientation, standing apart from the business district centered on East Bay Street (Figure 2). Each of the corner sites was occupied by a building housing the headquarters of a major cultural control mechanism. Although the population of antebellum Charleston was diverse in class, race, and ethnicity, the normative power structure recognized only the three groups that constitute the popular caricature of Southern society: a white planter-professional elite, poor whites, and black slaves. The focus of the city symbolized the elite's perception of the social order (Radford 1979), and in many ways expressed its world view.

The most prominent building in this antebellum nucleus was St. Michael's church. Its spire aided navigation and was a reference point for the whole city. It offered both the symbolic reassurance of ecclesiastical authority and a much more tangible resource as watch tower. The watch kept from the spire was less intent on divine intervention than on alerting the populace

Figure 1. The intersection of Broad and Meeting Streets.

to fire and slave revolt. In the minds of upper class prewar Charlestonians, slave revolt and arson were inseparable (Channing 1970). Fires were attended by the city guard, with a reserve held behind in the event that arson was used by free Blacks or slaves as a diversionary tactic. The watch began with "retreat—at a quarter past ten in the summer and a quarter past nine in the winter," after which the guard paraded in the street to the sound of St. Michael's bells (Smith

1950, 63). While the watch was posted in the spire, sections of the guard marched off to different parts of the city, where they patrolled until daybreak. The size and organization of the guard varied during the antebellum period, but it often numbered as many as 100 men and operated in addition to a regular police force. Its role was racial control. Blacks found on the streets at night would be taken to the Guard House, directly opposite St. Michael's on the southwest corner of the intersection. Those charged with offenses would appear, usually the next day, across Broad Street at the County Court House. The authority exercised from City Hall completed the formalization of the Four Corners as the symbolic control center of a city in a slave society (Radford 1976).

After the Civil War and Reconstruction, the center of Charleston became a symbol of the Lost Cause, a celebration of the defeated ideals of the Confederacy. Many tangible expressions of what in some eyes had been a noble cause can be found throughout the Southern states. Preserved battlefields and war graves, monuments large and small, and statues of the Confederate soldier abound. Most date from the turn of the century when the Lost Cause found currency in all regions of the nation (Winberry 1983). The pedigree of Charleston's symbolic center, by contrast, reached back to the colonial era, before the development of Southern sectionalism, and it embodied a continuity unmatched by such post-Reconstruction creations.

Against all odds, the Four Corners survive almost intact today. The only building replaced is the Guard House, its site now occupied by the U.S. Post Office. The other three buildings have endured a succession of catastrophes. Since the mid-18th century five major fires have come within two blocks of the intersection. The most disastrous swept across the peninsula in December 1861, leaving a persistent "burned district." Its effect was greatly compounded by destruction during the Civil War, which left the city largely abandoned and substantially ruined. True, General William T. Sherman did not carry out his threat to level the cradle of secession to the ground. But few buildings were left untouched by the combined effects of Union bombardment and vengeful Confederate arson as the soldiers gave up the city in the latter stages of the War. Matthew Brady's photographers who arrived with the Union troops in 1865, have left us images of ruin and desolation.

As well as enduring fire and warfare, Charleston has also had to face more than its share of natural disasters. There have been several minor earthquakes, and one major tremor in 1886, which extensively damaged buildings near the heart of the city. Three tornadoes and at least 10 hurricanes are on record.

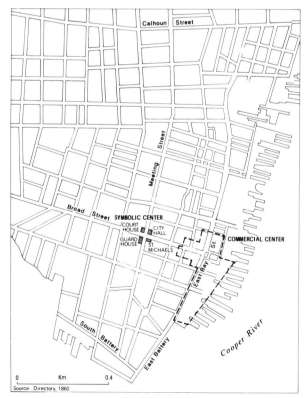

Figure 2. Major foci of antebellum Charleston.

None was more damaging than Hurricane Hugo, which came ashore on 21 September 1989, leaving in its wake an astonishing trail of destruction in the South Carolina low country. Sections of the roofs of the County Court House and City Hall were ripped off and St. Michael's church was damaged.

Countering these destructive forces throughout the years has been the role of the white elite in maintaining a morphological continuity. One easy explanation for the preservation of Charleston's anachronistic morphology is its relative stagnation and lack of affluence. By-passed by the dynamic forces that transformed most U.S. cities during the modern period, Charleston was peripheral even to developments within the New South. Its middle class evolved slowly, even by Southern standards, and the old elite all too often found itself "too poor to paint and too proud to whitewash." Much of central Charleston remained rundown beyond Reconstruction and even through the Depression and World War II.

While it is true that Charleston grew very little in the years after the Civil War, this stagnation can be explained by the white elite's continued disdain for trade and industry (Doyle 1981). Although its affluence paled in comparison with that of cities elsewhere in the nation, the elite maintained local preeminence, and continued to project a vision of Charleston that had no place for rampant commercialism. Its role in the city's morphological evolution has extended beyond mere preservation to a deep psychological need for restoration. Its first response to disaster was to put the pieces back together, attempting to perpetuate the social order by reestablishing the morphological *status quo ante bellum*.

The elite have long had a coherent vision of what Charleston "was meant to be." The vision originated early in the 19th century with the planter ideal of a summer resort, and was either opposed or suffered by the city's other interest groups. It emerged from the Civil War and Reconstruction with renewed intensity. Subsequently, although supposedly vested in Charlestonians living "South of Broad," it became part of a moral order that extends to much of the area below Calhoun Street, the pre-1850 city boundary. The intensity of this vision of the city has fluctuated over time, but its essential nature has never altered. It has at various times been denounced as regressive, racist, and bad for business. For a century after the Civil War, those who did not subscribe to its quiet tenacity, who were not part of its agenda, either conformed or went elsewhere.

Robin Datel (1990) has recently explored the impact of this profoundly conservative viewpoint on historical preservation in Charleston between 1920 and 1940, linking the movement to contemporary ideas of Southern regionalism. Sociologists, particularly those at Chapel Hill, North Carolina, explored the South's distinctive sense of regional consensus, which emphasized traditional social relationships and looked to the past as a source of inspiration for the present. Southern society possessed a traditional view of the family, an attachment to the land, and a preoccupation with locality. These values permeated the preservation movement in Charleston, and explain why visitors often found it the most "European" of American cities. Between the two world wars, therefore, white elite Charlestonians continued to endow their urban environment with an aura of continuity that belied rapid social change.

In recent decades the moral order has, with some ambivalence, gradually accommodated an acceptable form of commercialism: mass tourism. Although there is still a tendency to regard tourists as a nuisance, it is increasingly recognized that carefully regulated tourism can enhance rather than destroy the urban fabric. "Historic Charleston" has been on the itineraries of increasing numbers of winter sun-seekers from the snow-belt states and Canada since the 1960s. More recently it has begun to appear as an option on package tours from overseas. In the spring and fall, European and Asian groups often tour the plantations, taking the air on the Battery, and walking or driving through the intersection of Broad and Meeting.

It is difficult to know what impact a visit to "Historic Charleston" has on the average tourist attracted by Disney World or the Atlantic beaches. But perhaps it is fitting that the ultimate re-integration of this long-recalcitrant bastion of secession should be achieved through the absorption of its ethos by American popular culture.

References

Channing, Steven A. 1970. *Crisis of Fear: Secession in South Carolina.* New York: Simon and Schuster.

Doyle, Don H. 1981. Leadership and decline in postwar Charleston, 1865-1910. In *From the Old South to the New,* eds. Walter J. Fraser and Winfred B. Moore, pp. 93-108. Westport, CT: Greenwood Press.

Datel, Robin Elisabeth. 1990. Southern regionalism and historic preservation in Charleston, South Carolina: 1920-1940. *Journal of Historical Geography* 16:197-215.

Fraser, Walter J. 1989. *Charleston! Charleston!: The History of a Southern City.* Columbia: University of South Carolina Press.

Radford, John P. 1976. Race, residence and ideology: Charleston, South Carolina in the mid-nineteenth century. *Journal of Historical Geography* 2:329-246.

—————. 1979. Testing the model of the preindustrial city: The case of Charleston, South Carolina. *Transactions of the Institute of British Geographers* 4:392-410.

Severens, Kenneth. 1988. *Charleston: Antebellum Architecture and Civic Destiny.* Knoxville: University of Tennessee Press.

Smith, D. E. Huger. 1950. *A Charlestonian's Recollections.* Charleston: Carolina Art Association.

Winberry, John J. 1983. "Lest we forget": The Confederate monument and the Southern landscape. *Southeastern Geographer* 23:107-121.

Santa Fe, New Mexico: Place of Style and Postmodern Transformation

Linda W. Mulligan
Ohio State University

Multiple perspectives are essential to understanding the cultural evolution and place importance of Santa Fe, New Mexico. This analysis centers on the socio-historic, cultural geographical process (Jakle 1987; Meinig 1979; Norton 1989) whereby Santa Fe and its landscape has been transformed, over time, into an extraordinary postmodern cultural display, a place understood and visualized as a cultural text. Santa Fe is archetypical, due both to its Southwest regional cultural dominance and to its strong identification with symbols and metaphoric structures (Grimes 1976). It has become a key reference point' in postmodern America—a "placepoint" in American culture (Harper 1989).

Postmodern Santa Fe: A Mixture of Genres

Santa Fe, with a current population of approximately 60,000, lies against the western slopes of the Sangre de Cristo Mountains in northeastern New Mexico. Founded in 1610, its evolution incorporates Native American, Spanish, and Anglo-American periods of power and influence; contemporary Santa Fe contains all of these population groups. The heritage of Santa Fe encompasses three centuries—a history replete with cultural, architectural, and language styles originating from the culture of the place and its landscape. Contemporary postmodern Santa Fe can be understood as an innovative center from which material and non-material aspects of social change are diffused to the ultimate adopters throughout American society and internationally. Examples include clothing styles, adobe architecture, art, and a particular type of ambience, or "ethos" (Figure 1). Grimes, in his study of the

symbols and rituals in Santa Fe, has stated that the "ethos" of Santa Fe is in its ambience—that is, what its symbols evoke in the visual and auditory imagination (Grimes 1976).

Santa Fe and its place imagery have become metaphorically structured in the national literate vocabulary of American culture. The social process whereby this has taken place involves: (1) the visual and technological reproducibility of the clothing styles, art, and architecture—thus, in essence, allowing for the transfer of "Santa Fe Style" to other places and cultural regions of the United States and elsewhere; (2) the dissemination of visual and literary images of the place and its landscape, using artistic, literary, and photographic images over a long historic period; (3) effective postmodern efforts by promoters to advertize the landscape itself as a place of style and spiritual transformation, "combining the sacred and the profane, the religious and the secular, the spiritual and the commercial" (Sears 1989, 211); and, most importantly, (4) "stylization" of the place and its landscape.

Stylization is a process in which the aesthetic expressions of Santa Fe life and culture become an identifiable, composite cultural style. These include its adobe architecture; the sunny clear climate and high desert landscape; the wide use of nativistic color; the array of Spanish, Indian, and Anglo events, fiestas, and holy days; and opera, chamber music, and historic performances and events. Santa Fe is as an excellent example of places referred to by Sears as "containing the contradictions of American society in concentrated visual form" (Sears 1989, 211). The historical heritage of the Native Americans has been used to evoke symbolically an aura of spirituality that coexists with the

Figure 1. Visual metaphors of Santa Fe include the clothing styles, the architecture, and the tourists themselves. This figure shows pedestrians on Palace Avenue (top) and Galisteo Street (bottom). The front of the Original Trading Post, a Santa Fe landmark, appears in the bottom photograph. Photographs by Linda W. Mulligan.

commercialism of the shops surrounding the ancient Plaza. Moreover, Santa Fe, as embodied in these diverse expressions is publicly promoted by the city and its visitor's bureau, which distributes these images nationwide. This "stylization" of the place and its landscape gives Santa Fe definition and signature; the metaphorical imagery associated with Santa Fe and "Santa Fe Style" is central to the creation of a postmodern mythological theme in American symbolic culture.

The Landscape of Santa Fe: A Place of Style and Postmodern Transformation

Postmodern descriptions of Santa Fe and its landscape include those of travel writers (for example, Morris 1984), tour books, and visitors' guides published by the Santa Fe Visitors and Convention Bureau. However, one of the most revealing accounts of how Santa Fe is described and understood by the culturally literate

public is found in *The New York Times*, which publishes as a regular feature, *What's doing in . . . ,"* with Santa Fe appearing regularly every year. *The Times,* referred to by the traveling public and by a large majority of travel agents, states that in postmodern America one may add out-of-town tourists to the cultural heritages of Santa Fe, the others being Indian, Spanish, and Anglo. The *Times* goes on to say, "While not altering the city's basic character, neither can so many visitors melt into the background" (Fleming 1989, 10). Thus, the metaphoric imagery of Santa Fe does, indeed, include the tourists and the out-of-towners.

Content analysis of additional materials further attests to the metaphoric structure of Santa Fe and its landscape—"Santa Fe-style decor and fashions, whether in Los Angeles, New York, or Europe . . . [are a product of this population mix]. In Santa Fe, the style expands beyond decor to include high desert scenery, music, a green chili cuisine, clean air, and a special light as well as the citywide chorus of pastel coyotes howling from shop windows" (Fleming 1989, 10). Such media images stylize the postmodern identification and anticipation of the place and its landscape, and give it an evocative meaning. Santa Fe thus can be understood as a North American place-point of innovative, place-conscious style, panache, and commercialism.

Santa Fe's Emergence as a North American Place-Point of Innovation and Style

The evolution of the social process includes five major factors: obsession with symbols and history; migration of prominent writers to Santa Fe; emergence of "Santa Fe Style" as a "cultural image"; integration of musicians and artists with enclaves of scientists and celebrities; and coexistence of Spanish and Native American Indian spiritual and mystical cultures with a nearby population of 20th-century nuclear scientists (in Los Alamos), who participate in Santa Fe life and culture.

First, consider the identification of contemporary Santa Fe as a place obsessed with symbols and history. Grimes has noted that "Santa Fe . . . [is] a uniquely symbol-conscious city" (Grimes 1976, 22). He further states that since Santa Fe is the capital of New Mexico and it has a small population, the result is that close to two thirds of the city's residents work for the city, county, state, or federal governments. Moreover, "Santa Fe is the city of "holy faith"; and faith is closely linked to civility and ethnicity. The official insignia of Santa Fe draws on symbols of three nations [Spain, Mexico, the United States] and fuses them into something belonging to the city as a legal entity, thus fostering civil piety while drawing on ethnic identity" (Grimes

1976, 26). Annual religious celebrations, such as the Fiesta in September, as well as the Spanish and Indian Markets held every August, receive financial and public relations support from the city government. Thus the sociohistorical importance of Santa Fe is unusually symbolic and metaphoric in its construction.

A second key factor to Santa Fe's metaphorical imagery has been the steady migration of prominent writers to Santa Fe (and Taos) between 1916 and 1941. M. Weigle and K. Fiore (1982), in their analysis of the emergence of Santa Fe and Taos as famous places, demonstrate that this is directly attributed to the reputations of well-known writers such as D. H. Lawrence, Willa Cather, Robert Frost, Carl Sandburg, and Sinclair Lewis, as well as the artists of later years. These writers promoted the landscape and saw it as exotic. Moreover, before coming to Santa Fe, these prominent American writers had already established a wide general audience for their literary works; struggling young writers might not have communicated as effectively the "spell of the landscape of Santa Fe" to the culturally literate American public. These writers gave cultural and artistic panache to Santa Fe by their presence as well as by describing Santa Fe to the general public as enchanting and exotic—a foreign land (Weigle and Fiore 1982). Santa Fe's history is remarkably singular in this regard; few places in the world have had such prominent writers who incorporate the place, its symbols, and the physical landscape into their works. Internationally, Paris, France, is an outstanding example; in the case of Santa Fe, writers and their works include Willa Cather, *Death Comes for the Archbishop* (1927), and Paul Horgan, *The Many Centuries of Santa Fe* (1956) and more recently *Under the Sangre de Cristo* (1985).

Third, to comprehend the social process whereby a particular part of the symbolic order assumes both cultural geographic prominence and metaphoric structure, it is necessary to examine the dominant cultural styles of geographic place description in the late 20th century, as this process explains how places are culturally understood (Darnton 1984). S. Ewen, in his insightful analysis of cultural images in postmodern society, argues effectively that surface images and representations— the "skins" of reality have become reality itself (Ewen 1988). The emergence of Santa Fe, and "Santa Fe Style," can be readily understood as the expression of an aesthetically stylized postmodern geographic landscape. The Atchison, Topeka, and the Santa Fe Railroad was an active participant in promoting the cultural imagery of Santa Fe with calendar art, advertisements, and promotional side tours to Indian country (McLuhan 1985). Santa Fe has become overlaid with historical images, films, photographs—a factual past that serves to focus 20th-century perception on convincing the reader of the authenticity of present day Santa Fe. Indeed, the imagery of the past frames and references our present experience; it evokes strong emotional responses to the place and the landscape (Hollander 1989).

The metaphorical images and symbols in the national literate vocabulary do not depict the reality of everyday life of all local residents in present-day Santa Fe. Instead, the visual and literal images and metaphors associated with Santa Fe (for example, St. Francis Cathedral, the Plaza, the Palace of the Governors) become, in fact, the place as it is culturally perceived to be; the national culture and its literate vocabulary create and sustain the place and its landscape out of these images.

Fourth, Santa Fe has experienced an integration of its artistic and musical entrepreneurs with its population enclaves of scientists and celebrities. This provides both artistic excitement and the population base for a ready audience to whom to display music and art. Moreover, this semi-resident (arriving and leaving) population swells real estate prices, and contributes to the aura of late-20th-century style and glamour, in the same manner as the famous writers did in the early decades of this century.

Finally, Santa Fe has the coexistence of Spanish and Native American spiritual and mystical cultures with the nearby population of 20th-century nuclear scientists. The discontinuities of this cultural experience lend a vibrancy and evocative element to Santa Fe. Thus, Santa Fe is a "valued landscape" and is one of the primary geographic place sources of national cultural innovation and behaviors in North American culture.

References

Cather, W. 1927. *Death Comes for the Archbishop.* New York: Alfred Knopf.

Darnton, R. 1984. A bourgeois puts his world in order: The city as a text. In *The Great Cat Massacre*, pp. 107-143. New York: Basic Books.

Ewen, S. 1988. *All Consuming Images: The Politics of Style in Contemporary Culture.* New York: Basic Books.

Fleming, J. P. 1989. What's doing in Santa Fe. *The New York Times* 16 July: section 5, p. 10.

Grimes, R. L. 1976. *Symbol and Conquest: Public Ritual and Drama in Santa Fe, New Mexico.* Ithaca: Cornell University Press.

Harper, C. L. 1989. *Exploring Social Change.* Englewood Cliffs, NJ: Prentice Hall.

Hollander, A. 1989. *Moving Pictures.* New York: Alfred Knopf.

Horgan, P. 1956. *The Centuries of Santa Fe.* New York: E. P. Dutton.

————. 1985. *Under the Sangre de Cristo.* Santa Fe: Rydal Press.

Jakle, J. A. 1987. *The Visual Elements of Landscape.* Amherst: University of Massachusetts Press.

McLuhan, T. C. 1985. *Dream Tracks: The Railroad and the American Indian, 1890-1930.* New York: Harry N. Abrams.

Meinig, D. W., ed. 1979. *The Interpretation of Ordinary Landscapes.* New York: Oxford University Press.

Morris, J. 1984. Capital of the holy faith: Santa Fe, U.S.A. In *Journeys,* pp. 144-151. New York: Oxford University Press.

Norton, W. 1989. Explorations in the understanding of landscape: A cultural geography. In *Contributions in Sociology,* No. 77. New York: Greenwood Press.

Sears, J. F. 1989. *Sacred Places: American Tourist Attractions in the Nineteenth Century.* New York: Oxford University Press.

Weigle, M., and Fiore, K. 1982. *Santa Fe and Taos: The Writer's Era: 1916-1941.* Santa Fe: Ancient City Press.

The Skyscraper: America's Building

William R. Code
University of Western Ontario

In 20th-century America, the urban form has run to both horizontal and vertical extremes. The continent's people have usually welcomed both, often simultaneously and without sensing contradiction. Even Frank Lloyd Wright, the champion of the horizontal, was lured by the mile-high building. Though many of the great new towers, and still more of the great plans, are in Hong Kong, Frankfurt, Tokyo, and São Paulo, the skyscraper, along with the freeway landscapes in outer realms of the metropolis, defines the American city. Even now, 84 of the world's 100 tallest buildings are in North America, still providing the focal points for the North American city. The skyscraper is something that should not be easily dismissed; it is, as Ada Louise Huxtable has commented, "this century's most stunning architectural phenomenon" (Huxtable 1982, 7). But, why do these buildings exist? They have always been the object of fashionable criticism and cities can clearly function without them. Most would argue that mid-rise Paris would be better without Maine Montparnasse or La Defence. However, it is somehow less convincing to argue that New York or Chicago or Toronto would be better off without their towers. The reason, I suspect, is that these structures "fit," to use James Graaskamp's term (1981, 10) and, for the most part, have been an honest expression of the processes that created the American city.

The Distribution of the Skyscraper

The most recent census by the Council on Tall Buildings and Urban Habitat found 50 North American cities housing 697 buildings over 30 stories (CTBUH 1986). Not surprisingly, almost a third of these are in the cities where the skyscraper originated—New York and Chicago—with 129 and 88, respectively. In what Jean Gottmann (1964) called the "hinge cities" of the Atlantic seaboard of Canada and the United States (including New Orleans) 203 buildings are over 30 stories. In the midwestern interior cities from Toronto to Atlanta to Omaha, there are 215. On the Pacific, from Vancouver to Los Angeles, there are 84 with almost 70 percent in San Francisco and Seattle (37 and 21, respectively). Given its population of almost 15 million, the Los Angeles metropolitan area has surprisingly few such buildings—16. Only in part does this reflect that city's height limitations that existed until the 1960s, which were prompted by seismic risk. This paucity of tall buildings is also observable in most of the new cities of the "Sunbelt," except in the oil towns. Phoenix has only two buildings over 30 stories and even the Latin financial center of Miami only six, but oil centers make up for them—Houston's 49 buildings over 30 stories, and even Calgary with a population of fewer than 700,000 has 49 such buildings.

Scattered across the United States are small cities with their isolated tower, usually the creation of that place's dominant institution—Albany and Baton Rouge with structures housing the state offices, Salt Lake City with the Church of Latter Day Saints Office Building, Rochester with the Xerox Tower, and many others with the First Bank of _____ Building. These kinds of exceptions extend to the suburbs. Here, they have even reached the extremes of Houston's suburban Transco building—64 stories. Suburbs are, of course, becoming the dominant location of office space in the continent's metropolitan areas. They house offices ranging from the local insurance broker, through regional sales offices, and suburban manufacturer's head offices to the central business district's "back offices," and the headquarters of powerful corporations that are capable of internalizing their business environment. Rarely, however, even when concentrations of decentralized office space are significant, do the buildings adopt a tall profile. For every case like Transco Tower in Houston, or the towers of Toronto's planned suburban nuclei, or Santa Monica, there are thousands of low-rise office concentrations in places like Walnut Creek, Concord, and the "silicon valley" in the outer realms of San Francisco, or in Detroit's Southfield, or along Phoenix's Camelback Road, or among the miles of regional offices in Atlanta's De Kalb County. The skyscraper is still largely a phenomenon of the central city.

The Skyscraper as Cultural Expression

One can contend, as Gottmann (1966) does, that skyscrapers express our culture—our substitute for cathedrals—and in a broad sense they are. They do ". . . romanticise power and the urban condition and celebrates leverage and cash flow" (Huxtable 1982, 11). Accounting for the skyscraper as fulfilling some cultural need may or may not be valid. But it can only provide meaning at the most general level. The argument is, at heart, functionalist and even teleological. For this argument to have much operational meaning one would have to find buildings that are taller than the economics of the project dictate; for this to happen presumes that either the search for grandeur was shared by all involved in developing, financing, and occupying such structures, or that a developer proceeds with his or her grand designs irrespective of pro formas and economic self-interest, or because he or she places unusual value on making a marketing statement. This, of course, does happen and it does account for the isolated towers of many small cities, but it is unusual, for decisions regarding skyscrapers are usually collective within broad consortia of interests. As well, vast amounts of money are involved, and these funds usually flow through typically conservative financial intermediaries. Thus, the skyscraper, with some exceptions, is most directly accounted for, within the positivist frame, through the linkage of agglomeration economies, rent, technology, and land value within a market mechanism.

Demand Concentration, Rents, and Land Values

Concentrated demand is a necessary, even though insufficient, condition for any substantial concentration of tall buildings. The workforce must be focused in a relatively small area, and this workforce concentration must be translated through increasing floor-space con-

sumption into demand for office space. The demand for office space can come about for a number of reasons, including the focusing of mass-transit routes in the large metropolitan areas and, of paramount importance, information exchange.

Since the installation of the telegraph lines in the mid-19th century, there has been an intermittent but one-directional shift in the mobility of information, resulting from the sometimes spectacular innovations in technology and organization. The quaternary functions, dependent upon intelligence and dealing with information of sometimes inestimable value, have been among the first activities affected by reductions of spatial constraints. However, reducing the walls of distance has been selective in its effects. While in a truly frictionless space, information may easily flow to the center, there would be no greater advantages in this center than elsewhere and the demand for offices would be scattered in the way of Melvin Webber's (1964) "nonplace community." However, as the friction of distance has declined, its first effect appears to have been that of enabling the quaternary activities with particular advantages to exercise them more easily over greater distances. Even today, some of these advantages are still rooted in the easily made face-to-face contacts facilitated by high-density office concentrations.

A substantial portion of the information exchange in central agglomerations is made electronically, but much of the electronic exchange of information is made on the basis of prior face-to-face contact. When stripped of its electronic mask, one finds that at the center of the communications system is a method of information exchange not substantially different from that of medieval financial centers. One reason for this is the need for trust. Vast amounts of money and goods are moved daily on the basis of advice, and equally vast amounts are transferred, without the benefit of a written agreement, on the basis of understandings and custom. In these centers, easily made assumptions that the mobility of downtown quaternary functions are comparable to the wholesaling and retailing functions that pioneered the suburban diaspora are, at best, premature.

It is commonly assumed that the skyscraper is driven by land values, with every ratcheting upward of land value necessitating dispersal of these values over still greater floor areas. This position only looks at part of the relationship however, and essentially ignores the nature of land value itself. As David Ricardo (1817) long ago pointed out, land value is a derivative of the uses and intensity to which one can put the land—and the most intense use has not always been that of the tallest building. If one goes back to Nirenstein's

Atlas of 1929 and plots land values and building heights for a sampling of American downtowns, one detects only a loose relationship (Nirenstein 1929). The tall buildings were usually displaced from the peak value lands by the then more profitable retailing, particular department-store, uses. While the role of the downtown department-stores has weakened, and in some cases such as Detroit, actually disappeared, we can, even today, not be certain that the tallest buildings will be built on, or generate, the highest land values in the city.

The Supply Side: Risk Avoidance in Real Estate Cycles

While demand concentration is the *sine qua non* of the urban canyons of many American downtowns, it is not the only factor influencing the development of this landscape. There is also a "supply side." The capital investment involved in developing a skyscraper or any other large real estate project is both immense and immobile. Any shifts in the nature of demand, as is likely to occur in small or isolated centers with a narrow economic base, magnifies the already heavy investment risk with which commercial real estate is fraught in the unusually competitive American markets. The downtowns of large American cities which have been able to maintain a range of business services, serving a national or global economy, are somewhat less likely to experience the major fluctuations in demand found in smaller centers, or those dependent on a narrow industrial sector. Thus, over the long-term, an investment in Manhattan—or the downtowns of San Francisco, Chicago, Boston, or Toronto—could pose less long-term risk of extensive periods of high vacancy rates than a comparatively sized project in Tulsa or the New Jersey suburbs.

The Skyscraper as a Reflection of Building Technology

It is common to see the skyscraper landscape as a triumph of engineering and in one sense it is. From their roots in 1868 in New York's Equitable Life Insurance Building (Wiseman 1970) to the latest developments in mass damping technology, the tall building is a triumph of structural engineering and without that discipline's talents, skyscrapers could not exist in anything more than a rudimentary (and dangerous) form. Nevertheless, the real limits on building tall buildings are not those of the technology. The

technology is permissive; it is not a determinant of the skyscraper. Myriad advances make the very tall skyscraper a physical possibility—including efficient use and arrangement of structural steel and reinforced concrete, advances in predicting and solving problems of vortex shedding and building vibration, new technologies for high-speed elevators that use "fuzzy logic" controls to increase rates of acceleration. Moreover, these advances assist in reducing the marginal costs of building high. However, the part of the feasibility equations that is most important, the long-term revenue flows, is a function of office space markets, and consequent rents and vacancy rates (Code 1987).

Planning, Floor-Area Ratios, and the Redistribution of Skyscrapers

Even in market-oriented America, land-use planning has had a significant effect on the form and distribution of tall office buildings. This influence often has the effects foreseen by the drafters of the plan, but as often as not the significant ones are inadvertent. Of major importance in the evolution of the skyscraper was the 1916 zoning plan of New York City, designed to address safety and health issues and the degradation of property values by overshadowing, and to keep the gains emanating from the construction of new mass transit lines from being depleted by the congestion induced by clusters of tall buildings. Rooted in this plan was the establishment of what came to be called the zoning envelope, an imaginary three-dimensional mold representing the maximum bulk to which a building might be developed under the regulations. The 1916 regulations established five classes of height districts based on street widths. Above a maximum height of 91.2 meters (300 feet), buildings were required to be stepped back from the street in prescribed ratios. However a part of the building equal to one quarter of the lot's size could be built to any height as long as the base of the building did not cover more than one quarter of the lot area. In this way it was believed that adequate amounts of air and light would be ensured. The results were initially the "wedding cake" buildings that became such an important feature of the Manhattan landscape. On small lots, the 25 percent of the site that could be built to any height above the setbacks would not provide for adequate floor-plates especially in an era of relatively inefficient elevators and the consequent need for many elevator shafts. However, if a big enough lot were acquired, such as happened with the Chrysler and Empire State Buildings, the true potential of the massive office tower could be explored—the Empire

State to the extreme floor-area ratio (gross floor area/lot area) of 30:1. In most of the North American cities, this ultimately led to the establishment of maximum floor-area ratios, and one of the most important factors influencing the size and distribution of tall buildings.

While advances in structural and foundation engineering have stretched the ability to build very tall buildings on a relatively small base, there are physical and economic limits to the aspect ratio of buildings. This means that a very tall building needs a relatively large site. As well, tight restrictions on maximum floor-area ratios mean that very tall buildings must be on still larger sites than that implied by the aspect ratio. The net result is that many of the unusually tall skyscrapers are forced to the perimeter of the existing office agglomerations to larger sites with lower land values. The World Trade Center on the fringes of New York's financial district, the Sears Tower on the edge of the "Loop," and proposals in New York for the world's tallest building, first of all at Columbus Circle, and then on abandoned rail yards on New York's far West Side, all provide examples. Tight restrictions on core development, such as Toronto's restrictive floor-area ratios and San Francisco's market place for limited development rights, can also encourage the decentralization the central city's more footloose back offices.

The Vertical Organization of Uses

Just as the processes of congregation and segregation work in screening the horizontal plane of large-city downtowns, a vertical variant can also be observed. Like land use across the downtown, vertical screening is also affected by rent. Here, however, the rent variable is the inverse of the horizontal—it becomes more expensive with distance skyward. Even when controlled for building age, the rents in buildings over 50 stories in the United States average more than 40 percent higher than those of 30-story buildings (BOMA 1984). This is clearly not related to accessability, for even in an era of high-technology, rapid-acceleration elevators, there is a significant time penalty with building height. The main driving force of the rent difference is the vistas provided. Rents grade downward from the corner windows with the most dramatic views. Buildings are now even designed to increase the number of building facets and the number of corner offices. One typically finds in the upper floors the smaller offices, particularly those of legal and management firms as well as the executive suites of corporate head-

quarters, while the lower floors are leased in larger blocks to house the mass operations of the larger consumers of space, such as the money center financial intermediaries.

Conclusions: The End of the American Skyscraper?

Tall buildings require an expanding, but spatially concentrated, demand for office space. They also need efficient engineering to provide the necessary condition of low construction-cost increments with height, and an urban infrastructure capable of handling the massed demands placed on it. Very tall buildings also require large (for a central business district) plots of land to accommodate the demands of the aspect ratio and the planners' restrictions on floor area ratios.

While structural engineers have clearly made possible the construction of very tall buildings (even Wright's mile-high version), a number of trends are working to tilt the odds against the skyscraper. Most of the continent's cities no longer possess significant concentrations of those office functions demanding close, partially non-electronic, contact. Those that do—in cities such as New York, Chicago, San Francisco, and Toronto—exude an uneasy sense that growth in the traditional heart of the downtown office economy, the financial intermediaries, may be reaching its limits. If so, prospective developers of tall office buildings will be in the unenviable position of having to empty neighboring buildings and these in turn would have to compete with cheaper suburban office space for tenants. The prospective skyscraper developer also increasingly faces density controls in the form of restrictive floor-area ratios such as in San Francisco and Toronto. This demands more land which, if available, can only be assembled on the fringes of the central business district.

Nevertheless, in New York and Chicago, there are always those eagerly anticipating the next turn of the "office market cycle." Despite the decentralist gloom, pervasive since 1945, fortunes have been made gambling on still another rebirth of Manhattan's and Chicago's downtown. In these cities building the next greatest skyscraper may still not be a bad gamble. Their downtown plans are permissive; with luck, an upturn in demand could absorb the necessary 278,000 to 371,600 square meters (3 million to 4 million square feet) to be thrown on the market at one time. The structural engineers and architects have plans waiting—and there is tradition.

References

Building Owners and Managers Association. 1984. *BOMA Experience Exchange Report.* Washington, DC: BOMA.

Code, William. 1987. The impact on development feasibility of containment policies in central business districts. *Papers of the Regional Science Association* 33:124-131.

Council on Tall Buildings and Urban Habitat. 1986. *Tall Buildings of the World.* Leheigh, PA: CTBUH.

Gottmann, Jean. 1964. *Megalopolis.* Cambridge, MA: MIT Press.

_____ . 1966. Why the skyscraper? *Geographical Review* 56:190-212.

Graaskamp, James A. 1981. *Fundamentals of Real Estate Development.* Washington, DC: Urban Land Institute.

Huxtable, Ada Louise. 1982. *The Tall Building Artistically Reconsidered.* New York: Pantheon.

Webber, Melvin M. 1964. Urban place and nonplace urban realm. In *Explorations into Urban Structure*, ed. Melvin M. Webber, pp. 79-153. Philadelphia: University of Pennsylvania Press.

Nirenstein, Nathan. 1929. *Preferred Business Real Estate Locations.* Springfield, MA: Nirenstein.

Ricardo, David. 1817. *On the Principles of Political Economy and Taxation.* [Reprinted with introduction by William Fellner. Homewood, IL: R. D. Irwin, 1963.]

Wiseman, Winston. 1970. A new view of skyscraper history. In *The Rise of an American Architecture*, ed. Edgar Kaufmann Jr., pp. 115-160. New York: Praeger.

P A R T   E I G H T

PORTRAITS

The character of a place is known but often not recognized by those who live there. From an insider's view, the fundamental character of a place simply is. We might as easily ask a resident to describe her own identity and personality as ask for a description of the character of the place in which she lives. It may be even more demanding for an outsider to know a place's character. Some would say it is impossible, because he cannot know a place that is not a part of him. Even so, insiders and outsiders attempt to portray places in many ways—with pen and ink, oils, or pastels, with maps or photos in black-and-white or color, with music, poetry, or prose. As with the portrait of a person, both the tangible and intangible nature of the place must be apparent if the place portrait is to be effective. What is there, and what does it mean? Explore the varied and subtle answers to these simple questions in each of the portraits set before you. Accompany insiders as they detach themselves from familiar places—an intersection in a Missouri town and a neighborhood street in Cincinnati—to paint word portraits laden with personal meaning. Listen as thoughtful outsiders trace the social history of built landscape forms in southern Ontario and Kentucky's Bluegrass region. Watch with an insider as he looks out across the landscape visible from a study window, and learn what lies hidden within the view. And contrast these personal and personally accessible places with Walt Disney World, Florida, and Quartzsite, Arizona—two places meant to be wishful creations, consciously surrogate places but no less real for their artificiality and their origin in someone's imagination. The character of these places and the character of those residing there are illuminated by these portraits.

—Stephen S. Birdsall, University of North Carolina

The Geographic Truths in an American Intersection

C. L. Salter
University of Missouri

City workers continually paved the streets: they poured asphalt over the streetcar tracks, streetcar tracks their fathers had wormed between the old riverworn cobblestones laid smack into the notorious nineteenth-century mud. Long stretches of that mud were the same pioneer roads that General John Forbes's troops had hacked over the mountains from Carlisle, or General Braddock's troops had hacked from the Chesapeake and the Susquehanna, widening with their axes the woodland paths the Indians had worn on deer trails.

—DILLARD (1987, 73)

The American urban intersection is at once both a point and a meeting of lines. It represents a centrality even as it links distant points that have enough business and social connection with the intersection to have their roads lead to that common point.

At the same time, the intersection represents a pattern of meeting points in the rectangular grids that house most of the built environment of the American city, town, and even village. If one selects an appropriate intersection, the patterns of urban design, architecture, efforts at urban renewal, use, and contemporary decline or renaissance all can be seen from this point. There is great geographic truth in the American intersection.

As Annie Dillard observes so accurately, we have been building steadily at these intersecting points and along these travel lines that feed into and emanate from these crossings. Our various developments in the technology of road surfaces all stand layered atop one another . . . and the initial line may not be random at all, but derived from an aboriginal travel route or even an animal path. These lines—these signatures of human effort and intent—have a persistence, especially when they intersect at points to which we have ascribed economic and cultural importance.

The exercise that has issued this paper is one of observation and consideration of the geographer's field of vision at a mundane but significant city intersection. By selecting four, distinct 90-minute observation periods, I was able to observe not only the human use of the lines and points that converge at the intersection, but also on the geographic elements that mark the community's declaration of interest in that particular spot. These human assignments of urban attention are so universal that I include no map. These geographic realities are very much a part of all of our landscapes, even though this particular intersection is in Missouri.

Setting and Signatures

Our corner is the intersection of Broadway (Columbia's main east-west artery) and Ninth (the major north-south link of the Missouri campus to the city of Columbia) in the center of a nicely traditional downtown (Figure 1). I built my observations around a quartet of hour-and-a-half watches at 0630 to 0800 hours, 1100 to 1230 hours, 1630 to 1800 hours, and 2300 to 0030 hours. It is not a particularly busy pedestrian spot by larger city standards (Figure 2), but it is one of the most active pedestrian intersections in this town of approximately 70,000. The number of people who crossed either Broadway or Ninth in these four periods came to 42 people during the dawn watch, 479 at the noon watch, 211 in the late afternoon, and 84 in the midnight observation.

What is the geography of this place? How does it have meaning to the geographer, or—more importantly—to the people who cross through it?

It stands fundamentally expressed in two-story buildings. On the northeast is the lone bank of this intersection, built of columned limestone and standing taller than any of the structures on the other three

Figure 1. Ninth and Broadway, Columbia, Missouri.

Land use along the east-west corridor of Broadway has a long history of development, for this is a length of the old United States National Road. In the 19th century, this intersection stood as one of the major stops on the Boon's Lick Trail—the approach road to the Santa Fe Trail just 12.5 kilometers (20 miles) to the west—and it lay just south of the National Road (later to be called U.S. Highway 40) that wound up from the floodplain of the Missouri River. A great portion, therefore, of all land transit from the farmlands of the Midwest toward the grazing lands of the Great Plains probably found its way through this Columbia intersection.

The buildings on these four corners speak of the past in an informal way. Some of the cornerstones and brass plaques on Broadway speak of 19th-century construction, and the proximity of the university (opened in 1839) is evident by looking down past the two defunct cinemas toward campus four blocks south of this corner. The university had a good sense of geography and human activity when it opened its doors not far from this intersection. Classic, but more recent efforts at maintaining the importance and economic significance of this point are evident in the new(er) façades that reach from the sidewalk to a height of about 3 meters (8 or 9 feet). New bricks, new tiles, even a concrete awning intended to create the impression of a weather-free pedestrian shopping district for the four blocks that merge at this intersection, come together to create a modernity on a main street that is more than a century and a half old.

These expressions of newness—whatever their covering material—generally stop short of the second-story windows. Standing at this intersection, you can

corners. The ground-level businesses are all retail shops. In common 1990s fashion, each has a different function, yet all exist in part as a response to the major economic threat represented by the large regional shopping mall nearly 2 kilometers (3 miles) away on the western outskirts of Columbia. This contest of traditional main street shops versus the newer franchises of the regional mall is common to hundreds of cities in the country. Broadway and Ninth are a classic manifestation of this contest, which is shown in this heartland intersection by some empty store space, relatively frequent turnover on Ninth and even in parts of Broadway, and increasing numbers of specialty shops, such as expensive non-franchise men's and women's clothing shops.

Figure 2. A pedestrian's view of the intersection.

cast your eye along all four blocks that meet here and you will see only a few signs of current life or utility above the ground level. On Broadway, east of the intersection, there is the Formosa Chinese Restaurant—one of the very few restaurant menus in an American city that can get people upstairs without an elevator.

On the same street, west of Ninth, is Victor's Garrett—a stylish and rather pricey gift and luggage shop—that is linked to the ground level by an external elevator.

Around the corner, on north Ninth, is Leo's, a second-floor, second-hand clothing store that stands as a good marker of Ninth's attempt to draw college students from the 25,000-student University of Missouri campus (as well as the two other local private colleges) into this city center. Beyond these three shops, the floors above ground level bespeak a common American downtown appearance of emptiness and abandonment.

Window after upper window has been partially bricked so that the original elegant arched casements of the first and second decade of this century's Broadway and Ninth construction can be replaced by much cheaper rectangular windows. Behind those windows lie storage space, empty space, a very occasional apartment, and now-vacant professional offices that had an earlier importance for these corners.

One 1920s building has been removed and replaced by a drive-in banking space adjacent to the major bank on the northeast corner of the intersection. This continuing courtship of automobile convenience is symptomatic of the American downtown's struggle with the mall.

One turn-of-the-century building has been replaced completely with a dark brick complex that attempts to deal with classic American pedestrian disinclinations. One two-story building that stood in line with the flow of storefronts on Broadway, west of Ninth, has been replaced by something quite daring. It is a building with a three-sided atrium that lets pedestrians easily see a well-glassed and bright downstairs; a street-level series of retail businesses that are open and inviting; and an upper story that also has revealing glass instead of a hiding wall. All of these three levels wrap around plantings, broad stairs, and an airiness that suggests welcome at all levels.

This effort to overcome our profound disinclination to go upstairs or downstairs from the main street of our cities is another sign of the times that is evident from this intersection.

At the ground level this intersection told me of additional recent efforts to make these corners work. There were four planters with evergreens. There were trash containers, mail boxes, and an outside pay phone.

Wooden banners had been installed at two corners and banners hung from light poles celebrating St. Patrick's Day.

Looking a little higher, awnings at the ground level, bright windows with varied and current displays spoke of currency in the products offered. The stone and brick at these levels suggested prosperity and accommodation.

These signs of hope are contested with a low-grade, continuing business turnover. One person explained to me that when President Reagan had come to Columbia some years ago, three stores were going out of business and the Chamber of Commerce had put enormous "Welcome President Reagan" signs over the top of the closing notices so that the outsiders would not take note of the downswing in the business around this intersection.

One cluster of four shops on east Broadway has donned a modern façade of new brick on their ground level. Inside the shops these merchants have breached walls and created arches so that a patron can drift from a men's clothing store to a women's store and beyond without ever stepping outside into the elements. This camouflage is intended to evoke the climate-controlled ambience of the mall, but does so along the (nearly) traditional downtown main street.

This intersection, then, is a pattern book of traditional economic and social growth and change and decline and further change in the American city. In modest but terribly important terms, both the city and its socioeconomic vitality are written in this locale.

The Watches

Observation of the population flow of these corners also helps to explain the nature and value of this intersection. In the midday and late afternoon, it is Broadway that is the dominant street. Before a 1700- to 1800-hour quietude creeps over the intersection, there is a mix of well-dressed professionals, casual walkers and shoppers, mothers (and occasionally fathers) with children in strollers, and some teenaged and college students.

Around midnight, however, it is not Broadway but Ninth that delivers people to and through the intersection. The college students "own" the night in this locale and it is a converted movie theater that brings them the four blocks and more from the campus to this corner. What was once the Varsity Movie Theater is now a night club with bar and live entertainment four or five nights a week.

The students walking up Ninth during the evening seemed hardly cognizant of the other roles that this city intersection play as they anticipate the night's rock, rap, or jazz group. Some walked against the traffic light, but most paused in animated conversation as they stood by the relatively empty corridors of Broadway and Ninth at midnight.

That same nighttime watch was characterized by the most prominent homeless and wino activity of this point. One Oklahoma drifter fixed on me as I was "reading" the corner and he argued with great energy—sometimes talking to me, sometimes to the planter on which I was seated as I wrote in my field journal—about the stupidity of the buildings' empty second floors and his inability to find a place to stay. The logic of his observations was disturbing as I took note of floors and floors of unused space.

At about 0645 hours, one man grumbled about living above this intersection for the past 12 years. When I asked if he had made a conscious decision to stay downtown, he said, "Hell, no—I'd rather live out by the golf course with the rich folk—but my job only supports my sleeping room on Ninth!" He, like so many, was conscious of other points of potential but nonetheless his daily pattern of activity took him along lines that emanated from this intersection and the heart of the downtown.

The struggle of the intersection seems to be to expand the number of people—especially employed people —who might be captured by an attraction to these four corners and the adjacent shops, restaurants, offices, and services. Like the dawn person, they may come there reluctantly but at least they come there and are a part of the life of the intersection.

Conclusion

As I walked toward campus at the end of my last early morning watch, I saw an old Chevrolet station-wagon turn up Ninth and park just beyond Broadway. Its license plate said "1 Tuf 56." I thought of that as a fine marker for the intersection: 1956 must have been just about the last of the post-World War II years when these four corners were uncontested as the commercial and style center of Columbia. The intersection has been tough, and it has taken a tough spirit and real dedication to keep the intersection attractive (in the primary sense of that word) and economically viable since then.

These lessons are written—one way or another— in most intersections of our American cities. The lines of movement toward these points of commerce, social interaction, and amusement have characterized the growth of our urban places.

Reference

Dillard, Annie. 1987. *An American Childhood*. New York: Harper.

Sauntering along Evanswood

Bruce Ryan
University of Cincinnati

I toss the 365th apple core of the year into the 2.5-hectare bird reserve, over the same tree fork as always, and wonder why no apple orchard has yet taken root, after all these years. I realize that my mid-afternoon saunter up and down either side of Evanswood Place is half over, that I have only 400 paces to go. Were I to wander into the two tag-ends, I would have passed all 34 single-family houses and both the apartment buildings which have comprised my neighborhood for the past 25 years. They are typical American dwellings of their vintage—the haunted manor house built in 1850, the others between 1890 and 1956. Most are four-square houses with Prairie allusions. All but one have two stories, an attic, a basement, surrounding lawns and gardens, and mature trees that probe the sky for 20 meters, but their decorations and details are as eclectic as the individuals who occupy them (Figure 1).

Evanswood lies 6.5 kilometers north of downtown Cincinnati in Clifton, the city's first commuter suburb. Now it is besieged by a reverberating expressway, the gaseous, industrial Mill Creek Valley, and the expansionism of half a dozen clustered hospitals and the University of Cincinnati. The Clifton Town Meeting perversely strives to retain a gaslit garden village in the midst of inner city dereliction. I pause on my saunter to contemplate its success, asking myself why Evanswood Place has survived as a bastion of diehard professional culture while hemmed in so stressfully and ambiguously between crippled ethnic slums and anemic exclusionist suburbs. The roadway itself was platted in 1891, and after 100 years, or so I conjecture, a street may well acquire a personality or tradition which it imposes on its residents, as Churchill thought buildings did, shaping their expectations of themselves, and of each other.

It is late autumn, and today's swarm of enormous black crows is roosting for the night in the woods, before scavenging elsewhere tomorrow. The beeches are turning from copper to gold. Just skimming the pin oaks and buckeyes, a rescue helicopter clatters back to University Hospital from some highway accident. High overhead, an airliner glides silently down toward the inclement weather runway of the airport.

I pause where Evanswood intersects Cornell Place (Figure 2), and distantly observe that indelible driveway between hedgerows where the Cincinnati Strangler committed his fourth and final atrocity one midnight, a quarter century ago. I tell myself that it was the neighborhood's only recorded murder, and that the street could scarcely be American without one. The house behind me has just sent its daughter to college, where she intends to become the fifth generation of physicians in her family. Nor does she forget that her father's cousin was Arthur Schnitzler, the Austrian author and dramatist.

Opposite lives the Head of Architecture, past-President of the Faculty Senate, and designer of the downtown Butterfield Senior Center, in a house once occupied by the Dean of the Medical College, and before him by a refugee from Hitler who published exegeses of *The Book of Job* (1922) and *The Prophets of Israel* (1914). My wandering thoughts recall the little triangular Jewish cemetery at the opposite end of Evanswood, dating from 1849, when Jewish corpses were more welcome in Clifton than Jewish residents. One Evanswooder still tells strangers that he lives "in a quiet Jewish neighborhood." There are seven Jewish householders on Evanswood now, but three of them have gentile wives who light Christmas trees and hide Easter eggs among the ground ivy and pachysandra.

Evanswood follows a level ridge for 300 meters but drops away steeply on both sides and at each end. I look down the narrow eastern approach to the house of the late Herman Schneider, Acting President of the University and Dean of Engineering, who gave

Figure 1. The Evanswood House. Architectural details from all 36 houses on Evanswood Place are incorporated in this fabrication. Illustration by Bruce Ryan.

the world the cooperative (work-study) system of education in 1906. In the 1920s, Herman and Louise Schneider had lived convivially amidst the presidents and managers of Mill Creek manufacturing plants which now remain, if they remain at all, only as the artifacts of industrial archaeology—ironworks, cut stone and brick masonry, chemicals, plumbing supplies, leaf tobacco, valves, boilers, lubricants, alloys, clocks, meats

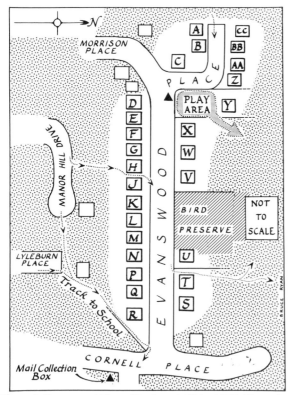

Figure 2. Evanswood Place, Cincinnati, Ohio, 1970. Illustration by Bruce Ryan.

(how we miss those malodorous stockyards), furnaces, elevators, veneers, beer, machine tools, rolling stock, radial drills, aluminum castings, furniture, and pharmaceuticals. The successors of these industrial magnates now tend to live near the peripheral, high-tech industrial parks, leaving Evanswood to the academics, the medical fraternity, and the besuited lawyers and business executives who commute downtown as brokers of money and power.

Not only black crows, I reflect, but human birds of passage have always roosted on Evanswood. Two of the last 20 valedictorians from the best public high school in Cincinnati have waved farewell to the street. From its Republican end came the present owner of the Cincinnati Reds baseball team (still basking in their 1990 World Championship), as well as the Keating brothers—one who went to Washington as a congressman, the other who went to jail for his commanding role in the Lincoln Savings & Loan fiasco. Not even Herman Schneider spread the notoriety of Evanswood so fulsomely, or feloniously.

For each farewell there has been a house-warming for new neighbors from all around the world (Figure 3). The Cincinnati school syllabus for first grade German was written by our native of Teheran, while native American neighbors have come to Evanswood from Saudi Arabia and Gadhafi's Libya during the bombing. Unofficial integration occurs every Glorious Fourth of July, when more than 100 Evanswooders picnic on the 140-meter sward behind the Siegfried house, mingling with an antique car collection which includes a 1947 French Delage, an even older camper-bus of mixed mechanical heritage, and some fin-bedecked Buck Rogers stegosaurs.

> Dyspepsia! Oh, Dyspepsia!
> Watch gluttony invade,
> Till each remotest neighbor
> Has passed out in the shade.

It is night on Evanswood, and the crashing rumble of trains in the classification yards wafts up from the valley bottoms, harmonizing with the ambulance sirens under an amber, sodium-illuminated sky. I pause outside the oldest house on the southern side of the street, a Queen Anne setting for Hitchcock mysteries, where a real estate lawyer (currently President of the Clifton Town Meeting and trombonist in the Clifton Community Memorial Day Band) has collected every Hungarian postage stamp ever issued. His wife, the epitome of charitable volunteers, has served several terms as President of the League of Women Voters and chairs the Clifton Senior Center. Other Evanswood volunteers serve just as diligently on community boards, from the

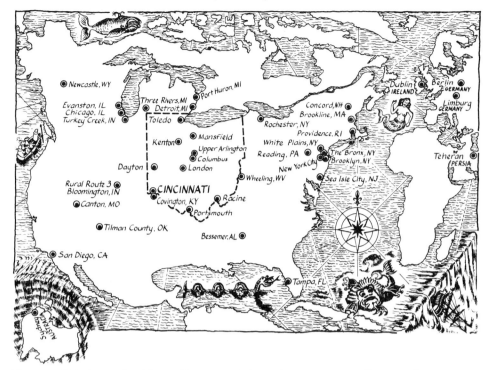

Figure 3. Birthplaces of residents of Evanswood Place, Cincinnati, 1988. Illustration by Bruce Ryan.

Christ Hospital Volunteers to the Crazy Ladies' Bookstore.

I saunter on to another house, its living room walls hung with an astonishing collection of ancient musical instruments (mostly plucked or brass), its former owner the associate conductor of the Cincinnati Symphony Orchestra. Next door lives His Honor, the present Mayor of Cincinnati, a Harvard lawyer whose dog-handling skills and Navy maneuvers have served him well in public life. Up the street, a navigation lawyer's widow inherited his 10 season tickets for seats right behind home plate at Riverfront Stadium.

Are we all eccentrics? Why should it be an Evanswooder who winds and maintains the clock in the tower of Clifton School? Why should out resident grief counselor specialize in helping the parents of murdered children, and inspire the morning television talk-shows? Where else does an engineer tune pianos for a hobby, or an architect carve violins, or an endocrinologist play piano with visiting chamber music ensembles? Surely others must breed raccoons and operate German beer gardens festooned with grapevines and build catamarans on their front porches.

It is winter, but my sauntering continues. The ghost of Dickens lurks at night where the gaslamps add haloes to the snow and the parked cars crouch like polar bears. Our pediatrician now jogs more gingerly along the ice-glazed sidewalks, while the hibernating writers

are hard at work behind their double-glazed windows and Christmas wreaths. Prodigious works greet the spring. Evanswood has nurtured Earl Case's *College Geography* (1932), the newspaper editorialist William Hessler's *Essays on Sea Power and Strategy* (1953), Leslie Martin's *Architectural Graphics* (1952), Helen Norman Smith and Helen Leslie Coops' *Physical and Health Education* (1938), Roy Blough's *International Business* (1966), Iola Silberstein's *Cincinnati Then and Now* (1982), Roy Hack's *God in Greek Philosophy in the Time of Socrates* (1931), Burtis Breese's *Psychology* (1917), Gaylord Merriman's *Calculus* (1954), and James A. Quinn's *Human Ecology* (1950). One classicist cataloged the Attic vases acquired by the Cleveland and Toledo Art Museums, while another excavated Lerna III (2450 to 2150 B.C.) in Greece. The creation on one very avant garde sculptress is now on permanent display in the Cincinnati Art Museum. Between deliveries, an obstetrician and gynecologist managed to write the history of Clifton and definitively ascertain the location of long-demolished Fort Washington, Cincinnati's original military garrison.

We are drawn together in our endeavors by the tritest acts of neighborliness—carpooling to the Playhouse-in-the-Park, joint ski vacations or expeditions to the Yucatán, baby-sitting, Scouts, church groups, house-watching, joint ownership of a yacht, hail and farewell parties for those who live nearby. Long before he became mayor, one of us had begun the tradition of candlelit Christmas caroling up and down the street. And it would not be New Year's Day without the traffic jam outside one open house. We play tennis and swim together at Clifton Meadows, we catch the same bus, we hike up the same hill to the university, we work for Procter & Gamble or Federated Department Stores, and our children ply the same houses each Hallowe'en.

True, we have to patch up occasional quarrels about straying pets, trespassing trees, inconveniently parked cars, encroachments across property lines, burglar alarms that sound for no good reason, kids who light campfires in the woods, errant trash cans on windy days, and sundry other nuisances, but there remains a determined and over-riding civility. We disagree currently about the proposed designation of Clifton as a historic district. Some medievalists crave it with a passion, wanting to wrap the glory days of the community in everlasting mothballs. Others oppose that hobgoblin of little minds, a single conforming architectural style, and hate to be characterised as old fogeys when the future of the inner city may require another direction entirely. We also disagree about building a new Clifton Library in Burnet Woods, the Tweedledums salivating for a park setting, the Tweedledees raving about the ecology.

But now it is a summer's evening on Evanswood, and the mothers are busy retrieving kids after the swim meet. How competitive we are, almost from infancy. Yet the fires of discord are barely smoldering now, smothered for a month or two by the oppresive humidity. A third of the houses are vacant, their owners away vacationing in France or Borneo, Tiber or Tennessee. Burglars are probably casing the street. I toss my lucky 366th apple core into the bird reserve, where the red-headed woodpecker and screech owl answer back, hoping that this time, at last, an apple orchard on Evanswood will be my reward.

The Fourth of Trafalgar:
Continuity along the Ontario Country Road

Thomas F. McIlwraith
University of Toronto

It is 1818, and an unheralded clerk of the British Colonial Office is dutifully ruling parallel lines on an outline map of Upper Canada. On this particular

day he is leaning over the westerly end of Lake Ontario, squinting at the Peel Plain forward of the face of the Niagara Escarpment. A decade earlier, his predecessor had boldly outlined Trafalgar and Nelson townships —no names could have been more fashionable—and surveyors had staked the perimeter, but then departed. The day's task is to draw in the farm lots ready for an expected rush of emigrants. Our clerk follows the "double-front" survey system: a rectangular grid of hundred-acre farms, each nearly square, with ⅗-kilometer (⅜-mile) frontage on a public right-of-way— a "Concession Line"—66 feet wide. After each five lots he rules a double line marking the cross-road allowance through the concessions, giving the linear landscape a second dimension. It is a simple mechanical exercise.

Farm lots 5 to 10 on the Fourth Line of Trafalgar Township took form on paper that day in 1818. With this directive, a surveyor strode through Trafalgar's dappled hardwood forest and drove stakes to mark the corners of the 12 lots. By 1830, settlers were opening little clearings, and a straggly cart-track was stretching out along the road allowance in front. Open spaces gradually merged during the next 50 years, the roadway became straighter, and fences appeared along boundaries. Rural services took up positions where Sideroad 5 and the Fourth Line crossed, and the spot was called Omagh. By the 1870s the squared cadaster had manifested itself as a farming landscape, and there the survey rests, immutable, to this day. The Omagh block, this unique 3-kilometer (2-mile) stretch of Trafalgar, tells a story repeated the full length and breadth of the Ontario countryside.

Trafalgar structures display the Ontario manner of building upon what exists with gentle touches, and

"Green Acres" house fairly shouts out its 150 years of steady adaptation. It is a story-and-a-half, the quintessential Ontario dwelling of the farm-making era: compact, upright, and symmetrical. Such houses have two floors of livable space in a timber-frame construction that had been introduced for simple one-floor houses. The story-and-a-half has been well suited to incremental increases in upstairs space as families grew and prospered, and by the 1880s tens of thousands stood on farms and in villages across the province. In the Omagh block, eight have kept up with changing needs and continue in family use.

The pedimental tops to the window-cases date "Green Acres" as a Classical Revival house of the 1840s. Already Gothic was appearing in the towns, but it would be years before the fashion trickled out to the Fourth Line. Meanwhile, the pediments gave simple grace to a vernacular building, marking a sophisticated populace intent on smoothing the rough edges of the civilized world. White clapboard siding is Classical too, but tell-tale corner caps show that it is aluminum panels no more than 20 years old: a worthy blend of continuity and convenience. The steeply pitched roof, punctuated by a front-wall gable, may have replaced a shallower slope common to Classical houses. The mixed farming era after 1870 was a prosperous time, and rural rebuilding was widespread. Raising the roof relieved the pressure and could explain the absence of cornice returns at the eaves, so common before 1860.

Handsome ridgeline chimney stacks, as old as the roof, once exhausted woodstove smoke. One has disappeared and the other's role has been taken by the furnace flue, a strictly functional and most unsympathetic column of cement blocks that rose along an end wall some time between 1925 and 1955. Elegant dwellings all over Ontario have been thus compromised, but this vernacular move modernized rural houses and made them attractive to the crowd escaping the cities and suburbs after 1960. The occupants of "Green Acres" have added a one-story family room to the rear, with a broad brick fireplace chimney filling most of one wall. Three chimneys recall a century and a half of refinement in domestic comfort and convenience.

Brick is another Ontario cliche well in evidence along the Omagh block. Several farmhouses in red with buff quoins and frieze decoration are joyous cel-

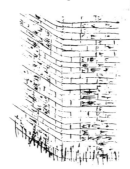

ebrations in a generally dour landscape. In lot 6, the harsh texture and improperly proportioned windows of the modern tail clash with the original, yet it is easy to overlook such shortcoming because someone made the effort to match the brick pattern. The 1840s Regency cottage in lot 7 is laid up in Flemish bond, a most subtle statement of pride to go along with the handsome sidelights to the doorcase, stone lintels with keystones, and polychromatic band below the eaves. This fine house and its contemporary next north set precedents that self-respecting Ontarians on the make in the next generation craved copying, often as a brick veneer hiding wood, stucco, or log.

Fencing has run full cycle, from the visible property markers of the emergent farmscape, through the unobtrusive and businesslike box-wire types of the 20th century, and then back again. Much along the Fourth

has completely disappeared, and more seems to have declined into picturesque reminders of where your place ends and mine begins. Panels, some mounted on old railroad ties, pick up and leave off casually; fragments of rusty wire dangle from stocky cement corner posts. In jarring contrast, a brick and wrought-iron wall brazenly announces the frontage of a 1980s house, while the absence of gates strikes a sympathetic chord with the weather-beaten fragments. A half dozen hobby sheep stray onto the road at their peril.

Trafalgar farm-makers imprinted the landscape with individualism. Detached farmsteads stand a hundred meters or so back from the roadside, occupying the slightest rise in what is essentially flat terrain. Houses thrust forward, barns defer behind, fields surround, and woodlots close the view. Neighbors see neighbors from afar, but are not intrusive. The quest for space, privacy, and access continues, and city refugees have found the most remote spots currently to be where farm-lot boundaries intersect the road. Nearly a dozen brick and aluminum houses have sprung up in the Omagh block since 1970, each on severances of half a hectare or so, farthest from the farmstead to which the land previously adhered. All line up squarely with existing borders, honoring a principle established by the old farm buildings.

The Fourth of Trafalgar is a showplace of farms and country living, and clearly in transition. Decorative farmhouse facades that once told passers-by of prosperity and self-esteem too often cower, their splendor diminished by a veneer of roadside debris. A gentle mix of obsolescence and nostalgia is evident in the broken wind-pump and tumbled barn foundation in lot 10. Both now function only as amiable subjects for local artists, but this modern circumstance has diminished the threat of complete annihilation. Again, the snow-plugged lane, the post missing its mail box, and the gaunt house behind an overgrown orchard add up to a serene case of decay, with good chance for a phoenix-like rejuvenation. One particularly fine farm view has been preempted by a roadside mobile home sharing its tiny lot with a sailboat, a school bus, and a packing-crate roadside shelter crudely labeled "bus stop"; the entire rootless mass could be driven away to Calgary or Newfoundland tomorrow. A satellite dish, heaps of unclean landfill, and plastic weather-sheeting over facade windows are certainly vernacular and honest, but likewise tacky.

Ontario countryside people are turning their backs on the road late in the 20th century. The tar stain on a brick front wall and the front door without steps are lingering signs of a verandah long since fallen away. Once families sat out on summer evenings and watched for a neighbor to pass, but even then they were using the remotest room in a backward-facing house. The enclosed front porchlet on "Green Acres"—an addition of the period before 1950 when residents walked—no longer invites. No steps or foot-path point to the road, and a side driveway leads behind to the garage and back door. The newest houses have attached garages facing sideways, and automatic door-openers permit occupants to slip home and be swallowed up, car and all, in a most unneighborly manner. Inward-facing people seem to be uncomfortable occupants of an outward-facing landscape.

The road reinforces the effect, arrow-straight and always stubbornly in tension against the rumpled carpet

that is Ontario's physical base. Four times in the Omagh block the Fourth crosses minor water-courses. Had the survey been twisted 30°, one crossing might have suf-

ficed; a deflection of a hundred meters in lots 6 and 7 could have eliminated two. Rather, the rivulets deviate a hundred meters along wide ditches to pass under the roadway at right angles. The 1929 topographic map identifies wooden bridges at these points, and surely they were landmarks for the commercial traveler in his Model "A." Today's broad cement box culvert, without railings and suitable for four lanes, has totally

separated motorists from a landscape with which they once were intimately associated. Only the largest stream holds its course against the road, which dips grudgingly while keeping its line, faithful to the cadaster and the speeding motorist.

Omagh once was a focus for country services, with shops and institutions nearest the corner and houses

beyond. As hamlet functions disappeared between 1930 and 1970, buildings not easily adapted vanished too. A mainline church has withstood the tide thanks to an evangelical congregation moving in, but otherwise Omagh is merely a scatter of houses. The hollow center is the legacy of this process, conveniently permitting road widening and rounded corners to hurry drivers onward to town or regional mall.

The evangelical church supports a feeling that revival may be coming to the Fourth, and in ways that are true to its rural-looking past. The 1980s house in lot 7 mimics a *Canada Farmer* pattern published in 1864, with its symmetry, thrust central bay, and fan-shaped window over the door. It announces renewed sophistication after 70 years of degeneration, and the persuasiveness of established models. The unlandscaped frontage speaks of more than aloofness, for the Regency farmhouse site is equally faceless. Each is at the start of a cycle of landscaping that may again yield roadside lines of maples, farm-lane avenues of basswood, hemlock windbreaks, and orchards, as well as foundation shrubbery and kitchen gardens around the house. Vegetation grows up and fades away inexorably, but the power of the old Ontario rural stereotype in the hands of a creatively reflective generation can will the cycle to be repeated. Just possibly the porchlet on "Green Acres" will be removed too, revealing the squared Classical Revival doorway that surely lurks beneath.

The Fourth of Trafalgar does not exist for the viewing pleasure of the passing motorist, and many would dismiss the current process of a commuter landscape overtaking a farmer landscape as an inevitable tragedy, or simply as banal. But the land bears its thoughtful and thoughtless users with equanimity, and to the discerning eye presents a rich palimpsest, ever renewed while not quite wiping out whatever went before. The Fourth of Trafalgar celebrates nearly two centuries of gradual accommodation among common features in ordinary places; this rich time depth is its irreplaceable legacy. The Fourth of Trafalgar is a measure of the pulse of the Ontario countryside late in the 20th century.

Further Reading

Blake, V. B., and Greenhill, Ralph. 1971. *Rural Ontario.* Toronto: Oberon.

Fram, Mark, and Weiler, John. 1981. *Continuity with Change.* Toronto: Queen's Printer.

Jackson, J. Brinckerhof. 1984. *The Necessity for Ruins.* Amherst: University of Massachusetts.

McIlwraith, Thomas F. forthcoming. *Beautiful Monotony: Reading the Ontario Landscape.*

Meinig, Donald W. 1980. *The Interpretation of Ordinary Landscapes.* New York: Oxford.

A Place and Point of View: Southwestern Wisconsin

Clarence W. Olmstead
University of Wisconsin–Madison

The windows of my home study overlook a landscape in southwestern Wisconsin (Figures 1 and 2). With this landscape I attempt to demonstrate that any place or landscape is a "window on the world," if examined in the context of the continuing physical and human processes that shape it and connect it to its region and the world.

Elements of the Land Structure

The view in Figure 2 is west-southwest. The horizon represents the eastern end of the Military Ridge cuesta (Figure 1). The cuesta ridge, parallel to the westward course of the lower Wisconsin River, is the northern outcropping escarpment edge of the Platteville-Galena dolomitic limestone formations, which once continued northward over the geologic dome of north-central Wisconsin (Martin 1974). At center right, the cuesta is surmounted by Blue Mounds. The southward-inclined top of West Mound, the highest elevation in southern Wisconsin, is an outlying remnant of the Niagara limestone formation whose outcropping edge forms cuesta ridges facing inward toward the geologic dome of Wisconsin, and outward around the Michigan basin to Niagara Falls. The view is across the headwater Pine Bluff basin of the Sugar River, a tributary of the Pecatonica. Although Military Ridge averages only 24 kilometers (15 miles) distance from the lower Wisconsin River, most of the area to the southward is drained by the Pecatonica down the southward-dipping slope of the Platteville-Galena formations to the Rock River in Illinois, and thence to the Mississippi.

The point of view is just beyond the terminal moraine of the Wisconsin glaciation or just within the Driftless Area, that large area, mainly in southwestern Wisconsin, which was missed by the several advances of the continental glaciation (Figure 1). The landscape here is transitional between glaciated eastern Wisconsin, with its moraines, drumlins, lakes, and marshes, and the heart of the Driftless Area with its steep-sided ridges and valleys. The floor of the Pine Bluff basin is covered by glacial alluvial outwash.

A transition even more significant for those who live from the land is that from the natural woodland environment of Wisconsin and the Great Lakes to the grassland environment of Iowa and the Great Plains (Curtis 1959). The transition occurs over a considerable area (Figure 1) with riparian woodlands formerly reaching southwestward and with fingers and islands of prairie penetrating northeastward, primarily on uplands. With its superior soils and longer growing season, the prairie environment is richly endowed for farming. The woodland region has a greater diversity of landform and water features. The transition zone, with elements of both, presents landscapes both productive and attractive.

Human Settlement and Use of the Land

People have occupied the area at least since the retreat of the last (Wisconsin) glaciation, approximately 11,000 years ago (Quimby 1960). In the centuries before first contact with Europeans, a sparse population of native Americans lived from hunting, fishing, gathering, and garden crops of maize, beans, squash, and tobacco. The coming of the Europeans precipitated change in the Indian economy and culture, as well as new conflicts and locational shifts among tribal groups. European

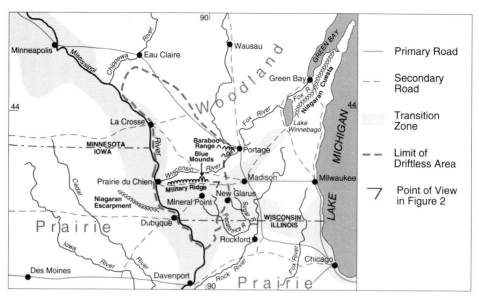

Figure 1. Southwestern Wisconsin and the prairie-woodland transition.

Figure 2. Views southwest across the Pine Bluff basin of Sugar River to Military Ridge and the Blue Mounds.

appropriation and settlement of Indian lands in eastern North America spurred successive westward migrations of Indian peoples as well as of the invading Europeans and their descendants. Iroquois Indians from the east, and Hurons, Ottawa, and others whom they uprooted, came into competition and conflict in Wisconsin with native Chippewa and Siouian tribes. They, in turn, were drawn into shifting alliances with the warring French, British, and Americans in their conflicts for control of the fur trade and the land during the century and a half after 1660.

After 1660, French traders from Québec entered Wisconsin at Green Bay to trade cloth, tools, guns, liquor, and trinkets to the Indians for cured fur pelts, especially beaver. Trade requires centers and lines of communication and transport. Waterways—the shorelines of the Great Lakes and the Mississippi, Wisconsin, Fox, and other rivers—had long served as routeways for the native Americans. Most of their villages were along lakes or streams. The Fox-Wisconsin waterway, with its short overland portage, became the major routeway of the Wisconsin territory. From the portage, at the great bend of the Wisconsin, the Fox River flows northeastward via Lake Winnebago to Green Bay (Figure 1). Trading posts were established at either end of the route, Green Bay at the mouth of the Fox in 1764 and Prairie du Chien at the mouth of the Wisconsin in 1781. In 1816 Fort Howard was built at the first and Fort Crawford at the second; a decade later, Fort Winnebago was built at the portage. The connecting route became known as Military Road. In southwestern Wisconsin it followed the sinuous crest and drainage divide of Military Ridge, thus minimizing grades and bridge-building (Smith 1973).

Sauk, Fox, and Winnebago Indians had used lead, found in the southwestern Wisconsin-northwestern Illinois area, mainly for decoration. After 1810 they began to mine and trade the ore. The new enterprise attracted white entrepreneurs and spawned a mining boom (Nesbitt 1973). By 1827, 6,000 miners, mainly migrants from the lower Ohio Valley, had moved up the Mississippi River onto Indian lands in what is now southwestern Wisconsin. In the next two decades they were joined by thousands more miners—many from Cornwall, England—and farmers. Quickly and inevitably, despite efforts such as the Black Hawk War of 1832, the native Americans lost their land, their way of life, and much of their culture and pride (Vogeler 1986). Following cessions of small areas at Green Bay and Prairie du Chien in 1815 to 1816, all of Wisconsin except small reservations was ceded to the white authorities from 1827 to 1848. Land offices for white settlement were established at Green Bay and Mineral Point (center of the lead-mining region) in 1834, and

at Milwaukee in 1836. White population exceeded 30,000 by 1840, 300,000 by 1850.

Before 1830, the spread of European-American settlement into Wisconsin was mainly from the south up the Mississippi River and into the lead-mining area. After 1840, a greater wave came from the east via the Great Lakes, especially through Milwaukee. Before 1850, most of the migrants were Americans, especially from New York, Vermont, and Ohio; the Swiss colonization near New Glarus in 1845 is an exception. Soon, immigrants from Germany and Norway far outnumbered the Swiss. Likewise, farmers outnumbered miners of lead, even within the mining area. Land was surveyed by the nationally-adopted rectangular system of townships and sections. Roads, farms, and fields conformed to the rectangular pattern except where steep slopes, streams, or marshes interfered. (Contour-strip fields were an innovation after 1940.) Most families acquired farms of one to three contiguous 16-hectare (40-acre) blocks.

Change in Land Use and Landscape

European settlement immediately began to change the character of the landscape, both natural and cultural. Wildfires, which frequently had burned unhindered through the prairies and savannas (oak openings), were curtailed or stopped, and scattered stands of trees rapidly thickened into forests (Ellarson 1949). Clearing of woodland and cultivation of uplands resulted in removal of soil from slopes to valleys (Knox 1977). On cleared land, farming was diversified but market-oriented from the beginning. Wheat was the major cash crop until about 1880 when impoverished soils, disease, and competition from newer farms in Great Plains grasslands forced experimentation and change (Ebling *et al.* 1948). During the decades just before and after 1900 there evolved on either side of the prairie-woodland line, a regionally dominant type of farming. On the prairies to the south and west, "Corn Belt" farms grew corn (maize), oats, and hay to feed farm-bred hogs and beef cattle. The woodland farms of Wisconsin were dominated by dairy cattle fed on alfalfa (lucerne), hay, corn, oats, and natural pasture.

Both systems of farming have continued to evolve and change in keeping with global, national, and local conditions. Most pervasive has been the substitution of capital inputs for labor and even for qualities of the land resource. The inputs fall mainly within three categories: machines powered by petroleum and electricity (replacing animal and human power, especially after 1940); chemicals for fertilizer and for control of weeds, disease, and pests; and genetic and managerial improvements, developed by state-sponsored and private research as well as by innovative farmers.

The external inputs have spurred increasing specialization and scale of operation, both on farms and in farm-service industries, and increasingly close and complex interconnections between the two. Mechanization and enlargement of farms proceeded faster in the Corn Belt where a typical family-operated farm has increased from 64 to 256 hectares (160 to 640 acres) during the 20th century. Yields of corn per hectare have tripled in the last half century. During mid-century decades many farms expanded their cattle-feeding enterprise by importing additional year-old cattle from western ranches. Now most of the ranch-produced cattle are "finished" in huge feedlots in western irrigated areas while more and more Corn Belt farms specialize in growing corn and soybeans for the world market (Hart 1986). A poleward expansion of such specialization into the dairy region, especially on islands of prairie soils, has been encouraged by genetic improvements, especially development of faster-maturing hybrid plants.

Dairy farms in Wisconsin have been highly capitalized from the beginning (Lewthwaite 1964). The need for animal housing and feed in winter prompted large barns, with stables at ground level and huge mows for loose hay above, and the development of upright, cylindrical silos for storing silage (ensilage) made of corn plants finely chopped before the grain has hardened. Since mid-century, low, one-story farm buildings are more convenient for storing machine-compacted bales of hay and farm implements as well as for animal housing. Concrete or steel silos of increasing size now dominate the farmsteads. Most dairy farms have eliminated once-common subsidiary hog and poultry enterprises. Pasture-feeding, likewise, has largely disappeared in southern Wisconsin. Oats, once important as feed for horses, as a nurse crop for spring-planted alfalfa, and for animal bedding, is dropping out of the crop rotation, as it did earlier in the Corn Belt. A decreasing proportion of the high-yielding corn crop is needed for silage; most corn is ripened for grain to be fed or sold. The typical family farm of 1940, with 32 hectares (80 acres) and 20 to 30 milk cows, now may have 96 hectares (240 acres) and 100 or more cows; production of milk per cow has tripled.

In the early 20th century, farmers hauled their milk in cans to small, crossroads cheese factories, a typical factory being owned by the dozen to 20 farmers it served. Wisconsin was unique at mid-century by having paved its dense, rectangular net of rural roads—daily delivery of fresh milk was imperative. The network of crossroads cheese factories was mirrored by a network of one-room, eight-grade public schools to educate

farm children. Both networks disappeared after mid-century, replaced by larger, more modern schools, and even fewer milk-processing factories, in towns and cities. Former cheese factories and many schools, like many former farmhouses, have been converted into non-farm residences.

Most of the population of southern Wisconsin is now concentrated in the Milwaukee-Chicago corridor, gateway to the east; in Madison, the state government and university center; and along inter-city routeways. The routeways, first used for travel by foot and water, then by wagon and rail, are now dominated by motor vehicles and airlines. Many railways that formerly served small towns and villages have been abandoned, some converted to bicycle and ski trails. The importance of the route from Chicago northwest to Madison and Minneapolis, via either Milwaukee or Rockford, now far surpasses that of the older but still important Military Road route from Green Bay through Madison to Prairie du Chien or Dubuque (Figure 1). Secondary routes facilitate Madison's function as a transport hub. Villages and hamlets, which once functioned primarily as farm-service centers, have lost much of that function to larger firms in larger places (Brush 1953). Those within a 64-kilometer (40-mile) radius of Madison or other cities now serve as residential centers for workers who commute daily to the cities, or for retired people. Selected routes in the rectangular grid of rural roads have been gradually improved to provide for the daily flow of workers to and from the cities as well as for city-based services to the fewer but larger farms. Once-farmed land in hilly moraines of the glaciated area, or in rougher, stream-dissected parts of the driftless area, now support second homes and recreation industries for people from the cities. Some changes formerly viewed as desirable are now being questioned or even reversed. Examples are the draining of wetlands, excessive use of chemicals in farming, abandonment of railroads, or unlimited urban expansion onto agricultural land.

Scales of Observation

The patterns and significance of the points, lines, and transition zones can profitably be observed at different geographic scales, for example, continental, regional, and local. The natural processes of atmospheric circulation, glaciation or geology, that produced the prairie-woodland transition, the driftless area, or Military Ridge, are of continental or world scale. So, too, are the human processes that brought new settlers and changing technologies to the land and evolved different systems of farming.

Other forces operate, or are expressed, at a regional scale. Fire, either induced or controlled by humans, affected the location of prairie-woodland boundaries. The mantle of loess, windblown during glacial times, thins gradually eastward for 96 kilometers (60 miles) along Military Ridge and its southward-dipping slope. The Corn Belt-Dairy Belt transition occurs over a comparable distance. The service areas of Milwaukee and Madison overlap in the 128-kilometer (80-mile) interval between them.

Finally, the patterns are most critical to decision-making, and most revealing of their secrets, at the local level. Many early settlers located their farmsteads at a prairie-woods edge to provide a diversity of resources: fertile prairie soils for grain crops and wood for building and fuel. A site near the head of a valley provided spring water, protection from wind, and natural pastures along the stream. In the landscape pictured here, near the northeastern margin of the prairie-woodland transition, farmstead locations were governed by the modified rectangular road pattern, but sites were also selected at the base or edge of wooded rock outcrops. In the Pine Bluff basin, a pre-settlement prairie, woodland now marks farmsteads, broken edges of the Platteville-Galena limestone uplands, or thin-soiled areas or outcrops of the underlying St. Peter sandstone formation.

The 1,280-hectare (5-square-mile) basin area partially represented in Figure 2 contains ten family farms, several operated by descendants of the original German and Irish settlers. Most support dairy herds but produce surplus grain and corn for market. In addition, a nursery and a fruit and vegetable farm, each 16 hectares (40 acres), have been established since 1970. Twelve former farmsteads, a cheese factory, and a school have been converted to non-farm residences. The 18 additional non-farm residences (excluding Pine Bluff hamlet), most built after 1970, some to house the older generation on an operating farm, others occupied by urban migrants lured by attractive homesites. The names of Mineral Point Road and Old Military Road suggest their original role as part of the route from Green Bay to Prairie du Chien and the lead-mining region. Their current function is commuter route to Madison.

The hamlet of Pine Bluff is marked by one of the largest farmsteads (with a subsidiary farm-equipment-manufacturing enterprise) and the steeple of the Catholic church whose adjacent school is now closed. The hamlet also contains the garage/office government "seat" of the town (township), two taverns, a former general store, a baseball field, and two dozen residences. The village of Mount Horeb, 13 kilometers (8 miles) distant on the horizon at left, is marked by its globular water tower. Its long-standing function as farm service center,

still with a milk-processing plant (two until 1991) and a cooperative farm store, is now secondary to that of a fast-growing bedroom community for Madison and, to a lesser extent, that of tourist center emphasizing its Norwegian heritage. The growing recreation industry is represented also by the state park on the summit of West Blue Mound.

The landscape, like the transition zone of which it is part, continues to change as people assess local, regional, and world conditions. As modern, mainly urban, influences continue to encroach, one hopes that neither the cues to the evolution of the landscape, nor the values and respect, which so far have maintained much of its bounty and beauty, will be lost.

Acknowledgments

I am grateful to James C. Knox and Thomas R. Vale for helpful suggestions and to the University of Wisconsin Cartographic Laboratory for Figures 1 and 2.

References

Brush, John E. 1953. The hierarchy of central places in southwestern Wisconsin. *Geographical Review* 43:380-402.

Curtis, John. 1959. *Vegetation of Wisconsin.* Madison: University of Wisconsin Press.

Ebling, Walter H., Caparoon, Clarence D., Wilcox, Emery C., and Estes, Cecil W. 1948. *A Century of Wisconsin Agriculture, 1848-1948.* Madison: United States and Wisconsin Departments of Agriculture.

Ellarson, Robert S. 1949. The vegetation of Dane County Wisconsin in 1835. *Transactions of the Wisconsin Academy of Sciences, Arts and Letters* 39:21-45.

Hart, John Fraser. 1986. Change in the Corn Belt. *Geographical Review* 76:51-72.

Knox, James C. 1977. Human impacts on Wisconsin stream channels. *Annals of the Association of American Geographers* 67:323-343.

Lewthwaite, Gordon. 1964. Wisconsin and the Waikato: A comparison of dairy farming in the United States and New Zealand. *Annals of the Association of American Geographers* 54:59-87.

Martin, Lawrence. 1974. *The Physical Geography of Wisconsin.* Madison: University of Wisconsin Press.

Nesbitt, Robert C. 1973. *Wisconsin: A History.* Madison: University of Wisconsin Press.

Quimby, George. 1960. *Indian Life in the Upper Great Lakes, 11,000 BC to AD 1800.* Chicago: University of Chicago Press.

Smith, Alice E. 1973. From exploration to statehood. In *The History of Wisconsin,* Vol. 1, ed. William F. Thompson. Madison: State Historical Society of Wisconsin and University of Wisconsin Press.

Vogeler, Ingolf. 1986. *Wisconsin: A Geography.* Boulder, CO: Westview Press.

Fence Lines on Kentucky's Bluegrass Landscape

Karl B. Raitz
University of Kentucky

Carolyn Murray-Wooley
*Architectural Historian,
Lexington, Kentucky*

Fences demarcate property boundaries, delimit land uses, and contain stock, or protect crops, gardens, and orchards from our own or our neighbors' animals. Historically, farmers erected fences using building materials obtained on their own lands. Fence form often followed models used along the developing frontier in America's eastern woodlands. In some places, the fence became more than a practical farmstead structure, as in central Kentucky's Bluegrass region, where the fence came to symbolize values and traditions.

Nineteenth-century Bluegrass farmers established a farming system based upon cattle grazing, which necessitated fencing from the earliest years of settlement. They imported blooded animals from Great Britain and sold breeding stock up and down the Ohio River valley. Some farmers drove cattle across the Appalachians to markets in large East Coast cities. Others sold cattle to Southern plantations (Henlein 1959).

Rail Fences

Bluegrass farmers built rail fences of local hardwoods using ordinary tools, following fence designs common in the east—post and rail, and Virginia worm being two common types (Hoffman 1835). The eastern hardwood forest supplied farmers in many sections with a seemingly unlimited supply of walnut, chestnut, and oak for rails. Bluegrass farms did not enjoy this same bounty. The region's woodlands were open; often centuries-old trees stood above an understory of cane, pea vines, and grasses. The old trees provided summer shade for stock and grass, and farmers reluctantly cut them for fencing materials.

Fences came to represent more than a simple enclosure. The well-made rail fence came to symbolize the considerate and law-abiding farmer. Neighbors and passersby used the condition of fences as a measure of moral character and social standing (Pocius 1977). Maintaining a strong, well-made fence, especially along a public thoroughfare, was requisite for continued community acceptance.

Rock Fences

Rail fences were not long-lived and a wood shortage in the region during the 1820s and 1830s prompted many farmers to consider an alternative fencing type and material, limestone (*Western Agriculturist and Practical Farmers Guide* 1830). The entire Bluegrass region—including its three subdivisions, the Inner Bluegrass, the Eden Shale Hills, and the Outer Bluegrass—is underlain by Ordovician-age rocks with some Silurian age near the region's edges (Figure 1). The regoliths that developed on the Ordovician limestones yielded fertile soils. Rock for fence construction could be quarried along streams or hillsides where ledges outcropped near the surface.

By the 1790s, some Scottish and Scots-Irish settlers had built rock fences to enclose gristmills along streams near the Kentucky River as well as some home yards and cemeteries. Large-scale rock-fence construction did not begin until the 1830s and 1840s, and it continued until after the Civil War. Some rock fences enclosed entire farms and therefore functioned as property boundaries. Others divided farmsteads into lots for buildings, gardens, and livestock. By the 1850s, turnpike construction crews were macadamizing road surfaces with limestone, and farmers began to build rock fences along the roads that fronted their property. On Inner Bluegrass farms where soils were deep, fence

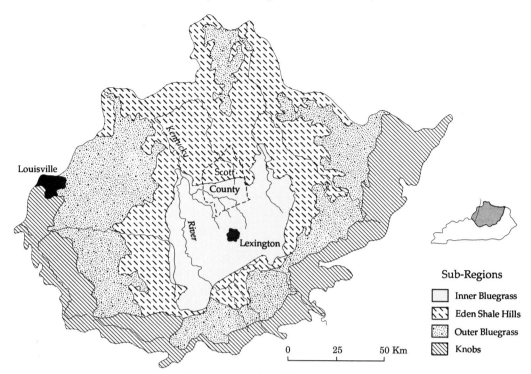

Figure 1. Kentucky's Bluegrass region.

rock had to be quarried and hauled to the construction site. On the stream-dissected land near the Kentucky River and the Eden Shale Hills, farmers used rock cleared from fields to build fences.

Local popular belief held that slaves built the Bluegrass rock fences. In addition to cattle, sheep, mules, and horses, the region's farms produced some grain crops and hemp. Many farmers owned a few slaves and some owned as many as a hundred or more. Slaves not needed for day-to-day farm operation were often hired out to neighbors or town residents, particularly after the crops had been harvested. But, neither his-

torical documentation nor fence morphology support the notion that they built the region's rock fences (Murray-Wooley and Raitz 1992).

Beginning in 1850, the federal census recorded occupation and place of birth. A count of all residents in several Inner Bluegrass counties revealed numerous persons whose occupation was stonemason or rock mason (Table 1). By 1850, 17 stonemasons lived in Scott County, for example, and 11 of them were born in Ireland. Many Irish laborers had been recruited to work on the Erie Canal in New York as early as 1817, and they worked at other construction projects such

Table 1. Stonemasons: Scott County, Kentucky

| | White | | | | | |
	Kentucky[a]	U.S.[b]	Foreign[c]	Ireland[a]	Black	Total
1850	1	2	2	11	1	17
1860	5	3	1	38	1	48
1870	9	0	1	12	16	38
1880	7	0	2	3	13	25
1900	5	0	1	0	20	26
1910	4	0	0	0	8	12

[a]Indicates place of birth.
[b]U.S. place of birth, other than Kentucky.
[c]Foreign place of birth, other than Ireland.
Note: The 1850 and 1860 censuses do not list slave occupations.
Source: Federal Census Manuscripts.

as canals, railroads, or in East Coast industrial mills. The potato famine of the 1840s lent considerable urgency to increased Irish migration to the United States, and many found their way to central Kentucky. Some stonemasons were skilled in preparing ashlar stone for building construction. Others had experience building common structures and fences using tools and techniques familiar to masons in Ulster, southern Scotland, and northern England.

The Irish-built rock fences had a common form (Figure 2). Starting with rock gathered from a field, a creek bottom, or a quarry, the mason laid large foundation rocks in a shallow trench dug to the frost line. Then in horizontal courses above the foundation the mason placed rocks without mortar into two faces so that the center formed an irregular void that was packed with small rocks and chips. The fence narrowed from a width of about 30 inches at the first course above the foundation to about 18 inches at the top horizontal course, some 4 feet above ground. This top or cap course spanned the fence's width helping tie the two faces together. Atop the cap course, masons placed a coping course of large triangular rocks set on edge and leaning downhill at a slight angle. The coping added height and the heavy rock gave stability to the entire structure. Sharp corners deterred stock. Many masons employed an additional feature to add strength to the fence, "through" rocks or "headers." At a height of about 18 inches above the foundation and spaced about 48 inches apart, the mason laid a "header course" of large rocks that spanned the entire fence width. Another such course was then laid about 18 inches higher and spaced so that the individual "headers" alternated with those in the lower course.

This fence was not simply a carefully piled stack of rocks but had a definite morphology that was duplicated over a broad area and through a half century of fence construction. To build this fence correctly, and with sufficient speed so that at a piece rate one could make a living at it, one would require the knowledge that could be obtained by apprenticing with an experienced mason. Therefore, this fence type diffused from Great Britain and Ireland directly to central Kentucky.

Some Irish masons boarded with farmers while building fences for them. Others lived in villages and walked to construction sites. Often senior masons started small companies that built fences and hard-surfaced roads, boarding a dozen or more Irish laborers in their homes. On the larger farms, owners delegated slaves to gather rock or haul it from a quarry to the place where a fence was to be built. Slaves also may have dug fence foundation trenches and some probably assisted the Irish masons in actual construction, thereby learning the Irish fence-building technique (Clay and Thornton 1844).

By the eve of the Civil War, the number of Irish stonemasons increased, and in Scott County, for example, their number had more than tripled (Table 1). Thereafter, their number steadily declined, and they were replaced by freedmen so that by 1900, few Irish-born stonemasons remained. Black masons were still working. This pattern of decline and replacement was repeated in the other Bluegrass counties. The process illustrates how black masons learned to build fences according to the Irish-Scottish-English model.

Travelers admired the Bluegrass rock fences as did the region's residents. The fence was substantial and permanent. It represented foresight, investment, and the polish that comes with settlement maturity rather than the rudeness represented by expedient frontier rail fences. Because it required skills and tools uncommon among most upland South farmers the rock fence represented the land-owner's wealth and status. In fact, the fence came to be a symbol of the region and when built elsewhere was known as a "Bluegrass Fence" (Riedl et al. 1976).

Plank Fences

By the 1880s, as sawn lumber became more readily available, Bluegrass stock-farmers began building plank fences. Farmers tore down many rock fences to enlarge pastures and fields, or to circulate more easily within farmsteads. Some farmers white-washed their fences to retard weathering, and soon the white plank fence could be found on many larger estates. During this same period, wealthy industrialists from the north bought land in the Bluegrass and established horse farms where they bred racing stock for the nation's

Figure 2. A fence built on an Inner Bluegrass farm by Irish masons using rock quarried on the property. The pre-Civil War fence encloses a woodland pasture.

thoroughbred and standardbred racing circuits. One of these persons used an asphalt-based paint on his farm's fences and so lent aesthetic credibility to the black plank fence. As the racehorse industry expanded during the early 20th century, new farms were usually surrounded by white or black plank fences. Few woven wire fences were built, in part because horses could become entangled in the open mesh, and barbed wire was not even considered for the potential hazard it posed.

A standard plank fence form gradually evolved. Long-lasting locust or cedar posts spaced about 16 feet apart supported four horizontal oak, pine, or cedar planks, about 1 inch thick and 6 inches wide. The bottom planks were spaced somewhat closer together than those at the top. A simple oak batt about 6 inches wide covered the joint where the planks were nailed to the posts. The owner of a large farm might build 20 miles or more of such fences. When pressure-treated wood became available during the 1950s, it gradually replaced traditional materials, although the treated planks and posts were usually painted white or black for aesthetic reasons (Stone 1971). Neither painting nor pressure-treating completely solved the problem of fence maintenance, and wood decay or faded paint prompts periodic rebuilding or repainting. Some farms maintain labor crews whose full-time task is painting and rebuilding fences.

Fences as Regional Symbols

The white or black plank fence has become a symbol of Kentucky even though the section that actually produces horses and builds white fences is a small area of a few hundred farms north of Lexington. Tourists often note the beauty of the white-fenced horse farms. This fence type has also become synonymous with the horse industry so that horse buyers and race fans tend to equate a prestigious horse breeder with an aesthetic Bluegrass farm surrounded by white fences, even though racing horses are bred in many other states. The white

fence is a key element of a stereotyped Bluegrass landscape whose image is often used in advertizing regionally or even nationally marketed consumer products, or in publicity materials prepared by the state to enhance tourism or recruit new business and industry (for example, Raitz and Van Dommelen 1990). Few new rock fences are built because of the expense, but frequently the owner of a large estate will erect a formal rock fence on each side of an entrance gate backed by a plank fence. This rock fence is merely a subaltern to the plank which actually contains the stock, and serves to symbolize the elite status of the farm owner, rather than acknowledge the 19th-century masons who brought the rock-fence technology to the region.

References

Clay, B. J., and Thornton, F. 1844. Agreement, April. Clay Family Papers. Lexington: Special Collections, University of Kentucky.

Henlein, P. 1959. *Cattle Kingdom in the Ohio Valley, 1783-1860.* Lexington: University of Kentucky Press.

Hoffman, C. 1835. *A Winter in the West,* Vol. 2. New York: Harper.

Murray-Wooley, C., and Raitz, K. 1992. *Rock Fences of Kentucky's Bluegrass.* Lexington: University Press of Kentucky.

Pocius, G. 1977. Walls and fences in Susquehanna County, Pennsylvania. *Pennsylvania Folklife* 26:9-20.

Raitz, K., and Van Dommelen, D. 1990. Creating the landscape symbol vocabulary for a regional image: The case of the Kentucky Bluegrass. *Landscape Journal* 9(2):109-121.

Riedl, N., Ball, D., and Cavender, P. 1976. *A Survey of Traditional Architecture and Related Material Folk Culture Patterns in the Normandy Reservoir, Coffee County, Tennessee.* Knoxville: Report No. 17, Department of Anthropology, University of Tennessee, Tennessee Valley Authority.

Stone, C. 1971. Fence construction. In *A Barn Well Filled,* ed. K. Hollingsworth, pp. 48-51. Lexington: The Blood Horse.

Western Agriculturist and Practical Farmers Guide. 1830. Fences, p. 55. Cincinnati: Robinson and Fairbank, for the Hamilton County (Ohio) Agricultural Society.

The Poor Person's Palm Springs: Quartzsite, Arizona

Barbara A. Weightman
California State University, Fullerton

"Snowbirding," the seasonal migration of mostly retired people from colder regions of the United States and Canada to warmer regions of the American Sunbelt, is a 20th-century phenomenon that is spawning a diversified network of facilities and services (Belasco 1981; Mings 1984). While snowbirds employ a variety of travel means, of increasing importance is the recreational vehicle, a self-contained home on wheels.

Recreational vehicles have expanded the possibilities of place-making because they foster mobility far beyond that of simple seasonal movement. Consequently, destinations for recreational vehicles experience wide variation in population, social, and economic activity.

In the Southwest, snowbirds in recreational vehicles journey from point to point along a well-established circuit including such places as Death Valley, Imperial Valley, and Palm Springs in California; and Phoenix, Tucson, and Yuma in Arizona (Mings 1984). One point on this circuit is Quartzsite, Arizona (Figure 1).

Quartzsite is both a retirement community and leisure landscape. By Karl Raitz's (1987) definition, Quartzsite is a consummate "leisure landscape ensemble" where people come, at particular times, for "quality experience" derived from activities engaged in as well as character of place. While Patricia Gober (1985) notes the geographical significance of retirement communities, both Karl Raitz (1987) and Peirce Lewis (1979) have called for geographic investigation of leisure landscapes in terms of the roles played by natural and cultural phenomena.

Quartzsite, however, is more than an assemblage of landscape features. As a retirement community, it is unusual because it is non-urban and largely impermanent. More importantly, its sense of place changes in context of its yearly cycle of activities.

Time is fundamental to the essence of this place. Several authors have revealed the criticality of time in the structuring of space (Lynch 1972; Parkes and Thrift 1978, 1980). As Don Parkes and Nigel Thrift (1978, 119) have noted: "Timed space is the essence of place." The experience of Quartzsite, as a retirement community and as a leisure landscape, is couched in the tempo of recurrent population movements and particular events associated with seasonal climatic change. Quartzsite epitomizes a "timed place."

Quartzsite is a landscape anomaly. Situated 32 kilometers east of the Colorado River, at the junction of Arizona State Highway 99 (SH-99) and Interstate 10 (I-10) in the Sonoran Desert, Quartzsite's midsummer population approximates 3,300 residents. However, by October this figure rises well above 50,000 and by February, it soars to an estimated 750,000 to more than 1,000,000 (Chamber of Commerce, 1991)! Explanation of this phenomenon lies in Quartzsite's evolution.

Evolutionary Landscape

As a watering-hole and stage-stop for travelers and the U.S. Army, Quartzsite was known as Tyson Wells or Fort Tyson in the 1850s. The settlement was named Quartzsite in 1896 in light of abundant quartz crystals in the local mountains. Sporadic mining activity boosted the population to 339 by 1910 (Feitz 1980). By the 1960s, recreational vehicles were becoming popular and Quartzsite was earning a reputation as a center for rockhounding (rock, gem, and mineral collecting) and swap meets, as well as a winter haven for snowbirds. Summer populations of about 50 souls

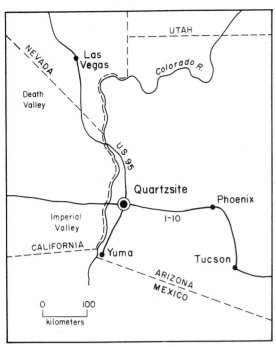

Figure 1. Quartzsite, Arizona.

were increasing to as many as 1,500 campers in recreational vehicles by January.

Some of the snowbirds decided to stay permanently and in 1965, 44 residents formed the Quartzsite Improvement Association to monitor and improve services and shape the destiny of the growing community (Allen 1981). Since this was an area already popular with rockhounds, the Quartzsite Improvement Association decided to hold a gem and mineral show and promote swap-meet activities.

More than 200,000 visitors attended the first large-scale show in 1974. The Quartzsite Improvement Association named this event the "Pow Wow." Today the Pow Wow is advertised as the "World's Largest Gem and Mineral Show" and "Largest Display of Lapidary Equipment in the Country." Subsequent growth in Quartzsite has evolved around seasonal patterns of mobility facilitated by the recreational vehicle, and a schedule of events anchored by the Pow Wow.

Contemporary Landscape

Year-Round

By capitalizing on the region's attributes and offering an array of unusual and interesting events and services, the Quartzsite Improvement Association has engendered a unique leisure landscape with its financial roots firmly in snowbird tourism, the leading dollar

generator. In 1985, 105 businesses and services were listed in the Quartzsite Improvement Association Directory. The town incorporated in 1989 and the 1991 listing totals 219 enterprises, including a McDonalds' built in 1988.

Quartzsite's commercial landscape is markedly unusual. *Time* (Ackermann-Blount 1989) characterizes the 4-kilometer (2.5-mile) commercial strip along Main Street (the business loop of I-10), as a "giant flea market." Ted's Truck Stop and Bull Pen Restaurant is the largest employer with 100 employees, while McDonalds ranks second with 20 regular and 20 seasonal workers. Most of the more than 200 remaining enterprises operate part-time and seasonally. Many are staffed by owners or volunteers and are run more for enjoyment than profit.

Permanent residents, for the most part, live north of I-10 where names such as Snowbird Lane, Lizard Lane, and Quail Trail mark the mostly unpaved streets. Camel Drive recalls the days (1857 to 1864) when the U.S. Camel Corps, created by Jefferson Davis, was training in the area.

Housing types tend toward the vernacular, exhibiting a plethora of adaptations and combinations of adobe, wood, and stone structures with aluminum trailers or mobile homes. Properties are demarcated by a variety of fence types, rows of rocks, or cactus hedges. Wooden windmills, coyotes, and roadrunners proliferate in carefully tended earthen yards. Many yards display accumulations of potentially valuable "junk" (in swap-meet context).

Winter

Winter is a time of great events that attract thousands of visitors (Figure 2). More than 3,000 "tailgaters" come to hawk their wares from various sized trucks and vans. Characterized by periodicity within a January-February time frame, these rock shows and swap meets are catalytic to the extent and intensity of social and economic activity.

These space-consuming activities border I-10, SH-95, and Main Street and the ensuing traffic congestion is monumental. The Chamber of Commerce comments that, "Newcomers are sitting in traffic, waiting their turn. . . . Seasoned veterans know better, you'll see them walking" (1991). According to the Arizona State Police, as many as 2,700 vehicles and 1,000 pedestrians an hour cross the intersection of I-10 and SH-95 during the peak period.

Quartzsite's rock, mineral, and gem shows draw dealers from Afghanistan, Brazil, Saudi Arabia, and Australia. Many of the exhibitors and dealers follow

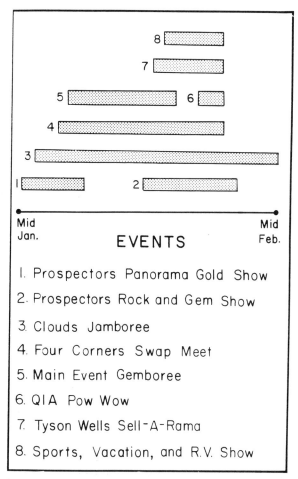

Figure 2. Timing of winter rock, mineral, and gem shows; swap meets; and hobby, sporting, leisure, and RV shows.

a show circuit. Fifty-three gem and mineral shows in 10 western states advertised at the 1991 Pow Wow. The other events exhibit, promote, and sell everything imaginable but most of the items are related to rock-hounding, lapidary, and other crafts, or the maintenance and enhancement of recreational vehicles.

Events are fundamental to the sense of community that prevails throughout the year. Weekly pancake suppers attract hundreds of participants and holiday dinners feed well over 1,000 people. Saturday night dancers pack the Stardusty Ballroom, foxtrotting and waltzing to the music of Desert Varnish or the Hi Jolly Musicmakers. There are Bingo nights, exercise classes, hobby and card-playing clubs, arts and crafts lessons, and organized excursions to shop in Mexico, gamble in Nevada, or fossick for gold.

During the Pow Wow, balloon and helicopter rides add to the festivities and rock-hounding and other special-interest field trips are conducted into the local hills. The success of these events derives from volunteerism. The Pow Wow alone calls for more than

1,200 volunteers drawn from snowbird as well as permanent populations.

Quartzsite's snowbird visitors are categorized in temporal context. Those who stay for most or all of the September-to-mid-April winter are termed "boondockers." Those who remain for a couple of weeks or less, typically during the "tailgater" period, are called "tourists."

Boondockers, tourists, and permanent residents, in concert with the tailgaters, seek social and economic satisfaction through the medium of "swapping" in "malls without roofs." Though all of the shows have ended by 15 February, many tailgaters keep their stands open until mid-March when the temperatures rise enough to precipitate the great snowbird emigration to the more northerly states and Canada.

While the key events are concentrated in two months, Quartzsite's population expands and shrinks over a stretch of eight months. From September to April, snowbirds take over the town and its environs. Thousands reside in their self-contained recreational vehicles, also referred to as "big-rigs." According to the Chamber of Commerce, Quartzsite has 72 recreational-vehicle parks (1991). However, these are insufficient for the massive snowbird influx.

On the outskirts of town, in the open desert dotted with saguaro cactus and other xerophytes, are three Bureau of Land Management 2-week recreational vehicle-camping areas. To ease the snowbird burden, the Bureau of Land Management has recently opened a 4,452-hectare long-term visitor area requiring only a $25 permit for an extended stay. These Bureau of Land Management campgrounds are staffed by Quartzsite volunteers.

There is one set of public telephones for all the Bureau of Land Management campers. Six kilometers south of Quartzsite, standing in the long-term visitor area's desert expanse, it is frequently marked by a cluster of vehicles and a line of campers waiting their turn to call friends and family across the continent. Perhaps they are explaining just how they are following through on one of the more popular bumper stickers: "We are having fun spending our children's inheritance."

Timed Place

Quartzsite is a periodic retirement community that responds to the demands of a recreational-vehicle lifestyle. It is a leisure landscape ensemble keyed to seasonal climatic change and recurring special events. This landscape of events is a metaphor for Quartzsite, the place, which a local author has dubbed the "poor man's Palm Springs" and a town "helped by arthritis and inflation" (Feitz 1980, 3). For thousands of snow-birds, it is the point of all points within the winter haven of the American Southwest.

References

Ackermann-Blount, Joan. 1989. Parked in the middle of nowhere. *Time* 22 May:108-109.

Allen, Mary. 1981. *Q.I.A. How Come? A History of the Quartzsite Improvement Association 1965-1980.* Quartzsite, AZ: Quartzsite Improvement Association.

Belasco, Warren J. 1981. *Americans on the Road: From Autocamp to Motel, 1910-1945.* Cambridge, MA: MIT Press.

Chamber of Commerce. 1991. *Quartzite, Arizona 1990/91 Visitor's Directory.*

Feitz, Leland. 1980. *Quartzsite, Arizona: No Ordinary Place.* Colorado Springs: Little London Press.

Gober, Patricia. 1985. The retirement community as a geographical phenomenon. *Journal of Geography* 84:189-198.

Lewis, Peirce. 1979. Axioms for reading the landscape. In *The Interpretation of Ordinary Landscapes*, ed. D. W. Meinig, pp. 11-32. New York: Oxford University Press.

Lynch, Kevin. 1972. *What Time Is This Place?* Cambridge, MA: MIT Press.

Mings, Robert C. 1984. Recreational nomads in the southwestern sunbelt. *Journal of Cultural Geography* 4:86-99.

Parkes, Don, and Thrift, Nigel. 1978. Making sense of time. In *Making Sense of Time*, eds. T. Carlstein, D. Parkes, and N. Thrift, pp. 119-129. New York: Halstead Press.

———. 1980. *Times, Spaces, and Places: A Chronogeographic Perspective.* New York: Wiley.

Raitz, Karl. 1987. Place, space and environment in America's leisure landscapes. *Journal of Cultural Geography* 8:49-62.

Walt Disney World, Florida: The Creation of a Fantasy Landscape

Morton D. Winsberg
Florida State University

In 1971 an 11,100-hectare amusement area named Walt Disney World was opened in central Florida near Orlando. Within it was the Magic Kingdom, the first of three large theme parks located there by 1991. The venture rapidly became an international success, and annual attendance today is approximately 30 million. It is the most visited privately owned tourist facility in the world. More than any other economic activity, including the Kennedy Space Center on Cape Canaveral, it has generated economic growth in central Florida.

Conceptualization of the Park

Walter E. (Walt) Disney was most responsible for a new concept in amusement parks which after World War II revitalized a dying industry (Mosley 1986; Thomas 1976). He proposed to build a park based upon themes. As in the older ones, there would be thrill rides, but it would lack many of the carnival-like amusements that earned the older parks such a bad reputation among middle-class American families. The ambience of his park would be one of scrupulous cleanliness, and be totally non-threatening. Employees, most of whom would be young, were expected to adhere to a rigid code of appearance and were to be cheerful at all times.

Disney was confident that if his park projected the ambience of wholesomeness, it would become highly profitable. He based this optimism on the post World War II national prosperity, greater use of automobiles, and the baby boom. He believed that many of these new families willingly would drive long distances to a park with themes to entertain young children. Disney

chose themes based on fairy tales, adventure stories, folklore, nostalgia, and future technologies. In sum, the goal was to create a friendly and wholesome environment in which visitors could have a pleasant escape from reality and exercise their imagination.

Disneyland opened near Los Angeles in 1955 (Bright 1987). It was divided into six sections, each built around a theme: Main Street U.S.A., Adventureland, Fantasyland, Tomorrowland, New Orleans Square, and Frontierland. Characters and themes from Disney's animated movies were employed throughout, particularly Mickey Mouse, who became the central figure of Disneyland, and later the Magic Kingdom in Florida.

Location of Walt Disney World

Disney conceived Walt Disney World shortly after he opened Disneyland. It was to be much larger in scale and would necessitate a huge tract of land. The purchase of a large tract was intended to provide ample area for expansion and to create a buffer between it and the numerous non-Disney attractions that would be drawn to compete for visitors (Thomas 1976). Important location factors were accessibility to a large number of people, proximity to a city large enough to provide labor, and a climate that would permit operation throughout the year.

Disney chose Florida. Although far from the nation's large cities, it was already popular with tourists and had a state government receptive to investments in tourism. The Orlando area was warm enough for year-round operations and was near the crossing of several limited-access highways, part of the nation's interstate highway system, then under intensive development.

Disney assembled this huge tract of land very cheaply by dealing with agents who had no idea they were working for him. He realized that had the owners known whom they were selling their land to, which was mostly poor palmetto pasture, they would have inflated the value greatly.

Disney desired as little interference from government as possible in the development of the property and was pleased that the Florida legislature awarded him special governmental status over the land purchase (Walsh 1986). The 11,100-hectare tract was named the Reedy Creek Improvement District. In effect it became the 68th county in the state, with authority over electric power, water, zoning codes, and fire protection. Police and judicial authority continued to rest with the two counties in which the land was situated, and the park pays county property taxes. Although Disney World today employs approximately 35,000 workers, there are only about 40 voting residents of the district, most Disney employees.

The special status of Disney World has led to tensions between it and the counties in which it is situated (Walsh 1986). Whereas other developers must pay impact fees to the county for roads and other public construction at the time they develop their property, Disney does not. Many believe Walt Disney Company is not contributing its fair share for the development of the area's infrastructure. Disney executives maintain that they have made substantial contributions, citing the company as the major force behind the growth of the region's economy.

There is no dispute that the Disney operation was the principal stimulant for growth in central Florida. In 1971, the year the Magic Kingdom opened, Florida tourism was much more uniformly distributed throughout the state than it is today. Most tourists vacationed along the beaches of the lower peninsula. Orlando, in the interior of the upper half of the peninsula, had comparatively little appeal to either tourists or retirees. In 1971, when the Magic Kingdom opened, approximately 450,000 inhabitants lived in the Orlando metropolitan area. Twenty years later, in 1990, the population had risen to slightly over 1 million. Among the nation's metropolitan statistical areas, including that of New York City, Orlando and Las Vegas, Nevada, continually exchange the lead in the number of hotel and motel rooms.

Creating a Sense of Place

Anthropologist Alexander Moore (1980) compares a visit to the Magic Kingdom to a religious pilgrimage. He believes that its success may be attributed to people being exposed since childhood to Disney images. Whether we can accept such an analogy, most who visit it become emotionally absorbed by it. They come to feel that, as Disney wished, "They are in another world" (Mosely 1986).

The Magic Kingdom is deep in the interior of Walt Disney World. It is reached by mass transportation either from Disney's own hotels within the park, or by car and bus from hotels outside the park. An especially large concentration of hotels outside the park has grown up along International Drive (Figure 1). Depending on the season of the year, between 25,000 and 75,000 guests are admitted to the Magic Kingdom each day. Many thousands more visit Disney World's two newer theme parks, EPCOT and Disney-MGM Studio Theme Park.

Guests to the Magic Kingdom (in Disney parks all visitors are referred to as "guests") go to its Transportation and Ticket Center. Here they purchase tickets to the Magic Kingdom, or a "passport" to visit all three parks (Birnbaum 1989). Although it is not nec-

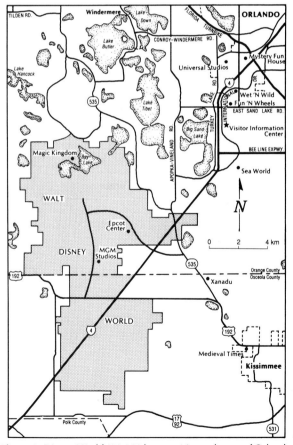

Figure 1. Disney World (11,000 hectares) is southwest of Orlando on U.S. Route 4.

essary, those who wish to heighten the illusion they are going into a different world can exchange U.S. dollars for one- and five-dollar Disney bills. From the ticket office, guests have a choice of reaching the Magic Kingdom by paddle-wheeled ferry boat across artificial Bay Lake or by monorail.

Even before arriving at the entrance to the Magic Kingdom all but the most calm begin to be swept into the spirit that Disney strove to generate. The sheer size of a typical day's crowd, which includes many children, heightens the excitement. People who visit the park marvel at crowd control. Although there are often long lines at popular attractions, the attendants move people through with great efficiency.

Upon entrance to the Magic Kingdom the first theme area or "land" entered is Main Street, U.S.A., the Disney concept of a late-19th-century small-town main street. Main Street, U.S.A., provides an excellent opportunity to appreciate the attention to detail that is the hallmark of all Disney attractions. It is used here as an exemplar of the other theme areas within the park.

The buildings are careful reconstructions of structures of the period (Goldberger 1972). Detractors argue that no main street at the end of the 19th century was as clean as Disney's. Draft horses pull streetcars down Main Street, but droppings or litter of any sort never remain on the ground for long, thanks to a crew of constantly vigilant sanitation workers. From a railroad station near Main Street one can take a train around the perimeter of the Magic Kingdom. The old steam locomotive that pulls the train is one of four salvaged in 1969 from an abandoned Mexican railroad system.

Illusion is vital in the creation of the Magic Kingdom's landscape. The kingdom seems much larger than it actually is. Buildings often are not built to true scale, or lower floors are, but the upper ones are not (Bright 1987). Many features are scaled down for children. Throughout the park are numerous examples of tricks of scale, to make buildings appear much taller or more distant through design. This technique, called "forced perspective," had earlier been refined by movie-set designers in Hollywood.

Since it is the visitor's first impression of the Magic Kingdom, to heighten the level of excitement, Main Street also is the venue for much of the Kingdom's live street entertainment. This entertainment includes a 19th-century barber shop quartet, strolling banjo players twice each evening during the busiest periods of the year, and a parade featuring a marching band and approximately 30 floats. From Main Street most move toward Cinderella's Castle (Figure 2), in the

Figure 2. Cinderella's Castle, in the center of the Magic Kingdom, is a point of orientation for visitors to the theme areas that surround it. The design of the castle was inspired by 12th- and 13th-century French castles as well as King Ludwig of Bavaria's Neuschwanstein. Source: Florida State Archives.

center of Magic Kingdom. Here roads lead directly to the other theme sections of the park.

Design and Construction of EPCOT

Walt Disney World's second theme park, EPCOT, opened in 1982. In design it bore little resemblance to what Disney himself had proposed in 1966, when the Florida property was purchased. In Disney's words Experimental Prototype Community of Tomorrow "will take its cue from the new ideas and new technologies that are now emerging from the creative centers of American industry" (Mosley 1986). It was never to be completed, but would always be introducing, testing, and demonstrating new materials and systems. It would be "a showcase to the world for the ingenuity and imagination of American free enterprise."

Instead of a community, what emerged was a park with two theme sections, Future World and World Showcase. Disney planners recognized that the demographic profile of the nation was changing as the baby boom children became adults (Lyon 1987). They wanted to build a park with themes that would appeal to this group. Future World was built with corporate financial support, including giants such as General Motors, General Electric, and Exxon. Within it are exhibitions, in the style of a World's Fair, that explain technological evolution. World Showcase is a joint development between Disney and a number of nations.

The exhibition areas of the countries were designed to incorporate popular visual impressions of them. On the "typical" English high street at the United Kingdom exhibition are half-timbered buildings that lean and have hand-painted smoke stains that make them appear to be centuries old. Some roofs are of thatch, but fire regulations demand that the material be non-flammable plastic. Mansard roofs are conspicuous at both the Canadian and French areas, and the French area is distinguished by a large model of the Eiffel Tower. Italy is depicted by a corner of Venice's St. Marks Square, including the campanile and the Doge's Palace, set on EPCOT's lagoon with a seawall that has been made to appear stained with age. Most national exhibitions have souvenir shops as well as restaurants where guests can sample the national cuisine. The United States is represented in World Showcase by a colonial style brick building where a 30-minute film panorama of American history is narrated by robots representing Ben Franklin and Mark Twain.

Detractors criticize Future World for too heavily relying upon technology to solve the world's problems (*American Heritage* 1987; Harrington 1979; Morison 1983). Those who disparage World Showcase, particularly the United States show, often say that history is put in too positive a perspective (Morison 1983). Disney officials make no apologies for their presentations in either section.

Physical geographers should find the landscaping throughout Walt Disney World of great interest, especially that of the national exhibits at EPCOT. The climate of central Florida is humid subtropical, while that of most of the exhibiting nations is colder or drier. Furthermore, the park is open all year, and designers did not want the landscape to differ radically between one season and another. Imaginative substitutions often were made by the park's horticulturists to create natural landscapes exotic to that of central Florida (Birnbaum 1989). For example, at the Canadian exhibit hemlock trees are represented by deodar cedars, a tree native of the Himalayas that is better adapted to Florida's hot summers. Geometrically sculptured bushes in the British exhibit are not yews but podocarpus. The European plane tree does not grow well in Florida. Where this tree is needed the western sycamore, of the same genus, has been substituted. At the French exhibit sycamores are pruned to about 6.1 meters (20 feet), and have developed the same characteristic knots found on plane trees along French streets.

Conclusion

No person in American history has more effectively imprinted on the nation's public the visual conception of a utopian and fantasy landscape than Walt Disney. Although he was a political and economic conservative, and some have accused him of both racial and ethnic prejudice, from his youth he had a utopian vision of community acquired from his father, an outspoken socialist (Harrington 1979; Mosley 1986). While EPCOT departed from his original conception, Disney would most likely approve of its attempt to bring the world together in a peaceful, orderly, and educational setting. He would have been delighted with the evolution of the Magic Kingdom, and Walt Disney World's newest park, Disney-MGM Studio Theme Park. At the inauguration of Disneyland, the progenitor of the Magic Kingdom, Disney said, "I don't want the public to see the real world they live in while they're in the park. I want them to feel they are in another world" (Mosley 1986). The Walt Disney Company is now successfully involved in joint theme park ventures in Japan and France, proving that the Disney landscape can appeal to the inhabitants of nations whose cultures are distinct from ours.

References

American Heritage. 1987. Disney: Coast to coast. 38:22-4.

Birnbaum, Steve, ed. 1989. *Walt Disney World.* New York: Avon Books.

Bright, Randy. 1987. *Disneyland: The Inside Story.* New York: Harry N. Abrams.

Goldberger, Paul. 1972. Mickey Mouse teaches the architects. *New York Times Magazine* Oct. 22:40ff.

Harrington, Michael. 1979. To the Disney station. *Harpers* 258:35-39.

Lyon, Richard. 1987. Theme parks in the USA: Growth, markets and future prospects. *Travel and Tourist Analyst* 9:31-43.

Moore, Alexander. 1980. Walt Disney world: Bounded ritual space and the playful pilgrimage center. *Anthropological Quarterly* 53:207-219.

Morison, Elting E. 1983. What went wrong with Disney's world's fair. *American Heritage* 35:70-79.

Mosley, Leonard. 1986. *Disney's World.* New York: Stein and Day.

Thomas, Bob. 1976. *Walt Disney: An American Original.* New York: Simon and Schuster.

Walsh, Matt. 1986. It's not easy living with the mouse. *Florida Trend* 29:70-75.

CLOSE-UPS / ZOOMING IN

Questions about geographic patterns may require the probing search for relationships at different scales of analysis and the need for attention to underlying causation. Thus, the climatology of a suburban residential area reveals different processes at work at different scales. Health hazards from seasonal temperature variations in American cities can be linked with broader climatological patterns. Scale of analysis is also important in geomorphological studies, as illustrated in this section for Mount St. Helens, Death Valley, the Carolina bays, and coastal shell reefs. The roots of physical landforms lie in subsurface rock structure and the dynamics of geo-tectonic forces, as well in the chemical, biological, and mechanical transformation of surface materials. Answers to explain form and process or even, for instance, to appreciate the distinctive character of wines produced in California's Napa Valley, may require interrelated studies at different scales. Geographers use satellite imagery and aerial photography for broad assessments. Confirmatory on-site field investigations provide "ground truth." Laboratory analyses with electron microscopes and chemical tests probe less visible but important indicators of landscape composition and change. Comprehension of physical processes at different scales also helps in evaluating the implications of human intervention. This is illustrated for canals in the development of Louisiana's wetlands and for protective levees along the Mississippi River. Assessment of the broad regional economic and local environmental consequences of the Grand Coulee Dam also benefits from zooming-in at different scales of analysis.

—Patricia F. McDowell, University of Oregon

Mount St. Helens

Amalie Jo Orme
California State University, Northridge

The cataclysmic eruption of Mount St. Helens on 18 May 1980 was but the most recent in a series of volcanic events in the Cascade Range of the northwest United States and Canada over the past several million years. The event, which was preceded by increasing seismic and volcanic activity, was triggered by a moderate earthquake that dislodged part of the mountain and allowed rising magma to be blasted violently northward from the side of the main volcanic pile. Formerly 2,950 meters high, the mountain lost 400 meters in elevation and 2.73 cubic kilometers in volume during the eruption. A mix of volcanic ejecta and melting glaciers in turn generated a series of lahars, or volcanic debris flows, which surged westward down the North and South Forks of the Toutle River and neighboring rivers; and in the days and weeks that followed, a vast cloud of tephra spread eastward over North America. The eruption was a dramatic reassertion of nature's primitive power. Fortunately, it occurred in sparsely inhabited country so that, although vast areas of timberland were leveled by the blast, loss of human life was small. Since 1980, Mount St. Helens has again begun to grow while the nearby landscape is being recolonized slowly by young forest trees and other plants.

The Cascade Volcanic Arc

Mount St. Helens is one of 19 major volcanoes in the Cascade Range of the northwest United States and Canada. This range is one expression of the many tectonic structures that have resulted from the subduction of the Farallon oceanic plate beneath the North American continental plate during Cenozoic times (Figure 1). As large slabs of the Farallon plate were subducted beneath California during mid-Cenozoic

time, some plate fragments lagged farther west to form, north of the Mendocino Fracture Zone, the Gorda, Juan de Fuca, and Explorer plates. An understanding of the behavior of these plates is essential to explanations of the Cascade Range volcanic province. Paleomagnetic evidence suggests that, prior to 7 Ma (million years ago), the Juan de Fuca plate approached Vancouver Island at 60 millimeters per year and the Oregon coast farther south at 70 millimeters per year (Riddihough 1984). Plate velocities later decreased and now approximate 20 millimeters per year in the north and 10 millimeters per year farther south. The Cascade Range volcanics are thus strongly associated with the subduction of the Juan de Fuca plate which dips 65° east to depths of 300 kilometers beneath the southern Washington Cascades. With deep and rapid subduction, high temperatures caused the subducted plate to melt, and unstable magma rose along pathways through the overlying North American plate to reach the surface as a volcanic arc. During the past 16 Ma, over 4,000 volcanic vents have developed in the region.

The Cascade Range consists of five distinct segments (Guaffanti and Weaver 1988):

(1) The volcanic front that extends north from Mount Rainier into British Columbia comprises a line of volcanoes and vents that lies west of the 60-kilometer-depth contour of the Juan de Fuca plate. Volcanism during the past 5 Ma is expressed in isolated dacite and andesite cones. Though seismicity is currently low, late Quaternary activity has given rise to lava, pyroclastic and mud flows, ash and pumice falls, and steaming fumaroles and explosions. For example, Mount Baker has had several episodes of steam exhalation with traces of ash and sulfur while Mount Garibaldi has produced lava flows and pyroclastics.

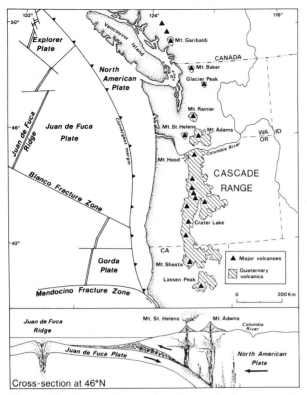

Figure 1. The Cascade Range volcanic arc and adjacent tectonic plates. Inset shows the subduction of the Juan de Fuca plate beneath the North American plate with subsequent volcano development.

(2)   Lying between Mount Rainier and Mount Hood, the Mount St. Helens area has produced large andesite and dacite composite cones in association with olivine-dominated basalt fields and vents. Quaternary volcanism consists of violent explosions, pyroclastic and mud flows, dome growth, steam, and ash and pumice falls. The Mount St. Helens area and its associated "hot spots" appear to lie east of the 60-kilometer-depth contour of the subducting Juan de Fuca plate and experience elevated seismic activity (Weaver and Smith 1983).

(3)   South of Mount Hood, the region has smaller andesite and basalt vents with extensive fumaroles, hot ground, tephra, and pyroclastic flows. The most recent activity at Mount Hood (A.D. 1760 to 1810) produced several episodes of extensive lahar inundation in neighboring drainage basins (Cameron and Pringle 1987).

(4)   The Mount Shasta and Medicine Lake region has a paucity of vents and occurs between the junction of the north Gorda and Juan de Fuca plates. These plates differ in subcrustal seismicity, age, and structure.

(5)   The Lassen Peak region lies well east of the 60-kilometer-depth contour and exhibits some historic (A.D. 1914 to 1921) pyroclastic flows, steam and ash episodes, and dacite dome emplacement.

A sixth segment of the Cascade volcanic arc may exist in the lava plateau of eastern Washington and Oregon, along the northern margins of the extensional Basin and Range province.

The Cascade volcanic arc thus varies both spatially and temporally—the region north of Mount Rainier being characterized by isolated older dacite and more recent andesite cones; the region south of Mount Hood being characterized by smaller andesite and basalt vents. Mount St. Helens lies within a geophysically anomalous region between the two—an area of large composite cones activated around 40 Ma with a more recent shift towards more andesitic dome growth and violent explosions.

Eruptive History of Mount St. Helens

Mount St. Helens has experienced at least nine eruptive periods over the past 40 ka (thousand years) (Lipman and Crandell 1981). Between 40 ka and 2.5 ka, volcanic events were dominated by dacite dome growth and pyroclastic and lava flows with hornblende, biotite, and hypersthene mineral assemblages. These earlier events include the Ape Canyon (40 ka to 35 ka) and Cougar (20 ka to 18 ka) periods, which produced large amounts of air-fall tephra, pyroclastic flows, and pumice lahars. The Swift Creek period (13 ka to 8 ka) yielded several tephra deposits found some 200 kilometers eastwards in central Washington and at least 20 kilometers west-southwest of the volcano. The Smith Creek period (4.0 ka to 3.3 ka) produced voluminous tephra which has been found 1,100 kilometers to the northeast, near Edmonton, Canada. Also, large pyroclastic flows and lahars moved northwestwards along the North Fork of the Toutle River nearly 50 kilometers west of the present Spirit Lake. A proto lake may have resulted when these flows plugged the main river drainage. The Pine Creek eruption (3.0 ka to 2.5 ka) expelled lithic pyroclastic flows and lahars which, with abundant fluvial deposits, created Silver Lake 50 kilometers west-northwest of the mountain.

The last 2.5 ka, however, have witnessed significant compositional changes as dacite materials alternate with andesitic and basalt flows. The mineral assemblages associated with this most recent period of activity are dominated by olivine, hypersthene, augite, and hornblende. These materials may be derived from multiple magma chambers or a heterogeneous single magma source. The Castle Creek period (2.2 ka to 1.7 ka) produced widespread complex assemblages of lava and pumiceous pyroclastic flows as well as two scoria deposits. During this time, the active building of Mount St. Helens as a composite cone commenced. The Sugar Bowl (1.15 ka) and Kalama (0.45 ka to 0.35 ka) periods witnessed thick extrusions of ash, lapilli, lahars, and scoriaceous tephra, with the basal Kalama tephra found 70 kilometers northwards at Mount Rainier (Smith *et al.* 1977). The Goat Rocks event (0.15 ka to 0.10 ka) produced explosions of tephra and dense smoke in 1857, with pumiceous ash extending 450 kilometers eastwards into Idaho.

The Eruption of 18 May 1980

Dormant since 1857, Mount St. Helens came to life 20 March 1980 when a shallow M4.1 earthquake rumbled northwest of the summit. Over the next eight days, hundreds of M3.5+ earthquakes occurred, a 1.5-kilometer-long crack developed along the north face, and a 75-meter-wide crater opened. Additionally, ash and steam were ejected 3,300 meters above the surface and sulfur dioxide-hydrogen sulfide gases, indicative of high temperatures and active magma, were expelled. The next seven weeks witnessed numerous harmonic tremors associated with migrating magma, a fivefold expansion of the crater, and the emergence of a prominent bulge along the north flank of the mountain.

On 18 May at 0832 Pacific Daylight Time an M5.1 earthquake signaled the catastrophic eruption that would ultimately remove the upper 400 meters of Mount St. Helens. Within 7 to 20 seconds of the tremor, a 2.8-square-kilometer portion of the north flank detached from the mountain, separating into three debris avalanches. One 195-meter-thick lobe of brecciated andesite, basalt, and dacite blocks, tephra, and colluvium rushed 7 kilometers northward at 50-70 meters per second, topped a 350-meter ridge, and ultimately entered the Coldwater Creek drainage. A second lobe of juvenile dacite, andesite, wood debris, and older Tertiary volcanics slammed into Spirit Lake, raising the water level 60 meters. A third 20- to 70-meter-thick lobe of glacial ice, basalt, andesite, alluvium, and tree trunks surged 22 kilometers down the North Fork Toutle River (Voight *et al.* 1981).

The detachment of large slide blocks and their transformation into massive debris avalanches released pressure on the subsurface dacite dome and led to substantial explosions as previously confined hydrothermal energy escaped. The result was a laterally directed blast as a hot (327°C) vapor-liquid-solid pyroclastic

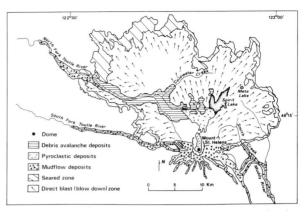

Figure 2. The distribution of 18 May 1980 debris avalanche, pyroclastic and volcanic mudflow (lahar) deposits, and the sear zone. Arrows indicate the direction of blast (surge) emanating from the north face of Mount St. Helens.

surge raced northward at an initial velocity of 100 meters per second, ultimately increasing to 325 meters per second with further gas expansion (Kieffer 1981). Extending in a 180° arc from the point of blast (Figure 2), the devastated area encompasses nearly 600 square kilometers with a distinct inner blow-down area (Figure 3) containing splintered, uprooted and downed trees blasted by supersonic air shock, phreatic water burst, and lithic fragments, and a more limited 4-kilometer-wide sear zone with charred and stripped trees. Associated with the surge were density flows containing silt and sand-size ash, accretionary lapilli, pumice, and organic debris (Fisher 1990). Other density flows contained coarse lithic materials and abundant wood fragments overlain by finer sands, suggestive of a turbulent high-velocity leading edge and a slower tail of finer deposits.

Within 4 to 10 minutes of the initial blast, unloading of the volcano triggered an explosive eruption of dacitic magma and a vertical hydroexplosive column of ash

Figure 3. View across Meta Lake in the blowdown area northeast of the summit. Photograph by A. J. Orme.

Figure 4. Mount St. Helens 18 May 1980, 1200 PDT. View northeast shows the vertical column of ash from the vent, pyroclastic flows, and melting snow and ice. Source: U.S. Geological Survey 80S3.

was ejected 25 kilometers into the atmosphere (Figure 4). A 20-centimeter-thick layer of pumiceous gravel tephra covered Smith Creek and neighboring watersheds. By 1545 Pacific Daylight Time, 7 hours after the initial explosion, the plume had reached Helena, Montana, some 750 kilometers to the east. Within several days ash reached as far as Oklahoma and Minnesota.

Within the first hour of the eruption, debris avalanche materials, which were 85 percent sand size or greater with saturation values of 21 to 71 percent, were transformed into lahars, which thundered through nearly all watersheds adjacent to Mount St. Helens. In particular, the Smith Creek, Muddy River, and Pine Creek drainages contributed over 13,500,000 cubic meters of rubble and water at rates of 17,000 cubic meters per second to the Swift Reservoir 40 kilometers to the south (Cummans 1981). The initial mudflow in the South Fork Toutle River emerged at 0850 Pacific Daylight Time, though within 90 minutes a 4- to 7-meter-high wall of hyperconcentrated flow with water, logs, volcanic debris, and human artifacts moving at velocities of 2 to 8 meters per second, was observed 43 kilometers downstream. The largest mudflow, however, occurred on the North Fork Toutle River. With the dewatering of 22,000,000 cubic meters of ash, pumice, rock, snow, and ice deposited in the upper 7 kilometers of the river, a massive lahar was generated at the distal portion of the debris avalanche some five hours after the eruption. Mobilizing through slumping, scouring, water bursting through depressions in the mass, and the melting of pulverized glacial ice, the lahar moved at rates of 1 to 3 meters per second with local velocities of 12 meters per second. The flow swept away bridge decks and supports, altered the channel pattern from meandering to braiding, and re-

shaped the valley through channel scour (Janda et al. 1981). Overbank deposits were 3 meters thick along several channel reaches, while bars of unconsolidated boulders, cobbles, and gravel, up to 4 meters high and 60 meters long, were fashioned under supercritical flow conditions. Approximately 0.5 kilometers upstream from the confluence of the North and South Forks of the Toutle River, the channel bed was raised over 5 meters for a distance of 300 to 400 meters. Ultimately the mudflow entered the lower Toutle and Cowlitz Rivers which accumulated nearly 27,000,000 cubic meters of debris (Lombard et al. 1981). The Cowlitz River waters rose 6 meters and fluid temperature approached 30°C. Some 1,300,000 cubic meters of debris were deposited 100 kilometers downstream in the Columbia River, requiring a 4-meter channel to be dredged to release 31 stranded ships (Swanson et al. 1989).

Significance and Implications

The eruption of 18 May 1980 reduced the height of Mount St. Helens from 2,950 meters to 2,549 meters and removed 2.73 cubic kilometers of rock from the summit. The new dome now forming lies more than 1,000 meters below the former summit dome. An assessment of damages following the eruption indicated that 43 bridges, 200 homes, 320 kilometers of road, and 25 kilometers of railway were destroyed (Schuster 1981). Episodic volcanism since the reawakening of Mount St. Helens has produced a lava dome nearly 300 meters high, together with several dozen lahars, avalanches, and ash expulsion (Pringle 1990). Hillslope erosion of tephra during the first year following the eruption injected 8,000,000 tons of sediment into the Toutle River (Collins et al. 1983), while the fluvial incision of debris avalanche materials produced 20,000,000 to 30,000,000 tons of debris annually through 1986. Despite subsequent erosion, perhaps 80 percent of the original volume of ejecta remains plastered across nearby hillsides, and is available for further mobilization.

The 18 May eruption of Mount St. Helens is just one reminder of the continuing hazards of volcanicity in the Cascade Range arc. As with the eruption of Mount Lassen, California, earlier this century and the presence of fumaroles on Mount Baker today, the volcanoes of the arc are capable of significant future activity. However, because the precise prediction of eruptions remains a difficult task, it is fortunate that these areas of potential volcanicity are distant from major urban centers. Mount St. Helens, however, has provided a valuable venue for advances in seismic monitoring and volcanology, and for an enhanced understanding of the mechanics of volcanic blasts, debris avalanches, and lahars.

References

Cameron, K. A., and Pringle, P. 1987. A detailed chronology of the most recent major eruptive period at Mount Hood, Oregon. *Geological Society of America Bulletin* 99:845-851.

Collins, B. D., Dunne, T., and Lehre, A. K. 1983. Erosion of tephra-covered hillslopes north of Mount St. Helens, May 1980-May 1981. In *Extreme Land Forming Events*, Zeitschrift für Geomorphologie Supplement 46, eds. S. Okuda, A. Netto, and O. Slaymaker, pp. 103-121. Berlin: Gebrüder Borntraeger.

Cummans, J. 1981. *Mudflows Resulting from the May 18, 1980 Eruption of Mount St. Helens, Washington*, U.S. Geological Survey Circular 850-B. Reston, VA: U.S. Geological Survey.

Fisher, R. V. 1990. Transport and deposition of a pyroclastic surge across an area of high relief: The 18 May 1980 eruption of Mount St. Helens, Washington. *Geological Society of America Bulletin* 102:1038-1054.

Guaffanti, M., and Weaver, C. S. 1988. Distribution of late Cenozoic volcanic vents in the Cascade Range: Volcanic arc segmentation and regional tectonic considerations. *Journal of Geophysical Research* 93:6513-6529.

Janda, R. J., Scott, K. M., Nolan, K. M., and Martinson, H. A. 1981. Lahar movement, effects, and deposits. In *The 1980 Eruption of Mount St. Helens, Washington*, U.S. Geological Survey Professional Paper 1250, eds. P. Lipman and D. R. Mullineaux, pp. 461-478. Reston, VA: U.S. Geological Survey.

Kieffer, S. W. 1981. Fluid dynamics of the May 18 blast at Mount St. Helens. In *The 1980 Eruption of Mount St. Helens, Washington*, U.S. Geological Survey Professional Paper 1250, eds. P. Lipman and D. R. Mullineaux, pp. 379-400. Reston, VA: U.S. Geological Survey.

Lipman, P., and Crandell, D. R. 1981. The eruptive history of Mount St. Helens. In *The 1980 Eruption of Mount St. Helens, Washington*, U.S. Geological Survey Professional Paper 1250, eds. P. Lipman and D. R. Mullineaux, pp. 3-15. Reston, VA: U.S. Geological Survey.

Lombard, R. E., Miles, M. B., Nelson, L. M., Kresch, D. L., and Carpenter, P. J. 1981. *Channel Conditions in the Lower Toutle and Cowlitz Rivers Resulting from the Mudflows of May 18, 1980*, U.S. Geological Survey Circular 850-C. Reston, VA: U.S. Geological Survey.

Pringle, P. 1990. Mount St. Helens—a ten-year summary. *Washington Geologic Newsletter* 18:3-10.

Riddihough, R. 1984. Recent movements of the Juan de Fuca plate system. *Journal of Geophysical Research* 89:6980-6994.

Schuster, R. L. 1981. Effects of the eruptions on civil works and operations in the Pacific Northwest. In *The 1980 Eruptions of Mount St. Helens, Washington*, U.S. Geological Survey Professional Paper 1250, eds. P. W. Lipman and D. R. Mullineaux, pp. 701-718. Reston, VA: U.S. Geological Survey.

Smith, H. W., Okazaki, R., and Knowles, C. R. 1977. Electron microprobe analysis of glass shards from tephra assigned to set W, Mount St. Helens, Washington. *Quaternary Research* 7:207-217.

Swanson, D. A., Cameron, K. A., Evarts, R. C., Pringle, P. T., and Vance, J. A. 1989. *Cenozoic Volcanism in the Cascade Range and Columbia Plateau, Southern Washington and Northernmost Oregon (Field Trip Guidebook T 106)*. Washington, DC: American Geophysical Union.

Voight, B., Glicken, H., Janda, R. J., and Douglass, P. M. 1981. Catastrophic rockslide avalanche of May 18. In *The 1980 Eruption of Mount St. Helens, Washington*, U.S. Geological Survey Professional Paper 1250, eds. P. Lipman and D. R. Mullineaux, pp. 347-377. Reston, VA: U.S. Geological Survey.

Weaver, C. S., and Smith, S. W. 1983. Regional tectonic and earthquake hazard implications of a crustal fault zone in southwestern Washington. *Journal of Geophysical Research* 88:10371-10383.

Spatial Organization in the Landforms of Death Valley

Ronald I. Dorn
Sandra L. Clark
Arizona State University

Death Valley, California, presents a classic site for the study of desert geomorphology in the Basin and Range province of North America. The harsh landscape reveals spectacular examples of faulted wineglass valleys, alluvial fans, shorelines of ancient lakes, weathering and erosion of brightly colored sediments, and a rich history of western frontierism and hardluck miners. A 25-kilometer trip from Badwater (−85 meters) to Telescope Peak (3,368 meters) traverses a diverse line of environments: from the evaporite salt polygons of the Devil's Golf Course on the valley floor; through aromatic mesquite (*Prosopis juliflora*) thickets; up extensive aprons of coalesced alluvial fans dotted with hardy xerophytes (especially creosote bush, *Larrea tridentata*); up steep hillslopes shedding debris at rates on the order of meters per thousand years; finally into conifers (such as *Pinus flexilis, P. longaeva*) which tenaciously stabilize fossil periglacial landforms atop the Panamint Range.

In an attempt to bring meaning and organization to diverse landforms in this stark setting, we treat the geomorphology of Death Valley as a hierarchy of spatial scales (*cf.* Chorley 1972; Fenneman 1916), from the regional overview provided by satellite. imagery at 1,000,000 meters to the intimate scrutiny of electron microscopy at 0.000 001 meter. We employ distance rather than area in our discussions, in accordance with the linear nature of fault lines, shorelines, and debris flows in Death Valley; and we use radiocarbon and potassium/argon radiometric clocks to measure time.

Process and Form

Different processes operate at different scales to organize landforms. Viewed as a morphotectonic entity at 1,000,000 meters, Death Valley is an incredibly active region of block and strike-slip faulting (Figure 1). This is in great contrast with neighboring provinces: the relatively stable Colorado Plateau to the east; the less active Mohave Desert to the south; and the Sierra Nevada batholith to the west.

Though still dominated by regional tectonism, satellite views at 10,000 to 100,000 meters begin to reveal a number of the elements that give Death Valley its own unique local character (Figure 2). Most striking are features such as major fault zones within a horst and graben setting, Death Valley's salt playa at the terminus of the Owens and Amargosa drainages, and extensive flanking bajadas. Eastward tilting of the Panamint Range and active normal faulting along the Black Mountains creates the great west-to-east asymmetry of these elements: long and complex west-side fans contrast with short and simple east-side fans (*cf.* Hooke 1972) (Figure 2).

At 1,000 meters, regional tectonic interrelationships no longer occupy center stage. In contrast to fault zones on satellite imagery, single faults become evident on aerial photos (Figure 3A). Individual alluvial fans (Figure 3A), channel patterns, and playa types are resolved. Human effects are apparent as straight roads cut across bajadas and playas.

Scales of 10 meters to 100 meters reveal details of this weathering-limited landscape. Individual sand dunes, springs, debris flows, and hillslope features predominate (for example, Figure 4). Paleolake features, such as wave-cut benches on Shoreline Butte, are distinguishable.

Landforms lose their character at meter and finer scales, but with the aid of magnifying tools microscopic units that compose individual landforms can be explored. Microfractures in rock minerals reveal tectonic

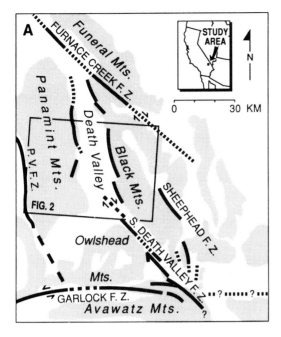

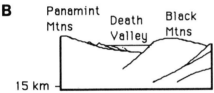

Figure 1. Morphotectonic setting of Death Valley: A. in plan; B. in cross-section. Modified from Serpa et al. (1988).

What distinguishes Death Valley from humid landscapes is the ubiquitous contrast of very stable land surfaces adjacent to constantly changing forms. As the observer zooms in, changes over time become more and more "catastrophic." If it were possible to revisit Death Valley 1 million years ago, its morphotectonic appearance at 1,000,000 meters would have changed little. At the scale of 100 meters, a change in process can create a drastic change in form. In Death Valley, if gullying starts, more exposure of bedrock generates more runoff; gullies continue to grow and permanently alter hillslope form. In contrast, in humid landscapes such as nearby coastal California, geomorphic changes tend to occur slowly; perturbations such as gullying are repaired by creep and biogenic transport of colluvium, and hillslopes again converge toward a convex form. At the weathering scale of 1 meter to 0.000 001 meter, some surfaces revisited during the Quaternary have not changed at all except for more layers of rock varnish (Figure 5). In contrast, other rock surfaces are completely remade by salt weathering (Goudie and Day 1980). We speculate that catastrophe theory (Graf 1988) is most viable as a tool to interpret geomorphic change in drylands at more detailed scales.

Integrating Process and Form with New Geomorphic Tools

A major difficulty in integrating the study of process and form in geomorphology is the inability to translate contemporary process measurements to landforms made in the past. This is well illustrated in Death Valley. In the late Wisconsin glacial period about 20,000 BP (radiocarbon years before present), a *Yucca* semidesert occupied the alluvial fans (Wells and Woodcock 1985) and, some 85 meters deep, Lake Manly filled Death Valley to sea level (Dorn 1988; Hooke 1972). About 10,000 to 13,000 BP, the lake dried up and sparse desert scrub species replaced more mesic vegetation. Geomorphic evidence of this climatic change is ubiquitous at several spatial scales, from alluvial fans (Figure 3B) to shorelines (Figure 3A) to talus flatirons (Figure 4B) and weathering phenomena (Figure 5).

Most of the age-determination techniques used to reconstruct geomorphic changes over time are based on dating materials found in a stratigraphic sequence, for example radiocarbon, thermoluminescence, and potassium/argon dating. These methods provide insights into changes in a vertical column at one site. However, if both spatial and temporal reconstruction of landscape change is desired, a key problem emerges: the inability

stresses. Soil horizons are replaced by the structure of individual peds. Salt crystals—potent rock-weathering agents—proliferate in this arid environment (Goudie and Day 1980). Currently, frontiers in weathering are explored at the 0.000 001-meter scale with electron microscopes (Figure 5).

A long-recognized aspect of self-organization holds true for Death Valley: as the scale of observation changes, so does the effectiveness of different geomorphic processes to control the appearance of a landscape (Chorley 1972; Fenneman 1916). For example, at a scale appropriate for evaluating morphotectonic history (1,000,000 meters), weathering plays a minimal role in the appearance of a landscape as tectonically active as Death Valley. In contrast, at the scale of an individual landform (10 meters to 100 meters), tectonism controls potential energy but has less impact than, for example, the root strength of a bristlecone pine (*Pinus longaeva*) holding regolith in place near Telescope Peak. Key insights linking process and form are lost without consideration for the interconnectedness of processes operating at different scales.

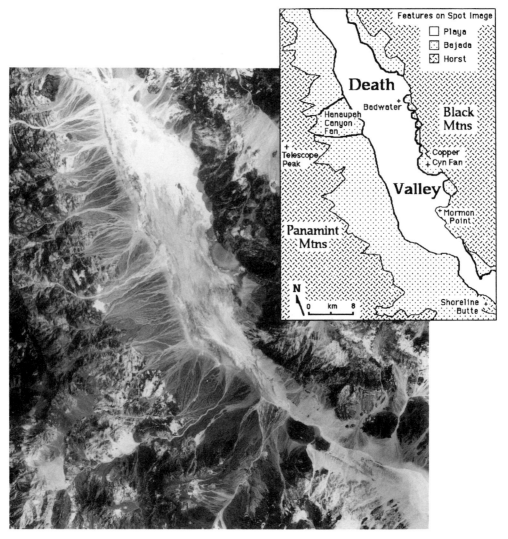

Figure 2. SPOT satellite image and corresponding map of southern Death Valley. The frame of the SPOT image is shown in Figure 1A.

to correlate in time spatially disjunct stratigraphic sequences.

New surface-exposure dating techniques such as
cosmogenic isotopes and rock varnish have the potential
to provide quantitative data to reconstruct landscape
history from place to place (Dorn et al. 1991; Phillips
et al. 1990). These methods permit the correlation of
landform development with available paleoclimatic
records (Hooke 1972; Wells and Woodcock 1985).
For example, the pronounced shorelines at Mormon
Point (Figure 3A) correlate with an early 180,000- to
130,000-year-old cycle of Lake Manly, and with a
major glacial pulse in the nearby Sierra Nevada (Dorn
1988; Phillips et al. 1990). Similarly, the 13,000- to
14,000-year-old talus relics at Artist's Drive (Figure
4B) correlate with a period when climate changed
from semiarid to hyperarid (Hooke 1972; Wells and
Woodcock 1985). The separation of talus from its

sources at this time of climatic transition, creating the
flatirons, is consistent with R. Gerson's (1982) model.

Climatic geomorphology has established correlations
between climate-environment and landform processes.
However, it is an untenable jump, at this stage, to
relate quantitative process studies with landforms developed in past climates. Still, correlating paleoland-
forms with paleoclimates is a vital ingredient in being
able to link geomorphic processes with the vast number
of fossil landforms beyond the reach of process geomorphology.

Summary

Three classic themes of geomorphology are stressed
in this examination of Death Valley, an archetype of

Figure 3. Landform associations in oblique aerial photographs: A. Mormon Point displays shorelines of Pleistocene Lake Manly (large arrow), normal faulting (smallest arrows), and a tectonic turtleback (medium arrows). B. Hanaupah Canyon alluvial fan. Note the distributary patterns on the active Holocene section of the fan (lower left), and the differential development of rock varnish and desert pavement on the older sections of the fan.

Figure 4. Individual landforms seen from the ground: A. A mid-Holocene alluvial-fan deposit from Hanaupah Canyon fan. The "bar and channel" topography is gradually evolving into a smooth desert pavement. B. Talus flatirons near Artist's Drive, radiocarbon dated by rock varnish to have formed about 13,000 to 14,000 years ago.

dryland geomorphology. First, processes that operate at different spatial and temporal scales combine to organize the landforms of Death Valley in a distinctive pattern. Second, gradual rates of geomorphic change that characterize landforms of humid climates contrast with the extreme dichotomy of rates of landscape change in Death Valley; landforms have remained virtually unchanged for hundreds of thousands of years are adjacent to landscapes remade yesterday. Third, the difficulty of integrating process and form in geomorphology is nowhere better illustrated than in Death Valley, where the contemporary arid climate could not have produced humid landforms left fossilized today. In Death Valley we illustrate the potential for understanding the link between process and form by correlating, in time, fossil landforms with the paleoclimates that produced them.

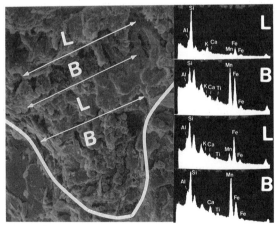

Figure 5. Two cycles of lamellate (L) and botryoidal (B) layers in rock varnish seen by a scanning electron microscope, collected from an alluvial unit of Hanaupah Canyon fan, about 170,000 years old (Figure 3B). Scale bar about 8 micrometers; line shows the boundary between varnish and rock. The corresponding x-ray analyses show the chemistry of the different layers. Botryoidal layers form in less alkaline, semiarid periods, while manganese-poor lamellate layers form during more alkaline, arid period (Dorn 1988).

Acknowledgments

Research on the geomorphology of Death Valley was made possible by grants from the National Geographic Society and National Science Foundation. We thank L. Cremis, D. Friend, S. Lichty, T. Liu, M. Pickup, and T. Wasklewicz for comments, D. Dorn for field assistance, and B. Trapido for graphical assistance.

References

Chorley, R. J., ed. 1972. *Spatial Analysis in Geomorphology.* New York: Harper and Row.

Dorn, R. I. 1988. A rock varnish interpretation of alluvial-fan development in Death Valley, California. *National Geographic Research* 4:56-73.

————, Phillips, F. M., Zreda, M. G., Wolfe, E. W., Jull, A. J. T., Donahue, D. J., Kubik, P. W., and Sharma, P. In press. Glacial chronology of Mauna Kea, Hawaii, as constrained by surface-exposure dating. *National Geographic Research and Exploration.*

Fenneman, N. M. 1916. Physiographic divisions of the United States. *Annals of the Association of American Geographers* 6:19-98.

Gerson, R. 1982. Talus relics in deserts: A key to major climatic fluctuations. *Israel Journal of Earth Sciences* 1:123-132.

Goudie, A. S., and Day, M. J. 1980. Disintegration of fan sediments in Death Valley, California. *Physical Geography* 1:126-137.

Graf, W. L. 1988. *Fluvial Processes in Dryland Rivers.* Berlin: Springer-Verlag.

Hooke, R. LeB. 1972. Geomorphic evidence for late Wisconsin and Holocene tectonic deformation, Death Valley, California. *Geological Society of America Bulletin* 83:2073-2098.

Phillips, F. M., Zreda, M. G., Smith, S. S., Elmore, D., Kubik, P. W., and Sharma, P. 1990. Cosmogenic chlorine-36 chronology for glacial deposits at Bloody Canyon, eastern Sierra Nevada. *Science* 248:1529-1532.

Serpa, L., de Voogd, B., Wright, L., Willemin, J., Oliver, J., Hauser, E., and Troxel, B. 1988. Structure of the central Death Valley pull-apart basin and vicinity from COCORP profiles in the southern Great Basin. *Geological Society of America Bulletin* 100:1437-1450.

Wells, P. V., and Woodcock, D. 1985. Full-glacial vegetation of Death Valley, California. *Madrono* 32:11-23.

Carolina Bays: Lines and Points on the Atlantic Coastal Plain

Thomas E. Ross
Pembroke State University

A landscape resembling a cratered moonscape surrounds travelers on Atlantic Coastal Plain highways. Although barely noticeable at ground level, from the air the spectacular view convinces many that they are witnessing the result of thousands of meteorite impacts (Figure 1). What they are seeing are land-surface features known locally and in the scientific community as "Carolina Bays."

The term "Carolina Bays" refers to shallow and elliptical depressions with parallel axes found primarily on the Atlantic Coastal Plain. They are concentrated in North and South Carolina, although some of the estimated 500,000 bays are found as far north as southeastern New Jersey and as far south as northeastern Florida (Johnson 1942; Prouty 1952) (Figure 2). They are called bays because of the abundance of several species of bay trees (*Magnolia virginiana, Persea borbonia, and Gordonia lasianthus*) in and around the bays, not because they are water bodies. Their parallelism and ellipticism distinguish the Atlantic Coastal Plain bays from other types of depressions and oriented lakes (Figure 3). Other physical characteristics of bays are their sand rims and large differences in size.

In general, the bays have either an elliptical or oval outline. Bays in North and South Carolina are usually elliptical while those further south are more ovoid. The oval bays' axes are oriented between south and east, with most lying between S10°E and S55°E. In contrast, the elliptical bays' axes tend to have a north-west-southeast orientation. There are, however, considerable departures from the prevailing directions of the axes of both the oval and elliptical bays.

All the bays are shallow and are typically lower than the surrounding plain. The deepest part is usually less than 2 meters below the adjacent Coastal Plain surface. Many of them are bordered by sand rims that rarely surround the whole bay. The rims range in height from less than 1 meter to about 3.5 meters, and in width from 5 meters to more than 100 meters. When a sand rim does exist, it is most evident at the south-eastern end of the bay. The composition of the rims is, in most cases, clean, fairly uniform white or buff quartz sand, which is not the same as the material composing the Coastal Plain. Some bays have multiple rims. There is no correlation between the size of the bay and the size of the rims.

The bays differ most in size range and in the absence, presence and/or extent of sand rims. Thousands of bays are less than 25 meters along the longest axis, while a few are more than 8 kilometers. They differ also in their surface cover. Some are shallow lakes or peat-filled swamps; others may be overgrown with wooded marsh vegetation and underlain by peat deposits. Many bays have long been edaphically dry (Johnson 1942). Forests are found in many dry bays, while many others have, over the past two centuries, been used extensively for cultivation of crops (Ross 1987) (Figure 4). The dry bays can be readily identified by the presence of dark carbonaceous soil, which is sometimes surrounded by a lighter colored sand rim.

The first mention of the bays in the literature appears in the 19th century by Michael Toumey (1848), the State Geologist of South Carolina. Since that time more than 300 books, articles, and scholarly papers have been written on the subject, most of which are related to the theories of origin (Ross 1987). At least 20 theories have been proposed to explain how the bays were formed. They include meteor showers (Mel-

Figure 1. Aerial view of Carolina Bays illustrating the unique land surface. Bays shown here range in size from less than 100 meters to 2.3 kilometers. All of the dark colored elliptical forms are Carolina Bays. The gray or lighter-colored ones have been cleared and cultivated for generations. Most of these are "relict" bays. About half of the large bay (2.3 kilometers) in the upper left (northwest) quadrant has been cleared for cultivation and house construction. This area is in northwest Robeson County, North Carolina, between Red Springs and Pembroke and is typical of the landsurface in the area of "primary occurrence" shown in Figure 2.

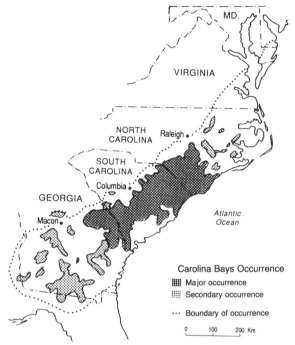

Figure 2. Primary and secondary occurrences of Carolina Bays with a long axis of at least 250 meters. Bays are abundant in the area of primary occurrence, covering as much as 60 percent of the total land area. They are less abundant in areas of secondary occurrence. An occasional bay is likely to be found in many parts of the remainder of the delimited region. Modified from Prouty (1952).

ton and Schriever 1933), fish nests made by giant schools of fish waving their fins in unison over submarine artesian springs (Grant 1945), sinks over limestone solution areas streamlined by groundwater (Legrand 1953), solution basins of artesian springs (Johnson 1942), shockwaves of cometary fragments (Eyton and Parkhurst 1975), and wind action on water bodies (Kaczorowski 1977; Thom 1970). Although no single theory has been universally accepted by geoscientists, it is very likely that bays were formed by a combination of terrestrial factors, including artesian springs, groundwater flows, ponding of surface water, and wind action.

The works of Bruce G. Thom (1970) and R. T. Kaczorowski (1977) are excellent examples of studies dealing with terrestrial causes. Both work from the assumption that the bays evolved from ancient lakes. According to Thom, the bays evolved as shallow ponds or lakes on the sandy Coastal Plain surface between 6,000 and 40,000 years ago (Thom 1970; Whitehead 1965). These water bodies included "deflation hollows within dune fields, or depressions on poorly drained interfluves, or lake basins at the contact between two river terraces" (Thom 1970, 806). Over time, southwest winds caused the bays to expand laterally and created

the northwest-southeast axis. The sand rims developed as beaches on the leeward side. Rims were formed along the shallow ends of the lakes. Kaczorowski also tied wind and water bodies together, but did not limit the geomorphic surface for evolution as strictly as Thom. He concluded that bays developed on any sur-

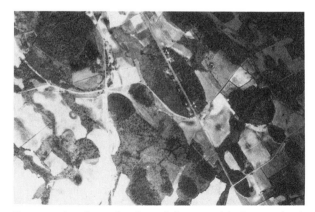

Figure 3. The elliptical and parallel nature of the bays is clearly evident here. Note the sand rims curving around the southeastern end of the bays. The large bay in the upper center, dissected by a highway, is approximately 1 kilometer long. This is near Red Springs, North Carolina.

Figure 4. Many bays, such as these in rural Robeson County, North Carolina, have been cultivated for several generations. The dark area is carbonaceous soil of the bay interior. The landowner put his residential, farm, and farm market buildings on the sand rim.

face, be it sand, clay, or silt, which had enough topographic irregularity for a pond to form (Kaczorowski 1977). The wind and water theory of origin as proposed by Thom and Kaczorowski is supported by findings reported by Carver and Brook (1989) in their study of Atlantic Coastal Plain paleowind directions. They write that "wind directions support the theory that Carolina Bays are oriented by winds" by R. E. Carver and G. A. Brook (1989, 214). The origin of the bays may never be known, but the wind and water connection advocated by Douglas Johnson, Bruce G. Thom, and R. T. Kaczorowski is most plausible and does command a substantial following in the scientific community.

Whatever the origin of the bays, their physical characteristics are important elements in shaping the present cultural landscape of the Atlantic Coastal Plain (Ross 1989). Since the earliest European settlements in the region, the bays have had a major influence on a variety of human activities and have imparted a subtle but distinctive pattern on the Coastal Plain's cultural landscape. There is even evidence that Indian life of 10,000 years ago was affected by Carolina Bays: Indian artifacts are much more common on bay sand rims than in the bay interiors (Knick 1988), which indicates that paleo-Indians camped on the rims around the bays.

Two characteristics of the bays important in shaping the region's cultural landscape are the parallel orientation and the sand rims surrounding poorly drained bay interiors. From the air, a curvilinear pattern of human features is clearly evident in association with larger bays (larger being defined as bays with a long axis of at least 500 meters). Many of the curves on the highways of the Coastal Plain are related to the

fact that the highways follow and are constructed upon the sand rims in order to avoid the poorly drained interior of the bays. Some of the highways in the region follow old Indian trails, which followed paths on the rims made by animals.

The siting of farmhouses and barns until recently was influenced by the presence of bays, or more precisely, the presence of a sand rim on the bay. Before 1970, almost all the buildings in the region where bays are abundant were built on the sand rim or adjoining Coastal Plain. Rare was a pre-1970 building sited in the interior of a bay. Population pressure and economic expansion, however, are leading to more construction within the bays, forever changing the pattern of housing sites in the Coastal Plain and diminishing the significance of the bays as a factor in the region's cultural landscape.

There is also a connection between bays and the patterns and shapes of other cultural features, such as cemeteries and orchards. Burial grounds are best located on well-drained land, and on the Coastal Plain of the Carolinas, an ideal site for a cemetery is a Carolina Bay sand rim. Thus many of the cemeteries are long and narrow, following the shape of the sand rim.

Bay lakes are used for recreational activities, and they retain the elliptical shape of the bay in which they are located (Figure 5). The sandy rims make ideal beaches for the water-filled bays and linear playgrounds, especially unpaved ones, are also found there.

The patterns of agricultural activities follow these elliptical bays. For example, the few orchards that are found on the Coastal Plain are positioned on the slightly elevated sand rims to take advantage of water and air drainage. Many bays have also been drained and planted with crops. Particular crops, such as blueberries, will

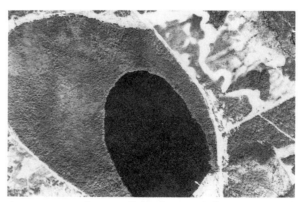

Figure 5. Jones Lake, near Elizabethtown, Bladen County, North Carolina, is a lake located within a Carolina Bay. Note that the lake is also elliptical, but its axis is not exactly oriented with the bay. The axis is more northerly than the axis of the bay itself. The bay is about 3 kilometers while the lake is about 1.5 kilometers along its axis.

be limited to the bays' interiors, where the soil is more acidic. Where corn or soybeans are planted, a distinct dichotomy occurs in the quality of the crop: the yield is less in the lower portions of the bay compared with that of the higher portions adjacent to the rim. In many of the bays, crops planted in the lowest part will sometimes be drowned by spring or early summer rains, or they will not grow as well as the plants on the higher ground because of excess moisture. After a few episodes of crop damage during a season, farmers ignore these areas and weeds are allowed to overtake the corn or beans. The result is a large field of corn or soybeans surrounding a plot of weeds.

Plant species are often delineated by their position with respect to the bay. In most cases, the species growing on the sand rim and surrounding Coastal Plain sediments vary greatly from those within the bay itself. According to a recently published report, the rims' vegetation tends "to be very easily differentiated from other bay community types" and a fairly distinct boundary exists between the rim and the bay proper (Bennett and Nelson 1991, 35). The northwest end of many of the water-filled bays is covered with marshy vegetation, whereas the water portion is found in the southeast end.

Thus, the boundary between the marsh and water portions of the bay is, in addition to the "point characteristic," another example of curvilinearity associated with the bays. And finally, forested bays are sharply defined on aerial photographs, and from higher altitudes appear as many small points on the landscape (Figure 6).

In summary, the Carolina Bays are unique in that they are both "points" and "lines" on the landscape of North America and that they are of interest to both physical and cultural geographers. Many cultural features in Carolina Bay country occur in a crescentic or curvilinear pattern because the bays' poorly drained interiors make them poor building sites: most building has been on the sand rims. Other activities, especially the cultivation of certain crops and trees that are greatly affected by the amount of moisture in the soil, occur as points in the landscape. In conclusion, the Carolina Bays are unique to North America, and are the most distinctive topographic feature on the Atlantic Coastal Plain of North and South Carolina and contribute much to the cultural landscape of the region.

References

Bennett, S. H., and Nelson, J. B. 1991. *Distribution and Status of Carolina Bays in South Carolina.* Columbia: South Carolina Wildlife and Marine Resources Department, Nongame and Heritage Trust Publications.

Carver, R. E., and Brook, G. A. 1989. Late Pleistocene paleowind directions, Atlantic Coastal Plain, U.S.A. *Palaeogeography, Palaeoclimatology, Palaeoecology* 74:205-216.

Eyton, J. R., and Parkhurst, J. I. 1975. *A Re-Evaluation of the Extraterrestrial Origin of the Carolina Bays.* Occasional Publication, No. 9. Urbana-Champaign: Department of Geography, University of Illinois.

Grant, Chapman. 1945. A biological explanation of the Carolina Bays. *Science Monthly* 61:443-450.

Johnson, Douglas. 1942. *The Origin of the Carolina Bays.* New York: Columbia University Press.

Kaczorowski, Raymond T. 1977. *The Carolina Bays: A Comparison with Modern Oriented Lakes.* Technical Report No. 13-CRD. Columbia: Coastal Research Division, Department of Geology, University of South Carolina.

Knick, Stanley. 1988. *Robeson Trails Archaeological Survey.* Pembroke, NC: Pembroke State University Native American Resource Center.

LeGrand, Harry. 1953. Streamlining of the Carolina Bays. *Journal of Geology* 61:263-274.

Melton, F. A., and Schriever, William. 1933. The Carolina Bays: Are they meteorite scars? *Journal of Geology* 41:52-66.

Prouty, W. F. 1952. Carolina Bays and their origins. *Bulletin of the Geological Society of America* 63:167-224.

Ross, Thomas E. 1987. A comprehensive bibliography of the Carolina Bays literature. *Journal of the Elisha Mitchell Scientific Society* 103:28-42.

———. 1989. Carolina Bays and the cultural landscape of the Atlantic Coastal Plain. *Abstracts of the Joint Meeting, Southeastern Division and East Lakes Division, Association of American Geographers*, p. 44. Charleston, WV.

Thom, Bruce G. 1970. Carolina Bays in Horry and Marion Counties, South Carolina. *Geological Society of America Bulletin* 81:783-814.

Toumey, Michael. 1848. *Report on the Geology of South Carolina.* Columbia: Geologic Survey of South Carolina.

Whitehead, D. R. 1965. Palynology and Pleistocene phytogeography of unglaciated eastern North America. In *The Quaternary of the United States*, eds. H. E. Wright and D. G. Frey, pp. 417-432. Princeton: Princeton University Press.

Figure 6. Forest in Carolina Bays. Observe how the trees emphasize the presence of bays. The large bay is about 3.6 kilometers long and has narrow multiple sand rims. Most of the smaller bays have much wider sand rims than the large bay. Observe the greater width of the rim of the bay in the lower left quadrant.

Linear Shell Reefs

Anthony J. Lewis
Louisiana State University

Although most shell reefs are irregularly shaped and have no preferred orientation to the coastline, under certain conditions some shell reefs exhibit pronounced linearity and a strongly preferred orientation. Such features usually are oriented at a high angle and are frequently perpendicular to the coastline. These linear shell reefs are geomorphically interesting and economically important. They are valued as an important source of young oysters and an effective dissipator of wave energy. At high tide they also represent a navigational hazard. Although linear shell reefs are not well known, nor often studied, the features were described and their importance recognized in the early 1900s.

Definitions

Shell reefs of this morphology and orientation found in the estuary of Newport River, North Carolina, were termed "oyster reefs" by Caswell Grave (1901) and were dominated by the common American oyster, *Crassostrea virginica*. Caswell Grave (1901) also speculated on their origin and discussed their economic significance. Subsequent studies have noted the same geomorphic feature but have used different terms. H. F. Moore (1907) referred to these features as "long reefs" in Matagorda Bay, Texas. The *Treatise on Invertebrate Paleontology* classifies them as "string reefs" (Stenzel 1971). A. J. Lewis (1971) described similar reef forms as "perpendicularly oriented shell reefs."

For consistency and simplicity the term "linear shell reefs" refers in this study to narrow linear accumulations of shells and shell fragments oriented perpendicular or at a high angle to the shoreline and situated on, but above, tidal mudflats. The reefs are exposed only during low tide, and although most of them are detached from the coast, several are attached to the coastline.

Linear shell reefs also exist as "buried oyster reefs" and have been identified at depths of over 4 meters below the bottom of several Texas bays (Norris 1953). Linear shell reefs are not confined to North America. A. J. Lewis (1971) described the same morphology in the Gulf of San Miguel, Republic of Panama. However, the reef composition represented different molluscs, mostly small clams, *Anomalocardia subragosa* and *Protothaca grata*, commonly found on mudflats along the Pacific Coast of Middle America (Keen 1958). Although the species composition is different, the form and, most likely, the process are the same.

Reef Composition and Environmental Conditions

Gulf Coast and East Coast linear shell reefs are composed primarily of living and dead shell and shell fragments of the common American oyster, *Crassostrea virginica*, although J. H. Coleman (1966) also reported high percentages of *Brachidontes recurvus* and *Crepidula plana* in Louisiana. The common American oyster is versatile, existing within a salinity range of 10 to 30 parts per thousand (Butler 1954). The *Treatise on Invertebrate Paleontology* classifies *Crassostrea virginica* as the most euryhaline oyster genus with an optimum growth salinity of 17.5 parts per thousand (Stenzel 1971). Exposure to cold winter air during low tide or temperatures below the freezing point of sea water ($-1.7°C.$) are fatal. Other parameters affecting their growth or existence are nutrients (Diatoma and Infusoria); water movement, which replenishes the food supply; and siltation. A firm substratum, natural or artificial, is a necessity.

Form and Process Hypotheses

Free-swimming oyster spat attach themselves to whatever is available. The most favorable location on a shell reef is at the tip of the reef where oyster shell and shell fragments provide a solid base and the swift current moving around the end of the reef keeps the oysters sediment-free and provides nutrients to the attached sedentary oyster (Grave 1901). Young oysters attaching to shell material on the flanks of the reef are likely to perish from excessive sediment or insufficient nutrients. With maximum survival probabilities at the tip of the shell reef and minimum survival probabilities along the sides or flanks, a long, narrow growth form is favored.

Caswell Grave (1901) hypothesized that the unique growth pattern was a response to the nearshore currents paralleling the shoreline (Figure 1). The growing edge of the shell reef deflects the current resulting in a higher velocity around the edge and thereby an optimum location for growth. In agreement with Caswell Grave (1901), H. F. Moore (1907), W. A. Price (1954), and R. H. Parker (1960) demonstrated that the alignment of the linear oyster reefs in Texas and Louisiana is related to the circulation or current direction—that is, the growth is always at right angles to the direction of flow. This is in response to the supply of nutrients. After the initial stage of oyster attachment to a coastal headland (Figure 1A), the shell reef grows away from the coastline and into the longshore or nearshore current. Upon reaching the current (Figure 1B), continued growth deflects the current, and the reef begins to bifurcate, thereby keeping the reef growth point perpendicular to the displaced current (Figure 1C). With increased deflection, bifurcation continues (Figure 1D), and the supply of nutrients reaching the landward part of the shell reef is depleted. The landward part of the reef dies and erosional processes dominate. The width of the linear shell reef decreases until it is finally breached by current and wave action. The linear

shell reef, although separated from the land (Figure 1E), continues to grow at the outer margins of the reef where nutrients are available and oysters are swept clean of sediments (Grave 1901).

Another hypothesis to explain the phenomenon of linear shell reefs combines the necessity of a firm substratum, for the attachment and growth of oyster reefs, and the linearity and firm substratum afforded by distributary levees of ancestral delta complexes. W. A. Price (1954) reported that the presence of paired linear shell reefs in central San Antonio Bay, Texas, and the lower James River, Virginia, is not explained by Caswell Grave's circulation model in Figure 1. He associated the location of these linear shell reefs with old channels (Price 1954). J. H. Coleman (1966) correlated the occurrence of linear shell reefs in coastal Louisiana with distributaries of ancestral Mississippi River deltas and concluded that abandoned distributary trends and the associated natural levees were the main controlling factors in determining the distribution of linear shell reefs in Louisiana.

Shell Reefs in the Gulf Coast

Linear shell reefs have been described in many of the bays, marsh lakes, and tidal channels along the Gulf Coast of Louisiana (Coleman 1966; Thompson 1956) and Texas (Moore 1907; Norris 1953). Linear shell reefs in Louisiana and Texas range in size from several tens of meters to tens of kilometers and are found as far as 16 kilometers offshore. The largest reef complex is nearly 42 kilometers long and is located east of Marsh Island, Louisiana (Figure 2). Individual linear shell reefs, such as those south of Marsh Island (Figure 3) attain a length of 4 kilometers and a width of 50 meters (Coleman 1966). Price (1954) reported that linear shell reefs in Louisiana and Texas are 6 meters deep. Tidal channels have also been noted on the up-current side of the shell reefs (Coleman 1966; Lewis 1971; Price 1954).

H. F. Moore (1907) and H. B. Stenzel (1971) report that linear shell reefs are characterized by straight or curved narrow crests or backbones that run the length of the reef. Many of the linear shell reefs are arranged *en echelon* (Stenzel 1971). Others are more loosely arranged and, according to W. A. Price (1954), form rounded or oval forms charted as "tow-heads." These characteristics are evident on the aerial photographs of Marsh Island, Louisiana, taken by the U.S. Air Force in February 1970 (Figure 4).

Photographic evidence of linear shell reefs off Marsh Island goes back to at least 1954 when black and white aerial photographs were collected by Jack Ammann

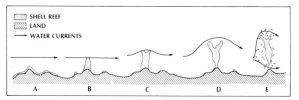

Figure 1. Linear shell reef growth stages. After Grave (1901).

Figure 2. Marsh Island, Louisiana.

Figure 4. Linear shell reefs located along the southwest coast of Marsh Island. Current deflection around the linear shell reefs is evident by lighter tone turbidity patterns. Dark tone, small length/width ratios, and greater distances from the coastline suggest a later development stage. Source: U.S. Air Force 1/20,000 color infrared photography, acquired February 1970.

Photogrammetric Engineers, Incorporated (Coleman 1966), although the reefs appeared on the 1921 edition of the U.S. Coast and Geodetic Survey Chart No. 1277. The February 1970 overflight by the U.S. Air Force provided the first large scale (1/20,000) color infrared photographs. National Aeronautics and Space Administration high altitude color infrared aerial photography collected after the passage of a cold front in December 1989, resulted in excellent regional coverage of Louisiana's linear shell reefs.

Analysis of the 1970 and 1989 aerial photographs reveals that although some evidence of reefal modification exists, the linear shell reefs have not moved during the last 20 years. Major displacement since the 1954 photography is improbable. Comparison with the 1921 coastal chart is difficult but suggests that the reefs are in the same positions. Tonal changes (more dark tones) and surface reef morphology changes (reef flank widening, lateral channelization parallel to and on the down current side of reef flanks, and decreasing crest continuity) are possible indicators of growth stages, as is size. Obviously the smaller reefs are probably younger. It is also probable that youth is indicated by

lighter tones, higher length/width ratios, the absence of lateral channels on the down-current side of the linear shell reef, and a continuous crest. When viewed over time, these interpretation keys could be used to determine growth trends of linear shell reefs.

Prospects of Future Research

Hopefully, future research on linear shell reefs will determine the effect linear shell reefs have on shoreline stabilization and suggest methods to promote the growth of linear oyster reefs. A better understanding of these linear shell reefs will have immediate and practical applications and could help reduce shoreline erosion in certain coastal areas and abate the depletion of oysters (Alford 1973). Accurate mapping of the global distribution of linear shell reefs would also be valuable for navigational and scientific purposes.

References

Alford, J. J. 1973. The role of management in Chesapeake oyster production. *Geographical Review* 63(1):44-45.
Butler, P. A. 1954. Summary of our knowledge of the oyster in the Gulf of Mexico. In *Gulf of Mexico: Its Origin, Waters, and Marine Life*, Fisheries Bulletin 89 (Fisheries Bulletin of the Fish and Wildlife Service, Vol. 55), pp. 479-489. Washington, DC: U.S. Government Printing Office.
Coleman, J. H. 1966. *Recent Coast Sedimentation: Central Louisiana Coast*, Coastal Studies Technical Report No. 29. Baton Rouge: Louisiana State University Press.
Grave, Caswell. 1901. The oyster reefs of North Carolina: A geological and economic study. *John Hopkins University*

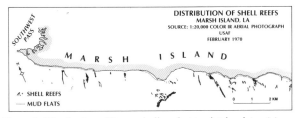

Figure 3. Distribution of linear shell reefs, Marsh Island, Louisiana. Source: U.S. Air Force 1/20,000 color infrared photography, acquired February 1970.

Circular, pp. 50-53. Baltimore: Johns Hopkins University Press.

Keen, A. Myra. 1958. *Sea Shells of Tropical West America.* Stanford: Stanford University Press.

Lewis, A. J. 1971. *Geomorphic Evaluation of Radar Imagery of Southeastern Panama and Northwestern Colombia*, CRES Technical Report 133-18. Lawrence: University of Kansas.

Moore, H. F. 1907. *Survey of Oyster Bottoms in Matagorda Bay, Texas*, Bureau of Fisheries Document No. 610, pp. 1-86. Washington, DC: U.S. Department of Commerce and Labor.

Norris, R. M. 1953. Buried oyster reefs in some Texas bays. *Journal of Paleontology* 27(4):571-574.

Parker, R. H. 1960. Ecology and distributional patterns of marine macro-Invertebrates, northern Gulf of Mexico. In *Recent Sediments Northwest Gulf of Mexico*, eds. F. P. Shepard, F. B. Pheleger, and T. H. van Andel, pp.302-333. La Jolla, CA: Scripps Institute of Oceanography.

Price, W. A. 1954. Oyster reefs of the Gulf of Mexico. In *Gulf of Mexico: Its Origin, Waters and Marine Life*, Fisheries Bulletin 89 (Fisheries Bulletin of the Fish and Wildlife Service, Vol. 55), p. 491. Washington, DC: U.S. Government Printing Office.

Stenzel, H. B. 1971. Oysters. In *Treatise on Invertebrate Paleontology*, Part N (Mollusca), ed. R. C. Moore, Vol. 3, pp. N953-N1224. Boulder, CO: Geological Society of America.

Thompson, W. C. 1956. Sandless coastal terrain of the Atchafalaya Bay area, Louisiana. In *Finding Ancient Shorelines*, Special Publication No. 3, eds. J. L. Hough and H. W. Menard, pp. 59-76. Tulsa, OK: Society of Economic Paleontologists and Mineralogists.

Canals and the Southern Louisiana Landscape

Donald W. Davis
Louisiana Geological Survey

Canals dominate the marsh-swamp terrain of southern Louisiana. Since construction of the first drainage projects in the early 1700s, dredged waterways have advanced relatively unchecked in the exploitation of wetland resources—fur, timber, agriculture, and hydrocarbons. Although aiding wetland transportation, dredging activity has exacerbated land loss and destroyed valuable estuarine habitats.

Louisiana's near-featureless marshes and adjacent water bodies span the entire coast. The low-lying marshes vary in width from 25 to 80 kilometers. Less than a 4-meter difference in elevation separates the marsh from its adjacent natural levees, cheniers, and beaches. The region accounts for 41 percent of the country's marsh ecosystems and is defined by elevation and absence of trees. Where the land is at least 45 centimeters above sea level, a cypress (*Taxodium distichum*)/tupelo gum (*Nyssa aquatica*) swamp may exist. The marsh, on the other hand, is a conspicuous lowland—literally a sea of grass. This alluvial wetland is a vast storehouse of renewable and nonrenewable resources.

Exploitation of these resources, along with numerous natural processes, has caused Louisiana's coastal margins and wetlands to vanish, and canal construction is an important contributor to the land-loss problem. Within Louisiana's coastal lowlands, wetland losses exceed 100 square kilometers per year (Gagliano *et al.* 1981; Penland *et al.* 1990; Turner and Cahoon 1987). This is a fluctuating rate that appears to be decreasing (Britsch and Kemp 1990). Nevertheless, if allowed to proceed unabated, in less than three decades, Louisiana's land loss would equal the area of Rhode Island. In half a century, the sea would repossess coastal property equal to all of the tracts reclaimed in the Netherlands during the last 800 years (Turner 1987).

While several factors influence Louisiana's land-loss rates (for example, change in the Mississippi's flow regime, local hydrology, sea level rise, subsidence, and fluid withdrawal), the conversion of land to water in developing wetland hydrocarbon reserves is conspicuous. For example, in one oil and gas field, there are 67 kilometers of petroleum-related canals, representing removal of at least 3.3 million cubic meters of soil. These waterways are part of an extractive site encompassing 178 square kilometers and are a visible indicator of land loss.

The environmental consequences of wetland exploration did not become part of the Louisiana vernacular until the early 1970s. Before that, many considered the marsh worthless, and land was sacrificed to capitalize on the subsurface mineral fluids. In the process, southern Louisiana was crisscrossed, ringed, cut, bisected, and otherwise dominated by an immense array of excavated access arteries. Until recently, new channels were added constantly, and old ones were rarely filled in. Once cut, a canal endures. Theoretically, its duration is finite, but some have enlarged into straight-channeled bayous. They have lost their human-induced identity and are now a part of the natural topography. They are, therefore, lines on the landscape, divisible into five categories: drainage and reclamation, transportation, trapping, logging, and petroleum (Davis 1973) (Figure 1). These canal types demonstrate the historical evolution of resource use in southern Louisiana.

The Canal Types

Drainage and Reclamation Canals

Louisiana's natural waterways are framed by broad, gently sloping natural levees that serve as south Louisiana's principal high ground. The region's early settlers established their homes, villages, and hamlets on these

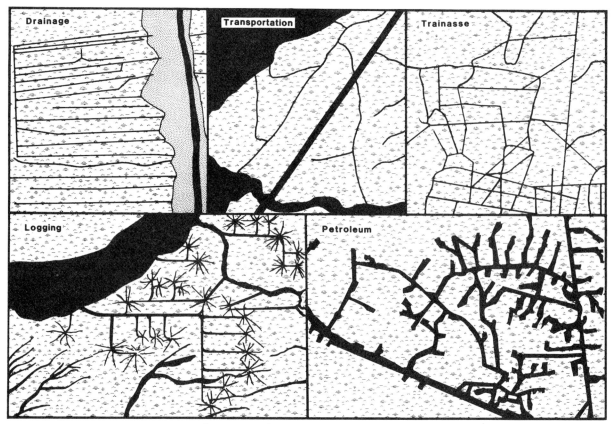

Figure 1. Canal patterns in coastal Louisiana.

protected and well-drained levees. These farmer-trap-per-fisher folks followed an annual-use cycle based on the "Bayou Country's" natural productivity. The province's economic history has, in fact, been dictated by access to high ground. Depositional landforms provided the locale for settlement. As early as 1720, two years after the founding of New Orleans, French colonists excavated drainage ditches to improve run-off from the natural levees. When New Orleans was surveyed, each block was ringed with canals that continue to reflect the "Crescent City's" dependence on a drainage network (Lewis 1976). These ditches were part of a great land reclamation enterprise, laid out to aid planters and urbanites in draining potential agricultural, pasture, residential, and industrial land. In 1770, Governor Alexander O'Reilly declared that every family settling in the province was to construct a levee and finish a highway with parallel ditches toward the levee within three years to guarantee Louisiana's lowland would be drained adequately (Martin 1827). Since then, particularly between 1880 and 1930 the rush to drain land has accelerated. In many cases, these original drainage canals evolved into vital transportation links, some of which remain as indispensable transportation arteries, even today.

Transportation Canals

Louisiana's natural hydrologic regime is predominately aligned north to south, making east-west movement difficult. To correct this navigational inconvenience, canals were excavated to increase the natural system's efficiency and to reduce travel time. Maps from the mid-1700s show transportation canals linking the wetlands to the Mississippi River. These pathways, often excavated by hand, served for moving goods to the market at New Orleans. French colonists also constructed canals to improve their contact with the Gulf of Mexico and to join the Mississippi with the bayous at the "rear" of their settlements.

By 1930 the coastal terrain was marked by many transportation channels. The system was small but adequate, and furnished local inhabitants with statewide connectivity. However, the United States' involvement in World War II, expanded the demand for petroleum, and Louisiana canals were integrated into a national coastal network. Congress approved financial support required to complete the Gulf Intracoastal Waterway. This watercourse in conjunction with the Mississippi River made Louisiana the center for wartime Gulf Coast petroleum traffic.

Evaluation of transportation canal expansion is difficult, because only a few were built exclusively for commercial traffic. Most were constructed to aid in developing agricultural, mineral, forest, and fur resources. They operated, secondarily, as transport couplings. Regardless, they furnished open-ended links between points and extended for kilometers along well-defined rights-of-way. Although the Gulf Intracoastal Waterway is the best known, many routes predate it, some as far back as the early 1700s, but all serve the same role—connectivity. Each transportation artery influences the economy, attracts industry, and alters property along the right-of-way. Through constant use, the sides of many canals have eroded such that they have become large waterways. In the process, their fabricated character has given way to a "natural" appearance.

Trapping Canals: A Trapper's Access to the Marsh

From prehistoric Indian sites to neo-modern settlements of French-speaking "Cajuns," Isleños, Austrians (Yugoslavian), Chinese, Malays, or Italians, the alluvial wetlands sustained a range of cultures and rural communities. Several thousand people, with nearly as many boats, were employed in seasonal harvesting that radiated out from each hamlet. Accessible only by water, these sites were linked to their resource base, markets, and sources of supplies by well-defined routes of circulation. With time, this dense, unorganized network of ridge and wetland communities represented a large, somewhat secluded, permanent population.

The natural and man-made waterways of the marshes and swamps were used extensively. To exploit wetland resources, *habitant* (peasants) dug *trainasse* with "crooked shovels." Trainasse, derived from the French means "to drag," but in the local *patois* it signifies "a trail cut through the marsh grass for the passage of a canoe [pirogue]" (Read 1937, 74). These paths were hacked out by trappers to gain access to their trapping leases. Though small, the 1.5-meter-wide and 15- to 30-centimeter-deep passages allowed the marsh dweller to systematically exploit the coastal zone's avian and fur- and hide-bearing animals, and aided in the harvesting process.

To trap marsh-dependent muskrats (*Ondatra zibethicus rivalicius*), nutria (*Myocastor coypus*), and alligators (*Alligator mississippiensis*), ditches (the trainasse) were methodically dug through the wetlands to create a massive array of watercourses and permanent landscape features. The elaborate network represents the wetlands'

earliest large-scale canalization projects (Davis 1976; 1978).

Logging Canals: The Spokes of a Wheel

Intensive bald cypress exploitation began after the Timber Act of 1879 repealed the Homestead Act of 1866. As a result of the act's repeal, vast cypress and tupelo tracts were sold for $0.60 to $1.25 per hectare, and by 1890, a sizable percentage of Louisiana's swamps was managed by the logging industry. Access problems were resolved by excavating canals to the logging sites, and the cut timber was removed using steam engines aboard pullboats to drag logs into a dredged channel.

To use a pullboat effectively, lumber companies dredged primary and secondary watercourses leading to logging sites, which provided an essential link to their reserves. To remove the harvested cypress, the canal configuration incorporated a series of intersecting channels with fan-shaped cable runs radiating out from points along the access routes. These radial designs were etched into the landscape by the cables that were required to "snake" the logs into the principal channel. This distinctive design can be detected on aerial photography and is an accurate record of intensive lumbering operations (Davis 1975).

Logging canals remain a part of the swamp's morphology. For over 50 years, the channels were essential to the forest industry, but as swamp timber was depleted and woodcutting operations terminated, only the canals and pullboat scars remained. They mark the once robust cypress trade and the near-complete depletion of virgin cypress and tupelo swamps.

Canals and Well Access

Concomitant with the logging era, the oil industry began to appraise the Louisiana coastal zone's hydrocarbon potential. Many favorable subsurface sites remained untested until drilling procedures, equipment, and geophysical expertise were developed to recover the mineral fluids. Once engineering and logistic problems were solved, the wetlands became a major hydrocarbon province. In 1934 successful use of a submersible drilling barge marked the beginning of expanded drilling activity. To maximize the floating unit's potential, canals were dredged into exploration and development sites. In most years, almost one third of the U.S. Army Corps of Engineers dredge and fill permits are issued in Louisiana, and it is now difficult to locate a stretch of marsh where canals are absent.

Because marsh soils presented no serious engineering impediments, powerful suction dredges, bucket dredges, spud barges, and marsh buggies easily excavated the required petroleum access, pipeline, and transportation waterways. In opening the coastal lowlands to hydrocarbon development, an extensive agglomeration of canals expanded into a complicated net of coalescing channels. Without any regard for changes in natural drainage, the long-range hydrologic effects have, in many cases, been disastrous.

To develop a hydrocarbon province, a petroleum contractor cuts a series of initial service routes, then adds supplementary passages as warranted. These tributaries filter into the central traffic corridor to guarantee well access. The assemblage ultimately dominates the surface topography, with each appendage representing a new well. It is a one-well, one-canal system. Constantly influencing larger quadrants, the geometric design grows rapidly. Once in use, the distinctive straight-sided canals often erode into a cuspated form. Some fields become so canalized that more than 20 percent of their surface morphology is devoted to these topographic elements. Along with the associated pipeline channels (Figure 2), the impressive number of these linear features resemble what may be described as a "spaghetti bowl." These interconnecting routes are now permanent features that represent 2.3 percent of

the present wetland area (Turner and Cahoon 1987). Oil or gas fields may evolve over many years with the exploration canals and appendages interlocking into a well-defined complex. These increased densities frequently unite to produce small lakes and bays. Unless sediment or fill is added, the land is permanently lost. With time, the array extends as tenacious features across the wetlands to create extractive patterns encompassing from less than 4 to more than 15,000 hectares. Land is lost, and open-water areas increase.

Conclusions

In no other part of the United States is there such a chronology of resource use depicted so clearly in canal systems. As a logical method for dealing with Louisiana's distinct wetland environment, canals have modified drainage patterns, changed flora and fauna, altered salinity regimes, and contributed to land loss. Whether the watercourses are the trapper's narrow trainasse, or the complicated assortment of petroleum access routes, they have a decisive and cumulative influence on the alluvial wetlands. Regulatory agencies have slowed the growth. However, once built, canals become permanent landscape features. Some canals in Louisiana are more than 200 years old. Past history indicates clearly the zealous activity of the canal builders. Canalization processes illustrate the transformation of wetland habitats through concentrated and uncontrolled dredging strategies. While it is an arduous task to evaluate total wetland change, canal construction emerges as one element in Louisiana's environmental modification. Exploitation patterns endure, even though resources are depleted; their longevity is a visible reminder of humankind's capacity to change unknowingly the natural system. The function may change, but the "lines" persist.

Figure 2. A pipeline is laid across the marsh to move hydrocarbons to market. Photograph by Donald W. Davis.

References

Britsch, L. D., and Kemp, E. B. III. 1990. *Land Loss Rates: Mississippi River Deltaic Plain,* Technical Report GL-90. Baton Rouge: U.S. Army Corps of Engineers.

Davis, D. W. 1973. *Louisiana Canals and Their Influence on Wetland Development.* Ph.D. Dissertation, Louisiana State University.

————. 1975. Logging canals: A distinct pattern of the swamp landscape in south Louisiana. *Forest and People* 25(1):14-17,33-35.

————. 1976. Trainasse. *Annals of the Association of American Geographers* 66(3):349-359.

———— . 1978. Wetlands trapping in Louisiana. In *Geoscience and Man*, Vol. 19, ed. S. B. Hilliard, pp. 81-92. Baton Rouge: School of Geoscience, Louisiana State University.

Gagliano, S. M., Meyer-Arendt, K. J., and Wicker, K. M. 1981. Land loss in the Mississippi River deltaic plain. *Transactions, Gulf Coast Association Geological Societies* 31:295-300.

Lewis, P. F. 1976. *New Orleans: The Making of an Urban Landscape.* Cambridge, MA: Ballinger.

Martin, F. X. 1827. *The History of Louisiana from the Earliest Period.* New Orleans: J. A. Gresham [reprinted, Gretna, LA: Pelican, 1975].

Penland, S., Roberts, H. H., Williams, S. J., Sallenger, A. H. Jr., Cahoon, D. R., Davis, D. W., and Groat, C. G. 1990. Coastal land loss in Louisiana. *Transactions, Gulf Coast Association of Geological Societies* 40:685-700.

Read, W. 1937. *Louisiana French.* Baton Rouge: Louisiana State University Press.

Turner, R. E. 1987. *Relationship between Canal and Levee Density and Coastal Land Loss in Louisiana.* U.S. Fish and Wildlife Service, Biological Report 85(14).

———— , **and Cahon, D. R.** 1987. *Causes of Wetland Loss in the Coastal Central Gulf of Mexico: Vol. 2. Technical Narrative.* Final report submitted to Minerals Management Service. Contract No. 14-12-0001-3252. OCS Study/MMS 87-0120. Metairie, LA: Minerals Management Service.

Levees

Warren E. Grabau
U.S. Army Corps of Engineers, Retired

H. Jesse Walker
Louisiana State University

The natural and artificial forms of levees are similar only in their linearity; natural levees develop during overflow stages of a river, whereas artificial levees are constructed to prevent bank overflow; natural levees provided the only suitable sites for occupation by early settlers along the river, whereas artificial levees made settlement possible in areas that were otherwise impractical; natural levees create unique ecosystems on floodplains, whereas artificial levees lead to the destruction of those systems. Here, we consider the levees of the Mississippi River.

The Levee as a Natural Feature

The sediment carried by the meandering Mississippi is primarily sand, silt, and clay, the proportions of each varying from place to place and from time to time. For example, the Arkansas River, with its sources in the Rocky Mountains, brings down large quantities of sand, whereas the Ohio River, deriving its sediments from the gradual erosion of the Appalachian highlands, carries clay-sized particles as a major component.

During the annual spring floods, before the Europeans settled here, these sediment-rich waters frequently overtopped the river banks and spread laterally over the floodplain. As flood water flowed over the banks, water velocity dropped, carrying capacity decreased, and suspended material was deposited. Grain sizes tended to be coarser near the river and became finer with increasing distance from it.

The result was the construction of natural levees along both banks of the channel (Figure 1). These levees were asymmetric in cross-sectional shape and grain-size distribution. They were highest near the channel, where most of the sand came out of suspension, and lowest at some distance from the channel, where

the finer grains eventually dropped to the bottom. As long as the channel remained stable, levees became higher each time they were overtopped by flood waters and the flood episode lasted long enough to fill the backswamps.

The latter condition is critical. The crest elevations of the natural levees were not uniform. They varied from place to place because all sections of the levee did not grow equally rapidly. Where floodplains were heavily forested along the bank, additional resistance imparted to the flow of water by the trees decreased current velocity, which consequently increased the rate of deposition.

Other mechanisms tended to keep low places low. When flood water overtopped the low sections of natural levees, erosion was likely to occur; erosion that resulted in breaks in the levee called "crevasses."

As the meandering Mississippi River changed course over the ages, section after section of channel was abandoned, new reaches formed, and new natural levees developed. With time, natural levees formed a reticulum that subdivided the floodplain into a landscape consisting of broad, shallow, waterlogged, and biotically diverse basins separated from their neighbors by relatively dry natural levees. In this wilderness, neither wholly land nor wholly water, humans were rare intruders, relatively powerless to alter the dictates of Nature. And all of this richness and diversity was maintained by the annual inundations.

The Levee as an Artificial Feature

All of this changed when European settlers appeared. The river, with its connecting maze of bayous and tributaries navigable to the small craft of the day, gave access to virtually every corner of the plain, and the

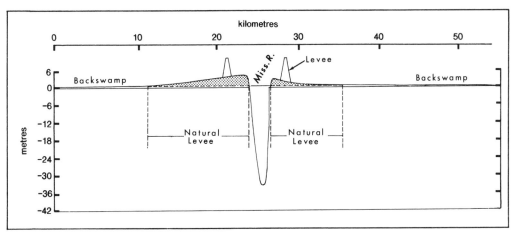

Figure 1. Cross section of levees along the Mississippi River.

natural levees, dry and well-drained except for the brief periods when they were inundated by spring floods, offered convenient sites for farms and plantations. Fertile soils produced abundant harvests, and the bayous and the river provided avenues to market.

This relation between farm and waterway created its own peculiar pattern. Each farm had to have access to the waterway, and each planter built his home on the highest—and therefore driest—land available, that is, near the front of the natural levee. The farms extended back from the river or bayou to the backswamp in the interior. This land division scheme was brought from northwestern France to Acadia as well as Louisiana. Known as the arpent system, it has left its imprint on the land to this day in the form of long, narrow lots (Figure 2).

The annual floods were an intolerable nuisance from the beginning. Indeed, in 1717, when New Orleans was being built, the city planner directed that "a dike or levee be raised in front, more effectually to preserve the city from overflow" (Mississippi River Commission 1940, 10). By 1727 the levee stretched for 1.6 kilometers, a length that grew to 70 kilometers by 1735 (Figure 3). The early levees were small embankments that bordered each planter's property. As one of the conditions of their grants from the King of France, planters had to build and maintain their own levees (Frank 1930). By 1812, levees extended along the right bank of the river as far north as the Red River (Figure 3).

The importance of levees to Louisiana was emphasized by J. B. D. de Bow (1847, 239):

> Levee is a French word, of primary importance within the State of Louisiana: It pervades its statute-book and is daily heard within its halls of justice. . . . Thus the Dutch are not the only people who have won their domain from the watery element.

During the first 128 years of the 275-year history of flood control along the Mississippi River, levee

Figure 2. Example of the influence of the arpent system along a Mississippi Valley bayou. Public domain photograph, provided by the U.S. Geological Survey.

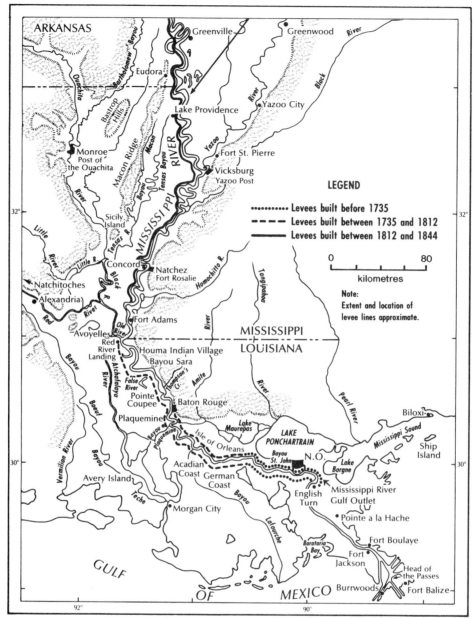

Figure 3. Map of levee construction along the Lower Mississippi River. Modified from Cowdrey (1977).

construction and repair were almost entirely in the hands of the planters. However, after a severe flood in 1844, county ("parish," in Louisiana) governments became actively involved in levee affairs (Harrison 1951). In 1849 and 1850, Congress passed the Swamp Land Acts with a condition that funds from land sales be used for levee construction.

In 1887 the federal government created the Mississippi River Commission, which was charged specifically to design and construct an integrated system for preventing the flooding of the lower Mississippi River floodplain.

The primary flood-control device was a system of levees so high and so extensive that even the highest floods would be confined to the channels of the river and its tributaries. Some of these are enormous, with heights exceeding 10 meters and bases exceeding 200 meters in width. Many have roads on top. Originally they were built along the river banks at the highest part of the natural levees, so that the artificial levee was nothing more than an extension of the front part of the natural structure. However, modern artificial levees cut indiscriminately across the country, in many places well away from the river channel (Figure 3). They are designed to create what is called a "floodway," a broad artificial channel that includes the natural channel as well as a narrow belt of floodplain on both sides. The floodway can carry a specific magnitude of flood.

The fill required for these structures is usually taken from just inside the floodway. Along some segments they contain as much as 3,000,000 cubic meters of fill per kilometer of floodway. Along hundreds of kilometers the artificial modern levee parallels long, deep lakes; the "borrow pits" from which the soil for the embankment was taken. These lakes now form an important component of the modern floodway ecosystem, alive with fish, birds, and small mammals that depend upon a semi-aquatic habitat. The levees themselves are kept free of woody vegetation and are heavily sodded; they are much in demand for cattle grazing (Figure 4).

Construction of levees had continued at such a pace that in 1926 it was stated that the end of flood-control

Figure 4. Artificial levee south of Baton Rouge. Note grazing cattle and road on top. Photograph by H. J. Walker.

fighting along the Mississippi River was near. However, in 1927 a major flood inundated nearly 50,000 square kilometers of the lower Mississippi River Valley and drove more than 700,000 people from their homes (Frank 1930). Crevasses were numerous, possibly as many as 225. However, New Orleans, where leveeing had begun 211 years earlier, was little affected. At the time, *The New York Times* reported (quoted in Cowdrey 1977, 36) that "New Orleans, sitting serenely between the river and Lake Pontchartrain, with virtually the entire city of half a million below river level, went calmly and unhurriedly about its ordinary work."

Nonetheless, the flood of 1927 was so destructive of levees that the "levee only" policy of the Mississippi River Commission was subsequently altered to include cut-offs in the middle reaches of the Mississippi River (Harrison 1951) and a system of floodways and spillways. The Atchafalaya River (Figure 3) was converted into a floodway that today, by law, carries one third of the discharge of the Mississippi River. New Orleans is further protected by the Bonnet Carré spillway that was built in the early 1930s about 50 kilometers north of New Orleans. It diverts floodwaters into Lake Pontchartrain (Figure 3) and has been opened seven times (1937, 1945, 1950, 1973, 1975, 1979, and 1983) since completion.

Because the modern system of flood control has been so successful (the last major flood in the lower valley was that of 1927), the character of the modern floodplain is utterly different from what it was a century ago. The great forests are gone, replaced by huge fields of cotton, soybeans, and corn. Bayous now exist as shallow swales carrying only a small flow of near-stagnant water. The complex ecosystem once nourished by annual inundations is gone, replaced by a landscape tamed and placid. So completely has the threat of flood been removed that many floodplain farms even find it necessary to irrigate their fields during dry summers.

References

Cowdrey, A. E. 1977. *Land's End.* New Orleans: U.S. Army Corps of Engineers.

De Bow, J. D. B. 1847. The Mississippi, its sources, mouth and valley. *De Bow's Review* 3(5):423-437.

Frank, A. 1930. *The Development of the Federal Program of Flood Control on the Mississippi River.* New York: Columbia University Press.

Harrison, R. W. 1951. *Levee Districts and Levee Building in Mississippi.* Stoneville: U.S. Department of Agriculture, Mississippi Agricultural Experiment Station.

Mississippi River Commission. 1940. *The Mississippi River.* Vicksburg: Mississippi River Commission.

Grand Coulee Dam

Keith W. Muckleston
Oregon State University

Grand Coulee Dam is a massive concrete structure on the Columbia River in north-central Washington state. The dam and its related resource developments reflect evolving strategies of resource management in North America during the 20th century. They symbolize five themes: stimulation of economic development in remote areas; government enhancement of agricultural production and settlement in dry areas; cooperative management of international water resources; increasing the scale of interregional trade; and growing concern for environmental and ecological values.

Resource Development and Regional Growth

Grand Coulee Dam is the epitome of society's earlier efforts to bend nature to its purposes by structural methods. Although started longer than five decades ago, the dam remains the world's largest single concrete structure, reaching more than 1.5 kilometers in length, containing more than 9,100,000 cubic meters of concrete, and requiring the excavation of almost 29.5 million cubic meters of earth and rock (U.S. Department of Interior n.d.). Rising approximately 168 meters from its foundations, this massive structure creates a reservoir almost 250 kilometers long that reaches to the United States-Canada border and contains approximately 6.4 cubic kilometers of useable storage water (Figure 1).

The Roosevelt Administration initiated construction of the dam in 1933, during the depths of the Great Depression. New Deal policy makers intended such bold action to reduce severe unemployment, stimulate economic growth in the region, and develop irrigation.

The initial commitment of funds represented an enormous and controversial expenditure during the Great Depression. Construction of the dam required six years of nonstop effort, employing more than 7,000 workers during peak periods. Construction did stimulate the regional economy, although the dam's many critics reviled it as "the new pyramids program," which squandered public funds to produce generating capacity in an area where it could never be used. It is estimated that more than half of the expenditures on the dam were in Washington and Oregon; and that total public expenditures in the region, of which Grand Coulee Dam represented a significant part, accounted for about 40 percent of the increased regional income between 1933 and 1939 (Johansen and Gates 1957). Over the last half century, abundant, low cost hydropower has attracted the aluminum industry and other producers oriented toward inexpensive electrical energy.

Irrigation and the Promotion of Settlement

Grand Coulee Dam is the *sine qua non* for the largest single irrigation project ever built by the U.S. Bureau of Reclamation. The Columbia Basin Project currently irrigates approximately 225,000 hectares, about half of the area originally authorized for development. Irrigation in the project area represents the third phase in a succession of agricultural uses: grazing, dry-land farming, and irrigation.

The irrigation project functions as follows: Between April and September, when the reservoir is generally high, water is pumped approximately 85 meters from the reservoir behind the dam and directed a short distance to the ancient course of the Columbia. A

Figure 1. The Grand Coulee Dam. Source: U.S. Department of Interior (1981).

deeply entrenched depression in the ancient course —the Grand Coulee—functions as a regulating reservoir more than 40 kilometers long from where more than 1.5 cubic kilometers of water per year are delivered by canals and syphons to the project lands lying 80 to 140 kilometers south of the dam.

Similar to most Bureau of Reclamation projects, the energy generated at Grand Coulee Dam subsidizes irrigation through its sale to nonagricultural users. Approximately 90 percent of the irrigation costs are carried by sales of hydropower to municipal and industrial users (Macinko 1963). Given this high level of subsidy and present policy of fiscal austerity toward federal water projects, it is doubtful that the unconstructed half of the Columbia Basin Project will be developed.

International Management of the Columbia River

Grand Coulee Dam plays an important part in joint United States-Canadian management of the Columbia. More than 3,500 of the 8,900 kilometers of boundary between the two countries are in international waters, along rivers and lakes. Moreover, dozens of rivers are transnational, flowing across the United States-Canada border (Muckleston 1980). The United States-Canadian Boundary Waters Treaty of 1909 is a recognition of the myriad hydrographic connections between the two countries.

The Columbia is a transnational river, rising in British Columbia and flowing into the United States some 250 kilometers upstream from Grand Coulee Dam. Therefore modification of the Columbia's river regime in British Columbia affects power production

at Grand Coulee Dam and the nine additional U.S. hydroelectric dams below it. Under natural conditions, prior to provision of upstream storage, discharge of the Columbia peaked during the period of rapid snow melt from May to July and then fell to between 20 and 25 percent of the June discharge during the period of August to April. This pattern of discharge is exactly opposite the seasonal demand for electric energy in the Pacific Northwest. Therefore upstream reservoir capacity was provided to store part of the spring snow melt for release in the autumn and winter periods of higher electrical energy demands.

Grand Coulee Dam was the first major storage dam to perform that function in the Columbia system. But as the demand for electricity accelerated during the 1940s and 1950s, energy shortfalls during the season of high demand made it clear that additional upstream storage capacity was required. A number of potential upstream storage sites in the U.S. part of the Columbia system were not used for a number of reasons, including disputes between public and private utilities over development of the sites, growing concern for environmental values, and strong objections from economic interests in Idaho and Montana over the prospects of losing productive resources to reservoir development. These concerns combined to suspend the provision of approximately 42 cubic kilometers of upstream storage (Marts 1954).

After years of study and negotiation, the need for large additional increments of upstream storage was finally met by the *Treaty between Canada and the United States of America Relating to Cooperative Development of Water Resources of the Columbia River Basin.* Finalized in 1964, the treaty provided among other things that approximately 19.1 cubic kilometers of storage in British Columbia be available for downstream power production in the United States. In return Canada receives one half of the downstream power benefits accruing in the United States that result from use of water stored in Canada. Thus, the treaty incorporates the principle of sharing downstream benefits, the only international treaty to do so (Sewell 1964).

A significant part of downstream power benefits in the United States is generated at Grand Coulee Dam. Its average output is almost one quarter of the total from 10 large hydroelectric dams situated below Canadian storage sites in the United States. Grand Coulee Dam's present morphology and its vastly increased generating capacity are also related to the cooperative development; markedly increased storage capacity in Canada made it feasible to almost triple the installed generating capacity at the dam, from 2,400 to 6,494 megawatts. This required extensive modification of the dam—construction of a 56-meter wing-dam and massive powerhouse, both of which attach at an angle to the main dam (Figure 1).

Interregional Trade

Just as the refrigerated railroad car allowed large and increasing volumes of fresh fruits and vegetables from California to be marketed on the U.S. east coast after the turn of the century, the introduction of extra-high-voltage transmission technology in 1968 made feasible the movement of large volumes of surplus electrical energy from the Pacific Northwest to major California population centers. This technology increased the maximum economic distance of transmission from approximately 500 to 1,600 kilometers. In the next 20 years of intertie operation the volume of electrical energy moving from the Pacific Northwest to California saved utilities there from purchasing 42,930,000 liters of petroleum for thermal electric production (U.S. Department of Energy 1990). In addition to the substantial monetary savings in both regions, this southward movement of electrical energy prevented about 2,450,000 metric tons of carbon dioxide from entering California's heavily polluted airsheds (U.S. Department of Energy 1990).

Salmon Restoration

Since 1983 innovative reservoir operations at Grand Coulee Dam symbolize the growing perception across North America that more attention must be directed to sustainable resource development. This view holds that the production of traditional utilitarian outputs from natural resources should no longer be emphasized without careful consideration of environmental and ecological questions.

A case in point is the coordinated release of water from Grand Coulee Dam to create partial freshets to speed juvenile salmonids downstream during the two-month period in the spring when they migrate to the ocean. This revised mode of reservoir operation contrasts sharply with the traditional management of storage at the dam. Before 1983 as much of the spring runoff as permissible was stored for release during the low-flow period of fall and winter when regional energy-demands were high. Thus, by "ironing out the hydrograph" at Grand Coulee Dam (and other upstream storage reservoirs), flood control was attained during May and June while energy production was optimized. The new pattern of releases from the dam means that less energy from the Columbia hydro system is produced during the high demand season and more during late spring. Because an excess of hydroelectric generating potential already exists in the spring, the additional energy produced from fish migration flows must be sold at prices considerably lower than those received during the fall and winter. The estimated yearly value of this tradeoff is about $50 million (Muckleston 1990).

It is perhaps fitting that spring releases for salmon migration come principally from Grand Coulee Dam because its construction and subsequent operations contributed significantly to the decimation of Columbia River salmonids. The dam permanently blocked bountiful salmon runs spawning above it, eliminating about 1,850 kilometers of salmon habitat.

The total volume of water reserved for release at Grand Coulee Dam during the 15 April to 15 June migration period is 4.25 cubic kilometers (Northwest Power Planning Council 1987). Since this is inadequate to sustain continuous flow conditions satisfactory to migration, releases are timed to correspond with short periods of heavy migration when river conditions would otherwise be unsatisfactory. Despite careful management, during some years reserved water for migration is exhausted before the migration period is over, resulting in poor passage conditions and high mortality rates.

Although annual spring releases of 4.25 cubic kilometers of water from Grand Coulee Dam have improved downstream migration conditions, since 1989 salmon interests have campaigned for larger and for continuous flows during the two-month period. Implementation would require larger tradeoffs of power sales. Whatever the outcome, operations at Grand Coulee Dam will be crucial to the outcome of the salmon restoration program.

Conclusions

The world's largest concrete structure and its evolving uses illustrate some of the resource-management strategies in North America. A massive monument to the structural approach in resource development, for almost six decades Grand Coulee Dam has been an instrument by which several water-related outputs are produced in prodigious quantities. Hydropower, irrigation, and flood control are the traditional outputs flowing from the dam, which were intended to increase the social and economic well-being of the region and country.

Although to many observers Grand Coulee Dam remains a prime example of successful resource development, it has long been a subject of great controversy due to its high initial cost, symbolization of New Deal philosophy, and heavy subsidy to irrigation. The dam's preeminent role in the decimation of Columbia River salmonids adds appreciably to its controversial nature. Innovative reservoir-management techniques to aid downstream migration of salmon below the dam have had limited success and may be further modified in the form of larger releases during the spring.

The increasing distance of efficient transmission of electric energy brings Grand Coulee Dam into closer relationships with western Canada and especially with the Pacific Southwest. While the United States-Canadian treaty has had a marked effect on the morphology and generating capacity of the dam, benefits resulting from Canadian storage are shared with the upstream neighbor. Construction and operation of Grand Coulee Dam have helped mold North America to its present dynamic relationship between society and environment. While continued production of the dam's water-related outputs is likely well into the next century, further operational modifications are expected to favor the environment over power and agriculture.

References

Johansen, Dorothy O., and Gates, Charles M. 1957. *Empire of the Columbia.* New York: Harper & Brothers.

Macinko, George. 1963. The Columbia Basin project: Expectations, realizations, implications. *Geographical Review* 53:185-199.

Marts, Marion E. 1954. Upstream storage problems in the Columbia River power development. *Annals of the Association of American Geographers* 44:43-50.

Muckleston, Keith W. 1980. International management of the Columbia. In *Conflicts over the Columbia River*, pp. 69-88. Corvallis, OR: Water Resources Research Institute, Oregon State University.

————. 1990. Salmon vs. hydropower: Striking a balance in the Pacific Northwest. *Environment* 32:10-15,32-36.

Northwest Power Planning Council. 1987. *Columbia River Basin Fish and Wildlife Program.* Portland, OR: Northwest Power Planning Council.

Sewell, Derrick W. R. 1964. The Columbia River Treaty: A landmark in international river development. In *Yearbook of the Association of Pacific Coast Geographers*, ed. J. F. Gaines, Vol. 26, pp. 15-22. Corvallis, OR: Oregon State University Press.

U.S. Department of Energy. 1990. *Bonneville Power Administration 1989 Annual Report.* Portland, OR: Bonneville Power Administration.

U.S. Department of Interior. n.d. *Grand Coulee Statistics.* Grand Coulee, WA: Grand Coulee Project Office, Bureau of Reclamation.

U.S. Department of Interior. 1981. *Bureau of Reclamation Project Data*, Water Resources Technical Publication, p. 377. Washington, DC: U.S. Government Printing Office.

Viticulture and Wines
of the Napa Valley

Deborah L. Elliott-Fisk
University of California, Davis

Teresa L. Bulman
Portland State University

Ann C. Noble
University of California, Davis

Because of the alleged effects of vineyard geography on the flavor of wine, geographers have long been interested in viticulture and wines. Many consumers claim they can identify a wine region by wine taste, and formal descriptive analyses of wines sometimes discriminate wines by region (Heymann and Noble 1987, 1989). We examine here the variability of vineyard geography, viticultural practices, and wine-making for the Napa Valley of California, North America's premier wine appellation.

The Physical Environment

Napa Valley (Napa County, California) has been world-renowned as a grape-growing area for decades, long before it became the second of the U.S. federally approved viticultural areas in 1981. The appellation includes most of Napa County. The Napa Valley viticultural area encompasses about 120,000 hectares, with more than 13,000 hectares planted in grapes (Figure 1). The diversity of viticultural environments has led to the establishment of several small viticultural areas wholly or partially within the Napa Valley. Vineyards grow at elevations ranging from 6 meters in the southern valley near San Pablo Bay to higher than 750 meters in the mountains above the northern valley floor (Vaca Range and Mayacamas Mountains), and in the intermontane tributary valleys to the east.

Geological History

The vineyard environments have evolved through geologic time, and past geologic processes now control grape production and wine composition (Noble and Elliott-Fisk 1990). Napa Valley is a synclinal valley of Cenozoic age, derived from Jurassic and Cretaceous marine sedimentary rocks. Faulting and minor folding initiated in the Miocene resulted in formation of the valley proper, with marine incursion resulting in the formation of a series of weakly consolidated, fossiliferous sandstones (Kunkel and Upson 1960). This was followed by a series of Late Cenozoic volcanic eruptions which deposited the Sonoma Volcanics, a series of basaltic to rhyolitic flows, tuffs, and tephras that blanketed the entire landscape. Further tectonic activity created terraces and a series of small hills on the valley floor.

Quaternary sea-level changes accompanying the glacial-interglacial cycles affected San Pablo Bay as part of the San Francisco Bay-Delta system. The bay has transgressed and regressed through at least the lower half of Napa Valley several times in the past. This resulted in the deposition of bay muds that vary from carbonate to organic-rich, and reduced to oxidized as soil parent materials. Variations in the Quaternary hydrology of the Napa River system have also resulted in the deposition of river and fan sediments of different ages, which vary in texture from fine clays and silts to coarse gravels and boulders (Noble and Elliott-Fisk 1990). All of these lithologies form residual and depositional soil parent materials.

Soils

The viticultural soils of Napa Valley are a function of the soil-forming factors of climate, soil parent materials, topography, organisms, and time (Jenny 1941). The role of each of these variables changes from site to site because of Napa Valley's active and diverse geological history. More than 40 soil series are found in several orders in the valley (Soil Survey Staff 1978). The depositional lowland soils vary in mineralogy, chemistry, texture, and structure, depending on variations in their parent material, landform age, and alluvial floodplain and fan stratigraphy. The uplands contain some of the oldest residual and depositional soils in Napa Valley at high elevations on terraces and ridges; some of these soils are deep, highly leached, largely pedogenic clays forming from colluvium and bedrock. Soils on other slopes and valley floors are of intermediate age, but immediately along the Napa River they are recent.

Although farming practices can modify the soil chemistry, structure, and horizonation to some extent, often the grape-grower must accept what the subsurface provides because vines root to great depths where possible. These soils interact with the climatic environment of the grapevine to control water and nutrient flux, vine phenology, fruit production, and fruit composition (Noble and Elliott-Fisk 1990).

Viticultural Environments and Viticultural Areas

A gradient of viticultural environments thus occurs from south to north, along the Napa Valley (Figure 1).

Carneros

The climate of the Carneros district (approved in 1983)—cool, subhumid (450 millimeters of precipitation annually), windy, and marine—results from the moderating influence of San Pablo Bay (Amerine and Winkler 1944). Carneros soils are predominantly clays to silty clay loams with moderate drainage and complex horizonation, having evolved from bay mud and estuary deposits. More than 2,100 hectares are planted in cool varieties of grapes, such as Chardonnay, Pinot Noir, Gewurztraminer, and Johannesburg Rie-

Figure 1. Napa Valley viticultural area.

sling (Cunha *et al.* 1989). A significant portion of the grapes grown here are also used to produce sparkling wines.

Mount Veeder and Wild Horse Valley

The cool mountain climates just north of San Pablo Bay and the low-elevation Carneros district include the Mount Veeder (approved in 1990) and Wild Horse Valley (approved in 1988) viticultural areas. The Mount Veeder appellation runs along steep mountain slopes in the Mayacamas Mountains of the west side of Napa Valley at elevations of 120 to 550 meters. Mount Veeder has a general exposure to the south and east, cold air drainage off the slopes, and increased orographic precipitation (as much as 875 millimeters annually). Soils are shallow and have developed on sandstone, rhyolite, and rhyolitic tuff. Cool-climate varieties dominate, especially Chardonnay, although substantial acreages of Cabernet Sauvignon have been planted across the 130 hectares of currently established vineyard (Cunha *et al.* 1989).

Wild Horse Valley is an upland plateau on the east side of the Napa Valley in the Vaca Mountains. This upland is blanketed with basalt and rhyolitic flows. The shallow and residual volcanic soils have low water-holding capacities but good nutritional status. Strong ocean breezes strike the mountain slopes and ridges directly off the bay, making this region cooler and more arid than Mount Veeder to the west. Cool varieties are planted here (25 hectares), including Chardonnay, Johannesburg Riesling, Gewurtztraminer, and Pinot Noir.

Stags Leap District

Between Napa River and the flanks of the Vaca Mountains on the eastern Napa Valley floor, lies the Stags Leap District (approved in 1989). It is distinguished by deep, gravelly valley-floor soils derived from old Napa River deposits and clay-rich colluvial footslope soils, both principally from volcanic parent materials. The dominant soil, sandy clay, here has a duripan at a depth of 0.5 to 1.5 meters, and defines an older channel of the Napa River. The climate is relatively cool, with precipitation of approximately 750 millimeters annually, but this area experiences strong afternoon sun and sea breezes, drawn up the valley by warm air rising off the eastern mountain escarpment. Cabernet Sauvignon is the principal variety of this area (570 hectares).

Oakville and Rutherford

North and west of the Stags Leap District along the valley floor are the proposed viticultural areas of Oakville-Oakville Bench-Rutherford-Rutherford Bench. This four-fold appellation scheme proposes two appellations that span the valley floor on both the west and east sides of the Napa River: Oakville to the south and Rutherford to the north, named after the communities located there. On the west side of the river, where two large alluvial fan complexes form the viticultural landscape, two proposed viticultural areas, which fall within the two appellations, are the Oakville Bench on the Oakville fan complex and the Rutherford Bench on the Rutherford fan complex. Both fans have very deep, gravelly to very gravelly sandy clay loam soils. The Oakville fan materials are derived from the marine sedimentary Great Valley sequence and the Rutherford fan materials are derived from the marine sedimentary and slightly metamorphosed Franciscan Formation (Noble and Elliott-Fisk 1990). The soils on the east side of the Napa River within the larger proposed Oakville and Rutherford appellations are primarily volcanic depositional soils of moderate to shallow depths over volcanic bedrock. The climate here is slightly warmer, with annual precipitation of approximately 1,000 millimeters and strong morning sun. The dominant variety is Cabernet Sauvignon, although other varieties are widely planted in this grape monoculture of more than 4,500 hectares.

Howell Mountain

The third viticultural area to be approved (in 1984) in Napa Valley was Howell Mountain in the eastern Vaca Mountains. This region is on a relatively high volcanic plateau along the Napa River's eastern watershed boundary. The climate is cool, days fog-free, and ripening period prolonged. Soils are volcanic lithosols. About 175 hectares are currently planted in grapes, evenly divided among Zinfandel, Cabernet Sauvignon, and Chardonnay. The area is most widely recognized for its concentrated, tannic Zinfandel and Cabernet Sauvignon wines (Noble and Elliott-Fisk 1990).

Spring Mountain and Diamond Mountain

North and west of the proposed Rutherford Bench viticultural area in the western Mayacamas Mountains is a high region, locally referred to as Spring Mountain and Diamond Mountain. This area receives the heaviest precipitation in Napa Valley (as much as 1,500 millimeters annually), and has clear, cool nights and clear cool to slightly warm days. Geologically, the mountains are composed of extensively faulted and fractured Sonoma Volcanics (dominanting to the north) over diverse Franciscan sandstones and conglomerates (dominating to the south). Some very old, leached soils occur on high terraces and interfluves. Approximately 250 hectares are planted, largely in Cabernet Sauvignon, Chardonnay, and Merlot.

Calistoga

At the north end of the valley, along the valley floor, temperatures are much warmer. This is the as yet undesignated Calistoga region. The Sonoma Volcanics rim all sides of the valley here, producing Bale series valley-floor soils that may be very bouldery deposits across alluvial fans or finer, gravelly deposits along the

Napa River. Cabernet Sauvignon and Zinfandel are widely planted, along with Sauvignon Blanc and Chenin Blanc.

Land-Use Regulations and Water-Resource Development

Although the physical geography of Napa Valley provides an ideal environment for growing high quality wine grapes, it was the development of favorable land-use regulations and adequate water resources that enabled the grape industry to become the dominant economic activity of Napa Valley in this century. Viticulture has been practised in Napa Valley for longer than 150 years, in spite of insect infestation, Prohibition, economic depression, and war.

Until the mid-1900s, the agricultural focus of the county was orchard crops, grains and hay, livestock, and tanneries. Vineyards began to appear on the hillslopes in the 1870s, when growers recognized that only in the thermal belt could they avoid the damaging spring frosts (Smith and Elliott 1878). The expansion of vineyards continued steadily until in 1890 more than 600 farmers were growing more than 7,000 hectares of vines, the fruit of which was processed in 100 wineries (Figure 2). However, the vine louse, *Phylloxera vastatrix*, had begun to attack Napa Valley vineyards by the 1870s; by the mid-1890s more than half the vineyards were infested (Verardo and Verardo 1986). At first, vineyard lands declined, but growers replanted with more resistant native and hybrid rootstocks. Just as the new, grafted rootstocks began to bear fruit, the 18th Amendment to the U.S. Constitution prohibiting alcoholic beverages was passed.

Despite economic depression, when Prohibition ended in 1933, grape-growers attempted to reestablish

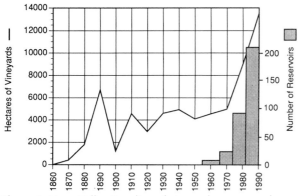

Figure 2. Napa Valley time line of vineyard acreage and construction of reservoirs for frost protection.

their foothold in Napa Valley. However, during World War II, the labor force was drafted, growers faced price ceilings and equipment shortages, and many of the fine wine grapes were requisitioned for raisins and industrial alcohol. By the end of the war, the grape acreage of 4,800 hectares was only a marginal increase over the 1929 acreage of 4,600 hectares. However, at this time, the price of wine grapes began rising and consumers increased their wine consumption, allowing wine grapes to assert their dominance over other agricultural industries in Napa Valley (Dowall 1984). By 1971, grapes surpassed both the livestock and the orchard industries in economic value in Napa County. However, the economic and cultural favoritism that the wine industry was experiencing could not protect the grape-growers against the pressures of urban growth and the whims of nature in the form of killing frosts. For these, land-use regulations promoting agriculture and water-resource decisions favoring growers were needed.

Land-Use Regulations

By 1950, Napa Valley was gaining a reputation for its beautiful landscape, relaxing atmosphere, and proximity to the San Francisco Bay area. Urban growth, from the annexation of unincorporated land around the cities of Napa, St. Helena, and Calistoga, was threatening the agricultural economy of the county by the mid-1960s. The California Land Conservation and Development Act (the "Williamson Act"), designed to preserve agricultural land throughout the state, was passed in 1965. The Williamson Act taxed farmers on the agricultural value of their land, rather than on its urban-use potential, so long as the farmers voluntarily contracted with the local government to forego non-agricultural uses of the land. To promote such voluntary contracts, Napa County planners rezoned unincorporated areas in minimum lots of 8- and 16-hectare parcels, which in effect made subdivision development difficult. As a result, Napa Valley land remained in agriculture and, by 1988, 20,000 hectares of Napa County land were under Williamson Act contracts (Sokolow 1989).

Water-Resource Decisions

To maintain the quantity and quality of the harvest, water is used to protect the vines from excessive heat, inadequate precipitation, and damaging frosts. The greatest climatological concern to grape-growers is late-winter and early-spring frost damage along the cold valley floor. Sprinkler systems began to replace orchard heaters and vineyard fans in the late 1960s. The sprinklers are better protection against frost and drought, and generate less air and noise pollution. An increase in the price of grapes to $400 per ton by 1969 justified their installation.

However, sprinkler systems can severely strain water supplies. Water-resource managers were concerned that vineyardists, who diverted water directly from the Napa River and its tributaries, would drain the Napa River dry during a frost event, and that steamflow would be insufficient for valley-wide frost protection. Water shortage in private reservoirs was a recognized solution, but implementing the solution was not possible under then-existing California water law, which limited reservoir use in the valley (Bulman 1990).

Although the county had built reservoirs at the turn of the century, the rate of reservoir construction for frost protection and related purposes such as irrigation began to increase in the late 1960s (Figure 2), resulting in conflicts between water users and violations of law (Bulman 1990). Lawsuits brought by the state and by individuals resulted in: the development of a Napa River Water Distribution Program designed to monitor and regulate water use for frost protection; an amendment to the California Code of Regulations, titled "Napa River, Special," which delineated provisions for water diversion and reservoir construction for frost protection within the Napa Valley watershed; and careful monitoring of all reservoir construction and use by the State Water Resources Control Board.

These unique water-resource decisions served not only to regulate water use in the vineyards of Napa Valley but, by specifying frost protection as a beneficial use of water resources, they promoted vineyard development in the valley. By 1989, more than 13,000 hectares of grapes, representing 90 percent of the grape acreage in the entire county, were frost-protected by reservoir water.

Although the viticultural environment of the valley has been preserved, competing demands for land and water resources have created conflicts. Urban expansion has pushed residential areas to the vineyard boundaries, resulting in complaints about agricultural dust, noise, and chemicals. The increasing number of wineries, many of which offer tours and tastings, has attracted growing numbers of tourists who stretch the valley's traffic and service infrastructure to its limits. The recent introduction of a tourist "Wine Train" caused extensive and heated debate in the valley, raising issues of traffic, pollution, and the general quality of life for residents. The county has recently imposed a ban on new wineries until a resolution can be found to the tourist and traffic

problems. Despite these conflicts, the Napa Valley wine industry remains strong and the quality and reputation of its wines continues to improve.

Grape-Growing and Wine-Making

Through the vicissitudes of *Phylloxera*, Prohibition, and World Wars I and II, grapes were considered to be the crop of an agricultural producer, and wine simply a way to preserve it. In the early history of California wine-making, because of problems with microbial stability, the majority of the wine produced was sweet dessert wine. However, Napa Valley has a long history of production of labeled, branded table wines, where even in the 1930s, 100,000 cases of wine were shipped out of Napa Valley as bottled, winery-labeled wine; less than 1 percent of this was sweet wine.

In 1936, the 4,100 bearing hectares in Napa Valley included approximately 20 percent Petite Sirah, 15 percent Zinfandel, 10 percent Carignane, and 9 percent Alicante Bouschet (*Wine Review* 1938). Of the remaining varieties, the major white varieties were 2 percent each of Sauvignon Vert (Trebbiano) and Golden Chasselas (Palomino). These varieties decreased over the next 50 years (Figures 3 and 4). With the exception of Zinfandel and Petite Sirah, none of the early varieties are perceived to produce premium wines.

Since 1970, the planting of grapes associated with the highest quality wines has increased dramatically, and the predominance of red grapes has shifted to more than 50 percent white. The rapid rise in the price of grapes in Napa Valley and reports predicting increased demand for wine contributed to this planting boom in the 1970s (Bank of America Economics Department 1970). Today, the total acreage planted in Napa Valley exceeds 13,000 hectares, of which the major varietals are by far Chardonnay and Cabernet

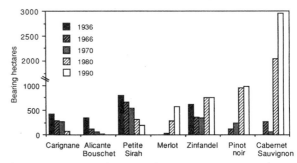

Figure 3. Bearing red varieties of grapevines (in hectares) for Napa Valley from 1936 to the present.

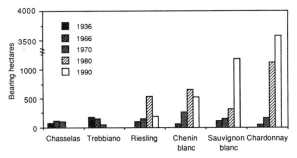

Figure 4. Bearing white varieties of grapevines (in hectares) for Napa Valley from 1936 to the present.

Sauvignon. New vineyards are being rapidly established in both the hillsides and the low-elevation, southern Carneros district, where land is available at somewhat lower prices than along the main valley floor.

Enological Techniques

In addition to the dramatic effect on the quality and style of wine that planting of premium wine-grape varieties afforded, several advances in wine-making techniques had an enormous impact on the wines, including use of refrigeration, stainless steel tanks, yeast and malolactic starter cultures, and improved sanitation. In the 1930s, white grapes were fermented on their skins in redwood tanks for several days without cooling, resulting in oxidized, golden-colored wines. Red grapes were fermented in redwood or concrete tanks, also without cooling. For both, fermentations were conducted relying on the yeasts present on the grapes. For the red wines, in which malolactic fermentation was desired, wineries waited until this secondary fermentation started in the spring "after the tanks warmed up." Now, fermentation can be controlled by addition of cultures of yeast and, when desired, malolactic bacteria. Most grapes are fermented in temperature-controlled, stainless steel tanks.

These factors changed the chemical composition and sensory properties of the wines enormously. With the use of cooling, white wines have more fruity, floral flavors, and are free of harsh, oxidized notes. Progressively, vineyardists and wine-makers have focused on details during every stage of grape-growing and wine-making to improve the quality of the grapes and wines.

Here we use the highest priced grape variety, Chardonnay ($1,400 per ton), to illustrate many of the changes that Napa Valley wines have undergone due to innovations in wine-making. Before 1960, Char-

donnay was not widely planted in Napa Valley because the clones that were widely available yielded low tonnage (less than 1.25 tons per hectare). The increase in acreage of Chardonnay in the last 20 years is extraordinary (Figure 4), and continues at a rapid pace in Carneros and other regions of the valley.

In the 1970s, wine-makers were producing Chardonnay wines with 13 to 14 percent ethanol. These "big" wines were aged for increasing intervals in new oak barrels, resulting in strongly flavored wines dominated by vanilla and spicy notes, lacking fruit, and referred to as "uni-dimensional." In the 1980s, wines were made from less-ripe fruit, which yield wines lower in alcohol. More often, oak aging was used in moderation so that the decreased contribution of oak no longer overwhelmed the "tropical fruit" aroma of the grape. In the 1990s, wine-makers are experimenting with the effects of many viticultural practices and wine-making techniques on Chardonnay and other wines. With Chardonnay, most wines undergo a malolactic fermentation and are fermented in oak barrels, to get a more subtle oak flavor than that achieved by aging in oak after the fermentation is complete. Wines are aged on the lees for as long as nine months to contribute the distinctive caramel, soy, or toasty notes that are products of yeast autolysis.

Viticultural Practices

A variety of vineyard microclimate and soil characteristics can be modified through viticultural practices to ensure grape maturation and optimize fruit production. In the past, vines were head-trained and widely spaced (3 to 4 meters apart), producing yields of 10 tons per hectare or more on deep, moist soils along the floor of the Napa Valley, or 5 tons per hectare or more along the hillslopes. Today, various types of vertical trellis systems and close planting (1- to 2-meter spacing) are employed to modify the surface and subsurface environment of the vine. Yield is increased or decreased by these techniques as deemed appropriate for the variety and site by the viticulturalist.

Various rootstocks increase or decrease root exploration of the soil system and modify vine vigor. Increasing leaf and fruit exposure through canopy management minimizes the development of the intensely vegetative flavor characteristic of Cabernet Sauvignon and Sauvignon Blanc. By finding climates in which the vegetative flavor is reduced and by manipulating the canopy of the grapevine to minimize its development, Sauvignon Blanc and Cabernet Sauvignon wines exhibit more fruity flavors, but are still varietally distinct.

Summary

Napa Valley has a firmly established international reputation for the production of high quality wines. Pressures of competition for land and water resources within Napa Valley and environmental concerns, such as reduction of the use of agricultural chemicals, have already slowed the growth of the industry and limited the potential for expansion. The Napa Valley wine industry is increasingly concerned with usage of every natural and technological resource for further increasing wine quality. Toward this end, recognizing the effect of geography on the quality of grapes and the importance of the quality of grapes on the quality of the wine, wineries are coordinating the work of the grape-growing and wine-making teams, each using cutting edge technologies and further exploring ways to produce the highest quality Napa Valley wines.

Acknowledgments

We appreciate the financial support of the National Geographical Society; the Stags Leap District appellation and quality control committees (including Chimney Rock, Clos du Val, Joseph Phelps, Pine Ridge, Robert Mondavi, Shafer, Stags Leap Wine Cellar, and Steltzner wineries); Caymus, Chateau Montelena, Inglenook, Newton, and Robert Keenan wineries; and the University of California, Davis.

References

Amerine, M. A., and Winkler, J. A. 1944. Composition and quality of musts and wines of California grapes. *Hilgardia* 15:493-675.

Bank of America Economics Department. 1970. *Outlook for the California Wine Industry.* San Francisco: Bank of America.

Bulman, T. L. 1990. *Water Management in Napa County: The Role of Private Reservoirs.* Ph.D. Dissertation, University of California, Davis.

Cunha, M. B., Elliot-Fisk, D. L., and Mendelson, R. P. 1989. *Appellation Napa Valley.* Davis, CA: Vine Arts Mapping.

Dowall, D. E. 1984. *The Suburban Squeeze.* Berkeley: University of California Press.

Heymann, H., and Noble, A. C. 1987. Descriptive analysis of commercial Cabernet Sauvignon wines from California. *American Journal of Enology and Viticulture* 38:41-44.

———. 1989. Comparison of canonical variate and principal component analyses of wine descriptive analysis data. *Journal of Food Science* 54:1355-1358.

Jenny, H. 1941. *Factors of Soil Formation.* New York: McGraw-Hill.

Kunkel, F., and Upson, J. E. 1960. *Geology and Ground Water in Napa and Sonoma Valleys, Napa and Sonoma Counties, California,* Water Supply Paper 1495. Reston, VA: U.S. Geological Survey.

Noble, A. C., and Elliot-Fisk, D. L. 1990. Evaluation of the effect of soil and other geographical parameters on wine composition and flavor: Napa Valley, California. In *Proceedings, 4ème Symposium International d'Oenologie, Actualités Oeonologiques 89,* pp. 51-58. Paris: Bordas.

Smith, C. L., and Elliot, W. W. 1878. *Illustrations of Napa County, California with Historical Sketch.* Oakland, CA: Smith and Elliot.

Sokolow, A. 1989. *Report of the Williamson Act Study Group, University of California. Part II: Preserving Agricultural Land in California: A Short History of the Williamson Act.* Davis: Agricultural Issues Center, University of California, Davis.

Soil Survey Staff. 1978. *Napa County Soil Survey.* Washington, DC: U.S. Department of Agriculture Soil Conservation Service.

Verardo, D., and Verardo, J. D. 1986. *Napa Valley: From Golden Fields to Purple Harvests.* Northridge, CA: Windsor.

Wine Review. 1938. Bearing and non-bearing acreage of principal wine grape varieties in California grapes. 18-19 March.

Scaling North American Urban Climates by Lines, Lanes, and Rows

H. P. Schmid
Swiss Federal Institute of Technology

T. R. Oke
University of British Columbia

Morphological Scales

The texture of a typical North American suburban residential area is dominated by linear elements, such as streets, alleys, and house-rows. Consequently, a hierarchy of morphological units can be identified: house lots; blocks; and neighbourhoods, consisting of several blocks. The size of these units is determined by the spacing of the houses, streets, and alleys that form their boundaries. Figure 1 is a schematic illustration of the dominant spatial scales in the city-block system of a residential area in Vancouver, British Columbia. A block consists of two rows of 10 to 15 homes with front and back lawns. These rows are separated by an alley and the whole block is bounded by two orthogonal sets of parallel streets. In the study area an average block is 100 meters wide and about twice as long. Every few blocks, wider thoroughfare avenues and streets group a number of blocks into a large city-block or "super-block." In reality this pattern is of course interrupted by parks, schools, and other land uses.

Our study shows that this pattern of linear elements in a suburban area governs not only the morphological units, but also the scales of variables controlling the surface climate and processes of surface-atmospheric interaction. The aim is to seek and identify objective scales of urban boundary layers based on dominant units of surface morphology, which may help to establish the validity of a scale-based classification of urban climates.

The need for such a scale-based approach arises from both theoretical and experimental evidence that, close to complex and rough surfaces, the well-known atmospheric scaling schemes for the surface layer and the mixed layer collapse, since homogeneity and fetch requirements are not sufficiently met (for example, Beljaars *et al.* 1983; Steyn *et al.* 1981). Additional scale lengths are needed that reflect the relevant scales of surface texture (for example, Garratt 1978).

Climatic Scales

Boundary-layer climates are largely determined by the interaction of the surface and the atmosphere through the transfer of heat and water vapor (latent heat) and momentum. The most important processes contributing to heat exchange with the surface in a suburban area are conveniently summarized in the surface energy balance equation:

$$Q^* + Q_F = Q_H + Q_E + \Delta Q_S + \Delta Q_A$$

where the individual terms (from left to right) represent the energy fluxes of net radiation, anthropogenic heat, sensible heat, latent heat, heat storage change, and net heat advection (units are $W \cdot m^{-2}$ = watts per square meter).

Schmid *et al.* (1991) show that the variability of the individual terms in this equation may be large at small scales (much less than 1 kilometer). Their findings indicate that this is most pronounced for ΔQ_S and Q_F, but also Q_H and Q_E are variable at scales smaller than previously suggested (for example, Ching *et al.* 1983) whereas Q^* is rather conservative. They show that variations of surface properties play an important role in forcing spatial modulations of energy fluxes.

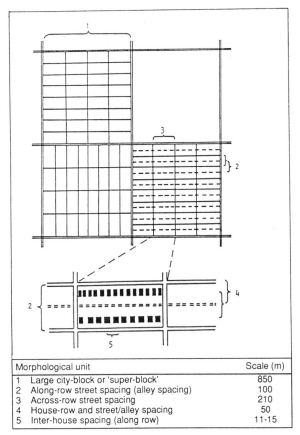

Morphological unit	Scale (m)
1 Large city-block or 'super-block'	850
2 Along-row street spacing (alley spacing)	100
3 Across-row street spacing	210
4 House-row and street/alley spacing	50
5 Inter-house spacing (along row)	11-15

Figure 1. Dominant spatial scales in the suburban city-block system of Vancouver, British Columbia.

Relevant surface properties that serve as control variables for energy and momentum fluxes are: *surface temperature* (for Q_H and Q_E); *geometry* (for Q^*, and also for the turbulent fluxes of heat, water vapor, and momentum because of turbulence enhancement); *thermal conductivity and heat capacity* of the surface cover (for ΔQ_S); *human activity* (for Q_F); and *surface moisture* (for Q_E) (Grimmond 1988; Schmid 1988).

Here we focus on surface temperature and geometry as the most important climatic surface controls. The distribution of radiation temperature over the suburban study area in Vancouver, British Columbia, was obtained by an airborne infrared scanner (pixel size 3.25 meters) in two flights: one in the early afternoon of a clear and hot summer day (25 August 1985) and the other just before sunrise (26 August 1985). The distribution of geometric elements (or roughness elements) was obtained graphically from air-photos and by site inspection, and was then digitized. Five categories of roughness elements were assigned average heights found to be characteristic of the study area: houses (10 meters), garages (3.5 meters), trees (15 meters), streets (0 meters), and grass (0.1 meter).

A portion of the resulting data base is shown in Figure 2 together with the corresponding section of an aerial photograph for comparison. Linear elements are distinct in the surface temperature and geometry visualizations, even more so than in the aerial photograph. The variability of surface temperature is more pronounced in the daytime image (b) than at night (c). It also appears to occur at smaller scales than at night: in (b) the "hot spots" are house-roofs and (slightly cooler) street surfaces, with an average spacing of about 25 meters (that is, by the combination of two overlapping 50-meter scales; see Figure 1). At night, the streets and alleys are warmest with little variability in between. In (d), the relative geometry of houses and trees, compared to the streets and grass areas, is dominant. Variability is therefore concentrated at very small scales and is virtually absent at larger ones.

Spectral Scales

The distribution of variance of surface temperature or geometry is conveniently analyzed using spectral methods, borrowed from texture analysis (for example, Chen 1982; Ford 1976). The variance spectrum decomposes the variability of a spatial process (such as surface temperature or roughness element height, in this case) into wave-number components. The wave-number, k, is the inverse of a length scale or a "spatial frequency" and thus, the spectral density curve may be plotted against the equivalent length scale or wavelength. If the original process is statistically stationary (spatially homogeneous), the integral under the curve over all length scales is exactly equal to the total variance.

Figure 3 shows a smoothed composite spectrum of daytime surface temperature plotted against the equivalent length scale. The spectral density, P, has been multiplied by the wave-number to preserve the variance/ area proportionality in this semi-logarithmic plot. In this representation there is considerable uncertainty in the magnitude of individual peaks, since the composite consists of only 10 linear spectra in the east-west axis, and the smoothing process is somewhat subjective. However, a number of relative maxima can clearly be identified and may be discussed in relation to the morphological scales in the study area.

The most prominent peak occurs at a length scale of 20 to 30 meters. A comparison with Figures 1 and 2b and shows that this distance corresponds to the overlapping spacing of the "hot lines" (house-rows and streets or alleys) in the direction of the spectrum (east-west). Minor peaks may be associated with the house-row spacing at 50 meters (marked as scale 4 in Figures

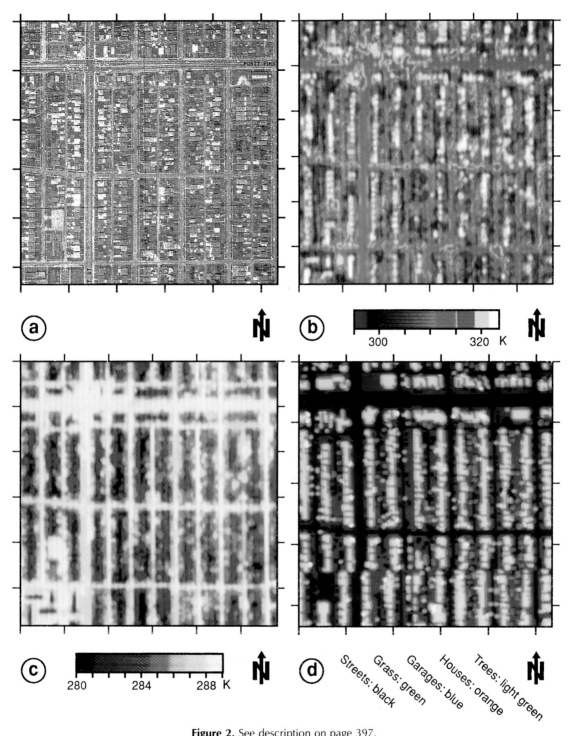

Figure 2. See description on page 397.

Figure 2. Visualization of a section of the study area and the distribution of climatic surface control variables. The grid distance is 100 meters: a. aerial photograph; b. daytime temperature; c. nighttime temperature; d. geometry. See accompanying color plate.

1 and 3), and with the along-road street-spacing, marked as scale 2. The small peak at approximately 10 meters is close to the inter-house spacing along the rows (scale 5). Some portions of the line transects used for this spectrum run through neighborhoods where the orientation of house-rows is at a right angle to those shown in Figure 2.

This distinct directionality of the pattern becomes apparent when a two-dimensional spectrum of the area is taken. In Figure 4 the variance contributions of daytime temperature, nighttime temperature, and geometry from 16 sectors are presented in rose-diagrams. The orthogonal layout of the city-block system is reflected in a contribution of the cardinal axes to over

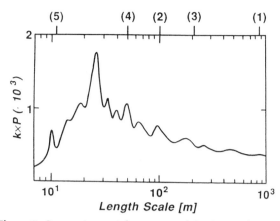

Figure 3. Composite spatial spectrum of daytime surface temperature. The orientation of the transects is north-south. The dominant morphological scales of Figure 1 are indicated at the top.

50 percent of the total variance. The east-west axis used in the composite spectrum of Figure 3 is orthogonal to most of the house-rows and clearly is the principal component of the variance distribution.

Objective Urban Climate Scales

The foregoing shows that urban morphological patterns and scales are reflected in the spectral representation of climatically relevant surface variables or climate forcings. The usefulness of spectral representations in this context lies in their ability to organize the surface variability according to length scales. This notion, together with the link to urban surface morphology, provides the basis for a scheme to identify objective scales of urban climates (Oke 1984).

An urban climate scale may thus be associated with the morphological unit that contains most of the surface textural elements contributing to the variability of climatically relevant features. In other words, if con-

siderable contributions of climatic forcings occur at scales larger than scale 3 in the spectrum of Figure 3, it would make little sense to associate the local climate with individual house lots on the order of scale 5.

A useful measure of the portion of variance associated with a given spatial scale is thus the scale-integrated variance spectrum. If it is normalized by the total variance, the resulting curve starts with zero variance contribution, R, at the smallest length scale and grows monotonically to unity at the largest scales. To avoid the directionality illustrated by Figure 4, the two-dimensional spectra have been integrated according to orientation, resulting in radial, one-dimensional spectra. From these, the scale integrated and normalized spectra have been calculated and plotted in Figure 5 for day- and nighttime temperature and for geometry. Thus, the result is a curve, showing the cumulative distribution of variance versus a length scale, irrespective of direction.

All three curves show the distinct character of a rather homogeneous texture. The curves rise quickly to a high value of R and level out at relatively small scales. However, as suspected intuitively in Figure 2, the nighttime temperature distribution is associated with larger scales than the daytime case and even more so than the geometry distribution.

Figure 3 is one example of a series of two-dimensional spectra, which H. P. Schmid (1988) took in the same suburban area. The similarity among them is striking. In all of the variance contributions of daytime surface temperature and geometry, R reaches a level well over 90 percent at a length scale of 200 meters. Larger scales are slightly more important at night; however, then the range of temperature variability is greatly reduced (Figure 2C).

These findings indicate that an average suburban block contains all the important surface elements to form a characteristic local suburban climate. This does not mean that one block in isolation would suffice to make its own climate. Advective edge effects would

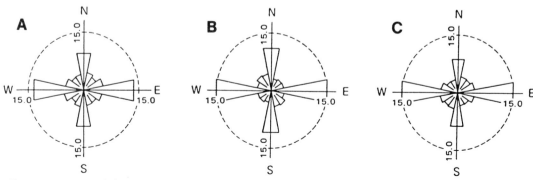

Figure 4. Directional distribution of variance: A. daytime temperature; B. nighttime temperature; C. geometry.

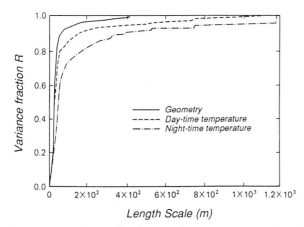

Figure 5. Cumulative distribution of variance versus a length scale (scale-integrated, normalized variance spectrum) for geometry, daytime, and nighttime temperature.

very likely erode any development in this direction. However, it is likely to be true when 10 or more such units in each direction form a more or less homogeneous region (as in the present study area), and then the present results have important implications for experimental design. The morphological unit corresponding to the objective climate scale is representative of the entire region and suitable averages of climatic variables, fluxes, and processes over this scale are expected to be characteristic of the local climate.

In principle this technique is applicable to any urban system. In practice it is likely to give the most clear-cut results for North American-type cities with a distinct grid pattern.

Acknowledgments

Janet Whiteside compiled the geometry data base. Financial support was provided by the Natural Sciences and Engineering Research Council of Canada to TRO; and by a University Graduate Fellowship (UBC) to HPS. This presentation was made possible by the Swiss National Science Foundation (grant 20-29541.90).

References

Beljaars, A. C. M., Schotanus, P., and Nieuwstadt, F. T. M. 1983. Surface layer similarity under nonuniform fetch conditions. *Journal of Climate and Applied Meteorology* 22:1800-1810.

Chen, C. 1982. A study of texture classification using spectral features. *Proceedings, 6th International Conference on Pattern Recognition,* p. 1074. Munich: IEEE Computer Society Press.

Ching, J. K. S., Clarke, J. F., and Godowitch, J. M. 1983. Modulation of heat flux by different scales of advection in an urban environment. *Boundary-Layer Meteorology* 25:171-191.

Ford, E. D. 1976. The canopy of a scots pine forest: Description of a surface of complex roughness. *Agricultural Meteorology* 17:9-32.

Garratt, J. R. 1978. Flux profile relations above tall vegetation. *Quarterly Journal of the Royal Meteorological Society* 104:199-211.

Grimmond, C. S. B. 1988. *An Evaporation-Interception Model for Urban Areas.* Ph.D. Thesis, University of British Columbia.

Oke, T. R. 1984. Methods in urban climatology. In *Applied Climatology, 18,* Zürcher Geographische Schriften 14, eds. W. Kirchhofer, A. Ohmura, and H. Wanner, pp. 19-29. Zurich: Verlag d. Fachvereine.

Schmid, H. P. 1988. *Spatial Scales of Sensible Heat Flux Variability: Representativeness of Flux Measurements and Surface Layer Structure over Suburban Terrain.* Ph.D. Thesis, University of British Columbia.

Schmid, H. P., Cleugh, H. A., Grimmond, C. S. B., and Oke, T. R. 1991. Spatial variability of energy fluxes in suburban terrain. *Boundary-Layer Meteorology* 54:249-276.

Steyn, D. G., Oke, T. R., Hay, J. E., and Knox, J. C. 1981. On scales in meteorology and climatology. *Climatological Bulletin* 30:1-8.

Heat-Related Mortality: What Might Happen in a Warmer World?

Laurence S. Kalkstein
University of Delaware

The impact of variable climate on human health and well-being is a topic of increasingly intensive study, partially because of the potential threat of an anthropogenic-induced global warming. The majority of the climate/human health evaluations to date concentrate on mortality, as these data are not difficult to obtain for the United States or Canada, and they are much easier to interpret than most sources of general morbidity data, such as hospital admission tallies or visits to the doctor.

Two findings involving weather/mortality relationships are relevant to geographical analysis and require further discussion in light of the large-scale climatic change foreseen by the scientific community. First, mortality from all causes appears to increase dramatically during very hot weather, and in some cases, daily deaths may be double mean or baseline values (Kalkstein and Davis 1989; White and Hertz-Picciotto 1985). Second, these mortality increases are not uniform throughout the United States, and results of previous research strongly suggest that people around the country exhibit a differential sensitivity to weather, especially in summer (Kalkstein 1989; States 1977).

This essay evaluates inter-regional differential human response to heat-related mortality and assesses its possible implications if the earth warms over the next century as many predict. It concentrates on results from 10 urban points located within various climatic regimes. An evaluation of this type requires an urban point focus for two reasons. First, densely populated urban centers with high daily mortality totals confined to small areas ensure that the daily variability of mortality is not primarily attributable to a single origin, such as a car accident killing six people. Such a single-source mortality cause would obscure the potential

mortality signal that is related to heat-stress. Second, an urban point evaluation maximizes the probability that the daily weather regime is homogeneous within the region. A larger-scale study raises the possibility that the weather could vary considerably across the region on a given day, further obscuring data evaluation.

If we consider each urban point as a microcosm of a particular climatological milieu, results from each point permit a broader interpretation of climate/mortality/global warming implications on a national scale.

Inter-Regional Mortality Responses to Hot Weather

Studies have shown that oppressively hot weather imposes great stress on the physiological responses of the human body, and this is manifested in increased mortality during these conditions (for example, Born et al. 1990; Jones et al. 1982). However, the degree of inter-regional response was vastly underestimated until recently, when a U.S. Environmental Protection Agency funded study of weather-related mortality for 50 U.S. cities indicated a systematic disparity across climatic regions (Kalkstein and Davis 1989).

The Environmental Protection Agency study demonstrated that many cities in the northeastern and midwestern United States show a sharp rise in total mortality during unusually hot weather, and in some cases, daily mortality can be more than double baseline levels when the weather is oppressive. But mortality rates rise only after a "threshold maximum temperature" is surpassed, and this threshold varies regionally based on the frequency of very hot weather. For example, the five cities that demonstrate the greatest rise in mortality during very hot conditions are New York, Chicago, Philadelphia, Detroit, and St. Louis (Kalkstein 1989). The threshold temperatures for these cities are: New York: 33°C; Chicago: 32°C; Philadelphia: 33°C; Detroit: 32°C; and St. Louis: 36°C. The magnitude of the threshold temperature appears to be related to its frequency across these cities; for example, a maximum temperature of 36°C in St. Louis occurs with approximately the same frequency as a maximum temperature of 32°C in Detroit. This strongly suggests that the notion of a "heat wave" is relative on an inter-regional scale, and depends upon the frequency of a given maximum temperature.

However, evidence also refutes this notion. Mortality rates in the southern and southwestern United States do not seem to be affected by weather in the summer, no matter how high the temperature. For example, Dallas, Atlanta, New Orleans, Oklahoma City, and Phoenix show little change in mortality even during the hottest weather. In fact, threshold temperatures for these cities are virtually impossible to define. One explanation may involve the variance in summer temperatures between the two regions. In the northern and midwestern cities, the very hot days or episodes are imbedded within periods of cooler weather. Thus, the physiological and behavioral "shock value" of a very high temperature episode is quite high. This point is further substantiated because most of the high mortality days occur during hot weather early in the summer season. Thus, the first or second heat episodes of June or early July are much more critical than a comparative episode in August. In the southern cities, the hottest periods are less unique and do not vary as much from the mean, possibly diminishing the impact of very hot episodes on human mortality.

The use of a new automated airmass-based synoptic procedure to evaluate weather/mortality relationships has supported and expanded the findings from the threshold temperature research (for a discussion of synoptic index development refer to Kalkstein et al. 1990). The synoptic procedure is designed to classify meteorologically homogeneous days into air mass categories. Thus, the synergistic relationships between numerous weather elements that comprise an air mass can be evaluated simultaneously, representing a significant improvement over an individual weather element approach.

The synoptic procedure was applied to 10 U.S. cities in different climates to determine inter-regional mortality/weather sensitivities: Atlanta, Boston, Chicago, Dallas, Memphis, New York, Philadelphia, St. Louis, San Francisco, and Seattle. For many of these cities, a single offensive summer synoptic category, which possessed a much higher mean mortality than the other categories was identified (Table 1). Although category 6 in St. Louis was only slightly cooler than category 9, category 9 alone appears to have exceeded an oppressive threshold associated with very high mortality.

A comparison of results from the 10 cities is instructive (Table 2). Most of the seven cities with a moderate or strong weather/mortality signal (as de-

Table 1. Mean Daily Mortality for Each Summer Synoptic Category for St. Louis[a]

Category	Whites	Non-Whites	Total	Elderly
1 Cool, continental	69	25	94	60
2 Transitional	71	29	100	63
3 Maritime	71	27	98	63
4 Overrunning	68	25	93	59
5 Maritime tropical, cloudy	72	28	100	64
6 Maritime tropical, sunny	73	30	103	64
7 Transitional to maritime	70	27	97	60
8 Frontal wave	70	25	95	60
9 Oppresive, tropical	88	41	129	84
10 Cold front passage	68	25	93	59
Overall mean, excluding category 9	70	27	97	62
Percentage that category 9 exceeds overall mean	26	52	33	36

[a] Values represent a one-day lag between synoptic category occurrence and mortality response.

Table 2. A Comparison of the Relative Impacts of Weather on Mortality for 10 Cities

Strong weather/mortality signal[a]	Boston Memphis New York City Philadelphia St. Louis
Moderate weather/mortality signal	Chicago San Francisco
No weather/mortality signal	Atlanta Dallas Seattle

[a] To have a moderate or high weather/mortality signal, a city must possess a synoptic category with unusually high mean daily mortality. In this analysis, this was determined subjectively.

termined by the existence of an oppressive synoptic category associated with unusually high mortality) were in the Northeast or Midwest (Memphis is a notable exception). The San Francisco area, which experiences infrequent but sometimes persistent hot weather, is also associated with this group. Of the three cities with no weather/mortality signal, two are in the South. The other, Seattle, possessed virtually no air mass types associated with very hot weather. Thus, this inter-regional disparity in response is supported within the synoptic evaluation.

Implications for a Large-Scale Global Warming

What do these mortality/weather responses imply for the warmer earth that many believe will exist in the 21st century? The answer depends primarily on how the population acclimatizes to this change in conditions.

Assuming that people react to weather in a warmer world much as they do today (implying little or no acclimatization), it is likely that deaths from heat stress-related causes will increase dramatically. Thus, the Environmental Protection Agency has reported to Congress that we might expect a sevenfold increase in heat-related deaths by the middle of the 21st century if acclimatization does not occur (Kalkstein 1989; Smith and Tirpak 1989). This would render heat-related deaths as a major killer, rivaling the current number of deaths from leukemia. The greatest brunt of this increase would be borne in the northeastern and midwestern United States, and it is estimated that the number of heat-related deaths in New York City would rise from the present 320 during a typical summer today to 1,743 during a typical summer in the mid-21st century (Smith and Tirpak 1989). Under such

circumstances, the number of days exceeding the threshold temperature would increase at least twofold, contributing to this dramatic rise.

Of course, a more likely scenario is that some degree of acclimatization will occur, as the gradual increase in warmth over the next 60 years permits time for certain social and physiological adjustments. A satisfactory means to account for this acclimatization has eluded the scientific community, although an attempt was made in the Environmental Protection Agency Report to Congress on the potential effects of global climate change (Smith and Tirpak 1989). Analog cities that possess present-day weather most duplicative of predicted mid-21st century weather for each evaluated city were established to account for full acclimatization. For example, the use of a climate-change scenario developed from a global circulation model to estimate weather in Atlanta in the mid-21st century will produce a regime that approximates the present weather in New Orleans (Hansen *et al.* 1988). Since New Orleans residents are fully acclimatized to this regime, the weather/mortality relationships that exist today for New Orleans can be used for the mid-21st century estimated climate in Atlanta to account for full acclimatization.

Employing this acclimatization procedure, new estimates were developed for the number of heat-stress-related deaths in the mid-21st century, yielding very different results from the unacclimatized estimates. Rather than the sevenfold increase developed for the unacclimatized model, less than a twofold increase results when accounting for acclimatization. In fact, many cities actually show a decrease in heat-related mortality if the population fully acclimatizes to a climate change. For example, the number of heat-related deaths that occur in an average summer today in Los Angeles is 84. If residents of Los Angeles fully acclimatize to the increased warmth expected in the mid-21st century, the number of heat-related deaths is estimated to drop to virtually zero (Smith and Tirpak 1989). Astoundingly, if residents of Los Angeles do not acclimatize at all, the number of heat-related deaths estimated for an average summer in the mid-21st century rises to 1,570! Similar disparities between acclimatized and unacclimatized estimates are noted for many northern and midwestern cities such as Detroit, New York, and St. Louis.

This wide variation has much to do with the technique used in this evaluation to account for acclimatization. Since the climate analogs for the northern cities are southern cities and, considering that southern populations respond much less dramatically to heat, it is not surprising that the fully acclimatized mortality estimates are so low. This points to some major shortcomings within our acclimatization procedure. First, the only similarities drawn between the target and analog cities relate to climate alone, with no consideration of possible urban structural or architectural disparities. For example, much of the low income urban population in the South, who are especially vulnerable to heat stress because they lack air conditioning or other amenities, reside in small frame houses often referred to as "shotgun shacks" (Kniffen 1965). Although far from luxurious, these residences are well-adapted to the rigors of a southern summer; they are often light colored with metal (reflective) roofs and well-ventilated with windows or doors on four sides. In contrast, most northern urban poor live in tenements constructed of dark-colored brick, black tar roofs, usually with only front and rear windows associated with a row house motif. During very hot conditions, these northern dwellings suffer from poor ventilation and absorb considerably more radiation than their southern counterparts. These types of factors might partially explain the differential mortality response between northern and southern populations.

Assuming that people can physiologically adapt to the predicted increasing warmth, we can expect only partial acclimatization to occur, because it is highly unlikely that the architectural makeup of the urban area, particularly dwellings of the poor, will change significantly over the next 50 to 75 years.

Demographic composition will counter acclimatization to varying degrees since analog and target cities may differ significantly. For example, our analysis does not account for the proportion of elderly (who are particularly vulnerable to heat-related mortality) and minorities. Furthermore, no attempt has been made to determine what these proportions might be in the target cities in the mid-21st century, although the percentage of elderly people (older than 65 years) is expected to increase over the next 75 years.

The report to Congress also includes an estimate of mid-21st century mortality assuming partial acclimatization (Smith and Tirpak 1989). The assumption here is that the apparent full acclimatization, which currently exists in the South, will not occur in a warmer North for two reasons: any architectural changes in dwellings will lag considerably behind the warming itself, rendering many residences unsuitable for the increasing warmth; and the demographic composition of these cities will be comprised of a larger proportion of people more vulnerable to heat-related stresses.

Finally, it is possible that the expected 1.5°C to 4.0°C warming by the mid-21st century might actually increase temperatures in many southern and southwestern cities to levels beyond the tolerance of our society. For example, the average annual number of days in Atlanta that exceed 32°C today is 17. This number is expected to reach 53 by the mid-21st century according to the Goddard Institute for Space Studies doubled carbon dioxide scenario (Hansen et al. 1988). Even more troubling is the number of days with temperatures exceeding 38°C. Currently, cities such as Memphis, Jackson, and Birmingham experience three such days during an average summer. This will increase to about 20 days (Titus 1989). Considering similar increases in the number of very hot days in Phoenix and other southwestern cities, certain areas might simply become uninhabitable in a warmer mid-21st century. Of course, this is highly speculative, as we have no examples to support this notion.

To improve upon present attempts to account for acclimatization, the Climate Change Division of the Environmental Protection Agency has developed a global warming/human health program that emphasizes an improved understanding of human acclimatization in a warmer world. The program will concentrate on expected social and cultural adjustments, including demographic alterations resulting from migration and other factors (Kalkstein 1990).

The extrapolation of this study to the larger scale will render this work more useful to the policy community, which ultimately determines the course of action to mitigate the potential global warming problem. Speculation in research of this type is extensive. Beyond the uncertainties of acclimatization, this evaluation assumes that estimates from global circulation models, which assume that the world will be considerably warmer, are reasonably accurate. This leap of faith has prodded many skeptics to consider climate impact research of this type virtually useless. However, there is certainly the possibility that humans in a warmer world will respond in a fashion somewhat similar to what has been described here, and global warming/impact research should be considered an "insurance policy" to be drawn upon if signs of a large-scale warming begin to appear.

References

Born, W., Happ, M. P., Dallas, A., Reardon, C., Kubo, R., Shinnick, T., Brennan, P., and O'Brien, R. 1990. Recognition of heat shock proteins and cell function. *Immunology Today* 11:40-43.

Hansen, J., Fung, I., Lacis, A., Lebedeff, S., Rind, D., Ruedy, R., and Russell, G. 1988. Global climate changes as forecast by the GISS 3-D model. *Journal of Geophysical Research* 93:9341-9364.

Jones, T. S., Liang, A. P., Kilbourne, E. M., Griffin, M. R., Patriarca, P. S., Wassilok, G. G., Mullan, R. J., Herricek,

R. F., Donnel, H. D., Jr., Choi, K., and Thacker, S. B. 1982. Morbidity and mortality associated with the July, 1980 heat wave in St. Louis and Kansas City. *Journal of the American Medical Association* 247:3327-3330.

Kalkstein, L. S. 1989. The impact of CO_2 and trace gas-induced climate changes upon human mortality. In *The Potential Effects of Global Climate Change on the United States: Appendix G—Health*, eds. J. B. Smith and D. A. Tirpak, pp. 1-12-1-35. Washington, DC: U.S. Environmental Protection Agency.

———. 1990. Climate change and public health: What do we know and where are we going? *Environmental Impact Assessment Review* 10:383-392.

———, and Davis, R. E. 1989. Weather and human mortality: An evaluation of demographic and interregional responses in the United States. *Annals of the Association of American Geographers* 79:44-64.

———, Dunne, P. C., and Vose, R. S. 1990. Detection of climatic change in the western North American arctic using a synoptic climatological approach. *Journal of Climate* 3:1153-1167.

Kniffen, F. B. 1965. Folk housing, key to diffusion. *Annals of the Association of American Geographers* 55:549-577.

Smith, J. B., and Tirpak, D. A., eds. 1989. *The Potential Effects of Global Climate Change on the United States*. Washington, DC: U.S. Environmental Protection Agency Office of Policy, Planning and Evaluation.

States, S. J. 1977. Weather and deaths in Pittsburgh, Pennsylvania: A comparison with Birmingham, Alabama. *International Journal of Biometeorology* 21:7-15.

Titus, J. G. 1989. Regional studies: Southeast. In *The Potential Effects of Global Climate Change on the United States*, eds. J. B. Smith and D. A. Tirpak, pp. 3-23-3-58. Washington, DC: U.S. Environmental Protection Agency.

White, M. R., and Hertz-Picciotto, I. 1985. Human health: Analysis of climate related to health. In *Characterization of Information Requirements for Studies of CO_2 Effects: Water Resources, Agriculture, Fisheries, Forests and Human Health*, ed. M. R. White, pp. 172-205. DOE/ER-0236, CO_2 Research Division, U.S. Department of Energy, Washington, DC.

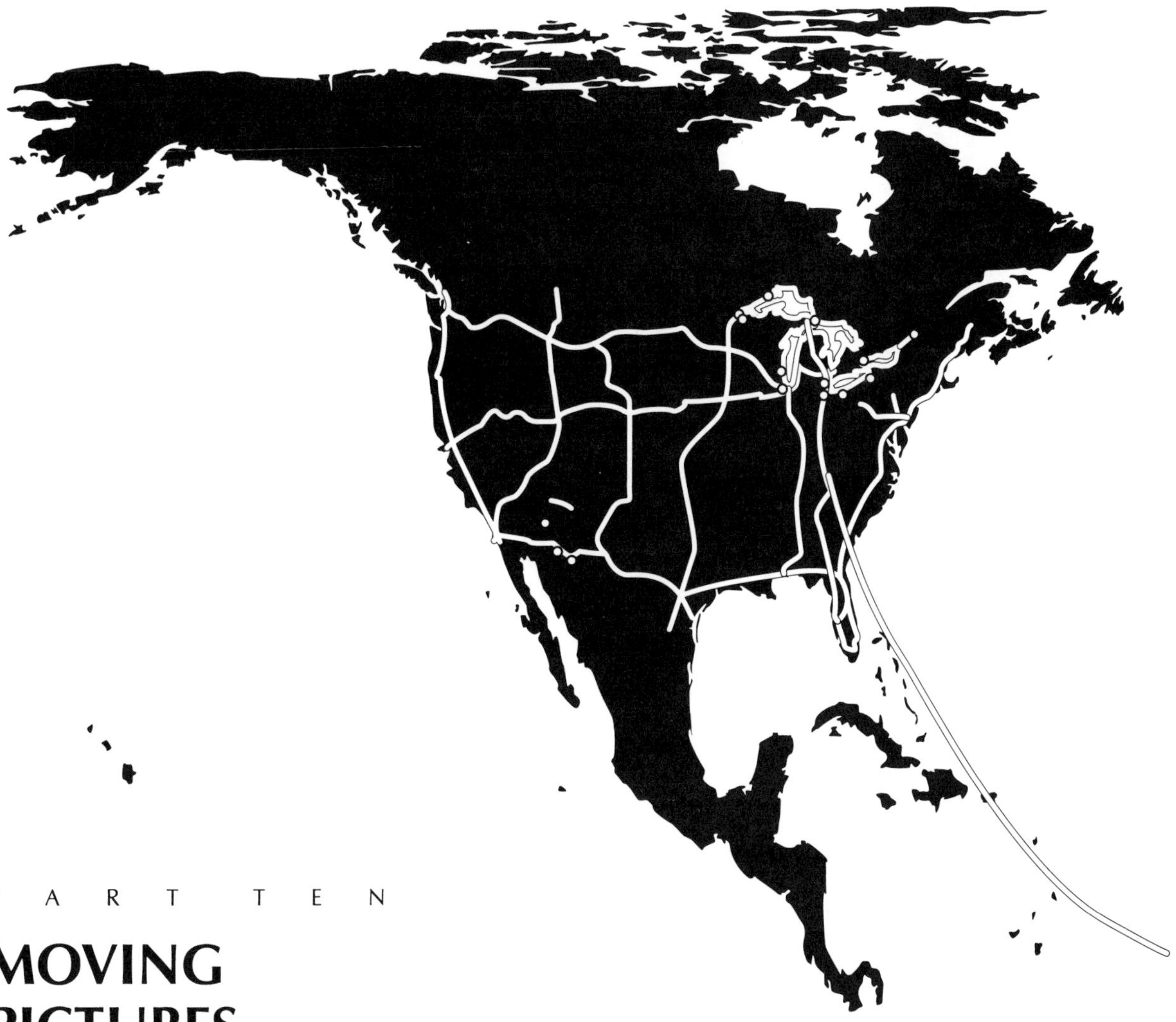

P A R T T E N

MOVING PICTURES

Geographic landscapes are seldom permanent—even continents drift into new locations over eons. But some movements, for example, hurricanes and dust storms, are more dramatic and sudden in their results. Winds, waves, and flows of water down hillslopes and through stream channels influence the long-term character of landforms and imbed their lines and points on the continent. In seeking to understand the action of these agents, physical geographers conduct empirical research, perform computer and field simulations, and develop predictive models. Human geographers employ similar tools to explain the transfers of people, goods, resources, and information along transportation and communication lines. These transfers not only reflect the day-to-day activities of life in North America, but structure the long-term economic linkages within and among regions. It is fitting to end this book on the road, with snapshots focused on highways. These concrete byways are arguably the dominant causal factor in the changing human occupance of much of North America and epitomize the entrenchment of technology and machines on the changing cultural landscapes of contemporary North America. Is the highway an icon?

—Athol D. Abrahams, State University of New York at Buffalo

Hurricane Hugo's Trajectory

James B. Campbell
Bonham C. Richardson
*Virginia Polytechnic Institute
and State University*

The global symmetry of hurricane tracks is an example of spatial order in nature. The distinctive, curving, poleward movements of these immense late-summer storms in both hemispheres illustrate a balance, even harmony, in the earth-atmosphere system. In North America, people (including possible hurricane victims) from Yucatán to Boston are at least vaguely aware of the general path a hurricane takes each autumn; they know that it usually enters the Caribbean region from the southeast, veers generally clockwise through the Gulf of Mexico, and then often parallels the southeastern coast of North America before dissipating its remaining energy and eventually disappearing in the North Atlantic.

Geographical awareness of North American hurricane tracks began when the aboriginal Caribs of the Lesser Antilles (whose word *furacan* is the origin of "hurricane") learned to predict a hurricane's arrival by observing cloud formations. Later, colonial planters of sugarcane, cotton, and tobacco throughout the circum-Caribbean realm all came to dread the possibility of a furious visitation by an autumn hurricane. Today, satellites track hurricanes westward from their genesis off West Africa and warn residents of Atlantic America of approaching storms. Now West Indians, Mexicans, and North Americans routinely monitor the paths of particular hurricanes through radio and television reports, their personal interests, of course, varying as to whether their own, or someone else's, locality will be affected.

Through modern storm-tracking technology, the route of a hurricane is information shared across national boundaries, thereby linking peoples of widely varying backgrounds and locations as victims and observers of a cumulatively catastrophic event. And along these routes, hurricanes often evoke the most vivid personal and collective recollections. Like declarations of war and presidential assassinations, the arrival of a hurricane throws into high relief one's otherwise mundane circumstances, especially if he or she has been caught in its path.

The path of Hurricane Hugo in September 1989, exemplifies the impact of large-scale hurricanes upon North America (Figure 1). The storm's origins lay, as do those of most tropical storms, in the atmosphere above the open ocean within a few degrees of latitude from the Equator. By late summer and autumn, tropical surface waters reach their maximum temperatures. Moisture evaporates easily, creating warm, moist, unstable air that is easily displaced upward. Then, as moisture condenses during ascent, release of the latent heat of evaporation powers further upward motion. This process sustains further vertical motion, provided the disturbance has continued access to a large body of warm water.

Hugo began in this manner in the Atlantic Ocean south of the Cape Verde Islands, 500 kilometers west of the African coastline. On 11 September the mass of cumulus clouds appeared on satellite images. As the depression moved westward it intensified; by the evening of 13 September Hugo's winds had reached hurricane velocities. Hugo's eye—the cloud-free region of descending air at the center of the storm—was recognizable on satellite images by 15 September. Hugo continued to intensify as it moved northwest, and reconnaissance aircraft measured sustained winds of 250 kilometers per hour. Thus, Hugo was near its maximum strength as it approached the Caribbean (Monk and Waters 1989).

Hugo's initial landfall occurred during the night of 16-17 September, when it hit the eastern half of Guadeloupe (Figure 2). The hurricane then headed

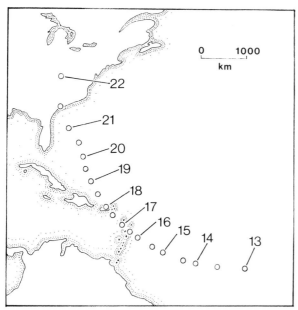

Figure 1. Hugo's path, 12-22 September 1989. Circles show positions of the eye at 12-hour intervals; numbers give noontime positions of the eye. Source: *New York Times* (24 September 1989:44).

across Grande-Terre, the low-lying limestone half of the French island. It ruined much of the recently planted sugarcane crop, downed power lines, and made roads impassable. Five persons were killed, 80 were injured, and an estimated 12,000 in Guadeloupe were left without shelter. Guadeloupe, an overseas department of France, received immediate aid from the metropolitan government. The French arrived shortly with emergency food supplies and relief workers aboard military aircraft. From Paris, the French Overseas Affairs Minister announced the immediate dispensation of the equivalent of U.S.$5.4 million in aid for Guadeloupe as well as a freeze on interest rates for local bank loans (Hevesi 1989a).

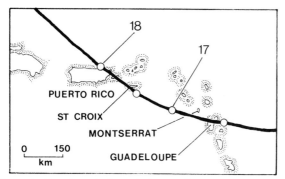

Figure 2. Hugo's path through the Antilles, 16-18 September 1989. Circles show positions of the eye at 12-hour intervals; numbers give noontime positions of the eye. Source: *New York Times* (22 September 1989:A22).

Next in Hugo's path was Montserrat, one of Britain's last Caribbean colonies. On the morning of Sunday, 17 September, Hugo's 250-kilometer-per-hour winds ripped through Montserrat, where residents had anticipated the storm. But warnings alone could not save Montserrat from being sent back to the "kerosene age"—as local islanders later described it—because nearly all the of tiny island's frame houses and public services were badly damaged by the high winds. Nearly everyone of the 12,000 Montserratians was left homeless, at least temporarily. The island's vegetation was similarly devastated, turning the landscape a dull brown and making Montserrat's nickname of "The Emerald Isle" a sudden anachronism. Amazingly, and probably because of the early warnings, Hugo killed only elderly Montserratians. As on Guadeloupe, metropolitan relief came quickly. Royal Air Force cargo planes brought supplies within a few days, and sailors and marines from the HMS *Alacrity* provided foodstuffs and also labor (Kifner 1989).

After departing Montserrat, Hugo churned relentlessly toward the northwest, heading for the U.S. Virgin Islands. There, as in neighboring British islands, the storm created enormous damage. On St. John, Hugo "ripped up" cottages on the island's two tourist campgrounds. St. John's main problems, however, were the temporary losses of electricity, telephones, and air service. Populous (53,000) St. Thomas was hit even harder. High winds blew out hotel windows in Charlotte Amalie, and some sailboats moored in St. Thomas's yacht harbors were thrown as far as 45 meters onto the shore (Hevesi 1989a).

But it was St. Croix, the largest of the U.S. Virgin Islands, that Hugo hit hardest, creating physical devastation that led to major social disruption. In the first hours of Monday, 18 September, the storm passed directly over Fredericksted, at the western end of the island. The town's pier was washed away and seafront stores were seriously damaged. Several major hotels sustained structural damage and they were unable to open during the ensuing 1989-1990 tourist season.

Even if all of St. Croix's hotels had remained open, it is unlikely that the island's ensuing tourist season would have boomed because of the looting in Hugo's wake, disturbances duly beamed into North American households by mobile television crews. St. Croix, the working island that describes itself as the "Connecticut of the Caribbean," also has marked racial and economic divisions that Hugo brought to a boil. Almost as soon as the storm arrived, crowds (mainly black) of Crucians began to raid stores and shops for food, clothing, and furniture. In a few instances they were thwarted by shotgun-wielding (mainly white) storeowners. On 20 September, U.S. President George Bush dispatched

1,000 military police to St. Croix to help restore order. Several U.S. Coast Guard cutters also headed to the island to evacuate terrified white residents and tourists (Joseph and Rowe 1990).

After daybreak on Monday, the storm was just east of Puerto Rico. It blew houses apart on the islet of Culebra and destroyed half of the 6,000 houses on Vieques. By late morning Hugo had descended on the northeastern corner of Puerto Rico, blowing down houses and destroying fishing villages' piers. Much of El Yunque National Forest, more than 11,000 hectares, was defoliated within minutes. Rural areas were hit hardest. Felled trees blocked dirt roads already made impassable by the mud-producing rains. Poles formerly supporting power lines were snapped in two by 210-kilometer-per-hour winds.

Continuing toward the north and west, Hugo's path created heavy damage in sections of San Juan (population 945,000). Directly east of the city, the hurricane dealt massive destruction to the shantytowns near the airport; it also demolished entire shops and took out windows in the fashionable Condado Beach area nearby. Hugo had left a path of destruction throughout northeastern Puerto Rico and had dealt heavy blows to much of the island's interior agriculture as well, notably the coffee and poultry industries. Following Hugo's departure, Puerto Ricans totaled the damage: 50,000 homeless, 20,000 houses destroyed, 4 Puerto Ricans killed (a tribute to the local warning system); total property damage was estimated at U.S.$1.4 billion (Schmalz 1989).

By noon on the 18th Hugo headed out to sea. The storm's passage over Puerto Rico's mountains had temporarily reduced its intensity and internal wind speeds to about 170 kilometers per hour, conditions that varied little during the next 24 hours. As it departed San Juan, the hurricane's path became the focus of intense scrutiny by the U.S. Weather Service. Forecasters watched via satellite as Hugo assumed a more northerly direction, passing east of the Bahamas at a reduced linear speed of about 20 kilometers per hour. They also noted an intensification of thunderstorms surrounding the eye and anticipated that Hugo would strengthen dramatically when it reached the warm Gulf Stream that parallels the United States' Atlantic coastline. As Hugo continued north and west during 19-20 September, the National Hurricane Center in Coral Gables, Florida, urged residents from the Florida Keys to Cape Hatteras, North Carolina, to begin preparations for possible evacuation (Hevesi 1989a).

By 0600 hours on the 21st (about 18 hours before landfall), Hugo's pressure began to fall and its winds to increase, as forecasters had anticipated, as it approached the Gulf Stream. The storm then intensified

even more than had been anticipated due to atmospheric factors identified in retrospect. First, tropical storm Iris, trailing behind Hugo, dissipated, allowing additional air and moisture to flow into Hugo. Second, Hugo's track was guided by mid-latitude pressure and wind systems that confined its circulation and tightened wind patterns near the eye, thereby strengthening the storm. These same pressure and wind systems are thought also to have then guided Hugo's path to the west (Stevens 1989).

Throughout the 21st, Hugo, becoming steadily more powerful, continued to move north and west, approaching the South Carolina coastline at about 40 kilometers per hour. By noon the low pressure center had intensified, bringing maximum wind speeds above 190 kilometers per hour. The U.S. Weather Service now classified Hugo as a category 3 hurricane. By 1500 hours, maximum wind speeds reached 200 kilometers per hour, bringing the classification to category 4, and by 1800 hours, maximum wind speeds attained 220 kilometers per hour. Although by these measures Hugo was short of category 5, the most severe category, it had now become one of the most powerful hurricanes of the 20th century (Michaels 1990).

Officials now ordered the evacuation of low-lying coastal regions. More than a half million people were evacuated from coastal regions of Georgia and the Carolinas. As it became clear that Charleston, South Carolina, was in special danger, more than half of Charleston's population of 350,000 was evacuated. Those who stayed the city filled bathtubs with water, boarded up windows, and prepared for the worst. Military aircraft within the projected danger zone were dispersed inland to airfields as far away as Ohio (Hevesi 1989b).

If Hugo's advance had slowed near the coastline, cooler waters west of the Gulf Stream might have weakened the storm. Instead, its speed increased in the final stages of its approach toward the continental United States, allowing the storm to maintain its full intensity. Maximum-ground level wind speed was estimated at 223 kilometers per hour, and instruments in Charleston recorded a low pressure of 936.5 millibars; the storm surge exceeded 5 meters, after allowing for usual tides in the region (Michaels 1990).

The surge produced dramatic storm damage. In Charleston, 20 boats (including one 15-meter craft) had been thrown on shore. In the harbor the *Narwhal*, a U.S. Navy nuclear submarine, was torn from its moorings and escaped damage only by submerging. Other naval craft sank during the storm. In smaller communities on the South Carolina coast, piers were destroyed and residences severely damaged; some were washed into the water (Gruson 1989; Hevesi 1989c).

By daylight on Friday, 22 September, damage to Charleston and its surroundings became visible. Hugo brought more than 12 centimeters of rainfall to Charleston; some streets were waist-high in water and damaged by water and mud. During the night, high winds had torn roofs from structures and broken power lines. Fires were created at broken gas lines. Eleven deaths were recorded. While basic services were being restored, government officials ordered a 1800 hours to 0700 hours curfew and established National Guard patrols (Hevesi 1989c). Although Hugo's eye passed over Charleston, the most severe winds were in the sector north of the eye. By Saturday morning, the 23rd, seven South Carolina counties had been declared disaster areas, 17 in the state had died from the storm, and damage in South Carolina had been estimated at $5 billion.

In the meantime Hugo had followed a path westward into western North Carolina, then north, parallel to the coastline but well inland through western Virginia, West Virginia, Pennsylvania, and western New York. The path inland was fortunate, as it avoided the populated coastal regions, dissipated Hugo's force in the rugged topography of the mountains, and took Hugo far from the Atlantic coastline. Beyond the region near Charleston, the coastline suffered little beach erosion.

Nonetheless, Hugo brought considerable damage to interior regions. In the Carolinas more than 700,000 people were without electrical power due to Hugo's winds. Charlotte, North Carolina, experienced 125-kilometer-per-hour winds, and received 8 centimeters of rainfall. Eighty-five percent of the city lost power, and one of Charlotte's television towers was blown over. In western South Carolina and North Carolina, heavy rains in rugged mountainous topography caused mudslides and flooding. By 0600 hours on 22 September Hugo had weakened sufficiently to be declared a tropical storm, and by 1800 hours this designation was withdrawn. Throughout Friday the 22nd the storm had brought 10 to 20 centimeters of rain to a broad band from the Carolinas to New York (Hevesi 1989c).

During its nearly two-week odyssey, Hugo killed 71 people and damaged billions of dollars of property.

Afterward, the giant storm was labeled "the most destructive storm in U.S. history" (Morrison 1990, 503). The superlative is questionable; only the year before hurricane Gilbert, which raked Jamaica and Yucatán, was described as the "mightiest storm to hit the Western Hemisphere in this century" (Browne 1988, A1). Hurricanes are like that. They inspire superlatives that may or may not be accurate. What is certain however is that this coming autumn, and every late-summer and fall in the future, as they have during every autumn in the past, people from Barbados to eastern Canada will monitor the tracks of individual hurricanes that accent the geography of a vast area of North America.

References

Browne, M. W. 1988. Meteorologists say the hurricane is similar to a monster tornado. *New York Times* 15 September:A1, A16.

Gruson, L. 1989. Amid howl, spark, and spray Charleston endures the wind. *New York Times* 22 September:A1, A20.

Hevesi, D. 1989a. Hurricane spares Bahamian islands; heads for the U.S. *New York Times* 20 September:A1, B9.

———. 1989b. Full force of storm hits South Carolina. *New York Times* 22 September:A1, A21.

———. 1989c. Carolinas reeling from devastation left by hurricane. *New York Times* 23 September:A1, A29.

Joseph, G. I., and Rowe, H. M. 1990. *Hell under God's Orders: Hurricane Hugo in St. Croix—Disaster and Survival.* St. Croix: Winds of Change Press.

Kifner, T. 1989. Hit hardest by hurricane, Montserrat starts to rebuild. *New York Times* 22 September:A22.

Michaels, P. J. 1990. The Hugo story. *Southeastern Climate Review* 1:3-9.

Monk, G. A., and Waters, A. J. 1989. Satellite photograph: 21 September 1989 at 0930 GMT. *Meteorological Magazine* 118:248.

Morrison, D. 1990. United States. In *1990 Britannica Book of the Year*, pp. 499-503. Chicago: Encyclopaedia Britannica.

New York Times. 1989. 22 September:A22.

New York Times. 1989. 24 September:44.

Schmalz, J. 1989. Puerto Rico fears for its recovery from wounds the hurricane left. *New York Times* 1 October:1, 32.

Stevens, W. K. 1989. Flood seen as storm marches up coast. *New York Times* 22 September:A20.

Summer Dust Storms in the Arizona Desert

Melvin G. Marcus
Anthony J. Brazel
Arizona State University

Dust storms are a fact of life in Arizona. They scrape paint from buildings and vehicles, damage crops, choke engines, disrupt communications, set people to coughing, and generally affect human health. Every Arizona driver learns to fear dust storms. In a matter of minutes visibility can drop from several kilometers to a few meters, and winds can rise from 10 to 64 kilometers per hour or more. Fatal multi-vehicle accidents are frequent under these conditions, particularly in the summer. By unhappy spatial coincidence, human land-use practices, such as farm abandonment, can exacerbate these roadway dust hazards.

The Nature of Summer Dust Storms

Strong local winds, synoptic cold fronts, broad-scale regional winds, and micro-scale tornado-like dust devils all contribute to the ubiquity of dust in Arizona. The major dust storms are usually associated with cold air outbursts from large cumulonimbus cloud systems (Figure 1A). Summertime cumulonimbus clouds rise in excess of 16,000 meters, accompanied by high vertical velocities, lightning, and thunder. Dust is lifted and suspended by shifting winds and strong downdrafts along the leading edge of the storm systems. The turbulence created by air temperature differences and by the mechanical effects of topography on winds also produce dust. The major portion of dust is carried by the pseudo-cold front flowing from the base of the thundercell (Figures 1A and 1B). After the dust has passed a given point, heavy precipitation often occurs,

moistening the surface and preventing further lifting of the silt-sized dust particles.

These dust storms in south-central Arizona occur primarily in summer and early fall when there is a coincidence of high temperatures (35° to 45°C) and water vapor imported from either the Pacific Ocean or the Gulf of Mexico. In Phoenix, for example, 86 percent of all dust-related thunderstorms with visibility less than 1.6 kilometers (1 mile) occurred during June through September, with the majority in July and August (Figure 1C).

The onset of dust generally occurs in late afternoon, peaking from 1700 hours to 2000 hours, and diminishing rapidly toward midnight (Figure 1D).

Regional Movement of Dust Storms

Thunderstorm systems in south-central Arizona develop and move east to west; however, variance in this pattern is evident. As dust clouds pass roughly from east to west, they intersect with the north-south trending roads. For Phoenix, the peak movement of thunderstorm-related dust events from the southeast (Figure 2A) relates to the prevailing upper level flow (with its influx of regional vapor flow known locally as the "Arizona monsoon") and to the open topographic basin between Tucson and Phoenix.

From weather satellite images, Figures 2B to 2D show a schematic sequence of cloud movements for a day when the major thunderstorm movement was from the east and northeast. It begins with a strong onshore

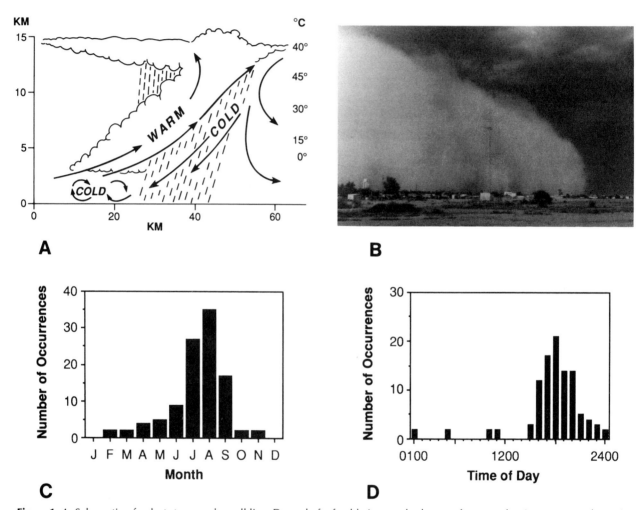

Figure 1. A. Schematic of a dust storm and squall line. Downdraft of cold air spreads along surface as a density current and entrains dust. B. Photograph of a dust wall front moving through the metropolitan Phoenix, Arizona, area winds left to right. C. Frequency of major dust storms by month at Phoenix Sky Harbor International Airport (1965 to 1980). D. Frequency of onset time of dust wall movement during the day past Phoenix Sky Harbor International Airport (1965 to 1980).

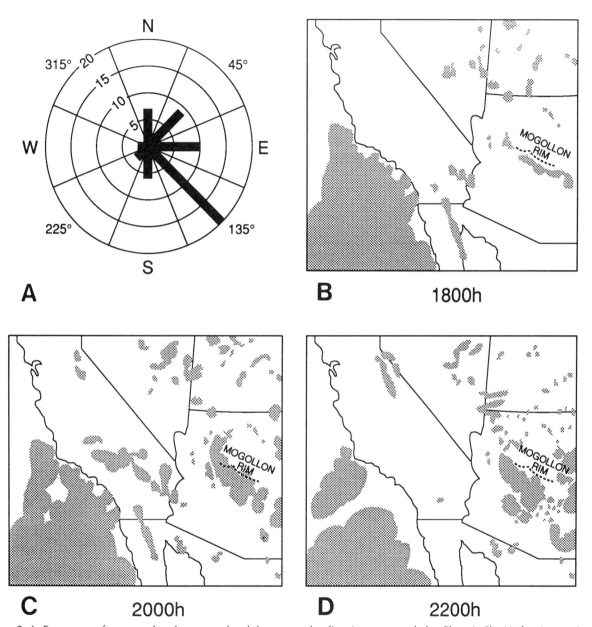

Figure 2. A. Frequency of summer thunderstorm-related dust storms by direction, as recorded at Phoenix Sky Harbor International Airport. B. Typical early evening cloud development pattern during summer in the Southwest United States. Note arc of clouds along the Mogollon Rim of Arizona. C. Mid-evening cloud thickening and further development in the uplands of northern Arizona. D. Late evening buildup and movement of massive thundercells away from the Mogollon Rim and across the lower desert region of southern Arizona.

flow of moist air from the Pacific Ocean, revealed by the classic offshore, cold current cloud cover. During the day, vapor moved across southern California and southwestern Arizona, pulled in by a surface low pressure over the southwest, and was lifted thermally and, as it approached the Mogollon Rim, orographically as well. By 1700 hours, an arc of developing cumulonimbus clouds bordered the Colorado Plateau (Figure 2B).

Once the sun was below the horizon, the southwesterly vapor flow subsided and the still-growing cloud systems moved both downslope and to the southwest under the control of upper air circulation (Figures 2C and 2D). The storms were mature by this time and cloud tops exceeded 14,000 meters. As thundercells merged and as associated high winds from downdrafts progressed across the heavily populated region of central Arizona in the evening hours, blinding walls of dust cross major highways. In Phoenix, visibility has dropped from 40 kilometers to less than 1.6 kilometers in a matter of minutes and winds have accelerated from 10 kilometers per hour to over 64 kilometers per hour. In similar circumstances during the summer of 1975, 35 people were injured and 2 killed south of Phoenix (Figure 3).

Dust Flow and the Human Factor

The dust-storm problem in southern Arizona is induced, in part, by human modifications of the desert landscape (Wilshire *et al.* 1981). This is illustrated for the highway corridor near Eloy, Arizona. In this area, cotton, alfalfa, and other crops have been grown for decades; but, in recent years, much of the farmland has been abandoned as groundwater drawn-down has become severe, and as regional and national economic factors have changed. The construction of interstate highways in the late 1960s, which interrupted irrigation systems, was also responsible for some of the abandonment. The frequency of accidents along this Phoenix-to-Tucson highway corridor prompted the Arizona Department of Transportation to initiate a Dust Storm Alert System in 1975.

In research funded by the Arizona Department of Transportation, M. G. Marcus (1976) found that severe dust storms are initiated by regional atmospheric flow across the highway network and that the severity of blowing dust escalated significantly along stretches of the highway bounded by disturbed and abandoned farmland. From meteorological effects alone, the probability of accidents along the highway corridor from Phoenix to Tucson would be nearly random; however, the accident distribution shown in Figure 4 is clustered and largely associated with nearby abandoned lands (Hyers and Marcus 1981).

A

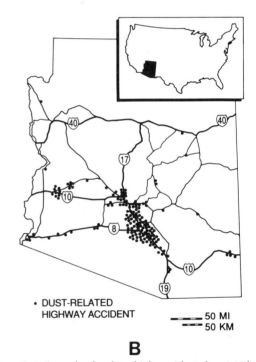

B

Figure 3. A. Example of multi-vehicle accident along I-10 between Phoenix and Tucson, Arizona, during a major dust storm event in summer. B. Dust-related accidents along Arizona highways (1960 to 1970).

Abandoned farmlands are prime areas for surface disturbance. They are flat and mostly devoid of vegetation cover. Thus, off-road vehicles, livestock, and other traffic easily disturb the surface and make it susceptible to aeolian transfer of dust. The quantity of dust blown across the highway is not only a function of regional wind potential, but is also related to antecedent anthropogenic disturbances of the landscape.

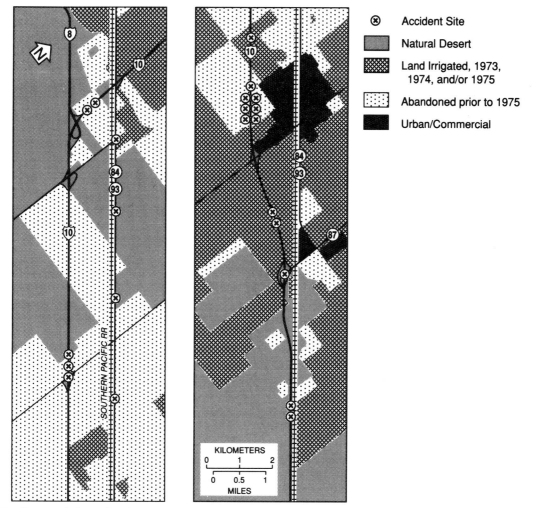

Figure 4. Land use and dust-related accident cluster distribution along two contiguous segments of I-10 between Phoenix and Tucson, Arizona. The right panel affixes to bottom of the left panel. Note accident clusters near abandoned land.

For instance, implementation of protective land-use practices correlates with a decrease of dust-related accidents along the Phoenix-Tucson corridor.

Local- and short-term problems are only part of the dust factor in Arizona. Seasonal and annual climate, soil moisture, wind frequency above critical thresholds, and variable vegetation coverage exert strong controls over the generation of blowing dust (Brazel and Nickling 1986, 1987). Against the short-term series of rapidly changing events, a backdrop of slower change in climate and landscape use must be better understood to address the expected dimension of the dust hazard in years to come.

References

Brazel, A. J., and Nickling, W. G. 1986. The relationship of weather types to dust storm generation in Arizona (1965-1980). *Journal of Climatology* 6:255-275.

_____ . 1987. Dust storms and their relation to moisture in the Sonoran-Mojave desert region of the south-western United States. *Journal of Environmental Management* 24:279-291.

Hyers, A. D., and Marcus, M. G. 1981. Land use and desert dust hazards in central Arizona. In *Desert Dust: Origin, Characteristics, and Effect on Man*, Special Paper 186, ed. T. L. Pewe, pp. 267-280. Boulder, CO: Geological Society of America.

Marcus, M. G. 1976. *Evaluation of Highway Dust Hazards along Interstate Route 10 in the Casa Grande-Eloy Region*, Research Paper No. 3. Tempe: Center for Environmental Studies, Arizona State University, Final Report for the Arizona Department of Transportation.

Pye, K. 1987. *Aeolian Dust and Dust Deposits*. London: Academic Press.

Wilshire, H. G., Nakata, J. K., and Hallet, B. 1981. Field observations of the December 1977 wind storm, San Joaquin Valley, California. In *Desert Dust: Origin, Characteristics, and Effect on Man*, Special Paper 186, ed. T. L. Pewe, pp. 233-252. Boulder, CO: Geological Society of America.

Overland Flow on Semiarid Arizona Hillslopes

Athol D. Abrahams
State University of New York at Buffalo

Anthony J. Parsons
University of Keele

On semiarid hillslopes virtually all runoff occurs in the form of overland flow generated when rainfall intensity exceeds surface infiltration rate. Overland flow gives rise to a variety of hydrologic and geomorphic phenomena, such as flooding and soil erosion, and, in the longer term, plays an important part in the geomorphic development of semiarid landscapes. This paper focuses on the hydraulics of overland flow and their implications for modeling hillslope runoff and soil erosion. The ability to model these processes is essential not only to understand the contemporary operation of these hillslopes but also to comprehend their past history and their future response to environmental change.

The hydraulics (that is, depth and velocity) of overland flow at a given rate of flow are a function of the resistance to flow offered by the ground surface. Because overland flow may be laminar or turbulent, flow resistance is best expressed by the dimensionless Darcy-Weisbach friction factor:

$$ff = 8gds/v^2 \tag{1}$$

where g is the acceleration of gravity, d the mean depth of flow, s the energy slope, and v the mean flow velocity. In overland flow ff may be divided into grain resistance exerted by soil particles and microaggregates and form resistance offered by microtopographic protuberances, stones, and vegetation.

Resistance to flow generally varies with rate of flow, which is represented by the dimensionless Reynolds number:

$$Re = 4vd/v \tag{2}$$

where v is the kinematic fluid viscosity. Laboratory experiments and theoretical analyses since the 1930s have shown that the power relation between ff and Re for shallow flow over a plane bed where ff is due entirely to grain resistance is a function of the state of flow. The relation has a slope of -1.0 where the flow is laminar and a slope between -0.25 (for a hydraulically smooth boundary) and 0 (for a hydraulically rough boundary) where it is turbulent.

Overland flow may be conveniently divided into rill and interrill flow, of which only the latter is examined here. Interrill overland flow is generally modeled as a sheet of water uniform in depth and velocity across the slope and increasing in depth and velocity downslope. This practice has permitted virtually all models of hillslope runoff to use the conventional relation between ff and Re (or surrogates thereof) for shallow flow over a plane bed, and most models of soil erosion to employ flume-based sediment-transport equations. Implicit in such models is the assumption that ff is due entirely to grain resistance.

In reality, however, interrill overland flow is not uniform in depth and velocity across the slope but contains threads of deeper, faster flow that diverge and converge around microtopographic protuberances, rocks, and vegetation. Clearly, form resistance is important and must affect the shape of the ff-Re relation. Yet no field studies have been aimed at measuring form resistance, and little field data exist on the shape of ff-Re relations (Dunne and Dietrich 1980; Emmett 1970; Roels 1984). To correct these deficiencies, two sets of field experiments were conducted on semiarid hillslopes in southern Arizona. This paper summarizes some of the results of these experiments and discusses their implications for modeling hillslope runoff and sediment transport.

Field Site

The field experiments were conducted at Walnut Gulch Experimental Watershed, Tombstone, Arizona (31°43′N, 110°41′W). Toward the end of the last century, the vegetation over much of the watershed underwent a transition from a grass- to a shrub-dominated community. A similar transition has been recorded over many parts of the northern Sonoran Desert (for example, Hastings and Turner 1965). As the shrubs grew and the understory grasses thinned, the A horizon of the preexisting soil was selectively eroded in the intershrub areas by a combination of rain-splash and overland flow, leaving behind a gravel lag. Under many of the larger shrubs, the A horizon has been preserved. In addition, under virtually all shrubs with moderate to dense canopies, sand-sized particles have accumulated as a result of differential splash—that is, the transport by rain-splash of more sediment from intershrub to shrub areas than in the reverse direction due to the shrub canopy protecting the ground surface beneath it from raindrops. The result is a mosaic of gravel-covered intershrub areas interspersed with shrub-protected areas of fine particles forming microtopographic highs (Parsons *et al.* in press).

The experiments were all performed on the interrill portions of shrub-covered, piedmont hillslopes underlain by well-cemented, coarse Quaternary alluvium. The shrub community is dominated by creosote bush, sandpaper bush, sotol, whitethorn, and yucca with a ground layer principally of desert zinnia and bitterweed. The vegetation typically covers less than half the ground surface, and almost all the shrubs stand atop mounds. These mounds contain a higher proportion of sand and larger quantities of organic matter than adjacent gravel-covered intershrub areas. They are also subject to more digging and burrowing by animals, notably rabbits and pack rats. These factors, along with the effect of the shrub canopy itself in dissipating the kinetic energy of raindrops, and thereby inhibiting surface sealing, cause infiltration rates to be much higher beneath shrubs than between them (Abrahams and Parsons 1991). As a result, overland flow is generated preferentially in the gravel-covered intershrub areas and travels downslope in well-defined threads, largely confined to areas that form swales between shrub-protected mounds (Parsons *et al.* 1990). The runoff plots used in the field experiments, therefore, were all located in intershrub areas, and overland flow was simulated by trickling water onto the plots.

Overland Flow Hydraulics

The first set of experiments was performed on six plots 1.8 meters wide and 5.5 meters long that ranged in gradient up to 33°. Although gravel mantled the plot surfaces, clipped plant stems occupied as much as 10 percent of their area, and the steeper plots had quite irregular surfaces. Analyses of 14 cross sections yielded *ff-Re* relations that were positively sloping, negatively sloping, and convex upward (Figure 1). These shapes are attributed to the progressive inundation of the roughness elements (gravel, plant stems, and microtopographic protuberances) that impart form resistance. So long as these elements are emergent from the flow, *ff* increases with *Re* as the upstream wetted projected area of the elements increases. However, once the elements become submerged, *ff* decreases as *Re* increases and the ability of the elements to retard the flow progressively decreases (Abrahams *et al.* 1986).

These field experiments show that the conventional *ff-Re* relation for shallow flow over a plane bed does not apply to the semiarid hillslopes under study. Furthermore, they suggest that the shape of the relevant

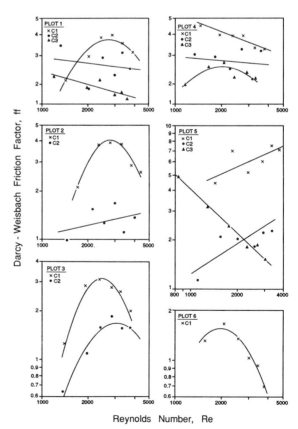

Figure 1. Graphs of the Darcy-Weisbach friction factor against the Reynolds number for up to three cross sections on each of six runoff plots. After Abrahams *et al.* (1986).

ff-Re relation is not a function of the state of flow, as is the case for the conventional relation in which *ff* is due to grain resistance, but of the surface form, which implies that ff is dominated by form resistance. The same is likely to be true on other semiarid hillslopes with gravel-covered surfaces. Consequently, such hillslopes might be expected to have *ff-Re* relations similar in shape to those in Figure 1, and the conventional relation should not be employed in runoff models for such hillslopes.

Modeling Runoff

There is, however, a more general problem with the use of *ff-Re* relations in runoff models for semiarid hillslopes. As Figure 1 illustrates, owing to the dependence of *ff* on surface form, the shape of the *ff-Re* relation changes as surface form changes across a hillslope, severely limiting the value of any *ff-Re* relation in such models. A possible solution to this problem is to replace the *ff-Re* relation with a multivariate relation that includes surface form variables. To explore this possibility, a second set of experiments was conducted on eight plots 0.61 meters wide and 1.5 meters long all with gradients less than 5° (Abrahams and Parsons in press). These plots were sited in intershrub areas with no plants and imperceptible microtopography to contribute to form resistance. Form resistance therefore derives entirely from the surface gravel, which was represented in the analysis by gravel size and concentration.

Stepwise multiple regression analyses based on data from 73 experiments on the eight plots yielded the equations:

$$\log ff = -5.960 - 0.306 \log Re \\ + 3.481 \log \%G + 0.998 \log D_G \quad (3)$$

with $R^2 = 0.61$, and

$$\log ff = 0.699 - 1.091 \log Re \\ + 3.151 \log \%G + 1.316 \log R_S \quad (4)$$

with $R^2 = 0.81$. In these equations D_G is the arithmetic mean size of gravel-sized particles (greater than or equal to 2 millimeters), %G is the percentage of the surface covered with gravel, and R_S is a measure of roughness size defined by $D_G \cdot d/A_B$, where A_B is the bed area over which D_G and d were measured. The product $D_G \cdot d$ represents the wetted upstream projected area of the average sized gravel particle, whereas the division by A_B makes R_S dimensionless.

Equations (3) and (4) have much higher R^2 values than the corresponding *ff-Re* relation ($r^2 = 0.032$)

and are therefore almost certain to predict hillslope hydrographs more accurately. Moreover, of the independent variables in these equations, %G is by far the single best predictor of *ff*, explaining 49.8 percent of the variance. These results confirm the importance of form resistance on the semiarid hillslopes under study and suggest that, wherever practical, multivariate relations should be used in place of simple *ff-Re* relations in runoff models for semiarid hillslopes.

Grain and Form Resistance

G. Govers and G. Rauws (1986) have recently developed a procedure for actually calculating the relative magnitudes of grain resistance *ff′* and form resistance *ff″* in overland flow. G. Rauws verified it in the laboratory (1988). Employing this procedure, *ff′* and grain resistance expressed as a percentage of total resistance %*ff′* were estimated for the 73 experiments on the eight plots. A relative frequency distribution of %*f′* for the 73 experiments is presented in Figure 2. The mode of this distribution occurs at 4.55 percent and the median at 4.53 percent. Thus, on the gravel-covered intershrub areas, grain resistance is typically about 1/20th of the total flow resistance. In other words, resistance to flow consists almost entirely of form resistance. This conclusion has major implications for sediment transport modeling.

Modeling Sediment Transport

As with resistance to flow, total shear stress $\tau = \rho gds$ where ρ is the fluid density, may be partitioned into

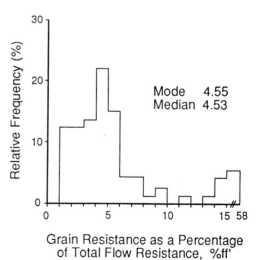

Figure 2. Relative frequency distribution of grain resistance expressed as a percentage of total flow resistance for 73 experiments on eight runoff plots.

grain shear stress τ' and form shear stress τ'' and grain shear stress expressed as a percentage of total shear stress $\%\tau'$ is equal to $\%ff'$ (Abrahams and Parsons in press). Therefore, if $\%ff'$ is typically about 5 percent for the gravel-covered intershrub areas, then so too is $\%\tau'$. This has important ramifications for sediment-transport modeling, for Govers and Rauws (1986) showed that, as in the case of river flow, sediment-transport capacity in overland flow is controlled by grain shear stress rather than total shear stress.

Almost all soil-erosion models developed in the past 20 years have contained a sediment-transport capacity equation, and many of these equations employ τ as a predictive hydraulic variable. As these equations were invariably developed or calibrated in flumes where form resistance was inconsequential, their use of τ was justifiable. However, when such equations are applied to overland flow on hillslopes where form resistance is substantial, τ must be replaced by τ' otherwise sediment transport capacity will be overestimated. The magnitude of the overestimation will increase as the exponent of τ' increases and as $\%\tau'$ decreases.

The magnitude of this overestimation for the present hillslopes may be estimated using the shear stress-based sediment-transport capacity relation recently developed by G. Govers (1990) for overland flow. This relation indicates that the exponent of τ' is 3.4 for the size of material transported during the 73 experiments. Given that τ is typically larger than τ' by a factor of 20, estimates of sediment-transport capacity based on τ will be larger than those based on τ' by a factor of $20^{3.4} = 26,500$! Inasmuch as $\%\tau'$ is likely to be very small on all gravel-covered semiarid hillslopes and the exponent of the relationship between sediment-transport capacity and τ' always exceeds 1 (Govers 1990), sediment-transport modeling on all such hillslopes must take into account the dominance of form resistance and be based on τ' rather than τ.

This finding is not only important for modeling contemporary soil erosion on semiarid hillslopes but also for modeling the long-term evolution of such hillslopes. Models of hillslope development must recognize the critical role of the surface gravel cover in controlling resistance to flow and, hence, the rate of erosion. Where gravel size or concentration varies in a systematic manner down a hillslope, form resistance and erosion rate will also vary, and this systematic variation may have a significant effect on the shape of the hillslope as it evolves through time. L. E. Band (1990) has shown how downslope changes in grain size can affect the evolutionary sequence in hillslope form. However, this study suggests that downslope changes in gravel concentration may be even more influential.

Acknowledgments

We acknowledge the assistance and cooperation of the U.S. Department of Agriculture/Agricultural Research Service Aridland Watershed Management Research Unit, Tucson, in conducting the field experiments and thank the many friends and colleagues who have helped with the field and laboratory work. The research was supported by North Atlantic Treaty Organization grant RG. 85/0066 and NSF grant SES-8812587.

References

Abrahams, A. D., and Parsons, A. J. 1991. Relation between infiltration and stone cover on a semiarid hillslope, southern Arizona. *Journal of Hydrology* 122:49-59.

_____ , and Parsons, A. J. In press. Resistance to overland flow on desert pavement and its implications for sediment transport modeling. *Water Resources Research.*

_____ , Parsons, A. J., and Luk, S.-H. 1986. Resistance to overland flow on desert hillslopes. *Journal of Hydrology* 88:343-363.

Band, L. E. 1990. Grain size catenas and hillslope evolution. *Catena Supplement* 17:167-176.

Dunne, T., and Dietrich, W. E. 1980. Experimental study of Horton overland flow on tropical hillslopes. 2. Hydraulic characteristics and hillslope hydrographs. *Zeitschrift für Geomorphologie Supplement Band* 35:60-80.

Emmett, W. W. 1970. *The Hydraulics of Overland Flow on Hillslopes,* Professional Paper 662-A. Reston, VA: U.S. Geological Survey.

Govers, G. 1990. Empirical relationships for the transport capacity of overland flow. *International Association of Hydrological Sciences Publication* 189:45-63.

_____ , and Rauws, G. 1986. Transporting capacity of overland flow on plane and on irregular beds. *Earth Surface Processes and Landforms* 11:515-524.

Hastings, J. R., and Turner, R. M. 1965. *The Changing Mile.* Tucson: University of Arizona Press.

Parsons, A. J., Abrahams, A. D., and Luk, S.-H. 1990. Hydraulics of interrill overland flow on a semi-arid hillslope, southern Arizona. *Journal of Hydrology* 117:255-273.

_____ , Abrahams, A. D., and Simanton, J. R. In press. Microtopography and soil-surface materials on semi-arid piedmont hillslopes, southern Arizona. *Journal of Arid Environments.*

Rauws, G. 1988. Laboratory experiments on resistance to overland flow due to composite roughness. *Journal of Hydrology* 103:37-52.

Roels, J. M. 1984. Flow resistance in concentrated overland flow on rough slope surfaces. *Earth Surface Processes and Forms* 9:541-551.

The St. Lawrence Seaway

Harold M. Mayer
University of Wisconsin–Milwaukee

Penetrating 3,540 kilometers (2,200 miles) into interior North America, the Great Lakes-St. Lawrence Seaway navigation system, completed in essentially its present form in 1959, was regarded as one of the great engineering accomplishments of the 20th century and an outstanding example of international cooperation. It was intended to remove much of the disadvantage of the continental interior in importing and exporting overseas. It was also designed to meet the increasing needs for electric power on both sides of the United States-Canada border.

Physical Characteristics of the Great Lakes

The seaway project did not create a new waterway but increased the effectiveness of a route provided largely by nature and previously improved in several stages during the 19th century (Figure 1).

The Great Lakes themselves constitute the world's largest freshwater surface with a total area of 245,000 square kilometers (95,000 square miles). Drainage is northeastward into the Atlantic Ocean through the St. Lawrence River. From Duluth-Superior, at the head of Lake Superior, the distance to tidewater at Montréal is 1,770 kilometers (1,000 miles); from Chicago, the southwesternmost penetration of the system, it is 161 kilometers (100 miles) less. Between Lake Superior and the sea, vessels must be lowered or raised 183 meters (600 feet), while Lake Michigan averages 177 meters (579 feet) above sea level. The difference in elevation is overcome by means of 16 locks between Lake Superior and Montréal; 15 between Lake Michigan and tidewater (Lesstrang 1976).

Tidewater Canals

After the War of 1812, canal projects on both sides of the international border were developed to capture the potential traffic of the Great Lakes region as the area was settled. To the north, a series of consecutive canals was developed circumventing the six rapids of the St. Lawrence River between Montréal and Lake Ontario, and Niagara Falls between lakes Ontario on the one hand and Erie, Huron, and Michigan on the other. These canals were enlarged several times, and by the end of the 19th century vessels of 4.32 meters (14 feet) draft transited the system. On the U.S. side of the border, the State of New York completed the Erie Canal, linking the Hudson River and hence tidewater with Lake Erie in 1825. By the late 19th century a modernized canal was completed, but its physical limitations made it available only to small specially designed craft (U.S. Army Corps of Engineers 1937). In recent years it has been used mainly by pleasure boats.

Internal Great Lakes Traffic

The Great Lakes above Niagara Falls were used during the last half of the 19th century for low-cost movement of bulk commodities, especially forest products, copper,

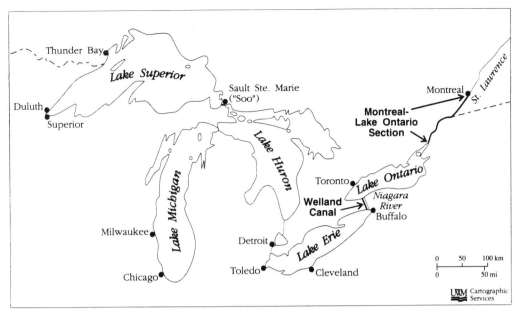

Figure 1. The Great Lakes-St. Lawrence Seaway route: Principal features and ports.

iron ore, and limestone from the northern lakes region, and coal from Appalachia. The iron and steel industry became dependent upon lake transportation. It and the electric power and cement industries constituted the principal sources of internal Great Lakes waterborne commerce during the 20th century. To handle these commodities, specialized vessels were developed. The opening of a consecutive series of locks at Sault Ste. Marie, Michigan (the "Soo"), overcoming the 6.1-meter (20-foot) difference in level between lakes Superior and Michigan-Huron, beginning in 1855 and culminating in the large Poe Lock in 1970, permitted ever-larger vessels to reach Lake Superior from the lower lakes. Until the opening of the present Welland Ship Canal in 1932, however, Lake Ontario was inaccessible to the large upper-lake vessels. The Welland locks are identical in dimension to the present locks in the Montréal-Lake Ontario section of the St. Lawrence Seaway, opened in 1959. The latter enabled vessels of former "maximum laker" size to reach lower St. Lawrence ports and, with certain design modifications, overseas ports.

Thus, between the opening of the present Welland Canal in 1932 and the Montréal-Lake Ontario portion of the seaway in 1959, the five Great Lakes formed a continuous navigation system for vessels as much as 222.7 meters (730 feet) length, 22.87 meters (75 feet) beam, and 7.78 meters (25.5 feet) draft. Between the easterly lakehead for these boats and tidewater at Montréal, cargoes were moved by smaller specially-designed "canallers" or by rail. The seaway proper, between Montréal and Lake Ontario, was designed to accommodate the then "maximum laker" which could carry approximately 26,000 tons, in contrast to the 3,000-ton "canallers" that were retired soon after the seaway was opened.

Pre-Seaway Direct Overseas Movements

Meanwhile small ocean-going ships carried cargoes between the lakes and overseas ports. In 1933 the first regularly scheduled cargo liners, with dimensions limited by the then-existing locks and depths of the St. Lawrence canal system, connected the lakes with overseas ports, especially those of northwestern Europe (Mayer 1954). The number and frequency of these services increased, interrupted by World War II, until 1959 when the newly opened seaway permitted larger, but still moderate-sized, ocean-going ships to enter and leave the lakes. However, the number of "salties" and the tonnage of their cargoes never approached that of the "lakers"—almost entirely Canadian-registered —which transit the seaway. The direct overseas cargoes were moved both on scheduled liners, which carry mainly high-valued manufactured goods, and tramps, which move bulk cargoes. The latter consist mainly of grain from both Canada and the United States outbound and iron ore from eastern Canada inbound.

Development of the Seaway

The discovery and prospective exploitation of the iron ore north of the Gulf of St. Lawrence made the seaway project politically practicable. Although enlargement of the earlier canals and locks had been proposed many years before, the project met with powerful opposition from the eastern U.S. railroads and the Atlantic ports on both sides of the border. With prospective low-cost transportation of the ore to the iron and steel plants of the Great Lakes region in the United States, the railroads, being good customers of the steel industry, withdrew their opposition.

The main bottleneck in the chain of waterways was the 177 kilometers (110 miles) above Montréal. Improvement of that stretch was authorized by the Canadian Parliament in 1951 and by the U.S. Congress in 1954. Construction commenced in August 1954, and the route was opened for through traffic in April 1959.

The principal features of the project were the canals circumventing the rapids in the St. Lawrence River above Montréal. They included seven locks—five in Canada and two in the United States—a power dam straddling the international border across the St. Lawrence, generators for 1.6 million kilowatts of hydro power distributed equally between the two nations, and deepening of stretches of the river and connecting channels between the lakes to accommodate vessels with a draft of 7.85 meters (25.75 feet). The entire seaway, including the previously built Welland Canal, was to be financed by the sale of power and by tolls charged the vessels and cargoes. Navigation benefits and costs, and hence toll receipts, were originally to be 71 percent Canadian and 29 percent United States; these proportions were slightly changed later. Lockage fees were and are charged additionally for Welland Canal transits, although the Great Lakes connecting channels, including the Soo locks, operated by the United States, are toll-free. The Canadian portion of the waterway is maintained and operated by the St. Lawrence Seaway Authority, and the United States portion by the St. Lawrence Seaway Development Corporation. The latter was originally an independent organization, but was later reorganized as a unit of the U.S. Department of Transportation, dependent upon congressional appropriations. A change from the original financial arrangement was made when the United States decided to rebate the U.S. portion of the tolls.

Commodity Movements

Seaway traffic consists of both bulk and general cargo, in contrast to internal Great Lakes traffic, which is almost entirely bulk cargo: grain, iron ore, coal, limestone, chemicals, and petroleum products. Seaway traffic is dominated by downbound grain, mostly for export from Canada and the United States, with lesser volumes of upbound iron ore. Downbound cargoes are carried overseas both directly in ocean-going vessels and in Canadian-registered lake vessels for termination or transfer to larger ocean-going ships in the lower St. Lawrence. While general cargo, carried principally in scheduled liners, increased rapidly until the early 1970s, it steadily declined in recent years, and—except for inbound steel products and special "project" or heavy-lift cargoes, which cannot move overland—has almost ceased.

During the first decade following the opening of the seaway, traffic matched or exceeded the pre-opening projections. The major lake ports—Toronto, Cleveland, Toledo, Detroit, Milwaukee, Chicago, Duluth-Superior, and Thunder Bay—witnessed annual increases in direct overseas traffic. Most of them benefited from numerous direct Great Lakes-overseas scheduled liner services. But in less than two decades traffic declined, and the decline continued into the early 1990s (Figure 2). From nearly 60 million metric tons moving through the Montréal-Lake Ontario section in 1977—the peak year—tonnage declined to about 38 million in 1990. Declines were similar in the Welland Canal, although in that waterway the trend was somewhat modified by movement of coal from U.S. Lake Erie to Canadian steel and electric plants on Lake Ontario.

The Future of the Seaway

Many factors contribute to the long-term decline of seaway traffic. An important trend is the shift in the direction of both U.S. and Canadian intercontinental trade from the Atlantic to the Pacific. Pacific ports, especially Los Angeles-Long Beach and Seattle-Tacoma in the United States, have surpassed New York and other Atlantic ports in annual tonnage, while Vancouver, British Columbia overtook Montréal, to lead in Canada.

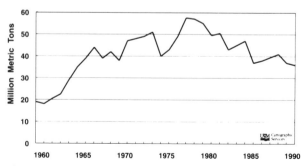

Figure 2. St. Lawrence Seaway traffic. Sources: Annual Reports of St. Lawrence Seaway Authority, and St. Lawrence Seaway Development Corporation.

Associated with the shift to the Pacific is not only rapid industrialization in east and southeast Asia, but also organizational and technological changes in long-distance transportation in North America, facilitating substitution of a land-bridge for many movements that otherwise would have used the Panama Canal. Some ship operators during the 1980s initiated intermodal railroad connections, using containers, including double-stacking. Improvements in port terminals have reduced both costs and delays in land-water transfers, while mergers of U.S. railroads have facilitated overland transportation.

An increasing proportion of the world's ocean-going ships exceeds the size limitations of the seaway's channels and locks. Whereas the typical ocean cargo ship of the time when the seaway was planned could pass through, most modern dry bulk carriers, as well as typical container ships, are much too large, both in draft and beam, to use the waterway. Furthermore, they represent so much capital investment that, even if they could physically enter the lakes, they could not afford the time to do so, seeking, instead, to maximize the number of voyages per year.

Modern highways, including the Interstate system in the United States, the Trans-Canada, and other express highways in Canada, and the growth of inter-city long-haul trucking, have further reduced the advantages of the seaway. JIT ("just in time") production, which is increasing in favor, requires dependable, regular arrival times. The seasonality of the seaway, which is closed for nearly four months each year, and the vulnerability of the route to blockage because of numerous bridges, locks, and narrow channels, produce uncertainties in transit time that many industrial and commercial shippers try to avoid.

In spite of these disadvantages, the seaway's future is not entirely negative. Even if it were not used, its existence as a competitive facility is an incentive for overland carriers—both rail and highway—to keep their rates lower than otherwise. Some industries within the Great Lakes region in both the United States and Canada produce and receive heavy and bulky items, which cannot be transported overland. Without the seaway such establishments would be tempted to relocate. Also, Great Lakes shipyards produce small- and medium-sized naval and commercial ocean-going vessels, many of which are exported.

A basic reason for development of the seaway was to reduce the transportation rate disadvantage of interior North America relative to coastal locations, where direct all-water routes are available. Improvement of the Mississippi River and its tributaries, especially after World War I, and the subsequent expansion of barge traffic on those waterways reduced competitive overland rates. The St. Lawrence Seaway was partially motivated by the same reasoning. Traffic on the Mississippi system continues to increase, currently amounting to about 10 percent of the U.S. freight transportation by all modes. At the same time, the Great Lakes-St. Lawrence system accounts for well under 6 percent, and a declining market share of the total domestic cargo movement by all modes.

In the U.S. Midwest, the two systems compete in a common hinterland, with the overland rates gibing the Mississippi system an advantage in many instances. Both of the seaway authorities agree that the initiation and protection of an advantageous inland rate structure and promotion of the seaway are essential to the future of that route, and vigorous efforts are being made in both respects.

Although the seaway maintenance and operation are carried out smoothly insofar as relations between the United States and Canada are concerned, the interests of the respective nations relative to the seaway are not identical. With increasing integration of the economies of the two nations and the formation of a common market, several issues relative to the future of the waterway have become increasingly timely. The trade treaty maintains the status quo with regard to cabotage—domestic maritime activity within each of the respective nations—but does not restrict traffic between them by "third flag" carriers: The Great Lakes-St. Lawrence Seaway system is an international waterway. Yet, in practice, international movements within the system, not involving overseas traffic, is almost entirely confined to American and Canadian registered vessels.

The trade agreement clearly upholds the cabotage principle: that domestic traffic within each of the two nations continues to be restricted to domestically registered vessels. There is, nevertheless, some apprehension among U.S. Great Lakes interests that, were the American-Canadian trade to be open to vessels of either of the two partners, Canada could dominate internal Great Lakes vessel operation, as it now does through the seaway, thereby virtually ending the need for U.S. Great Lakes commercial craft. An analogous situation exists in western Europe with the creation of an economic union. Can each nation, over the long term, continue to maintain a monopoly of its domestic transportation solely with its domestic carriers? Another concern relates to the fact that the seaway route traverses two cultural realms within Canada, with potential for conflict that may affect seaway operations. On the other hand, the Rhine is a multinational waterway that has been operated relatively smoothly for many decades, and, indeed, its operating agency is a prototype in some respects for the seaway itself.

Regardless of these concerns, the seaway presents both physical and economic challenges to both nations: seasonal closure with limited prospects for extension of the navigation season; vulnerability to interruption due to accidents, or, in troubled times, sabotage; inability because of physical dimensions to handle an increasing proportion of the world's ocean-going ships; diversion of traffic to alternate routes with improvements in their technology and organization, and shifts in the economic base of the Great Lakes region. Both nations are challenged to cooperate not only in physical and economic planning to enable the waterway to meet these challenges, but also to continue—as they are doing—to market effectively the use of the waterway to the maximum possible extent.

References and Further Reading

Heilmann, R. L., Mayer, H. M., and Schenker, E. 1986. *Great Lakes Transportation in the Eighties.* Madison: University of Wisconsin Sea Grant Institute.

Lesstrang, J. 1976. *Seaway.* Seattle, WA: Superior.

Mayer, H. M. 1954. Great Lakes overseas: An expanding trade route. *Economic Geography* 30(2):117-143.

———. 1957. *The Port of Chicago and the St. Lawrence Seaway.* Chicago: University of Chicago Press.

Schenker, E., Mayer, H. M., and Brockel, H. C. 1978. *The Great Lakes Transportation System.* Madison: University of Wisconsin Sea Grant College Program.

U.S. Army Corps of Engineers. 1937. *Transportation on the Great Lakes.* Washington, DC: U.S. Government Printing Office.

The Interstate Highway System

Henry Moon
University of Toledo

The interstate highway system represents the largest public works project ever undertaken in the United States. It is also a network through which the national government has interconnected and in effect homogenized its constituents. This system, whose mileage exceeds the earth's circumference by 50 percent, symbolizes much of North America's political, social, and economic ambition during this century.

Politically, the network came about as a result of a coalition of Congressional leaders of military, pro-development, and engineering mind-sets. This group was prodded by a powerful association of nearly 250 pro-highway lobbyists and industrialists branded by many as the "Road Gang" (Flink 1990). Socially, interstate highways have served for more than 30 years to segregate on a small scale and integrate on a large scale. In many U.S. cities interstate highways are viewed as barriers cutting through neighborhoods and isolating rather than connecting. Economically, the network is frequently associated with urban decentralization and sprawl, rural-urban fringe and suburban development, industrial relocation, and free-flowing capital transfer. Migrations of all shapes, sizes, and forms occur via interstate highways. If the District of Columbia is this nation's heart, then the intestate highway system is its circulatory system—far reaching, interconnected, and at times clogged.

The System as a Network

The origin of the National System of Interstate and Defense Highways can be traced to a formal report made to Congress in 1939. Prior to the report, evidence was increasing that the transportation network of the nation was rapidly becoming inundated with high traffic volumes. Automobile manufacturing had boomed fol-

lowing World War I and the truck and family car had become necessities in American daily life. The findings of the report, however, revealed an inadequate traffic flow to support a tollroad system and congress proposed a 42,960-kilometer, non-toll network. The government would share 50 percent of construction costs with individual states. A 1944 report addressed the specific details of the system as to purpose, design, and size. Acting on the proposals of both reports, Congress enacted the Federal-Aid Highway Act of 1944 and created the interstate highway system. Congress limited the network to 64,360 kilometers and sited it "to connect via routes, as direct as practicable, the principal metropolitan areas, cities, and industrial centers, to serve the national defense, and to connect at suitable border points with routes of continental importance . . ." (Federal Highway Administration 1976, 468). Postwar funding was delayed and initial construction began only after enactment of the Federal-Aid Highway and the Highway Revenue Acts of 1956. J. Flink (1990, 372) writes that ". . . the 1956 Interstate Highway Act ensured the complete triumph of the automobile over mass-transit alternatives in the United States and killed off, except in a few large cities, the vestiges of balanced public transportation systems that remained in 1950s America." Officially, the National System of Interstate and Defense Highways began on the morning of 30 June 1956. Today a 68,383-kilometer network of limited access highways crosses the continental United States, connecting more than 90 percent of all cities with more than 50,000 residents (Figure 1). The network comprises slightly more than 1 percent of the country's road and street mileage yet carries more than 20 percent of all traffic (Federal Highway Administration 1984).

While the system is a national network passing through all of the continental states, it is part of the

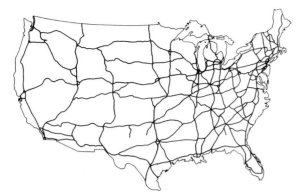

Figure 1. The National System of Interstate and Defense Highways.

state highway system in each of the states in which it is located. Federal highway funds from highway-user taxes were combined with state funds for construction. But the ownership of the system—the entire responsibility of maintenance, administration, and regulation—rests with individual states and localities. To date, the federal share of the total system's cost is approximately 90 percent with the local share at 10 percent. The system has been under construction longer than three decades and approaches 100 billion dollars in cost.

Interstate Routes as Lines

Although the purpose of interstate arteries is to connect larger cities, they pass through nonmetropolitan areas stimulating large-scale interaction and change. Not the least of the potential changes is that of an altered economy. The effect of interstate highways on non-metropolitan areas is being felt on a national scale as more urban residents, commercial establishments, and industrial enterprises find exurban location profitable. Reasons for relocation along interstate lines are numerous and varied but little doubt exists that the sprawl of urban America into the countryside could not occur at such a pace without them as medium and motivation. K. Patrick (1991) points out that interstate routes are often part of "linear bundles" also consisting of other federal routes and railroads stretching across nonmetropolitan areas connecting urban centers and diffusing change and influence.

Recent research on urban-commuting fields illustrates the significance of capital transfer into nonmetropolitan areas from larger urban centers (Mitchelson and Fisher 1988, viii). Apparently, peripheral regions benefit from the extended commuting fields of larger urban areas. By closely documenting the nature of commuting fields

in Georgia and New York and by zeroing in on evolving peripheries, they found that "results clearly point to the permissive role that implantation of limited access highways plays in development via commuting" (Mitchelson and Fisher 1988). Interstate highways facilitate commuting and commuters certainly transfer capital into rural areas. Commuters join local residents and passers-through in forming rural markets for goods and services.

Interchanges as Activity Centers

The type and intensity of the land-use impact of interstate highway construction varies from one region to another and within regions depending on a variety of site and situational characteristics. These characteristics may be in place before construction or may result from highway location. Among these characteristics are historic, social, economic, site specific, population, and geographic variables that necessarily influence the form and function of the region's transportation network and its local influence (Moon 1988). Both direct and secondary highway effects are concentrated at interchanges where access to and from the network takes place. The effect, while substantial in modifying traffic-flow patterns, is equally consequential in influencing the way in which land is used in areas adjoining system routes. While thousands of hectares of property have been removed from other uses by the structure of the highway itself, even more land has been drawn into the realm of highway-related development by its newly acquired connection with previously distant places.

New Urban Places

Some interstate highway interchange areas are evolving into central places. Recent research shows that about 10 percent grow into "interchange villages" (Moon 1990). These findings document the high level of importance that interchange villages carry, fostering local and regional economic development. They represent the convergence of distinct capital-bearing populations at critical points in space and the creation and growth of new nonmetropolitan central places— interchange villages. These evolving urban places are sophisticated economic centers—in part reflecting the demands of a transient highway population while responding to the demands of the region, nation, and world.

When an interchange community exemplifies a large and diverse mixture of transport-related and community-specific establishments, its role is that of a central place. Interchange villages are centers of commerce, manufacturing, distribution, and administration. An interchange village often serves as a hub, the focus of a community's economic, educational, cultural, and social activities. Different interchange village types perform specialized functions but exhibit similar form largely because most firms concentrate near the intersection to maximize access and visibility. This location behavior is consistent with urban history where important crossroads and downtown intersections were viewed as prime sites for small town commercial and administrative development.

As might be expected, a significant proportion of the firms located at interchanges are transport-related (service stations, restaurants, and motels) but a nearly equal number of structures are dedicated to local, regional, or multiple functions. Structure usage is often inconsistent with that associated with through travel. For example, nonmetropolitan interchange location is frequently chosen by local, state, and federal government agencies for local and regional offices. Interchange villages customarily feature a relatively high proportion of institutional land users such as schools and churches oriented toward smaller, local populations. Commercial businesses such as fast-food restaurants or gasoline stations that are dependent on large threshold populations also seek to locate at high traffic-volume interchanges or at interchanges near cities and to capitalize on two distinct populations. Some businesses and industries are dependent on interstate truck traffic as clients or as shippers to move their products or raw materials and may find interchange locations cost effective. On the other hand, residents of remote areas may build homes near an interchange to improve their accessibility to commuting opportunities. Since interstate highways have limited access, a single interchange village may be the accessibility focus for an entire region.

Conclusion

No mode of transportation has altered the geography of the United States more than the automobile. From the advent and distribution of the family car to the implementation of high-speed expressways, individual automobile travel has continually reshaped both urban and rural North America. Nationally, highway use as a mode of transportation dominates all others. Roughly 84 percent of intercity passenger travel occurs via automobiles and 23 percent of intercity freight tonnage is transshipped by trucks (Federal Highway Administration 1984). The interstate highway system reigns as a symbol of this impact. More than a collection of concrete and asphalt links and nodes, it is a composite of American political will and technical ingenuity. It segregates and integrates, serves as a facilitator and a barrier, decentralizes and coalesces, transcends and limits.

References

Federal Highway Administration. 1976. *America's Highways 1776-1976.* Washington, DC: U.S. Department of Transportation.

———. 1984. *America on the Move.* Washington, DC: U.S. Department of Transportation.

Flink, J. 1990. *The Automobile Age.* Cambridge: MIT Press.

Mitchelson, R., and Fisher, J. 1988. *Population Growth and Income Growth Multipliers for Extended and Localized Commuting.* Washington, DC: U.S. Department of Commerce.

Moon, H. 1988. Modelling land-use change around nonurban interstate highway interchanges. *Land Use Policy* 5:394-407.

———. 1990. Interchange villages as urban places. *Small Town* 19:4-14.

Patrick, K. 1991. *Transportation Corridor: Examining the Development and Spatial Structure of Linear Cultural Landscapes.* Presented at the Annual Meeting of the Association of American Geographers, Miami.

The Pennsylvania Turnpike

E. Willard Miller
Pennsylvania State University

The Pennsylvania Turnpike introduced the era of the American superhighways. It was dubbed the world's greatest highway in 1940, for this 160-penny-a-mile road across the Appalachian Mountains had no intersections, no road or railroad crossings, no traffic lights, and not even a speed limit. The traveling time between Pittsburgh and Harrisburg was reduced from nine hours to five and a half. It provided not only a psychological rallying point for a nation that was in the midst of a great economic depression, but it demonstrated the great advances technology was making. The turnpike has withstood the test of time for it is one of the few existing highways incorporated into the United States interstate (I-76 and I-70, Breezewood to New Stanton) highway system. The development of the Pennsylvania Turnpike as the major east-west transportation route across southern Pennsylvania holds an important place in the modern-day economy of Pennsylvania.

Origin of the Turnpike

Initial interest in seeking a land route with low grades across the Appalachian barrier of southern Pennsylvania began during the railroad-building era of the 1880s when William H. Vanderbilt, owner of the New York Central, decided to construct a railroad to compete with the Pennsylvania Railroad's Main Line from Philadelphia to Pittsburgh. Construction of the railroad began in 1883 when contracts were awarded for nine tunnels and for piers for a bridge across the Susquehanna River at Harrisburg. For almost two years construction on the tunnels and the right-of-way was pursued energetically, but in 1885 New York financier J. Pierpont Morgan persuaded Vanderbilt to stop construction on the South Penn Railroad. He warned that a rate war

with the Pennsylvania Railroad would reduce foreign investments, and eventually both railroads would suffer economically. The abandoned right-of-way and the half finished tunnels that had cost $10 million and 26 lives came to be known as "Vanderbilt's Folly" (Shank 1964).

By the 1930s, automobile and truck traffic was providing stiff competition to the Pennsylvania Railroad. Southern Pennsylvania once again found the Appalachian barrier a deterrent to movement of people and goods from the Midwest to the Atlantic coast. U.S. Route 30, the Lincoln Highway, the predecessor of the turnpike, wound through the ridges and valleys, in places rising as much as 2.7 (9 feet) for every 30.5 (100 feet) horizontal meters, making driving extremely hazardous. As a response to the changing mode of transportation, interest was kindled to provide a low-level highway across southern Pennsylvania.

To expedite construction of the highway, the state legislature created the Pennsylvania Turnpike Commission on 21 May 1937, with a simple instruction: Build a toll turnpike from Carlisle in Cumberland County, about 29 kilometers (18 miles) west of Harrisburg, to Irwin, about 39 kilometers (24 miles) east of Pittsburgh, a total distance of 257 kilometers (160 miles) (Cupper 1990). Route surveys were once again initiated. The Turnpike Commission concluded that the abandoned South Penn Railroad route was "the best ever devised between the Ocean and Ohio."

To finance the highway, the commission issued bonds totaling $60 million. However, there was much skepticism concerning the importance of the highway. The U.S. Bureau of Public Roads predicted that only 715 vehicles would use the highway daily. Bankers and investors were equally skeptical and a market for the bonds did not develop. To salvage the project, President Franklin D. Roosevelt, who not only recognized the employment potential in a depressed region but saw the need for a high-speed route for military purposes from the Midwest to the East coast if war developed in Europe, ordered two federal agencies to provide the initial funds. The Reconstruction Finance Corporation, headed by Jesse Jones, underwrote a $35 million issue of turnpike bonds, and the Public Works Administration, under Harold Ickes, provided an outright grant of $29.1 million to the commission (Patton 1990). The decision to build this superhighway was thus related not only to domestic economic conditions, but future national and world events.

Construction of the Turnpike

To begin the project, the commission secured title in September 1937 to the right-of-way and nine abandoned tunnels of the South Penn Railroad. After a survey it became evident that two tunnels were not needed and could be converted into open cuts. In the more than half a century since work had stopped on the South Penn Railroad, many of the portals had caved in and the tunnels filled with water. Only one tunnel, the Allegheny, had been completed. The undriven tunnels varied in length from 108 to 1,030 meters (355 to 3,379 feet).

Construction on the turnpike began 27 October 1938 in Cumberland County. About 18,000 men worked on the turnpike for an average hourly wage of 75 cents. The contractors were required to give men on relief preference in hiring. To meet a federally mandated deadline for completion of the turnpike on 29 June 1940, the men worked under conditions of a near-military emergency, using floodlights at night on three seven-hour shifts (Jones 1950). Although the official deadline was missed, the highway was opened to traffic on 1 October 1940 (Figure 1).

To construct a low-grade road in areas where local relief exceeded 214 meters (700 feet), new and difficult construction problems had to be solved (Cleaves and Stephenson 1949). In the eastern portion of the turnpike, in the highly folded Ridge and Valley Province, rock structure was a major consideration in tunnel construction. In the three eastern tunnels—Blue, Kittatinny, and Tuscarora—the rock strata dipped steeply, standing nearly on end. Because of this, horizontal tunnels proved most satisfactory. In the Tuscarora Tunnel the core boring was 442 meters (1,450 feet) long, believed to be a record at the time for a horizontal bore.

To achieve a maximum grade of no more than 3 percent meant that hills had to be cut through and deep valleys filled. Of the five major hill cuts, the deepest was through Clear Ridge in Bedford County, about 0.8 kilometer (0.5 mile) east of Everett. This cut is 754 meters (2,475 feet) long, 116 meters (380 feet) wide at the top, and 47 meters (153 feet) deep. More than 864,000 cubic meters (1,130,000 cubic yards) of rock were moved by trucks. In total, to achieve this low-grade route, 53.3 million cubic yards of earth and rock were moved.

A unique construction problem was encountered in Somerset and Westmoreland counties at the western

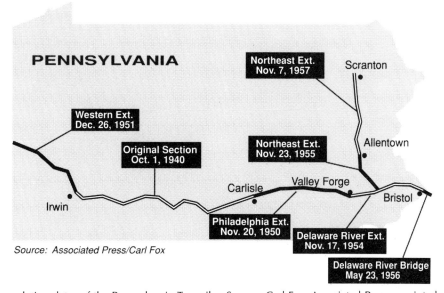

Figure 1. The completion dates of the Pennsylvania Turnpike. Source: Carl Fox, Associated Press; reprinted with permission.

end of the initial Turnpike (Cleaves and Stephenson 1949). Locally, coal mines, both producing and abandoned, occurred near the surface on the turnpike right-of-way. These mines presented a major construction problem because past experience had shown that roads built over coal mines are frequently damaged by surface subsidence. To solve this problem, the open galleries of the mines were packed with stone, and the balance of the mine was "slushed." Slushing involves the introduction of fine-grained rock debris, cinders, or crushed slag into the mine to fill the voids. Water was used to transport the slushing material and facilitate its distribution.

There were also a number of unique problems to solve in servicing highway users. This was the first highway to have service plazas constructed specifically for a single route. For the gasoline concession, Esso (Standard Oil of Pennsylvania) secured the contract. To operate the plaza restaurants, Esso turned to Howard Johnson, famous for his modestly priced restaurants, who decided to gamble that the turnpike would become the premier highway of the state. For decades he operated the only food concessions on the turnpike.

Developing a rapid toll collection system presented another problem. Finally a system adapting IBM's already famous punchcards was selected. Thousands applied for toll-taker jobs. The toll plazas exhibited dramatic functional architecture with their enameled blue, angled compartments, sheltered by great overhanging canopies called "marquees."

Completion of the Turnpike

After the opening of the original section in 1940, World War II delayed additional work on the turnpike by nearly a decade. To complete the turnpike, the 160-kilometer (100-mile) eastward extension to Valley Forge was begun in October 1948 and completed on 20 November 1950. The 108-kilometer (67-mile) westward extension to the Ohio line was begun in October 1949 and opened to traffic on 26 December 1951. The Delaware River Extension from Valley Forge to New Jersey was completed on 17 November 1954 and the Delaware River bridge on 23 May 1956. In the mid-1950s the northeast extension was planned. It was built in two stages, the first stage, from the turnpike near Valley Forge to Allentown, was completed on 23 November 1955 and its extension to Scranton was opened on 7 November 1957. The total system thus includes the 579-kilometer (360-mile) main line across southern Pennsylvania from New Jersey to the Ohio border and the 177-kilometer (110-mile) Northeast Extension that links Philadelphia and Scranton.

Traffic and Economic Importance

When the original section of the turnpike was opened, *The New York Times* (1940, 1 October) reported, "The Pennsylvania Turnpike is a road such as has been hitherto seen only in miniature at the Futurama of the World's Fair—superspeed, supersafe, the longest route of homogenous design in the country." When the road opened on its first Sunday, 6 October 1940, the early predictions of failure were dispelled for mile-long traffic jams formed at every tollbooth. More than 27,000 vehicles drove on the highway the first Sunday, and more the second Sunday. In an era when most roads were unpaved, narrow, and high crowned, highway driving was laborious and slow. The turnpike provided a revolution in mode of travel. It provided the model for the interstate highway system that began in the 1950s.

The road's attraction for America's motorists continued after the first novelty wore off. Turnpike planners expected 1,200,000 trucks and cars the first year, but they totaled 2.4 million, providing an income of $3.5 million (Annual Report 1989). Traffic has continued to increase steadily so that by 1989 traffic volume exceeded 78,000,000 passenger cars and 12,000,000 trucks, producing an income of $105,000,000 from passenger cars and $96,000,000 from trucks. In 1989, the turnpike carried its two billionth vehicle.

As a transit route across southern Pennsylvania (Miller 1986), the turnpike has affected the economy of the region in a number of ways. It took traffic away from the Pennsylvania Railroad and the railroad employment has declined. Highway U.S. 30, as a cross-state transit route has essentially disappeared, and now serves the local population. The most evident economic influence of the turnpike is at the intersection of the eastern and northern routes. This has created an economic expansion around King of Prussia and Valley Forge. Business and industrial parks have found it advantageous to expand at this major node of transportation.

At each of the toll plazas, service functions have developed such as gasoline stations, restaurants, gift shops, and especially motels. While most of these centers are small, Breezewood, at the intersection of I-70 and the turnpike, has become a "city of motels." It is not uncommon to see a 100 or more mammoth trucks parked near this toll plaza. In contrast, the turnpike bypasses most cities and towns as it crosses the state and has had little influence in their economic well-being. The population of most of the places along the turnpike has remained static or declined since the 1930s. Modest recreation facilities have been developed at such places as Bedford and Somerset.

Modern-Day Problems and Some Solutions

Although the turnpike represented the state of the art in 1940, its design is viewed today as primitive (Turnpike Design 1950). The 0.3-meter (10-foot) medial strip, once considered so wide that early travelers held picnics in it, is today 22 meters (73 feet) narrower than the average rural interstate medial strip. To provide a measure of safety a medial guardrail has been constructed along the entire length of the turnpike.

Unlike modern highways that have gradual turns to relieve driver's boredom and sleepiness, the turnpike is as straight as topography permits. The turnpike, built on the old railroad bed, frequently gives the driver an impression of being hemmed in on all sides. It is sometimes called a corridor of claustrophobia. Entry and exit lanes are too short and tight by modern standards, and the service ramps are points of many accidents. There is still no direct connection between the turnpike and Interstate 95 in the Philadelphia area, although this may change by 1995, depending upon agreements with federal highway engineers. Few mileage signs tell drivers distances to their destinations. In the 515 kilometers (320 miles) between the Ohio border and Valley Forge, only three signs indicate the distance to Philadelphia.

But modernizing efforts are being made regularly. In the 1960s, all tunnels were enlarged to four lanes. The interchanges are being rebuilt to accommodate larger trucks. A new toll-collection system has been installed in which ticket issuance is fully automatic for both cars and trucks. This is the only system in the United States where vehicles are weighed, classified, and issued toll tickets by a dispenser at the proper height without human intervention. A Sonic Nap Alert Pattern was installed in 1988 along 10 kilometers (6 miles) of highway berms. These trial rumble strips are designed to alert inattentive drivers who start to drift off the road. An emergency call-box system is in operation along 80 kilometers (50 miles) of the turnpike system between New Stanton interchange and 8 kilometers (5 miles) east of the Allegheny Tunnel, enabling travelers to reach the communication center in Harrisburg for aid.

The services along the highway have also been upgraded. The plazas now have a variety of fast-food vendors, including McDonald's, Hardee's, Bob's Big Boy, Burger King, Popeye's, Roy Rogers, Arby's, King's Family, and Sbarro Italian restaurants. At several locations a combination of facilities include separate snack bars, fast-food, and table service restaurants. Howard Johnson restaurants have long since disappeared. As Loren Martin, executive director of the Pennsylvania Turnpike Commission states, "Our concern ultimately is to satisfy the motoring public, because they pay extra to use the Turnpike."

Reflections

The Pennsylvania Turnpike, while a marvel of its day, is now, in many ways, a museum piece—a transportation antiquity. It is both loved and loathed by its users. But to many Pennsylvanians it is the equivalent of Chicago's Wrigley Field—a well-tended and dependable piece of living nostalgia. It is a relic that still functions as a major transportation corridor for millions of travelers each year.

References

Annual Report. 1989. Harrisburg, PA: Pennsylvania Turnpike Commission (annually since 1941).

Cleaves, A. B., and Stephenson, R. D. 1949. *Guidebook to the Geology of the Pennsylvania Turnpike: Carlisle to Irwin,* Bulletin G24. Harrisburg, PA: Topographic and Geologic Survey.

Cupper, Dan. 1990. *The Pennsylvania Turnpike: A History.* Lebanon, PA: Applied Arts Publishers.

Jones, P. R. 1950. *The Story of the Pennsylvania Turnpike.* Mechanicsburg, PA: Camelot Farms.

Miller, E. W. 1986. *Pennsylvania: Keystone to Progress.* Northridge, CA: Windsor Publications.

New York Times. 1940, 1 October.

Patton, Phil. 1990. The Pennsylvania turnpike: A "dream" of a road. *Smithsonian* 21(October):96-111.

Shank, W. H. 1964. *Vanderbilt's Folly.* York, PA: Buchart-Horn Consulting Engineers.

Turnpike Design. 1950. Dillsburg, PA: Capital Engineering Corporation.

America's "Just-in-Time" Highways: I-65 and I-75

James M. Rubenstein
Miami University

A large truck exits Interstate 75 in rural Ohio and heads for the nearest small town. Large frame houses line the main street, leading to a compact 19th century downtown area. On the other side of town, the truck passes strip shopping centers and fast-food restaurants surrounded by parking lots. Soon, the landscape is dominated by extensive fields of corn and soybeans. The truck turns onto a narrow two-lane road. At the end of the road, in the midst of the fields is a Japanese-owned factory making automotive parts.

The geography of automobile production within the United States changed during the 1980s and early 1990s (Figure 1). Auto makers clustered new plants within 100 kilometers of two major north-south interstate highways—I-65 and I-75—while closing a large percentage of the facilities elsewhere in the country. I-65 and I-75 have become known as the "kanban" highways, after the Japanese term for "just-in-time" delivery. The term kanban reflects both Japanese ownership of many of the new plants and the diffusion of Japanese-style production methods, including just-in-time delivery.

Between 1980 and 1991, construction started on 20 motor vehicle assembly plants in the United States. Eleven of the 20 new assembly plants were designed to build automobiles, four to build trucks, and five to replace older plants nearby that were then closed. Communities in the I-65 and I-75 corridors between Michigan and Tennessee were selected for 10 of the 11 new automobile assembly plants, three of the four new truck assembly plants, and three of the five replacements. Five of the 20 new plants were located in Michigan, four in Ohio, three in Kentucky, and two each in Indiana and Tennessee.

The only new automobile assembly plant opened in the United States outside the I-65 and I-75 corridors during the 1980s and early 1990s was a joint Chrysler-Mitsubishi facility in Illinois, less than 200 kilometers west of I-65. A new truck assembly plant was located in Louisiana and replacement plants in Kansas and Missouri. None of the new plants is in an east- or west-coast state. Japanese companies or joint ventures between Japanese and U.S. firms were responsible for building eight of the new automobile-assembly plants, while U.S. firms—General Motors, Ford, Chrysler, and American Motors (acquired by Chrysler in 1987)—built the remaining three new automobile assembly plants, four new truck-assembly plants, and five replacement plants.

Meanwhile, 18 U.S. plants ceased automobile assembly during the 1980s and early 1990s. Twelve closed entirely, while the other six stopped producing automobiles but remained open for assembly of trucks. Three of the 12 plants that closed completely and only one of the six plants converted from car to truck assembly were within 100 kilometers of the I-65 and I-75 corridors. Six of the 12 closures and four of the six conversions were in east- and west-coast states, while the remaining three closures and one conversion were in interior states outside the I-65 and I-75 corridors. As a result, the number of automobile-assembly plants in the I-65 and I-75 corridors increased from 16 in 1980 to 21 in 1991, while the number elsewhere in the United States declined during the period from 28 to 15.

Historical Pattern of Automotive Production

The concentration of assembly plants along the I-65 and I-75 corridors represents a change from a spatial pattern that dates from the World War I era. In the

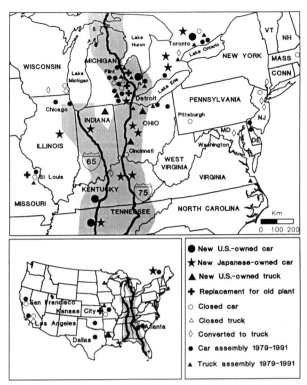

Figure 1. Changes in location of motor vehicle assembly plants, 1979 to 1991. All but one new U.S. assembly plant built during the 1980s and early 1990s has been located within 100 kilometers of Interstate Highways 65 and 75. Most plants located near large east- and west-coast population concentrations were closed during the 1980s.

early days of the automotive industry, Detroit was the center of production; in 1912, for example, more than 80 percent of all automobiles were assembled in the Detroit area (Boas 1961). Assembly plants for low-volume luxury cars, such as Cadillacs and Lincolns, remained clustered in Detroit, but by the 1920s, auto makers had decentralized assembly operations for their best-selling models, such as Ford and Chevrolet, to branch plants near major population concentrations, including New York, Atlanta, Los Angeles, and San Francisco (Boas 1961). Meanwhile, manufacturers of the thousands of components that go into automobiles remained clustered in southeastern Michigan and areas of adjacent states near the southern Great Lakes.

The primary motivation for the spatial distribution of automotive production within the United States has been to minimize freight costs. Components were manufactured in the southern Great Lakes region to take advantage of proximity to the main source of inputs—midwestern steel mills—as well as to the main customers—the major automobile companies' corporate headquarters and distribution centers in southeastern Michigan. Parts were jammed into box cars for the long rail journey to coastal assembly plants.

Truck carriers then hauled finished automobiles the relatively short distances from the assembly plants to dealers' showrooms.

Minimizing Freight Costs

Automobile producers are still selecting plant sites that minimize freight costs, but their optimal locations are now in the I-65 and I-75 corridors rather than near coastal population concentrations. The clustering of plants in the I-65 and I-75 corridors has resulted from the fragmentation of the U.S. motor vehicle market. The principal factors contributing to this fragmentation are the construction of Japanese plants in the United States and the increasing variety in the products sold by U.S. firms (Rubenstein 1986).

Prior to 1960, virtually all U.S.-made automobiles were of similar size, approximately 215 inches long (North American motor vehicles are measured in inches); models varied primarily by body shape and trim design. Beginning in the 1960s, U.S. producers introduced several new sizes to counter the increasing popularity of smaller European cars. Since the oil shocks of the 1970s, the North American market has comprised four sizes of automobiles—full-sized (longer than 200 inches), intermediate (approximately 190 inches), compact (approximately 180 inches), and subcompact (shorter than 170). The "Big Three" also produce a variety of sporty and luxury models. In addition, light trucks, including minivans, sport utility vehicles, and small pickups, now comprise one third of all North American motor vehicle sales.

Because building several sizes of vehicles on the same assembly lines is inefficient, U.S. companies have converted assembly plants that formerly produced identical models for regional distribution to specialized facilities that produce one model for distribution throughout North America. U.S. auto makers have calculated that to minimize aggregate distance to North American consumers, plants should be located in the I-65 and I-75 corridors. Consequently, they have constructed new plants in the I-65 and I-75 corridors, while closing most coastal ones. Similarly, Japanese companies have calculated that their first U.S. assembly plants should be located along the I-65 and I-75 corridors to maximize access to their customers (Hamilton 1985).

Just-in-Time Delivery

Manufacturers of automotive components increasingly have concentrated production in the I-65 and I-75 corridors, as well, because of the introduction of just-in-time or kanban delivery systems at both American-and Japanese-operated assembly plants (Rowand 1982). Suppliers receive schedules of needed components perhaps 10 days in advance and then more precise schedules showing the hour when components will be used on the assembly line during the next five days. To meet tight timetables, suppliers may need to locate some production facilities near final assembly plants, typically within 150 kilometers (Rowand 1982).

More than 200 Japanese manufacturers of automotive components built plants in the United States during the 1980s and early 1990s. Because all but one of the Japanese assembly plants are located in the I-65 and I-75 corridors, approximately 80 percent of the Japanese parts suppliers have also been attracted to the corridors. Approximately 20 percent of the Japanese parts manufacturers have located in Ohio, 15 percent each in Indiana, Kentucky, and Michigan; 10 percent in Tennessee; and 5 percent in Illinois (Figure 2).

Components suppliers have other reasons to seek greater proximity to final assembly plants. Traditionally, producers—concerned more with expense than the quality of parts—awarded annual contracts to the lowest-cost supplier. Quality suffered in part because some suppliers cut corners in the production process to reduce costs. Under the just-in-time system, suppliers

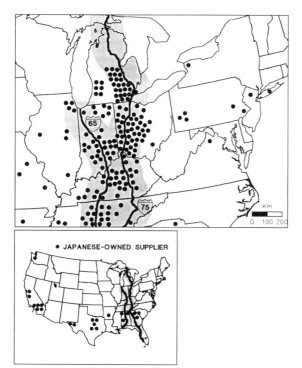

Figure 2. Location of Japanese automotive parts manufacturers in the United States. Excluded are already existing plants that Japanese firms have acquired. Plants constructed by U.S.-Japanese joint ventures are included.

are selected according to quality (McCormick 1983; Rowand 1982). To encourage development of high-quality components, producers give suppliers contracts that extend over the life of the model for which the parts are designed, for as long as 10 years. Armed with the stability of long-term contracts, suppliers can invest more in design and engineering to both improve the quality of parts and reduce production costs (McCormick 1983). Suppliers also have incentive to invest in new facilities, many of which are situated in the I-65 and I-75 corridors to minimize delivery time to nearby assembly plants.

Suppliers to North American assembly plants are now organized into a hierarchy, emulating a long-time Japanese practice. A manufacturer that delivers components to a final assembly plant is called a direct or first-tier supplier. A first-tier supplier provides producers with modules like seats, instrument panels, and suspension systems. Inputs needed to manufacture the modules are obtained from firms now known as second-tier suppliers, many of whom previously sold products directly to the assembly plants.

Rather than obtain individual parts, such as frames, foam, and cloth, from several suppliers, producers can deal with a smaller number of first-tier suppliers. Ford and General Motors were able to reduce their total number of first-tier components suppliers by one third during the 1980s. Because modules are relatively bulky to transport, first-tier suppliers have built facilities near their customers, the assembly plants in the I-65 and I-75 corridors.

Local Labor Climate and Site Selection

Within the I-65 and I-75 corridors, automotive firms—especially Japanese—have avoided communities such as Detroit, Cleveland, and Dayton, where motor vehicle production had traditionally been concentrated (Jensen 1985). Instead, most have selected small towns in the I-65 and I-75 corridors in west-central Ohio, southern Indiana, and central Kentucky and Tennessee (Rubenstein 1986).

Japanese firms have been especially eager to locate in small towns within the I-65 and I-75 corridors where local residents are likely to hold unsympathetic attitudes towards unions and positive attitudes towards adoption of Japanese-style flexible work rules (*Automotive News* 1981). Key elements of flexible work rules include elimination of most job classifications and organization of workers into teams to perform a variety of operations.

To assure that the employees are willing and able to adopt flexible work rules, Japanese firms screen job applicants meticulously. The application process typically requires five types of tests. First, workers take a battery of general aptitude tests to determine reading, writing, and mathematical skills. Then, their mechanical skills and performance capabilities are examined. Third, applicants are tested for work-place attitudes; for example, do they work well with others and adapt well to change? Fourth, applicants must pass medical tests. Finally, applicants are interviewed to determine ability to communicate. Public agencies may administer some of the tests (Kertesz 1989).

The selection of sites on the basis of non-union environments has adversely affected employment of blacks and other minorities. Because the local labor market area around most Japanese plants is predominantly white, relatively few blacks are hired at Japanese-owned plants. Japanese car makers choose small towns outside the industry's traditional core region, in part, to avoid hiring large numbers of blacks, whom they regard as less well-educated and less amenable to adopting flexible work rules (*Automotive News* 1981; Schlesinger 1988).

Faced with the need to close plants, U.S. companies consider local labor climate to be a critical factor. Unions charge that U.S. car makers use the practice of "whipsawing" to select plants for closure. Whipsawing means that an employer plays off the workers at one plant against those at another to extract the best package of concessions. The plant offering fewer concessions is then closed (Mandel 1988). Whipsawing helps to explain the closure of some of the plants in the I-65 and I-75 corridors, while a handful of coastal ones have remained open (Rubenstein 1986).

Conclusion

Further locational changes are likely because the most fundamental problem dominating North American automotive production has not been solved, namely excessive plant capacity. American and Canadian plants can turn out 2 million more vehicles per year than consumers in the two countries are willing to buy. To bring productive capacity in line with projected demand, auto makers, especially General Motors, must close approximately 10 assembly plants. Remaining coastal plants are especially vulnerable, but auto makers may close plants in the I-65 and I-75 corridors where the labor force is uncooperative. In the meantime, the steady stream of trucks enter Interstates 65 and 75 from dozens of small midwestern American towns, relentlessly distributing finished automobiles and automotive parts "just-in-time."

References

Automotive News. 1981. Nissan doesn't want union at Smyrna, Runyon says. 57(21 September):42.

Boas, C. W. 1961. Locational patterns of American automobile assembly plants. *Economic Geography* 37:218-230.

Hamilton, K. 1985. McCurry details Toyota site hunt. *Automotive News* 56(2 September):8.

Jensen, C. 1985. Mitsubishi-Chrysler ventures eye S. Ohio. *The Cleveland Plain Dealer* 16 April:1-E.

Kertesz, L. 1989. Injury, training woes hit new Mazda plant. *Automotive News* 64(13 February):1.

Mandel, D. 1988. One plant dies; one blossoms. *Automotive News* 63(28 November):1.

McCormick, J. 1983. Suppliers adjusting to just-in-time. *Automotive News* 59(22 November):40.

Rowand, R. 1982. New "just-in-time" systems called boon for Midwest. *Automotive News* 58(11 October):1.

Rubenstein, J. M. 1986. Changing distribution of the American automobile industry. *Geographical Review* 76:288-300.

Schlesinger, J. M. 1988. Fleeing factories: Shift of auto plants to rural areas cuts hiring of minorities. *The Wall Street Journal* 12 April:1.

Finding America on the New Jersey Turnpike

Briavel Holcomb
Rutgers University

The New Jersey Turnpike is quintessentially American. A line epitomizing mobility, communication and efficiency, the turnpike represents those characteristics that inhabitants of the United States ("Americans") are both renowned for and pride themselves on. The turnpike is built for the speed Americans love and the safety our risk-averse late-20th-century desires. Its utilitarian design echoes the modern period of construction, yet the pastiche of contradictory landscapes through which the road carries us reflects the postmodern state. The New Jersey Turnpike is a monument to mobility in a nation where travel is often preferred over arrival, where the grass (or greenbacks) are always greener elsewhere, where a fifth of the population moves to a new home each year, and where the fluidity of the economy is such that, to maintain employment totals, an average locality must replace 50 percent of its job base every five years. In its northern section the turnpike is at the heart of one of the largest transportation and communication hubs in the world (Figure 1). Here, air, sea, rail, road, and pipeline transport parallels electric, telephone, fiber-optic, and television cables in a cybernetic spaghetti (Figure 2), and the area is replete with the invisible messages of radio and television waves of a babel of competing stations.

The New Jersey Turnpike is the busiest toll road in the nation and probably in the world. In the late 1980s, it carried more than 190 million vehicles a year. It is the main artery connecting two of the great cities of the first megalopolis (New York and Philadelphia), and is the symbol of a state that Benjamin Franklin once described as "a barrel tapped at both ends." New Jersey's spinal column, the turnpike was an almost inevitable product of the state's corridor location. Its construction has significantly affected the geography of the state. Many people's mental images of New Jersey conform to its rightful position as the most densely populated and one of the most highly industrialized states. But the turnpike also passes through part of the 50 percent of New Jersey that is forest covered, over some of its 8,000 lakes, and near some of its truck farms, which gave rise to the state's moniker "The Garden State" (though some quip that today "The Garden Apartment State" would be more apt). New Jersey is statistically the most urbanized state in the United States. Every county in the state is classified by the federal government as being part of a metropolitan area. However, it is really the most *sub*urbanized and the turnpike is part of an interstate highway system that was one of the most powerful suburbanizing forces in the post World War II period.

Planning for the New Jersey Turnpike began in the late 1940s and the first construction contracts were awarded in December 1949. Several hundred engineers and 10,000 workers were employed. Money was raised by bond sales; not until 1956 did President Dwight Eisenhower sign the Interstate and Defense Highways Act, precipitating the largest civilian construction project in history. For the New Jersey Turnpike some 3,400 parcels of real estate had to be bought. The turnpike was designed as a high-speed highway (120 kilometers per hour [75 miles per hour] in the south and 100 to 115 kilometers per hour [60 to 70 miles per hour] in the northern section) although its official speed limit in these conservation-conscious years is 90 kilometers per hour (55 miles per hour). The Turnpike was built with lanes 3.7 meters (12 feet) wide, at a maximum grade of 3 percent. It was one of the first highways to have 370-meter-long (1,200-foot-long) exit and entry lanes; the first to use laser visibility sensors in fog areas; the first to have a median guardrail for its entire length; and the first to install radio controlled speed limits and emergency warning messages (Briggs 1987). The New Jersey Turnpike today connects

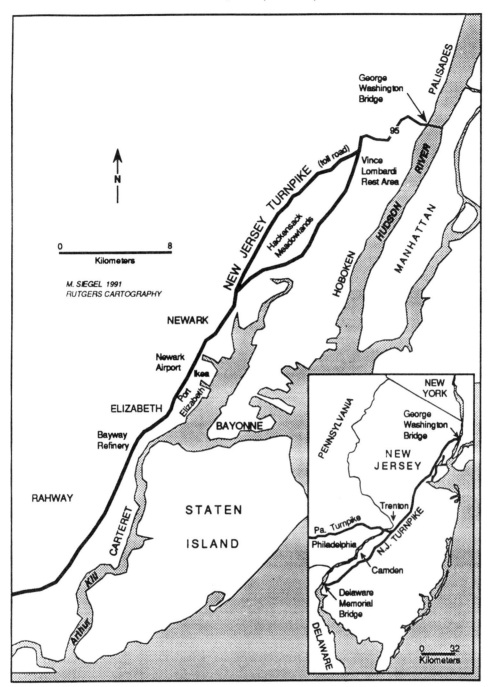

Figure 1. The New Jersey Turnpike.

with the Pennsylvania Turnpike—America's first superhighway—which opened in 1940 with no speed limit, no center guardrail, and traffic "so light that travelers sometimes pulled off and ate picnic lunches on the median without fear" (Cupper 1990, 9).

The partially completed New Jersey Turnpike was officially opened at the end of 1951 and the final section to the George Washington Bridge to Manhattan opened just over a year later. The speed and efficiency

with which the turnpike was constructed reflected the "can do" attitude that prevailed in the American 1950s. It was "another fulfilment of the American Dream . . . [and] tangible evidence of the success of the American Way of Life" (Gillespie and Rockland 1989, 38). Not until the 1960s was there widespread protest against the evisceration of urban neighborhoods by freeway construction (which the turnpike did in Elizabeth); not until the 1970s did environmentalists protest

Figure 2. Traffic along the Turnpike. Note the electricity pylons and the Bayway refinery, in the background. Photograph by Thomas A. Suszka, courtesy of the New Jersey Turnpike Authority.

the destruction of ecological niches (which the turnpike construction fomented in the Hackensack Meadowlands). If the turnpike were to be proposed today it would undoubtedly take years longer, be millions of dollars more costly, follow a different route, and perhaps never be built.

New Jersey boasts that it is a microcosm replete with many different physical and human environments. The turnpike crosses two physiographic provinces, the New Jersey Lowland and the Atlantic Coastal Plain. The Precambrian hills of the Reading Prong are visible from the turnpike's northern reaches, and the highway cuts through a terminal moraine of the Pleistocene glaciation and bridges rivers where fluvial processes are depositing silt laced with heavy metals and pesticides.

Similarly, all stages of human livelihood are represented. There are a few people whose livelihood is hunting and trapping (mostly muskrats) in the Hackensack Meadowlands within sight of the Turnpike bridges as well as the skyscrapers of Wall Street where people sell debt and junk bonds for a living. But most of the turnpike traverses land devoted to neither pre-agricultural nor post-industrial uses. In its northern section the landscape is one of manufacturing, heavy industry, and energy production, while the southern section passes through countryside both agricultural and uncultivated. The economic transformations in both these sectors are visible in the abandoned factories and derelict docks of the north and in the sod farms and horse pastures in the south. Although chemicals and pharmaceuticals remain the state's largest manufacturing industry, the tourist industry has outstripped it in terms of value of production. Thirty years ago

poultry and eggs were the state's biggest agricultural commodity, followed by dairy products and vegetables. Today the "farmers appear to have beaten their plowshares into lawnmowers. . . . The biggest share of agricultural income these days . . . comes from nurseries and greenhouses and sod, to green up the state's rolling tide of lawns, village greens and golf courses" (King 1989, B1). New Jersey now produces more horses (for show and racing) than it does dairy cattle. Due mainly to rising taxes on increasingly valuable land, the state lost an average of six farms a day in the early 1980s, a rate slowed by tax reform to one farm every six days by the end of the decade. Land adjacent to the turnpike, particularly near highway interchanges, is valuable for development, and the consequences of "land-banking" are visible in the uncultivated fields awaiting the earthmovers.

From the turnpike a mosaic of community forms is visible. As the highway cuts through the igneous intrusion of the Palisades, on one side are motels offering hourly rates, videos, and waterbeds, while on the other are classy quarter-million-dollar condos with skylights, balconies, and chimneys for wood-burning fireplaces. One assumes that the denizens of these communities commute to Manhattan to produce financial instruments and to sell the accoutrements of preferred lifestyles. Farther south the turnpike transects or skirts the old, industrial, depressed cities of Newark, Elizabeth, Bayonne, Cartaret, and Rahway. These are working-class, multi-ethnic, multiracial places. The residents' forbearers were immigrants and the neighborhoods over which the turnpike flies on its often elevated route are home to stoic metalworkers, hopeless crack addicts, enduring matriarchs, and children with dreams and nightmares. Their homes are dilapidated terraces and six-story walk-ups. But from the turnpike the conspicuous spires of churches loom: their original congregations suburbanized, they run outreach programs for addicts, child abusers, and simply lonely people.

The racism and discrimination with which many of the inhabitants of these cities (and an increasing proportion of Americans) cope is illuminated by statistics from an office of the Public Defender's recent study of complaints of harassment by state troopers on the turnpike. A week-long survey of traffic on a section of the turnpike showed that 4.7 percent of the sample cars were occupied by black people and had out-of-state license plates. However, 80 percent of 271 arrests for contraband from February to December 1988 that were referred to the Public Defender's office involved black motorists driving out-of-state vehicles. The difference between the typical traffic pattern and the arrests of blacks from other states "is dramatically above thresholds used to establish prima-facie evidence of racial discrimination" (Naus, quoted in Sullivan 1989, B3). There is evidence of homophobia as well as racism on the turnpike. A recent case involves charges brought to court by the American Civil Liberties Union and the Lambda Defense Fund charging that an undercover operation, which resulted in 540 arrests for lewdness over 20 months at the Vince Lombardi rest area on the turnpike, involved plainclothes state troopers encouraging men using the restrooms to make sexual advances. Such entrapment is illegal, but the case illustrates the spatial constraints imposed on gay lifestyles in America.

The turnpike is part of an interstate highway system first authorized under a Defense Department act to facilitate the movement of military personnel and materiel in times of national emergency. But it had the much more profound effect on the geography of the United States of greatly stimulating suburbanization and hastening the decline of central cities. New Jersey epitomizes this pattern and the turnpike passes distressed central cities such as Newark (the most poverty-stricken big city in the country in 1990), Camden (the New Jersey city with the highest poverty rate in 1990), and Trenton (the state capital where most state employees live outside the city), as well as thriving suburbs such as East Brunswick or Cherry Hill where a landscape of retail malls, hotel-convention centers, office condominia, and planned residential communities speak of the aesthetics of consumerism and the ethics of capitalism.

The most geographically fascinating section of the turnpike is probably the area around Exits 13(A) and 14(A) (Newark Airport, Elizabeth and its seaport), about which there is already a tradition of largely derogatory songs and stories. This is the hub of one of the most highly developed transportation systems in the world. Adjacent to the 20 lanes of freeway is Newark International Airport. Opened in 1928 on 28 hectares (68 acres) of reclaimed swampland, Newark is the oldest airport in the New York metropolitan region. It was greatly expanded in 1973 and in recent years has captured an increasing share of international travellers to the New York region. With a slogan "New York's best-kept travel secret" the airport won contracts from foreign airlines bringing business people visiting subsidiaries of the approximately 1,300 international businesses in New Jersey.

On the eastern side of the turnpike across from the airport is Port Elizabeth, the largest container seaport in the world. At a complex of docks built on reclaimed marshland, giant cranes load and unload thousands of containers filled with cars, grain, chemicals, machinery, or refrigerated tropical produce onto waiting flatbed trucks and rail cars. Containerization and mechanization of the docklands has eliminated thousands of longshoremen's (there were no longshorewomen) jobs and has radically changed the ambience of the social and physical environment. The Arthur Kill (*kill* is the Dutch word for stream) is one of the most heavily trafficked waterways in the world and carried more tonnage a year than the Panama Canal (Briggs 1987). Arthur Kill was one of the most polluted waterways in the United States until a few years ago, with untreated sewage, chemical wastes, landfill leachates, and DDT which "came down the river like soapsuds." Efforts by environmentalists and government to reduce pollution have transformed the kill into "a model of the coexistence possible between nature and urban life" (Golden, 1990, B4). Rising oxygen levels have allowed grass shrimp, fiddler crabs, killifish, and ribbed mussels to multiply and to attract back herons, ibis, and egrets. Similar improvements in environmental quality have been achieved farther north along the turnpike in the Hackensack Meadowlands where the responsibilities for zoning and environmental protection for parts of six municipalities lies with a Meadowlands Development Commission. Regulation in the Meadowlands has conserved vital wildlife habitats while enabling the construction of a large sports complex accessible from the turnpike. Here, major league teams compete in the state that hosted the first baseball game (in Hoboken in 1846) and the first American football game (between Rutgers and Princeton, in 1869) (Briggs 1987).

A little to the south of the airport and seaport, the turnpike passes the Bayway refinery, the largest such complex on the East Coast. Here are "fifteen hundred acres of tubes, tanks and pipes, steam, and flames venting from tall towers, all overlaid with a suffocating odor" (Briggs 1987). It has been suggested that this view would be a suitable model for a contemporary Hieronymus Bosch depicting hades. "As far as the eye can see, flames perform a dance macabre above ghostly white holding tanks. Skeletal towers rise against perpetually leaden sky. Steam hisses angrily from miles of black pipes. Corroded drums stand like sentries at the edge of Morses Creek. It is a world of twisted metal and high tension wire, in which any green, growing thing would be an intrusion" (Sarver 1988, 70). Yet this is a landscape of great fascination and beauty to some. A local diner sports a drawing of the scene on its placemats.

Recently, into this landscape of production an oasis of consumption has sprouted. The Swedish home furnishings firm, Ikea, opened a huge store adjacent to the turnpike and airport. The unconventional advertising brought thousands to the store's opening and reportedly irritated turnpike officials. Prominent billboards read "On May 23 find a place to make love on the New Jersey Turnpike," "On May 23 pick up a beautiful Swede on the New Jersey Turnpike," and "On May 23 children will be left to play on the New Jersey Turnpike."

The New Jersey Turnpike is a line that has radically changed the landscape of the state. While it was built initially with great speed and efficiency, the expansion and extension of the highway in recent years has required much greater attention to environmental and social impacts and has been much slower and more contentious. The Environmental Impact Statement for the Turnpike widening in the late 1980s filled many volumes and controversy over its funding continues. Yet, as Gillespie and Rockland argue, the turnpike is not merely the machine in the garden (state), "the garden has been obliterated, has disappeared. Here [in the stretch just described] the machine is the garden. . . . This is the ultimate technological landscape. Here is where you come to see the twentieth century. Here is where we would lead a Martian if we had to choose one place, one artifact, to explain our civilization" (1989, 17).

References

Briggs, Charles A. 1987. *Ride with Me on America's Interstates: I-95 New Jersey to Delaware.* Bethesda, MD: RWM.

Cupper, Dan. 1990. Pennsylvania Turnpike celebrates its 50th Anniversary. *Pennsylvania* 10(3):6-14.

Gillespie, Angus Cress, and Rockland, Michael Aaron. 1989. *Looking for America on the New Jersey Turnpike.* New Brunswick: Rutgers University Press.

Golden, Tim. 1990. Oil in the Arthur Kill: Publicity and peril for urban marsh. *New York Times* 18 January:B1,B4.

King, Wayne. 1989. When farming was the soul of the Garden State. *New York Times* 1 December:B1.

Sarver, Patrick. 1988. Off the beaten track. *New Jersey Monthly* 13:9,17.

Sullivan, Joseph. 1989. New Jersey police are accused of minority arrest campaigns. *New York Times* 16 February:B3.

Index: Regions and Topical Specialties

Page numbers refer to opening pages of relevant chapters.